Reader's Digest

HOW Was it Done?

THE STORY OF HUMAN INGENUITY THROUGH THE AGES

Published by The Reader's Digest Association Limited

LONDON NEW YORK SYDNEY CAPE TOWN MONTREAL

Contributors

The publishers wish to thank the following people for their valuable contributions in creating HOW WAS IT DONE?

HOW WAS IT DONE?
was edited and designed by
The Reader's Digest Association Limited,
London.
Editor: David Gould
Art Editors: Mavis Henley, Iain Stuart

Colour separations
Dot Gradations Limited,
South Woodham Ferrers
Paper
Smurfit Townsend Hook Limited,
Snodland
Printing and binding
Jarrold & Sons Limited, Norwich

Printed in England

ISBN 0 276 42192 2

WRITERS WHO CONTRIBUTED MAJOR SECTIONS OF THE BOOK

Angus Hall · Nigel Hawkes · Dr Christopher Lawrence MB, ChB, MSc, PhD
Antony Mason · Charles Messenger, MA · Dr Richard Walker BSc, PhD, PGCE

OTHER CONTRIBUTORS AND SPECIALIST ADVISERS

Charles Allen · Dave Batchelor, English Heritage · John Booth · Professor Keith Branigan
Dr Allan Chapman · Nancy Duin · Charlotte Evans · Dr Michael Fischer · Caroline Fraser-Ker
Thomas and Joan Gandy · Peter Goodwin, Keeper & Curator, HMS Victory
Dr Elizabeth Hallam · Colin Harding · Richard Harris · Avril Hart, Victoria and Albert Museum
Amy de la Haye, Victoria and Albert Museum · Dr Robert Headland · Jay Hornsby
Jim Lang · Louise E. Levathes · Jane Lewis, BA (Oxon) · John W. Lord, flint knapper
Dr Kevin McDonald · Rowland J. Mainstone · Dr Frank Meddens
Dr Anne Millard, BA (Hons), Dip Ed, Dip Arch · Michael Moody, Imperial War Museum
Anthony North, Victoria and Albert Museum · Dave Parry · Charles Phillips
William S. Pretzer, PhD · Jasper Rees · Jenny Rees-Tonge · Christina Rodenbeck
Dr Ronald A. Ross · Dr David Sim · Robert Stewart MA, DPhil (Oxon)
Paul Swart · Richard Tames · Michael Taylor · Professor Richard Tomlinson
John Trenouth, Science Museum · Captain Simon Waite, MNI

ARTISTS

Andrew Aloof · Ian Atkinson · Barbara Brown · Terence Dalley · Tony Forster
Ivan Lapper · Malcolm McGregor · Jonothan Potter · Paul Weston

INDEXER

Laura Hicks

The typefaces used in this book
are Ellington, designed in 1990
by Michael Harvey, and
Univers, designed in 1957
by Adrian Frutiger.

How Was it Done?

Introduction

This is a book for everyone who has ever looked with wonder at the world around them. It tells how people have overcome what often appear to be impossible odds to help to create that world.

There are tales of ingenuity and toil on a grand scale: monumental buildings and great cities, created by armies of labourers using a few simple tools. The Pyramids and the Colosseum, built thousands of years ago, are imposing reminders of our predecessors' determination and vision; the 15th-century octagonal dome of Florence Cathedral remains one of the most complex pieces of architectural engineering ever attempted. More recently, engineers have cut mighty canals and laid railways across continents; their epic struggles to overcome all obstacles exemplify the human drive to succeed.

There are epic stories of single-minded determination. Polynesian and Viking sailors crossed the oceans in small boats, without navigational instruments. Hannibal took an army of elephants over the Alps, and Samurai dedicated their whole lives to learning the art of war. Scientists like Galileo and Darwin challenged the established order with their ideas. All such achievements are based on perseverance, intelligence and unshakable self-belief.

HOW WAS IT DONE? is not solely concerned with the great breakthroughs in technology, art or thought. It also features the daily challenges which people have always had to meet: eating, keeping clean and getting dressed. How did people of the past light their homes at night, and how did they find out the date and the time? What did they do when they were ill, and how could they seek justice if they felt wronged?

The answers to all these questions – and many more – are given in this lavishly illustrated and vividly written reference volume, which the entire family will find useful and informative. Above all, HOW WAS IT DONE? brings the past to life as it charts the progress of civilisation, saluting the unsung heroes and heroines as well as the geniuses behind the human story – from the time of the first hunters to the modern age.

Contents

CHAPTER 6

Law and Order

CHAPTER 7

Feats of Science

CHAPTER 8

The World at Work

On the Move

HUMAN BEINGS ARE A RESTLESS SPECIES. DRIVEN ON BY A STRONG INSTINCTIVE CURIOSITY, EXPLORERS HAVE SCOURED THE EARTH TO DISCOVER WHAT LIES BEYOND THE FAR HORIZON. OVERCOMING EVERY OBSTACLE TO REACH THEIR GOAL, THEY HAVE CONQUERED THE FEAR OF THE UNKNOWN WORLD

In tune with the ocean's secrets

HOW POLYNESIANS FOUND THEIR WAY AROUND THE VAST PACIFIC

THE NIGHT BEFORE STARTING WORK, THE CANOE-BUILDER lodged his stone axe in the sacred enclosure. A feast of fatted pig dedicated to the gods was followed by sleep before he rose at dawn to begin assembling the wood to build the boat.

Over 1000 years before Europeans first sailed into the Pacific Ocean in the 16th century, Polynesians had been exploring the 25 000 islands in its waters. For a long voyage, they would build either a *vaka*, a boat with two hulls, or an outrigger – a boat with a float fixed to arms extended on one side, which gives extra stability. A crew of six sufficed: two steersmen, a sail-man, a bailer, a spare hand so that rests could be taken and – most important of all – a navigator whose years of training enabled him to find his way, without instruments or a fixed star, in the vast ocean.

The travellers took coconuts, vegetables, dried fruit and fish, and a cooked paste made of breadfruit with them, although stowage was limited and the crew must have gone hungry when they could not catch sufficient fish on the journey. To supplement rainfall and coconut milk, some water was carried on board in gourds or hollow bamboo.

Getting the feel of the sea

To keep the vessels on course, navigators could correct a few degrees variation in the wind direction by checking it against that of the long-range swells generated by the trade winds. Some would lie down on the outrigger to 'feel' the sea as it rose and fell, and swells were considered significant enough to be recorded on schematic reed-maps, some of which survive. The navigators also built up a prodigious knowledge of the ocean's currents as they gained experience at sea.

The Polynesians judged the latitude of their boats by the Sun and monitored their exact course by the stars – the movements of 16 groups of guiding stars were committed to memory by means of rhymes. They could associate stars with particular destinations accurately enough, according to a Spanish visitor of 1774, to find the harbour of their choice at night.

NORSE SAGA *This romantic portrayal of Eric the Red captures the defiant bravery of the Vikings as they sailed fearlessly into the unknown. Eric the Red was exiled from Iceland for murderous feuding, but on his voyages westward across the ocean in AD 982 he discovered a vast new land, where icy mountains hid fertile grassland valleys. Eric named the place Greenland, and when he returned to Iceland he persuaded his family and friends to emigrate with him. Of the 25 boats which set out, only 14 reached the new lands, but the descendants of their crews lived at Bratthalid and Godthabsfjord for the next 400 years.*

HEADING OUT *A newly built Polynesian vessel – already blessed by the head man and ceremonially launched – sails out of the safety of the lagoon. The villagers have gathered on shore to bid its crew farewell.*

UNIQUE DESIGN *A smooth-hulled vaka slips through the water at the start of another journey of exploration. The boat is made of horizontal planks sewn with fibre and sealed by sap from the breadfruit plant. It is steered by a paddle at the stern and a 'dagger-board' plunged into the sea near the bow to turn into the wind or at the stern to swing downwind. The central hut is for sleeping and shelter from the Sun as well as for storage. The coarse sails are made of fibre from the pandanus plant or coconut palm.*

Caroline Islands

Marshall Islands

Hawaiian Islands

Fiji

NEW ZEALAND

OCEAN ISLANDS *The majority of the 25 000 Pacific islands were first found and settled by daring Polynesian sailors.*

Masters of the oceans

HOW VIKINGS CROSSED THE ATLANTIC WITHOUT MAGNETIC COMPASS OR SEA CHART

THE WIND WHISTLING THROUGH THE RIGGING HAD TO BE a wind from the gods. It was freakish, blowing Eric the Red ever westward – further into exile. For days now, he had been sailing into the unknown and had given up hope of seeing land again when he saw the mass of Greenland heave into sight over the horizon.

The story of that first sighting of Greenland, in the 10th century, is a typical Icelandic folk memory, as is its subsequent colonisation by a hero cast out on stormy seas. The true heroism of the Vikings was their willingness to follow the winds and currents using 'environmental' navigation – which combined their observations of the Sun, the stars, the winds and the sea swell – to set their courses. When they approached land, they read the cloudscape or followed the flight of homing birds – like the legendary discoverers of Iceland in the 9th century, who released ravens at regular intervals and followed them to land.

Ships built of oak and iron

Inherited wisdom and personal experience also steered the settlers who followed Leif Ericsson, Eric the Red's son, to Newfoundland in the early 11th century. The settlers' ships were not the slinky 'serpent-ships' used by Viking raiders, nor the 'gold-mouthed, splendid beasts of the mast' celebrated in song by Norse poets, but broad, deep vessels similar to those unearthed by archaeologists at Skuldelev, Denmark, in 1962.

These ships were about 50 ft (15 m) long and 15 ft (4.5 m) wide. Their overlapping planking of pine, or sometimes ash or oak, was made watertight with a caulking of animal hair soaked in pine tar and then fastened with iron rivets. The keel, stems and ribs of the boat were made from oak, with the ribs fastened to the planking by wooden nails of willow.

The central mast had a square sail of coarse woollen cloth which was at its most efficient in a following wind. The ships were steered by a rudder which fitted over the starboard side of the boat, towards the stern, and three or four oars kept in the front of the boat were used to manoeuvre the vessel in confined spaces. Although there were decks at the bow and stern of the boats, the hold in the centre was left open to the sky and held cargo and livestock.

Stores – salted provisions, sour milk and beer – were also stowed amidships in skins or casks which were almost impossible to keep dry. No cooking could be done on board, but all the ships were provided with huge cauldrons for use on shore whenever possible.

To navigate without instruments into unknown regions thick with icebergs seems a strange undertaking, but the Vikings' reasons for going sound familiar. According to a Norwegian book of 1240: 'One motive is fame, another curiosity and the third is lust for gain.'

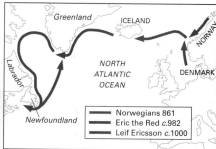

CHAIN OF CHANCE *A series of currents, winds and island-hops links Norway to Newfoundland across the Atlantic Ocean. The last stage of the crossing from Greenland to America is a short current-assisted haul. The Vikings' ability to cross open sea without charts or technical aids seems miraculous to modern sailors.*

Defying demons to follow the Silk Road

HOW MEDIEVAL EUROPEANS TRAVELLED OVERLAND TO CHINA

THE MERCHANTS' HANDBOOK

THE *PRATTICA DELLA MERCATURA*, WRITTEN IN ABOUT 1340, aimed to prepare European merchants for a business trip to China. The author, Francesco Balducci Pegolotti, had never been to China himself, but gathered information from those who had while working for one of the biggest merchant bankers of the day, the Bardi family of Florence.

'You must let your beard grow long and not shave,' advised the traveller's guide. 'And if the merchant likes to take a woman with him from Tana, he can do so.'

Helpful hints and tips
The guide suggested acquiring a good dragoman – a local interpreter – at Tana, now in the Ukraine, regardless of expense, and noted that on leaving Tana a merchant only needed 25 days' supply of flour and salt fish with him to see him through to Hangzhou, a city on the Pacific coast of China.

The route between Tana and Hangzhou is described in detail in the *Prattica* in terms of days' journeys between towns under the protection of Mongol police. It was important to travel with a relative, as a merchant's property would otherwise be forfeit to the local authority if he died *en route*.

Suitable conveyances were suggested for each stage: ox-cart or horse-drawn wagon to Astrakhan on the Caspian Sea; and then a camel or pack-mule as far as the river-system of China. Silver was the currency of the road, with rates of exchange specified for each stop, but it had to be exchanged for paper money on arrival in China.

OPEN TRADE *Mongol rule in central Asia ensured peaceful trade routes from the Pacific Ocean to the Black Sea. Out on the steppes, merchants camped overnight next to their camels.*

'THEY WERE HARD PUT TO COMPLETE THE JOURNEY IN 3½ years, because of snow and rain and flooded rivers and violent storms in the countries through which they had to pass, and because they could not ride so well in winter as in summer.'

Towards the end of the 13th century, the traveller Marco Polo returned to his native Venice after an extended stay in China. Although this summary of his journey overland from Europe to China implies considerable hardship, Polo never complained of robbers, official extortion, tough terrain or bureaucratic delays to the other merchants that he primed to travel the route. Instead, he focused his anxieties on the treacherous deserts.

One of these was Taklimakan, the desert north of Tibet. Here, caravans paused on the edge of the desert for a week's refreshment and stocked up with a month's provisions. As a rule, the bigger the caravan, the safer, but no more than 50 men and their beasts could hope to be sustained by the modest sources of water they would find over the next 30 days.

The voices of spirits across the sand
The worst danger of desert travel was getting lost – 'lured from the path by demon spirits' – and sandstorms were seen as a demon's device to disorientate the travellers. The Mongols recommended warding them off by smearing the horses' necks with blood. 'Even by daylight men hear these spirit voices and often you fancy you are listening to the strains of many instruments, especially drums, and the clash of arms,' wrote Polo. 'Bands of travellers make a point of keeping very close together. Before they go to sleep they set up a sign pointing in the direction in which they have to travel. And round the necks of all their beasts they fasten little bells, so that by listening to the sound they may prevent them straying off the path.'

Although individual travellers favoured horses, merchants relied on camels to carry their wares, each laden with bundles weighing up to 500 lb (225 kg). The camels could travel much farther than horses without regular food and water, and their soft hoofs did not sink into the sand. As the route was long and arduous, merchant caravans carried small quantities of high-value goods, and kept to routes between the Tien Shan and Kunlun mountains, where there were oases and settlements.

Independent travellers to China – who included merchants' couriers,

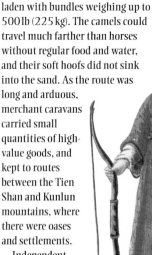

TRAVELLER'S TALES *Marco Polo's own account of his travels appeared in many editions. This 15th-century version focused on the more outlandish of his encounters.*

BEASTS GALORE *Much of Polo's travels took him through country 'abounding with... wild beasts'.*

ISLE OF DOGS *Polo wrote of many peoples met on his way, some more believable than others. Among them were the dog-headed inhabitants of the island of Andaman.*

FITTING IN *Marco Polo dressed in the costume of the Tatars, a tribe from central Asia. It was quite usual for travellers to adopt the local dress.*

MERCHANT ADVENTURERS *In another 15th-century edition of Polo's book, Marco Polo, his father and his uncle are shown saying goodbye to friends, rowing out to their ships and setting sail from Venice. Polo wrote the original version of his book while being held prisoner by the Genoese. Called* Description of the World, *it told of the story of Polo's 24 years of travelling through the empires of Kublai Khan, leader of the Mongols.*

missionaries and ambassadors – could travel safely under Mongol protection along a sparsely inhabited steppeland route to the north. The discomforts, however, were daunting: extremes of temperature, long periods of hunger and thirst, and stretches where the only fuel for fires, if any, was horse dung.

The great obstacle on this route was the Altai Mountains, crossed in two-day stages between *iams* – military way-stations – where travel-weary guests could sleep and the horses changed for fresh ones.

BY LAND OR SEA *Whether taking the route from Venice via Baghdad or the route overland from the Black Sea, suggested by Pegolotti, the road to China was dangerous.*

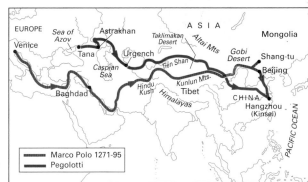

Bandits, mosquitoes and knee-deep quagmires

HOW EUROPEAN PILGRIMS REACHED SPAIN'S GREAT SHRINE OF ST JAMES

A STEP CLOSER TO GOD *The 'way of St James' was a pilgrimage of penance, undertaken in poverty after travellers had paid their debts, begged forgiveness of wronged neighbours and handed over property to the care of a religious order.*

MOSQUITOES INFEST THE MARSHY PLAIN SOUTH OF Bordeaux, where pilgrims straying from the road will sink up to their knees in mud. At Sorde, pilgrims are ferried across the river in hollowed-out tree trunks, in peril of drowning. After a steep, 8 mile (13 km) climb through the pass of Cize, bandits flog those who cannot pay the illegal toll. Spanish food should be avoided 'and if anyone can eat their fish without feeling sick, then he must have a stronger constitution than most'.

Preying on pilgrims

A 12th-century French guidebook listing the hazards of the pilgrims' route to Compostela in north-west Spain goes on to warn that by the salt stream at Lorca, two Basque tanners live entirely on profits made on the hides of horses poisoned by the water, while in Rioja, the locals poison the streams to increase sales of their wine. Travelling companions – unavoidable on popular routes – should be chosen with care, warns the guide, as a common ploy of robbers is to attach themselves to the unwary in pilgrim disguise. All along the roads, beggars exploit the pilgrims' obligation to give alms, deliberately bloodying themselves or faking leprosy.

Despite these hazards, Compostela – home of the shrine of St James, patron saint of Spain – remained a powerful attraction. Pilgrims were accorded special privileges. Their property was immune from civil actions in their absence. In theory, they were protected against molestation and depredation 'by the harshest penalties of the church', as the Archbishop of Lyon declared in 1096. To ease the journey, hermits built roads and bridges for pilgrims' use in northern Spain as works of charity.

Dubious hospitality

A universal obligation of hospitality rested on those who encountered pilgrims: 'Whoever receives them', says the guidebook, 'receives St James and God himself'. In practice, innkeepers were generally denounced as extortionate ('Judas lives in every one of them', warned a contemporary) and dishonest.

Fortunately, the roads to Compostela were studded with hospices run by religious orders, whose only charge was alms at the pilgrims' discretion. In the Pyrenees, the Hospice of St Christine was warmly recommended, but its main rival, that of Roland at Roncesvalles, went to great lengths to attract the passing pilgrims, promising attendance by 'comely and virtuous women' who washed feet, combed beards, cut hair and took 'more care than you can say'.

No miracles of healing were recorded at the shrine of St James, although pilgrims with ailments could visit curative shrines along the way. Their rewards included the chance to lessen a spell in purgatory, and a reputation for sanctity on arrival home.

Playing chess on the back of a camel

HOW MEDIEVAL MUSLIMS MADE THE PILGRIMAGE TO MECCA

IT IS THE DUTY OF EVERY MUSLIM, AT LEAST ONCE IN A lifetime, to make the *hadj* – the pilgrimage to Mecca. Crowds of the faithful, clad in penitential white, converge from the far corners of the globe, eager for a sight of the *Kaaba* – the holy meteorite cast down to Adam by God, on which the Old Testament prophet Abraham was told to sacrifice Isaac.

Today, the pilgrims arrive in chartered jets, but in the Middle Ages, getting to Mecca was rather more difficult. In 1325, Ibn Battuta, the great Arab travel writer, set out from Tangier – in what is now Morocco – on his first pilgrimage to Mecca. He reached Tunis in time to join the annual caravan.

Staying out of the war zones

At Cairo, other pilgrims joined the flow. Some had got there by sea, usually in Genoese merchant ships. From here, the numbers would be too great to travel in one company and parties were only allowed to depart by the Egyptian authorities at 24-hour intervals. Each elected a caravan-master whose authority was absolute, including the power of life and death. During the Crusades, the favoured route included an 18-day voyage up the Nile and across the eastern desert to the Red Sea. On the desert leg, a pilgrim could travel in relative comfort in one of the

two sides of a double pouch called a *saqadif*, slung over a camel's back. A traveller could snooze in the oppressive heat, or, 'if he considers chess lawful', play a game with his makeweight.

Danger on the Red Sea

According to Ibn Jubair, a writer who made the pilgrimage in 1183, the crossing of the Red Sea was 'a terrible danger, save for the few whom our God in His power and greatness, preserves', owing to the winds that might drive them south to arid or hostile coasts. The shipowners, moreover, 'treat pilgrims like demons', seating them 'one on top of another like chickens packed into a cage'.

The pilgrims had to arrive in Mecca on foot, but sometimes the feet could be those of bearers. Ibn Battuta was able, by the favour of an old friend who was a high official, to secure a half-share of a palanquin – a covered litter carried by four men – from Baghdad, with four men's rations of food and water all for himself. He suffered from diarrhoea all the way.

TRAVELLING IN STYLE *Surrounded by an entourage of archers, horsemen and servants, a wealthy lady makes the pilgrimage to Mecca concealed in a covered compartment.*

HOW CHRISTOPHER COLUMBUS CROSSED THE ATLANTIC

HOW DIAS FOUND THE GATEWAY TO THE INDIES

IN 1488, JOÃO II OF PORTUGAL SENT ONE OF HIS BEST navigators, Bartolomeu Dias, to find out if the Atlantic was connected to the Indian Ocean. He hoped to establish a direct sea route around Africa to the Indian subcontinent.

No shore, but an empty sea
South of the Canaries, Dias – in command of two caravels and a store ship – kept close to the African coast until he reached the mouth of the Orange River, in the south-west of the continent. There, storms drove him south, far out of sight of land. After the storms finally died down 13 days later, the prevailing wind was westerly. But when the ships turned to port, expecting to sail into shore, all they found was empty ocean, so they altered course to the north.

When Dias finally made landfall at what is now Mossel Bay, he found that Africa was now to the north, and concluded that he had reached the lower limit of the continent. He continued along the coast until he reached the Great Fish River – far enough east to be certain he had found the passage to India.

The Cape of Storms
Dias consulted his fellow officers. The crew was becoming mutinous – no other European ship had ever sailed so far – and they decided to turn back. On the return journey they sailed close enough to the African continent to see the cape – two windswept promontories capped by peaks rising to around 800 ft (244 m) – which Dias named the Cape of Storms.

Just over 16 months after starting out, Dias dropped anchor in Lisbon harbour. João II promptly renamed the cape 'Good Hope' to encourage future explorers.

SOUTHERN CROSS *On June 6, 1488, Bartolomeu Dias raised a limestone cross, or* padrão, *at the present Cape Maclear, west of the Cape of Good Hope. The cross bore the Portuguese royal coat of arms and marked the southernmost point reached by Dias, although there is no trace of the marker today.*

COLUMBUS REJOICED WHEN THE WINDS TURNED AGAINST him in the middle of the Atlantic Ocean. For weeks, since leaving San Sebastian de la Gomera in the Canary Islands on September 6, 1492, the ships had been blown west into the unknown. It was a good wind for discovery, but the crews had begun to get restless, wondering if they would ever find their way

IN SEARCH OF A NEW LAND *Most of the 90 Spaniards who crewed the three ships on Christopher Columbus's first transatlantic crossing were recruited from Palos, the setting for Emanuel Leutze's* The Departure of Columbus, *painted in 1855. A confident Columbus points the way to the fabled wealth of the Indies, while relatives and friends bid their last farewells to the sailors of the* Niña, Pinta *and* Santa Maria.

back home. Then, without warning, the pennant on the mast of the leading boat drooped as the wind turned. Progress would be slower, but at least, as Columbus wrote later, the crew now believed that there were winds in those seas by which they might return to Spain.

Following the trade winds

The success of Christopher Columbus's first voyage of discovery was due entirely to his boldness in sailing with the prevailing wind behind him. None of the attempts earlier in the century to penetrate the far Atlantic had succeeded, mostly because the ships had tried to guarantee their passage home by sailing into the prevailing westerlies, which blew across the Atlantic to the Azores. The strength of the westerlies defeated all efforts to sail westwards into the wind, but by starting his Atlantic crossing from the Canaries, farther south, Columbus avoided the problem. He hoped to find the westerlies for his return journey.

When Columbus's ships left from Gomera, the most westerly of the Canary Islands, all three were rigged with square sails. This rig was the most efficient when sailing with a wind from behind, but was less effective when sailing against the wind; in this case, the sails had to be turned until they were nearly parallel to the keel, allowing the ships to tack slowly back and forth.

On this first of his three transatlantic voyages, Columbus anticipated a journey of about a month – an exceptionally long spell on the open sea, but one for which ample supplies could be provided for the crew. Ship's biscuits – with flour to bake more if needed – salt meat and fish, wine, oil, vinegar, pulses and a hard cheese especially made in Gomera for sea-bound vessels were brought on board, as were small quantities of honey, rice, raisins and almonds, used only for those who fell sick.

Cooking on board was discouraged because of the risk of fire, although there was a hearth built among the ships' stone ballast. Sleeping arrangements were similarly primitive, and most men would curl up on the deck until Columbus saw hammocks in use in the West Indies. Thereafter, hammocks were provided for the crew.

Columbus claimed that his success in the art of navigation was due to 'a kind of prophetic vision', but at the same time he was always trying to impress his men by brandishing a chart and handling new-fangled gadgets, such as a quadrant, on deck. In fact, his methods were simple.

'A very big mainland . . . a new world . . . of which until this day nothing was known.'
CHRISTOPHER COLUMBUS ON DISCOVERING AMERICA

The first task was to keep course. Mariners' normal practice was to steer by the compass, but Columbus seems to have preferred to set his course by the sun and stars. Although he claimed to check his latitude by regular quadrant sightings, in practice he worked it out by timing the hours of daylight and comparing the results with a printed table listing estimated hours of daylight by latitude. The errors in his log match the errors printed in a contemporary table.

Columbus also needed to measure time as accurately as possible in order to estimate how far he had travelled. He calculated the ship's approximate speed by watching the water break over the bows, and multiplied it by the time elapsed. The ships' boys had the job of turning the hourglasses, but Columbus found it more reliable to watch the progress of stars around the Pole Star, which describe a complete revolution in exactly 24 hours.

A cry from the rigging

Columbus's courage and driving vision kept the men from despair as the days wore on without sight of land, but on October 7, he altered the ships' course to the south-west, writing on October 10 that the men 'could endure no more'. The very next day, sightings of flotsam multiplied and as night fell everyone was excitedly anticipating land. At two o'clock on the morning of the 12th a lookout up in the rigging peered into the darkness and sent up an excited cry. A shot from a small cannon rang out – the agreed signal for land – and the ships echoed with loud, heartfelt thanks to God and the holy Virgin.

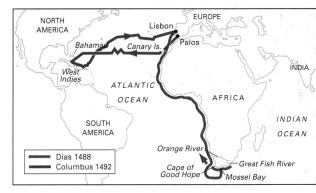

STRAIGHT INTO HISTORY *By sailing due west from the Canary Islands in 1492, Columbus managed to avoid the Atlantic winds, the westerlies, which blew to the north of him. He used the powerful winds for his return voyage to Spain.*

Fresh fish for the mountain-top ruler

HOW INCA COURIERS CROSSED THEIR VAST EMPIRE

ACCORDING TO INCA MYTH, KON TIKI WIRA KOCHA, a creator god, travelled straight – and as far as possible the Incas imitated him. Their road network covered some 12 500 miles (20 000 km), and probably more, connecting what is now southern Colombia with Chile. The roads spanned the world's longest mountain chain, the Andes, and at their highest point they reached 16 700 ft (5 090 m), cut out of the rock in steep places to form a stairway. Built without iron tools but hewn by the action of harder stone on softer, these roads were travelled by a people with no knowledge of wheeled transport.

A local venture
Routes were laid out by government engineers and then each section was built by gangs of local people. Mountain-top shrines, aligned as straight as possible, each visible from the last, were vital for finding direction, for the Incas used no writing and no maps. Although llamas were available as light, slow pack animals, imperial communications depended on professional runners who carried government orders around the empire and brought fresh fish every day from the coast to the Supreme Inca's table at Cuzco – more than 11 500 ft (3 500 m) above sea level and 200 miles (333 km) from the coast, as the crow flies.

On well-used routes, runners travelled between way-stations about ½-5 miles (1-8 km) apart. Two runners lived at each station, on permanent alert for the next message. For urgent messages, a courier would run flat out to the next way-station and shout the message to the next runner before arriving.

Travelling light
Locals recognised the royal messengers' uniform – a headdress of metal foil decorated with feathers and sometimes a square cape – and allowed royal runners to pass unchallenged. The messengers carried with them a shell horn with which to announce their arrival, coca leaves to chew for energy and a *quipu*, the knotted cord used as a memory aid by the Incas.

Permanent bridges made of fibre cables as thick as a man's torso were used to cross the precipitous gorges of the Andes. The famous bridge across the Apurimac river in present-day Peru spanned 148 ft (45 m), at a height of 118 ft (36 m) above the river.

PERILOUS PATH *A royal messenger pulls himself across a remote gorge high in the Andes, balanced on a wooden swing and suspended from long ropes strung between the cliff sides.*

HOW THE HORSE GOT TO THE AMERICAS

ALTHOUGH THE FIRST HORSES reached the New World in the 15th century, the first arrivals were more instrumental in transporting conquistadores than in establishing the herds that later roamed the plains. Perhaps this was no bad thing – Columbus had bought the best horses he could find in Seville, but their grooms and riders traded them in for inferior animals before sailing.

A grassland haven
Horses played a vital part in conquering the Americas, and more arrived with subsequent fleets. These fared so well on the pampas in the south and the prairies of the north that from the mid 16th century, Spanish *rancheros* moved into the American south-west.

Indians were forbidden to ride by the Spanish, but by the 1680s, local people could buy horses from ranches and missions. Herds of wild horses established themselves on the plains, descended from stock which had escaped.

ROUGH RIDE *By lashing two canoes together, two Spaniards transport horses across the Yucatan River.*

MESSENGERS OF THE GODS *Mythical counterparts of the empire's fleet-footed couriers prance across a pair of Moche ear studs.*

SUITED TO THE PLAINS *An ambitious expedition to find gold north of Mexico depended largely on the use of horses brought from Spain. The animals' hardiness and speed made them ideal for the vast spaces, extreme climate and rugged terrain.*

RICH PICKINGS *As the Spanish finished their plunder of the Mexican Aztecs, they turned their attention to rumours of gold farther north.*

KING OF THE PRAIRIE *The Spanish were amazed by the buffalo, which was described as a cross between a large goat, a sheep, a camel and a lion. As they travelled ever deeper into the heart of the continent, the expedition left mounds of buffalo dung along the route to show the way home.*

A search for plunder ends on the Great Plains

HOW CONQUISTADORES PUSHED INTO AMERICA'S MIDWEST

SOMEWHERE TO THE NORTH THERE WAS GOLD – GOLD IN unthinkable quantities. Over the mountains, legend had it, lay seven cities ripe for plunder, cities whose treasures rivalled the wealth of the Aztec empire that the Spanish conquerors had brought low.

The first aim of the expedition of Francisco Vázquez de Coronado, the young governor of the Mexican state of New Galicia, was to find these rumoured 'Cities of Cíbola'. The second was to find an overland route to Asia for the Spanish rulers of Mexico. Coronado intended to return to Mexico a rich man and planned his expedition on a large scale. He took a force of 300 Spanish horsemen and hundreds of Indian auxiliaries ahead of a huge support column driving pack mules and herds of horses and livestock.

The lure of gold

The expedition set out due north from Mexico's Pacific coast, leaving Culiacán, at the limit of Spanish control, in April, 1540. Cíbola was said to lie 'beyond mountains', so the expedition travelled upstream to the Colorado Plateau, and then followed the rivers down the other side. Leaving the support column far behind, Coronado and his advance party suffered extremes of hunger in the high plateau, and

some of them died after eating a mysterious poisonous vegetable. Finally, after two months, they came across well-worn trails and saw their first 'city of Cíbola' – the dusty Zuni pueblo of Háwikuh, hardly the El Dorado they had been expecting.

The Spaniards battered the locals into submission, and they then wintered with their supply column, which had now caught up with them, at Tiguex on the Rio Grande, near modern Bernalillo, New Mexico. There they heard tales told by a Plains Indian slave of a gold-rich city called 'Quivirá', said to be situated farther to the north.

A fruitless endeavour

After five frustrating weeks in 'lands as level as the sea' – the Great Plains of the American West – Coronado suspected that his Indian guides were trying to mislead them. Other Indians met on the route, however, waved the expedition northwards when asked about Quivirá.

Coronado took a bold decision. He sent most of his force and all the camp followers back to Tiguex and headed north from Palo Duro Canyon, near today's Amarillo in Texas, with only 30 horsemen. The pared-down expedition crossed the Arkansas river near the Great Bend in what is now central Kansas, and continued north-east to the Smoky Hills river before turning back. Coronado returned to Mexico empty-handed, having wasted two years in a fruitless search for riches, although he had found the Wichita and Texas Indians and possibly even the Pawnees. The Plains nations had had their first encounter with the European.

The lure of the fabled land of spices

HOW VASCO DA GAMA ROUNDED THE CAPE OF GOOD HOPE

BRIEF ENCOUNTER *As Calicut merchants were suspicious of the Europeans, da Gama only managed to purchase small samples of the products available.*

FANTASTIC VOYAGE *Exaggerated reports of exotic creatures were brought back by the sailors on da Gama's ship and soon spread throughout Europe.*

ORAL HISTORY *When da Gama arrived in Malindi 80 years after the Chinese admiral Zheng He, locals dressed in Chinese silks told of a fleet of immense ships which had visited several times but one day departed, never to return.*

VASCO DA GAMA'S HEART SANK. ALL THE YEARS OF preparation seemed to count for nothing when, on November 8, 1497, nearly three months at sea from his last port of call, he saw the coast of Africa dead ahead. He had meant to bypass the continent altogether, avoiding the treacherous storms that blew around its southern shores. Instead he made landfall just north-west of the Cape.

Searching for a new route

Ten years earlier, an expedition under Bartolomeu Dias had discovered that in the far south of the Atlantic there were westerly winds to carry adventurers around the Cape of Good Hope. Since then, a Portuguese information-gathering mission had approached the Indian Ocean via the Red Sea and been told by local merchants that it was not landlocked, but could be entered from the Atlantic.

Da Gama's four square-rigged ships combed the ocean in search of westerlies, making a detour of 6200 miles (9980 km) south-west into the Atlantic – by far the longest journey ever recorded up to that time on the open sea. Eventually he found the west winds that he was seeking, but he caught them too soon and they landed him short of the Cape. He had no option but to struggle around it, first checked by headwinds, then hurled forward by storms.

He entered the Indian Ocean by a route never, so far as is known, sailed before, and then crossed it guided by a Gujerati Muslim known to da Gama as Molemo Canaqua – a corrupt form of the Arabic for 'pilot-astronomer'. This pilot used the accumulated lore of centuries, recorded in the writings of Ibn Majid, a master navigator who was still alive at the time and who provided complete written sailing directions from Sofala (in Mozambique) to the Bay of Bengal, used well into the 19th century.

Thirty lives lost

The monsoon carried da Gama across the Indian Ocean from Malindi, in what is now Kenya, to Calicut, where he bought samples of all the goods available. Attempting to return without a pilot, and knowing nothing of the violent winds and rains of the monsoon, da Gama chose the wrong season. The reverse crossing of the Indian Ocean took three months and cost 30 lives to scurvy. Yet da Gama sailed into Lisbon in triumph. He had established a direct trade route from Portugal to India which was to bring his country huge wealth.

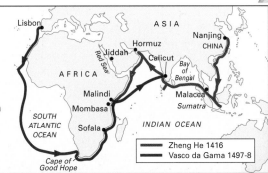

A SHIP TO SUIT THE WIND

UNTIL VASCO DA GAMA'S VOYAGE, PORTUGUESE exploration of the seas was dominated by the caravel, a small ship with triangular sails which was streamlined – up to 98ft (30m) long and 20ft-30ft (6-9m) wide – to give it extra speed. As da Gama expected to sail with the wind, he converted his sole caravel to a rig which had square sails on two masts and a triangular sail at the stern – more efficient in a following wind than triangular sails alone.

The other three vessels of his expedition were merchant ships, with broad hulls to give more storage space, deeper keels and square sails. The Cape run is notoriously dangerous, but only one of da Gama's ships was lost: the merchant ship *São Rafael*, scuttled when disease left him with too few men fit to run it.

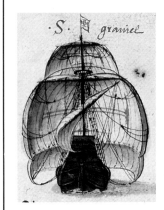

BUILT TO LAST *Two of the merchant ships were especially built for da Gama's trip; unusually for a voyage of exploration, they were armed with 20 guns each.*

A giraffe in the Forbidden City

HOW AN IMPERIAL FLEET SAILED FROM CHINA TO AFRICA

ON AN AUTUMN DAY IN 1416, THE EMPEROR OF CHINA, Zhu Di, accompanied by a gaggle of courtiers, made his way to the magnificent gates of his palace in the imperial capital at Nanjing. A distinguished foreign arrival was to be greeted, brought by a Chinese fleet from Malindi, on the east coast of Africa; it was a creature the Chinese believed was divine – a giraffe.

The admiral whose far-flung ventures brought such rare exotica to China was a Muslim eunuch known as Zheng He. His appointment in 1403 to lead an ocean-going task force was a triumph for those at court who sought commercial and imperial expansion. Before his last voyage in 1433, Zheng He led six expeditions to survey the farthest shores of the Indian Ocean and to impose Chinese control over

trade in the area. He brought back incense, cat skins, tortoise shells, hard wood, precious stones and pearls, as well as ingredients for medicine, such as sulphur from Sumatra.

Planned on a grand scale

The scale of Zheng He's expeditions was as impressive as their range. The first fleet comprised 62 junks, 225 support vessels and 27870 men. The support included water tankers, troop ships, special ships for carrying horses, war ships and speedy eight-oared patrol boats for chasing down pirates. The manpower was overwhelmingly military, though there were seven eunuch ambassadors, ten deputy eunuch ambassadors, 53 supervisory eunuchs,

DROPPING ANCHOR *A flag with the slogan 'lasting tranquillity' flutters over the giant junk as it lowers its sails and prepares to unload its exotic African cargo. Zheng He's vessels were immense, up to 390-408 ft (119-124 m) long and 160-166 ft (49-51 m) at their broadest beam, with hulls made up of three or four layers of pine planks. Specially grown trees were used for the curved bow and stern ribs. The junk's rigging was extremely efficient; its sails, some made of red silk, were stiffened at the foremast and mainmasts with battens of bamboo, and could be turned parallel to the keel.*

several experts in divination and 180 medical staff. The Chinese court was, in part, a large harem and also consisted of a number of eunuchs, who were often foreign prisoners castrated in childhood.

In all, Zheng He's expeditions reported on at least 30 countries around the rim of the Indian Ocean and up the Red Sea as far as Jiddah. The farthest voyage went as far south as the port of Mombasa in East Africa. On average, the expeditions lasted two years each, and the boats travelled quickly. The 3327 miles (5354 km) between Hormuz, an island off Persia, and Malacca, on the Malaysian Peninsula, was covered in only 44 days.

Zheng He's expeditions were brilliantly organised and were triumphs of shipbuilding and navigation. To find his way, Zheng He had sailing directions for most of the routes, compiled by Chinese merchants and the Asian shippers who had served them for centuries. Coasts had been roughly mapped from mariners' reports, and the shape of southern Africa, for instance, was known to a Chinese mapmaker 200 years earlier.

'How can such dissimilarities exist in the world?'
MA HUAN , INTERPRETER WITH ZHENG HE'S FLEET

One of Zheng He's tasks was to improve the maps, whose sailing directions relied on compass bearings and took the form: 'Sail on such-and-such a bearing for so many watches.' His crew supplemented them with celestial naviagation far in advance of anything Europeans were capable of at the time. Navigators carried diagrams of star-positions at various times and latitudes, which were compared with day-by-day observations.

Soon after Zheng He's final voyage in 1433, infighting at court led to the start of China's long years of self-imposed isolation from the rest of the world. By the mid 16th century the building of boats with more than two masts had been made illegal and merchants who traded by sea were put to death.

BLESSED BEAST
The Chinese believed Zheng He's giraffe was a qilin, or sacred animal, which was only seen at such auspicious moments as the birth of Confucius.

A magic stone shows the way home

HOW TRAVELLERS USED EARLY NAVIGATIONAL AIDS

ON NIGHTS WHEN THE MOON AND STARS ARE HIDDEN, sky and sea come together and shroud the way home in darkness. Yet even in the 12th century, a sailor need not despair. With the help of a lump of naturally magnetic iron ore – a lodestone – he could navigate without the stars. A treatise written in about 1190 advises navigators to take a pin rubbed with this stone and float it inside a straw in a basin of water. The straw automatically pointed north.

To use this compass as more than a rough guide, however, it had to be possible to make allowances for drift. Not until the 13th century did seamen have written tables to spare them the arithmetic. Experience, intuition and scanning the heavens for a guess at latitude were the basis of navigation until the end of the 15th century, when Portuguese trade along the African coast created a demand for latitude-finding devices. The height above the horizon of the Pole Star, or of the Sun at a given time of day, varies with latitude, which can therefore be measured as the angle between the horizon and the celestial body. Generations of instruments refined the principle, incorporating sights and scales of ever-greater refinement.

GETTING A BEARING *On this 16th century compass from Marseilles, north is indicated by the fleur-de-lys, emblem of the kings of France.*

TELLING TIME BY THE STARS *The nocturnal was a 13th-century invention for measuring time by observing the movement of the Little Bear around the celestial pole.*

PRECISE WORKMANSHIP *The astrolabe was originally designed for use on land, to calculate the angle of elevation of a heavenly body above the horizon and so determine the time of day. Simplified mariners' versions were made in the 15th and 16th centuries to determine latitude: the Pole Star was aligned with a pair of sights and its angle of elevation above the horizon read on an attached protractor scale.*

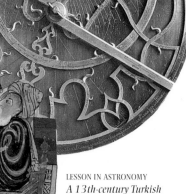

LESSON IN ASTRONOMY
A 13th-century Turkish sage demonstrates the use of an astrolabe to his three pupils.

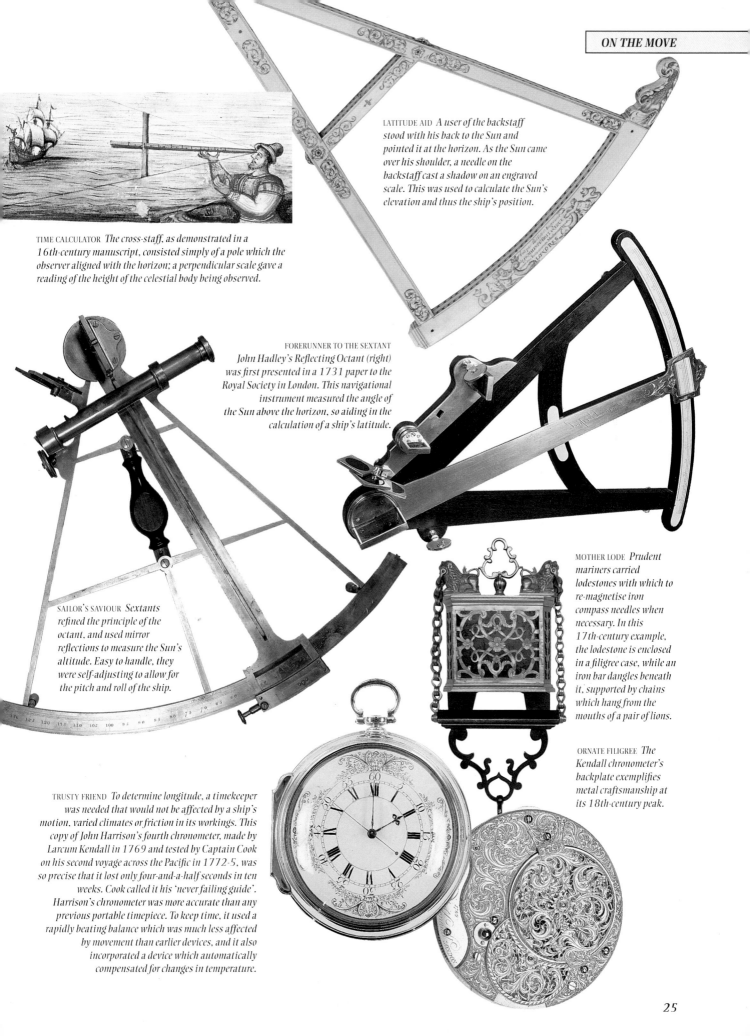

LATITUDE AID *A user of the backstaff stood with his back to the Sun and pointed it at the horizon. As the Sun came over his shoulder, a needle on the backstaff cast a shadow on an engraved scale. This was used to calculate the Sun's elevation and thus the ship's position.*

TIME CALCULATOR *The cross-staff, as demonstrated in a 16th-century manuscript, consisted simply of a pole which the observer aligned with the horizon; a perpendicular scale gave a reading of the height of the celestial body being observed.*

FORERUNNER TO THE SEXTANT *John Hadley's Reflecting Octant (right) was first presented in a 1731 paper to the Royal Society in London. This navigational instrument measured the angle of the Sun above the horizon, so aiding in the calculation of a ship's latitude.*

SAILOR'S SAVIOUR *Sextants refined the principle of the octant, and used mirror reflections to measure the Sun's altitude. Easy to handle, they were self-adjusting to allow for the pitch and roll of the ship.*

MOTHER LODE *Prudent mariners carried lodestones with which to re-magnetise iron compass needles when necessary. In this 17th-century example, the lodestone is enclosed in a filigree case, while an iron bar dangles beneath it, supported by chains which hang from the mouths of a pair of lions.*

ORNATE FILIGREE *The Kendall chronometer's backplate exemplifies metal craftsmanship at its 18th-century peak.*

TRUSTY FRIEND *To determine longitude, a timekeeper was needed that would not be affected by a ship's motion, varied climates or friction in its workings. This copy of John Harrison's fourth chronometer, made by Larcum Kendall in 1769 and tested by Captain Cook on his second voyage across the Pacific in 1772-5, was so precise that it lost only four-and-a-half seconds in ten weeks. Cook called it his 'never failing guide'. Harrison's chronometer was more accurate than any previous portable timepiece. To keep time, it used a rapidly beating balance which was much less affected by movement than earlier devices, and it also incorporated a device which automatically compensated for changes in temperature.*

Charting the great ocean

HOW RENAISSANCE CARTOGRAPHERS MAPPED THE ATLANTIC

IMAGINARY ISLANDS

ALTHOUGH MEDIEVAL CHARTMAKERS strove for accuracy, their maps of the Atlantic Ocean were often filled with non-existent islands.

Many mistakes resulted from a tendency to take literary sources as hard evidence of a place's existence. The Hesperides, for example, visited by Hercules in Greek myth, found their way onto many maps, as did six legendary islands to the west of the British Isles allegedly conquered by King Arthur.

In addition, the Atlantic was – and still is – notorious for false sightings of islands. Deceptive formations of cloud sometimes reinforced by the presence of a 'sea' of weed – giving the impression of land ahead – were generally responsible.

Delusions of land

As mapmakers copied each other without checking the information on earlier charts, false or misplaced islands tended to accumulate. Early 15th century maps, for example, commonly show the Azores twice or more in different places.

The temptation to fill empty space on the maps seems to have been irresistible, especially in the mainly decorative charts intended for landlubbers' libraries.

Maps were usually drawn on a single whole sheep or goatskin and cartographers would draw islands and monsters to fill the awkward triangular space where the neck had been.

SEA MYTH *The legendary adventures of St Brendan – a 6th-century Irish hermit who camped on a whale's back, mistaking it for land – inspired a spate of imaginary isles.*

TWO CANNON SHOTS RANG OUT ACROSS THE WAVES, giving the signal for a conference on Christopher Columbus's flagship. His second-in-command was rowed across and joined Columbus in the cabin of the *Santa Maria*, poring over a map of their route across the Atlantic. The map must have been highly speculative, but the solemnity with which they contemplated it reflected the reverence with which late medieval seamen regarded the mapmakers' art.

Scale maps were unknown in Europe. Lines of longitude and latitude were known only in theory, first appearing on charts and maps between 1500 and 1520, and no even approximate method was available for measuring distances at sea. Marine charts could therefore at best give a rough idea of direction, without ever showing an exact position.

Turning vellum into a map

When constructing a chart, a mapmaker began by drawing a number of compass, or wind, roses onto vellum made of sheep or goatskin. These roses gave him the directions of the main winds. Once he had added rhumb lines to indicate the courses taken when sailing with these winds, he filled in the coastline, starting with his home port.

Information used on maps of the Atlantic coasts of Europe and Africa came chiefly from mariners who

GLOBAL EXPANSION *In this world map of about 1500, the mapmaker's chief aim was to show the islands newly discovered in the West Indies and what was known of the rest of America. Compass lines give sailing directions. The dividing line between Spanish and Portuguese spheres of influence in the New World, agreed at Tordesillas in 1494, runs from north to south. Africa is dominated by a depiction of the Portuguese gold trading station of Sao Jorge da Mina, founded in 1482 and shown here as a sub-Saharan Camelot.*

Os montes claros em affrica.

Caraboa. Castello damina.

Linha equin

Moles Inne

Trop

compiled sailing directions for their own reference. A surviving work of 1508 includes remarkably accurate Portuguese navigators' directions for the entire coast of Africa down to the Cape of Good Hope.

The rhumb lines on the charts had destinations positioned along them, according to the best available information. A navigator would set a course by laying a ruler between the point of departure and the destination and then finding the rhumb line most nearly parallel to it. He could follow the progress of his voyage along the rhumb line with the aid of a pair of dividers, or compasses, and a ruler.

Fanciful outlines

Maps based on navigators' directions look wildly wrong to a modern eye, but it must be remembered that they showed coasts as they might appear to a sailor steering by compass bearings alone, without checking for longitude or for the difference between true and magnetic north. To map the open sea, mapmakers had to rely on seamen's often vague reports in order to locate the islands which were both the vital supports and lethal hazards of the sea lanes.

The best hope was to get the islands in roughly the right latitude relative to each other, for there was no accurate method of measuring latitude at sea, and most mariners relied on educated guesswork and observations of the Sun or Pole Star, with the naked eye, to work it out.

Several surviving 15th-century Atlantic maps include the mapmaker's statement that he has added the latest practical information brought home by explorers – although he had no means of checking it.

DESKBOUND SCHOLAR *Until the 15th century, no known maker of marine charts had actually served as navigator on board ship. Like Petrus Vesconte of Genoa in his self-portrait on an atlas of 1318, they presented themselves as scholars. As well as the earliest signed and dated sea chart of 1311, Vesconte produced the first known European maps to use a north-south grid.*

Braving scurvy, mutiny and starvation

HOW MAGELLAN'S EXPEDITION CIRCLED THE GLOBE

FREEBOOTING CAPTAIN *Ferdinand Magellan was an adventurer first and a master seaman second. The fact that he was a Portuguese soldier who had changed his nationality made him deeply unpopular with his aristocratic Spanish officers and his increasingly disgruntled crews.*

THE SAILORS TORE STRIPS OFF THE LEATHER WRAPPED around the yardarm, but even that was almost impossible to chew with scurvy-swollen gums. The ship's biscuits had crumbled to a worm-ridden powder that stank of rats' urine, and the drinking water was yellow and putrid. Rats were sold on board for half a ducat each, 'and even at that price', according to one crew member, 'we could not get them.' But at last land was in sight.

The first men to circle the globe could only explain their success by a miracle. No recorded voyage had ever lasted so long or gone so far.

Sailing west to reach the East

In March, 1518, when Ferdinand Magellan was commissioned by Charles V of Spain to lead an expedition from Seville to the Spice Islands, now part of Indonesia, his first problem was choosing a route across unknown seas. The eastern approach to the Spice Islands, across the Indian Ocean, was dominated by Portuguese shipping, but Magellan believed he could open a route approaching the islands from the west, around South America.

In common with most geographers at the time, Magellan seriously underestimated the size of the globe. He assumed that the 'sea of the south', as the Pacific was then called, would prove to be a narrow ocean and that once a way was found around the New World, the limits of Asia would lie not far beyond. He set out with five ships, instead of the two or three

usual for voyages of exploration, so he must have expected losses. The expedition carried raisins and quince jelly, which may have helped to control scurvy, and Portuguese casks, which could keep water – laced with antiseptic vinegar – drinkable for months.

Mutinous stirrings

Magellan's fleet departed from Spain on September 20, 1519, and, after spending Christmas at Rio de Janeiro on the east coast of South America, took leave of the known world at the end of the year. Dwindling food, hostile locals and the bitter cold as they sailed farther south along the coast made the crews mutinous. Magellan quelled them by resolute action – the ringleader was killed first and tried afterwards.

He decided to spend the winter in Patagonia, near the southern tip of the continent, to build up supplies. When the ships at last rounded Cape Horn and emerged into the Pacific on November 28, 1520, it was the ideal season to make a fast crossing with the south-east trade winds. Even so, the distance proved almost too vast. By the time they reached Guam, about 800 miles (1280 km) east of the Philippines, they had been without landfall or fresh food for 3 months and 20 days.

Worse was to come. On April 17, Magellan was killed during a battle between two local tribes. The surviving crew members of the two remaining ships had no choice but to carry on around the globe. On November 8, 1521, they reached Tidore, the first of the Spice Islands.

'Had not God and His blessed mother given us such a good wind [across the Pacific], we would all have died of starvation.'
ANTONIO PIGAFETTA, SHIP'S GUNNER

The two surviving ships separated. One searched for a route back across the Pacific but was captured by the Portuguese. The other, the *Victoria*, sailed home via the Cape of Good Hope, its starving crew eluding Portuguese marauders. When the *Victoria* reached Spain on September 6, 1522, only 18 hands out of the original crew of 265 were still alive, and the ship was barely kept afloat by continuous pumping.

FLIGHT OF FANCY *The crew's tales of flying fish – a vital addition to their rations on the open sea – were exaggerated by European artists, who imagined them to be like flocks of birds, swooping through the ships' rigging.*

SUDDEN DEATH *Six months into Magellan's attempt to find a westerly route to the Spice Islands, three of his captains mutinied. The reaction was swift. When the ringleader, Luis de Mendoza of the* Victoria, *refused to report to the flagship* Trinidad, *Magellan's messenger slit his throat.*

A MONK DISCOVERS THE WAY BACK

AT 52, FATHER ANDRÉS DE URDANETA WAS BEGINNING TO feel old age approaching. He had retired from his profession of navigator eight years previously to take up a teaching post at the friary in Mexico City. Yet early in May 1560, sealed letters arrived for him, bearing unsettling orders from Philip II in Madrid. Father Andrés was to lead an expedition across the Pacific and find a route back – something no one had ever done before.

Riding the monsoon winds
Magellan had shown a way across the ocean from America to Asia. But the expeditions that followed him found themselves stranded in the Moluccas or the Philippines with no favourable winds to carry them back the way they had come. Five attempts to find the return route had ended in failure, but it was generally agreed this was because they had not gone far enough north. The Pacific remained frustratingly uncharted.

Friar Andrés knew that exploring the northern route would be a long task. On reaching the Philippines, he stocked his ships with provisions for eight months – rice, biscuit, chick peas, oil, wine and 200 water casks: more than any ship had ever carried before.

The winds around the Philippines change direction according to the season, so he had to time his departure carefully. In June 1565, he caught the southerly summer monsoons, which swept him clear of the islands in record time. These and the Kuroshio current carried the fleet north by east to the middle of the Pacific almost as far north as the latitude of Oregon. From there, Friar Andrés struck east and zigzagged, searching first for wind, then for land.

A landfall at last
Working day and night, Friar Andrés served as a seaman as well as navigator. The ship made landfall in California and then cruised down the west coast of America on the California Current. By the time they got back to Mexico, scurvy had reduced most of the crew to toothless, crawling wraiths covered in purplish-brown patches. A companion ship, which had become separated from that of Friar Andrés on the outward voyage to the Philippines, claimed to have gone so far north that olive oil stored in the ship's galley froze.

Friar Andrés had made the longest voyage yet recorded without landfall – 11 600 miles (18 668 km) in four months and eight days – and established the trans-Pacific route for the rest of the age of sail.

CONTEMPORARY RECORD *A well-to-do couple (left) from Mindanao contrast with their backwoods equivalent (right) from the remote Zambales mountains in drawings from the Boxer Codex, c.1596.*

A new land in the southern seas

HOW CAPTAIN COOK CHARTED THE COAST OF NEW ZEALAND

THE BRITISH ADMIRALTY HAD ITS DOUBTS. IN AUGUST 1768, when it sent Captain James Cook to search the southern seas for the supposed continent of *Terra Australis* – the 'land of the South' – it allowed for the possibility that it did not exist. If he failed to find it, he was to go to New Zealand, which had been discovered in 1642 by the Dutch mariner, Abel Tasman, and chart that instead.

The Admiralty chose Cook, a Yorkshireman, because he proved himself an outstanding navigator in Newfoundland during the Seven Years War against the French. In fact, Cook both charted New Zealand and claimed New South Wales for Britain.

Cook's ship, the *Endeavour*, came within sight of New Zealand on October 7, 1769. Over the next six months he charted 2400 sea miles (4500 km) of coastline. Data for the general chart of the coast was gathered over 117 days' sailing at about 20 miles (32 km) a day, and altogether Cook and his crew spent 58 days at anchor, carrying out detailed surveys from boats and on shore. It was a labour of unprecedented accuracy and efficiency.

Measuring with accuracy
Cook produced two sets of charts: an outline of the entire coast and what was called a 'running survey' – detailed maps of all the harbours and anchorages which the *Endeavour* found, compiled in order to save ships from going aground in unpiloted places. The charts were generally assembled by sailing along the coast at a reasonable distance offshore while taking compass bearings and making rough sketches of the shoreline and its prominent features.

Cook had a passion for the latest technology and liked to use the most up-to-date equipment available. He measured the angle between selected points on shore using one of his many sextants – instruments that represented late 18th-century technology at its best. The distance between the various points was measured by log-line – a knotted cord which was trailed through a seaman's fingers at the stern of the ship. The ship's speed was calculated by counting the number of knots that slipped through his fingers in

the course of a minute. The ship then sailed along the coast to a point where several landmarks could be seen at once, and the procedure was repeated.

Where it was possible to make observations on land, greater accuracy was guaranteed by using 'triangulation'. This surveying technique divides an area into triangular segments and the angles of these are then used to calculate distance.

Keeping time by the skies
Cook and his men seized the opportunity to take a reading of the *Endeavour*'s latitude and longitude whenever they could. Calculating the longitude was a problem, as it depended on knowing the time at Greenwich, the fixed point from which it was counted. As Cook had no accurate timekeeper on board on his first voyage, he worked out the longitude by measuring the angle between the Sun and Moon, or a small number of stars, and comparing it with predicted angles listed at three-hourly intervals in the Nautical Almanac, first issued by the Commissioners of Longitude in London in 1766.

Observing celestial phenomena could also help to establish the ship's longitude, as these events took place at a predicted time. One such was the passing of the planet Mercury across the face of the Sun, which Cook and Charles Green, the ship's astronomer, observed at Mercury Bay in New Zealand's North Island on November 9, 1769. Even so, Cook placed New Zealand well to the east of its true position, and it was only by the time of his next voyage in 1772 that Harrison's chronometer – a timepiece which kept track of Greenwich time – was available.

Cook transferred his results daily to a 'compilation sheet' plotted to scale; surviving examples use 10 or 16 in (25 or 40 cm) to equal one degree of longitude. In cramped and unstable shipboard conditions, converting these sheets to graduated charts was a slow business, and they were completed only after the *Endeavour* returned to Britain.

LOCAL COSTUME *This portrait of an intricately tattooed local Maori chief, wearing greenstone ear-pendants and an ornamental* hei-tiki *round his neck, was painted in 1769 by Sydney Parkinson, the ship's artist.*

SCARE TACTICS *A Maori warrior uses a gesture understood by all races to show the sailors they are unwelcome.*

WELCOMING PARTY *A 70 ft (21 m) Maori war canoe pushes out to sea to meet the* Endeavour. *On board, some of the men brandish weapons. Cook's expedition had several clashes with local warriors and shot ten Maoris on the trip.*

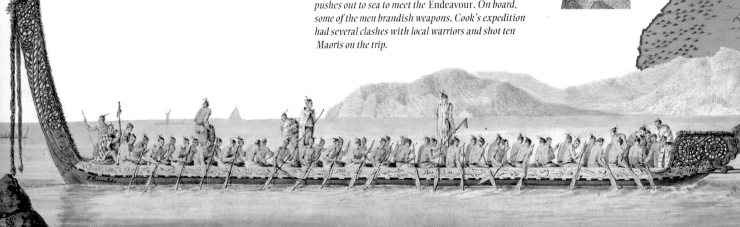

THE FIRST MAP *In his original charts of New Zealand, Captain Cook labelled the islands with their Maori names – Aeheinomouwe and Tovypoenammu.*

SELF-MADE MAN *The son of a farm labourer, Captain James Cook started his career as a ship's boy, became a brilliant navigator and rose through the ranks of the Navy to take command of his own ship.*

READY FOR BATTLE *A chief wearing a cloak of dog skin holds a fearsome battleaxe and carries a heavy truncheon at his waist.*

NATURAL BARRIER *This arch, carved by the sea from solid rock, impressed ship's artist Parkinson. A Maori* paa, *or refuge, stands on the arch.*

USEFUL SCENERY *Spectacular mountains on the west coast of New Zealand were used as landmarks to help in mapping the island.*

HOW SAUERKRAUT SAVED COOK'S CREW FROM THE WORST OF SCURVY

ONE OF THE MOST INTRACTABLE PROBLEMS THAT FACED Captain Cook, or any 18th-century ship's captain, on long ocean voyages was combating scurvy, the condition caused by a lack of vitamin C. It ravaged whole crews, while the cause remained a mystery. Wounds did not heal, gums and joints bled freely, and the victim's skin became horribly blotched with internal bleeding. If a crew member did not die of haemorrhaging, his weakened immune system left him open to other infections.

A dry and limited diet
Cook carried experimental rations to combat the condition. Fresh food was known to be effective against 'scorbutic and other putrid diseases', but it was hard to obtain on a long sea voyage. The usual rations were salt beef, salt pork, salt fat, hard biscuit, oatmeal and dried pulses, relieved only by raisins and sugar. The *Endeavour* also carried 1000 lb (455 kg) of 'portable soup' – cakes of meat essence that could be boiled up – and large quantities of tangy citrus fruit syrup, which was of little use, for all the vitamin C had been boiled away.

Cook's secret weapon against scurvy was sauerkraut – fermented cabbage – enough for 2 lb (1 kg) a week per man for 12 months. At first the crew would not touch it, but Cook forced his officers to set an example by eating the sauerkraut in full view of the crew. After that it had to be strictly rationed. Cook attributed his crew's exceptional health to it.

Fresh green vegetables in Madeira
In addition, Cook took every opportunity to make the men eat fresh victuals. At Madeira, where he had two men flogged for refusing their fresh beef ration, he distributed 30 lb (14 kg) of onions per man and bought all the green vegetables he could find. In Tierra del Fuego at the tip of South America, he had wild celery picked and brought on board for use as an antiscorbutic. The ravages of the disease could not be escaped completely, but on Cook's first voyage, for the first time on a Pacific crossing by European ships, no man died of scurvy.

REWARD FOR BRAVERY *Extra rations of rum were handed out after storms, and alcohol played a large part in keeping the crew happy. On Christmas Day, 1768, they were 'all good Christians', wrote naturalist Joseph Banks, and 'all hands got abominably drunk'.*

By boat, carriage and mule to Italy

HOW ENGLISH GENTLEMEN MADE THE GRAND TOUR OF EUROPE

'A MAN WHO HAS NOT BEEN IN ITALY,' ACCORDING TO Dr Johnson, 'is always conscious of an inferiority from his not having seen what it is expected a man should see.' The 'Grand Tour' of the cities of Europe was the culmination of an English gentleman's education in the second half of the 18th century.

A long and hazardous journey

With its treasure trove of Renaissance and classical art, Italy was the ultimate goal of any Grand Tour. Of course, it was convenient that France – with its own complement of historic cities – was en route, although extra hazards attached to travelling there when, as for much of the 18th century, the English were at war with the French. The most popular route from the French Channel ports was through Paris, by public mail coach or private carriage, to Châlon-sur-Saône in central France. From there the best and speediest way forward was by river boat to Lyons.

Once in Lyons, the Grand Tourists could choose to follow a hair-raising trail through the Alpine passes, or continue to Marseilles and sail from there – running the risk of capture by pirates. Most travellers chose to take the land route. The passes were unfit for carriages until the Turin carriage-road was opened in the 1780s; in the meantime, travellers went up the Alps by mule and were whisked down in a chair by chatty porters. Good roads resumed along common tourist routes as far as Rome and on to Naples and the new archaeological discoveries at Pompeii.

Shopping was high on the travellers' agenda – and art forgers and other swindlers made a small fortune from the more gullible English milords. But along with fake antiques, the Grand Tourists brought back works of art that grace many modern British galleries and form the bulk of the national art collection. The milords spent heavily – it was claimed in 1786 that English tourists spent over a million pounds a year in Paris alone. To cover expenses, bankers at home arranged for drafts on their foreign counterparts, or issued bills of exchange – but these were rarely honoured until, late in the 18th century, bankers created networks of guarantors. The bill of exchange guaranteed the bearer for a certain amount and each time money was drawn, the sum would be adjusted.

Wine – and other attractions

Many young gentlemen, sent abroad to polish their French or Italian, acquired only the words necessary for entertaining the fairer sex. A supply of contraceptives was an essential part of the baggage. British-made sheepgut condoms were so superior that Louis XV of France ordered his own supply.

Like their descendants, English tourists of the time had a reputation for drunken brawling. Being still few in number – 40 000 a year, according to an estimate of the 1780s – and always well connected, they could invoke diplomatic protection.

The Grand Tour might last as long as five years, especially if the young man's family would rather he stayed away. After all, as Dr Johnson said: 'If a young man is wild and must run after women and bad company, it is better done abroad.'

READY FOR THE OFF *With a fresh lick of paint and brasses polished to a winking shine, coaches of the Royal Mail trot past the Angel, Islington – one of London's great coaching inns – following the 1828 annual procession in celebration of King George IV's birthday. On this day, the 28 mail coaches that leave London every night have wound their way to Islington through the capital's West End. At St James' Palace, porter beer was distributed freely to the drivers and guardsmen, who drank the health of the king. As a mark of respect to King George, the guards and drivers wore new livery in scarlet and gold, the horses had new harnesses, and the coach boxes were covered with cloths bearing the royal coat of arms. But now all the pomp is over and it is time for the real work to begin. Cheered on by the watching crowd, the northern mails embark from Islington on their long night journeys to Manchester, Liverpool and Holyhead.*

CLASSICAL MUSINGS *Pen in hand, English poet Percy Bysshe Shelley stares soulfully into space outside the picturesque Baths of Caracalla in Rome – one of the essential stops on the Grand Tour. Shelley left England in 1818 and lived in Italy until his death – aged 29 – four years later.*

DEAR DIARY *George Scharf – later the first director of the National Portrait Gallery in London – kept a careful record of the sights of the Continent when on his tour. In Florence, he made a delicate sketch and wrote a detailed description of the Ponte Vecchio. Scharf's travels took him all the way to Asia Minor.*

Speed and split-second timing

HOW PEOPLE TRAVELLED BY MAIL COACH IN THE 1830s

BREATHLESS WITH THE EFFORT, GUARDS IN SCARLET uniforms fitted passengers and mailbags onto the roof. Meanwhile, with fresh straw underfoot, the four inside passengers settled onto the lightly padded benches. At the cry 'All ready inside and out!' the horse-handlers whisked off the animals' blankets and held the bridles until the coachmen called, 'Let 'em go!' The scarlet wheels spun and the guards blew a salute of three rising notes on their horns. If it was a mild evening, a crowd would have gathered to cheer in reply. In the 1830s, the six coaches of the Western Mails left London's Hyde Park Corner every evening at half-past eight exactly.

Fast trip to the south-west
The mails were a well-organised and reliable service. Four passengers travelled inside. Extra passengers at half-fare went on the roof of the coach – as long as space was available after loading the mailbags. Luggage was stowed in the front compartment; passengers were allowed up to three items each.

The coaches clattered along at an average of 8 or 9 miles (13 or 14 km) an hour. With 20 minutes for breakfast and about an hour for other meals, the 213 miles (340 km) to Devonport on the south-west coast – the fastest run – could be covered in 24 hours with 20 changes of team. It was a great improvement on the 'Exeter Flyer' of the previous century, which had taken three days to reach London.

At each coaching inn grooms and staff would be ready and waiting to tend to travellers and change the horses with the maximum speed and efficiency. The inns were a place where people from all walks of life met, though those who could afford to often hired private dining rooms. Beds were available but sometimes bedrooms had to be shared.

As he went off duty, the departing coachman would enter the public bar and call loudly, 'Gentlemen, I leave you now' as a signal for his tip. For the drivers, timekeeping was an obsession, with the exact minute of arrival and departure marked on printed forms kept at each station.

Full speed over the hill
By the 1830s, Britain had a network of fairly well-made roads. Even on the longest journeys, passengers rarely had to get out and push. Down steep hills, the guard put a skid under the nearside rear wheel as a brake. Experienced coachmen could surmount a double hill by 'springing the team' – descending without a skid and whipping the horses up to a canter just before reaching the bottom of the first hill to carry them over the next.

On the 42 hour trip to Edinburgh, full fare was just under £10. About another £2 would go in meals and tips. This made travelling by mail coach cheap by comparison with the £35 or so it would cost to hire a private post chaise for the same journey, but it still represented a month's wages for a skilled labourer, so for the majority of people walking was the only way to travel long distances.

On a fashionable run, passengers might prefer the privately operated coaches that competed for speed. These were often driven with alarming panache by gentlemen-amateurs. The 60 mile (97 km) London-to-Brighton run was notorious for accidents, but could be completed in five-and-a-half hours.

Ill-prepared and dogged by misfortune

HOW BURKE AND WILLS CROSSED AUSTRALIA'S HARSH INTERIOR

ROAD TO NOWHERE
Burke and Wills' observations at the north coast were so poor that they miscalculated their position by about 50 miles (80km). They never quite reached the sea, as they were defeated by impenetrable mangrove swamps.

BRIGHT HOPES *Robert Burke points the way forward as the expedition departs. He set out with 22 other men, including two scientists and four camel-drivers, as well as 25 Indian camels, a range of other pack animals and 21 tons of provisions.*

THE GROUP OF ABORIGINALS BACKED AWAY AS EDWIN Welch eased his horse down the sandbank. They pointed to a single lonely figure half-covered with scarecrow rags and wearing the remains of a hat.

'Who in the name of wonder are you?' Welch exclaimed.

The hoarse reply came: 'I am King, sir... The last man of the exploring expedition.'

'What! Burke's? Where is he – and Wills?' demanded Welch.

'Dead! Both dead long ago!' cried the scarecrow and collapsed in a faint.

It was September 15, 1861, and the end of a five-month search for a clue to the fate of the expedition led by Burke and Wills – a doomed attempt to be the first to cross Australia. The expedition had been a financial disaster. By the time it left in August 1860, it had already cost £12000 – more than five times the cost of the most expensive expedition to date.

Plagued by incompetence

'No expedition has ever departed under more favourable circumstances,' declared Robert O'Hara Burke as he and his men prepared to set out from Melbourne for the north coast. In fact, things had already begun to go wrong. The appointment of this impetuous, inexperienced 40-year-old inspector of mounted police as the expedition's leader was itself a mistake: when the party reached its first base at Menindee, in western New South Wales, the original second-in-command and the botanist both resigned.

Burke went ahead to establish a base camp at Cooper Creek. In October, the rearguard of the expedition should have set out north from Menindee to join him, but was delayed by incompetence and misunderstandings. Burke lost patience. On December 16, he decided

on a 'dash to Carpentaria', leaving behind a garrison. He set out with 27-year-old William Wills, the surveyor he had appointed as the new second-in-command, and two others. But they took with them neither tents nor experienced guides, and carried only three months' provisions.

The elusive north coast

They reached the Gulf of Carpentaria on the north coast on February 11, 1861. At least they tasted salt water in the mouth of the Flinders River, but they were unable to hack a way through the mangrove swamps to the sea. They turned back the following day, with just one month's rations left for a two-month journey. During March, they ate or lost most of their camels and their sole horse. Discarding everything except firearms, the survivors mounted the last two camels and rode for Cooper Creek. When they arrived on April 21 they found it deserted. Only a few hours earlier the garrison, in a state of despair – with no news either of the 'dashers' or of the rearguard, which was still stuck 75 miles (120km) to the south – had left for Menindee.

Even so, ample provisions remained at the camp and Burke could have followed the trail to Menindee or awaited rescue. Instead, he tried to trace a new route to a police outpost at Mount Hopeless in South Australia. But the group lost their way and then fell terribly ill, perhaps because of experimenting with seeds from the nardoo fern, which the Aboriginals used to make a filling cake. Unless prepared properly, the seeds are poisonous and can cause beriberi, a wasting disease. By the beginning of July, both Burke and Wills were dead. The only survivor, camel expert John King, was adopted by Aboriginals and lived with them until his rescue.

LAST HOURS
Burke and Wills both succumbed to a sickness that left them gaunt and weak.

SOLE SURVIVOR *19-year-old John King was the only man to survive the dash to the coast.*

34

Bloody flight to freedom

HOW THE BOERS MADE THE GREAT TREK ACROSS SOUTHERN AFRICA

LONG TRAIL *By the 1830s the eastern frontier had become crowded as Xhosa, Boers and British settlers jostled for farmland. Yet only a small proportion of Boers moved inland.*

DAWN RAID *Boers desperately defend their wagon encampment against Zulu warriors. On the morning of February 17, 1838, the Boers were waiting at Blaauwkrantz River in Natal for their leaders to return from a conference with the Zulu king. Instead, the Zulu army descended on them. That day the Zulus killed 500 trekkers in the camp and seized many thousand head of cattle.*

Straining teams of oxen, encouraged by the sharp crack of whips, hauled the wagons up the steep, rocky hillside. At the top of the slope the wagon drivers unharnessed the animals for the night and, with their families, set about manoeuvring the vehicles into a rough defensive circle, or *laager*, in case of attacks by locals or wild animals.

The travellers were Boers – descendants of the original Dutch settlers in South Africa – who had set out from Cape Colony in a bid to escape British rule. Their epic wagon journey across southern Africa in search of land – later known as the Great Trek – led to the settlement of Natal, Transvaal and the Orange Free State, but cost thousands of lives. The adventure began in 1835, when the first Boers set out.

Touring the high country

Trekking was already a way of life to frontier Boers whose wagons were their homes as they moved between watering holes. They had evolved the ideal vehicle with which to colonise the 'highveld', the high-lying plateaus in the interior of southern Africa. It was a narrow, tented wagon, 12-17 ft (3.7-5 m) long and 3-4 ft (1-1.2 m) wide. With a full team of 16 – and sometimes even 20 – oxen, the whole stretched an awesome 70 ft (21 m) and could carry almost 2 tons of cargo. Good oxen were the key to success: the best were Afrikaner cattle, a cross of European and local breeds.

The wagons had no springs, but the axles fitted smoothly with the other parts of the sub-assembly to soften the ride slightly. For security and to limit spillage, the sides were built up to 2 ft 6 in (76 cm). The upper structure was of layers of reed matting and canvas stretched over wooden arches.

The driver sat on a chest at the front of the wagon. To give direction to the wagon team, servants and family members – usually young children – took turns to walk ahead of the front pair of oxen, gently leading them by a leather thong. On the level veld, in two daily stages of three hours each, a wagon could cover an average of 12-15 miles (19-24 km) a day, rising to 20 miles (32 km) in the cool season. But the speed of migration was limited by the grazing of the huge numbers of livestock the trekkers brought with them. One group of 29 small family parties had 6156 cattle and 96000 sheep.

Defying the mountains

Sloping terrain was a great enemy. The emigrants had to cross the fords of the middle Orange on a front 100 miles (160 km) broad. Those who then turned east to Natal had to traverse the Drakensberg mountains, which the Zulus called *Quathlamba* – 'heaped up and jagged'. As many as 1000 wagons were coaxed up and down the steep defiles.

Hazards of war and disease had to be faced. One of the earliest parties was massacred near the junction of the Olifants and Limpopo rivers. One man, Louis Tregardt, was determined to find a route to the sea that would guarantee independence from British rule. He and his followers reached the Portuguese settlement on Delagoa Bay, but there, in Lourenço Marques (modern Maputo, in Mozambique), most of them died of malaria. Eventually, the British moved into Natal and many trekkers set off once more to join their compatriots in the Transvaal. By the late 1840s there were 35000 colonists, whose identity had been forged in an epic trial.

Trials of the Oregon Trail

HOW THE FIRST WAGON TRAINS CROSSED THE ROCKIES

'A JOURNEY ACROSS THE CONTINENT MIGHT BE performed with a wagon,' declared the *Missouri Gazette* in 1813, 'there being no obstruction in the whole route that any person would dare call a mountain.' The writer was more of an optimist than a realist, however. A great natural barrier to American westward expansion did exist in the form of the Rocky Mountains, a range which runs for 3000 miles (4800 km), from Alaska down to central New Mexico, and rises to 14 431 ft (4399 m).

Although the Rockies had been crossed in 1805, the opening up of the Far West would never go far without a viable wagon route. This would involve considerable work, as the Spanish friars who built the first road across the similarly inhospitable Sierra Nevada had found in 1775; they had to cut a stairway into sheer rock.

Dreams of a better future

The first target of the great pioneer drive to the West was the rich Oregon territory on the north-western seaboard. The lure was cheap land. Propagandists advertised 'a level open trail… better for carriages than any turnpike in the United States', and shrewd utopians such as the American Society for Encouraging the Settlement of Oregon Territory bought up land to sell to settlers with the promise of a 'city of Perfection' in the west. Misleading maps were available, which suggested that there was an easy river passage from Great Salt Lake along non existent waterways.

Even as the *Missouri Gazette* printed its over-optimistic pronouncement, a party of fur traders stumbled on the pass through the mountains that would one day be the vital east-west link of the Oregon Trail. South Pass, a fairly flat plain about 20 miles (32 km) wide around the Wind River range in Wyoming, remained unpublicised until 1824, however, when Crow Indians showed it, on a deerskin and sand-pile map, to a fur trapper and keen explorer, Jedediah Smith.

Wagons through the mountains

The first trading wagons were led through the South Pass as early as 1832, but without iron axles, the wagons were simply too fragile for unbroken trails of this kind. In an arid climate where, in the words of a missionary's wife in 1836, 'the Heavens over us were of brass and the earth of iron under our feet', dried-out spokes sprang out of place and iron rims slipped off shrunken wheels.

One solution to the problem of shrinking wheels was to remove the iron rims and enlarge the wheels by nailing on extra pieces of wood – or tying them on with rawhide cords, when the nails ran out. The iron rims would be heated as much as possible, eased into place and then shrunk by being doused with cold water. Wagons had to be manhandled up steep climbs and then lowered down cliff-faces with ropes – a feat which might take 80 men to one wagon – and ravines had to be filled with rubble.

Breaking new ground

Although fur traders were known to have travelled from Independence across the mountains and into Oregon in 1832, the first settler wagons to make the trip set out in 1840. The party of missionaries hired a trapper, Robert 'Doc' Newell, to lead them and their two wagons from Green River, on the western edge of the Rockies, over the mountains to Fort Vancouver on the Willamette river.

FAMILY QUEST *Among the many families who braved the trip to Oregon were the Applegates. Jesse (above) wrote to his brother Lisbon of his 'wild' decision to go, with his other brothers Lindsay (top left) and Charles (top right).*

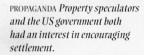

PROPAGANDA *Property speculators and the US government both had an interest in encouraging settlement.*

ROUTE WEST *The most popular route for those looking for land was the arduous Oregon Trail. Settlers joined the route in Independence, Missouri, and then travelled over the Rockies to the Willamette river in Oregon. The first settler wagons completed the trip in 1841, and at the peak of migration in 1850, as many as 5000 people overcame the hazards of the journey in the hope of finding a new life.*

A NEW DAWN *Pioneers gaze down from the western edge of the Rockies after a long, hard climb and pause for a moment in triumph. Their epic journey is almost over. Promises of cheap, fertile land for all in a settler's paradise were soon put to the test as the work involved in building up a homestead began.*

The party left Green River on September 27, and travelled as far as they could in the wagons, clearing a road as they went. When they could go no farther, they shifted their goods to packhorses and the wagons were stripped to their chassis. Eventually, one was broken up to make spares for the other. When they reached Fort Walla Walla – which is now in Washington State – Newell built a barge and hauled the remaining wagon along the Columbia river to the mouth of the Willamette. The 2000 mile (3200 km) trail from Independence, Missouri, to Oregon country had been born. Newell was jubilant, if wrong, when he wrote in his journal on April 19, 1841, 'that I, Robert Newell, was the first who brought wagons across the Rocky Mountains'.

'You who have never seen what is called a prairie country can form but a faint idea of the beauty of its scenery... a diversity of oak-covered hills and plains covered with the most luxuriant verdure.'
A SETTLER ON COMPLETING THE OREGON TRAIL, 1842

The first large train of settlers to complete the trail from Independence to Fort Vancouver passed through South Pass in 1841, led by Elijah White and with Thomas 'Broken Hand' Fitzpatrick, a long-time companion of Jedediah Smith, as guide. As with Newell's party, the settlers switched to packhorses at Fort Hall, this time abandoning their wagons, but it was not long before an annual trek of settlers was making the journey along 'the Oregon Trail', as it soon became known.

The wagon train which followed the next year, in 1842, was known as the Great Migration. It was powered across the mountains by sheer weight of numbers: with a complement of 1000 people – 200 families in 200 wagons, pulled by 694 oxen – there was manpower enough to keep the Indians at bay and to cope with the terrain. The trek took place between April and October, to avoid the winter snow in the mountains and slushy spring weather on the plains, and was the first to use wagons from start to finish.

A corridor of gravestones

Once over the Rockies – whether in Oregon country or Mexican California – the settlers claimed their land, planted the seeds and young fruit trees they had brought with them and built log cabins. The 1846-8 Mexican War led to the United States annexing the states of California and New Mexico, a move which encouraged yet more settlers.

Those filing along the route were all too aware of the sacrifices of previous pioneers, according to Francis Parkman, a young New Englander who went West in 1846 to chronicle the migration and became the leading historian of early exploration and settlement in North America. The Oregon Trail was marked by graves and memorial stones from one end to the other, and migrants that Parkman encountered told him of cattle driven off by wolves, of oxen scattered all over the hills, and of men slaughtered by the Pawnee and horses seized by the Dakota.

Disease and accident took more lives than the Indians ever did, however. Although the Oregon Trail had been conquered, it exacted a harsh revenge.

CANNIBAL CAMP

IN 1846, A PARTY OF 89 SETTLERS, led by the elderly George Donner, left the California Trail for an untested short-cut south of the Great Salt Lake. By the time they reached the foot of the Sierra Nevada mountains, the party had lost most of its animals. As they climbed the Truckee Pass in late October, early snow fell, and while they sheltered near a lake, snow drifts 20 ft (6 m) high closed the pass. Starvation set in.

On December 16, 15 of the men left to find help on the far side of the pass, but 25 days later, half of them stumbled into an Indian village; eight had died on the way and had been eaten by the rest.

On February 19, 1847, a rescue party found those left from the original party. In a scene of rampant cannibalism, stew pots of human flesh cooked next to the half-butchered bodies of the dead, and tending both were emaciated survivors. In all, 45 members of the original Donner party had died on the way to a better life.

CHILD SURVIVOR *A wedding portrait of Patty Reed, rescued at the age of eight, half starved, with others of her party.*

The umbrella that led to Timbuktu

HOW A SCOT AND A FRENCHMAN EXPLORED THE WEST AFRICAN INTERIOR

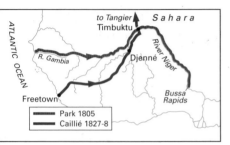

INHOSPITABLE TERRITORY *European explorers attempting to penetrate the heart of west Africa encountered local hostility and illness on the way.*

THE COURSE OF THE RIVER NIGER, DEEP IN THE WEST African interior and sandwiched between malarial swamps along the coast and the barren Sahara, was one of the great mysteries of Africa which lured European explorers. But the dangers of disease and local hostility defeated all but the toughest – or the most crafty. Two men, the Scottish doctor Mungo Park and the French scholar René Caillié, tried very different approaches.

Exploration on the grand scale

Young Dr Park's account of his first journey in the Niger region, *Travels in the Interior Districts of Africa*, was a best seller in 1799. Despite being abandoned by his men and held captive by Muslim tribesmen, he confirmed early reports by the Greek historian Herodotus and medieval Arab travellers that the River Niger flowed from west to east. The government asked him to return to west Africa to trace the course of the Niger from the interior to its outflow.

Park mounted a huge expedition – including 35 soldiers who had to wear the red woollen uniform of the British Army, even in tropical heat – and set out up the valley of the River Gambia. The expedition started badly: all but four of his 40-man expedition died of fever before Park even reached the Niger in November 1805. There they built a boat, planning to follow the river's course. 'If I could not succeed in my mission,' Park wrote, just before setting sail, 'I would at least die on the Niger.'

This indeed was to be his fate: one of the slaves in Park's expedition later made his escape to Freetown, on the Atlantic coast. There, he reported how the party had been ambushed by local inhabitants in the

Bussa rapids, about 500 miles (800 km) up the Niger from the coast – Park had perished trying to swim for the bank.

Travelling in disguise

Nearly 30 years later, a Frenchman, René Caillié, was lured by the 10 000-franc reward offered by the Geographical Society of Paris to any westerner who could reach Timbuktu. It was not a 'forbidden city' in the same sense as Mecca or Lhasa, but the Muslim rulers who controlled Timbuktu and the regions around it would not let Christians in – or, if they reached that far, let them out.

Caillié created an elaborate alias, claiming to be an Egyptian called Abdallahi, who, kidnapped in childhood by the French, was now returning to his family and faith. He learned Arabic and with his savings bought 100 lb (45 kg) of trade goods and an umbrella. Caillié set out from Kakondy, north of Sierra Leone, on April 19, 1827. Travelling among people who had never heard of Egypt, he was obliged to change his story, becoming 'a real sherif of Mecca' – a descendant of the Prophet.

He passed from caravan to caravan and guide to guide, but in August he fell ill with foot sores and scurvy. He took almost a year to reach the River Niger, where a local chief arranged transport by boat to Timbuktu in exchange for Caillié's umbrella. The barge – made of planks bound together with vegetable fibre, caulked with straw and clay and then covered with mats – needed constant bailing with empty gourds, or calabashes. It carried a cargo of rice, honey, textiles, and nearly 50 slaves.

On April 20, 1828, Caillié arrived in Timbuktu itself, at first sight appearing to be no more than a mere scattering of earth houses. Yet, 'there was something imposing in the aspect of a great city,' he wrote. Caillié's return journey took him across the Sahara to Tangier and then Paris, where he collected his reward. He was the second European to visit Timbuktu, and the only one to return alive.

BOLD SCOT *A 6 ft (1.8 m) Scotsman with flaming red hair, Mungo Park cut a conspicuous figure on his travels in west Africa.*

GRAND STYLE *Dr Park's ill-fated 1805-6 expedition was manned by soldiers and local guides.*

DESERT CITY *René Caillié (left) described Timbuktu as 'a mass of earth houses in a bleak plain of sand'.*

'Dr Livingstone, I presume?'

HOW AN AMERICAN REPORTER TRACKED DOWN A LOST EXPLORER

HISTORIC MEETING *'I pushed back the crowds and . . . walked down a living avenue of people . . . As I advanced slowly towards him I noticed he was pale, looked wearied, had a grey beard, wore a bluish cap with a faded gold band on it, had on a red-sleeved waistcoat and a pair of grey tweed trousers,' wrote Stanley. 'I would have run to him, only I was a coward in the presence of such a mob – would have embraced him, only, he being an Englishman, I did not know how he would receive me; so I did what cowardice and false pride suggested was the best thing – walked deliberately to him, took off my hat and said, 'Dr Livingstone, I presume?'*

IN 1871, JAMES GORDON BENNETT, PROPRIETOR OF THE New York Herald, recalled his toughest reporter from the war in Spain. 'Draw a thousand pounds now,' he told Henry Morton Stanley, 'and when you have gone through that draw another thousand, and when that is spent, draw another thousand . . . and so on; but FIND LIVINGSTONE . . . Good night and God be with you.'

In search of a scoop

Two factors made possible the most famous meeting in the story of African exploration: the ruthlessness of reporter Henry Morton Stanley and the wealth of his patron. David Livingstone – missionary and explorer – was a national hero, famed, among other things, for having been the first European traveller to reach the Victoria Falls on the Zambezi during his exploration of the river in 1855. He stayed deep in the heart of Africa for so long that most people believed he was dead, but Bennett had other ideas – he wanted a scoop for his newspaper.

Where Livingstone used the Bible, Stanley used a bludgeon. He forged his way forward, ruthlessly sacrificing life and principle on the way. When he arrived in Zanzibar, the main port for the east African slave trade, he bought the services of 157 porters, excellent interpreters, two inexperienced white lieutenants, a dozen of the native auxiliaries who had served with John Speke's Nile expedition of 1860-3 – and an enamel bath and a Persian carpet for his own tent.

Vitally for Stanley's strategy, Bennett's money also bought an arsenal of weapons and 18 bales of fine textiles, for he intended to fight and bribe his way to the town of Ujiji on Lake Tanganyika, which was Livingstone's last known station.

'Forward' was Stanley's motto. His demonic energy and the brutal pace he set reduced many of his men to mutinous stirrings, which he quelled with floggings – later in his career, as he fought his way up the Congo river in Zaire, his men nicknamed him *Bula Matari*, meaning 'smasher of rocks' in Swahili. He survived two murder attempts on the way to Ujiji, and nothing could stop him until his path was blocked by an army led by Mirambo, a slave-trader's porter who had appointed himself a chief.

Taking the back route

Stanley's instinct was to attack, but as his allies ran away, he was forced to embark on a long detour to the south. This diversion and recurrent bouts of malaria threatened the momentum of the expedition. His diary became feverish: 'No living man shall stop me . . . But death? Not even this . . . I shall find him – and write it larger – FIND HIM! FIND HIM!'

By November 6, 1871, Stanley despaired of reaching Livingstone 'without being beggared' by the extortions of local chiefs. Buying up as much food as could be carried, he bribed an expert local guide to lead his party away in small groups west, off the beaten track, under the cover of darkness. Four days later, they emerged within sight of Ujiji. The champagne and silver goblets Stanley had carried with him to toast Dr Livingstone had not, after all, travelled in vain.

'Ships that walk in the water'

HOW STEAMBOATS RULED THE MISSISSIPPI

THE RESIDENTS OF THE SMALL MISSISSIPPI RIVERPORT heard the distinctive, high-pitched whistle of the brightly painted steamboat *J.M. White.* long before it sailed majestically into view. It was the summer of 1880, and steamboats were the main means of transport in the American West. Faster, cheaper and more comfortable than overland stagecoaches, they churned along some 7000 miles (11 000 km) of the Mississippi and its tributaries. They were the main link from St Louis in the north to New Orleans in the south, travelling at a stately rate of about 15 mph (24 km/h) and calling at centres such as Cairo, Memphis, Natchez and Baton Rouge.

Christened the 'ships that walk in the water' by the American Indians, the steamboats had an average working life of four to five years, barring accidents such as collisions or fires. Measuring up to 350 ft (107 m) long and 58 ft (18 m) wide, they had low-lying hulls. Their slender smokestacks stood up to 95 ft (29 m) tall and were hinged to tilt so that the boats could pass under bridges. The engine rooms were on the main deck, level with the water line.

'Aladdin built a palace/He built it in a night/And Captain Tobin bought it/And named it J.M.White.'
ANONYMOUS

On either side of the boat, twin engines were connected to the paddle wheels by thick wooden levers. These turned the wheels as the boilers built up pressure of at least 150 lb per sq in (10.5 kg/cm²). Many of the boats had boilers which burnt wood, often using so much that the vessels docked twice a day to take on more logs. Those which were cheaply built and poorly looked after tended to explode.

Mississippi river sidewheel steamboats used two paddle wheels mounted on separate shafts so that they could work independently of each other. This made for great manoeuvrability. By reversing one wheel, for example, and going ahead with the other, a steamboat could be turned in its own length.

The captain acted as overall manager of the boat, while river pilots, who worked in shifts, were responsible for safe navigation. The pilots' expert knowledge of the river made the difference between success and disaster, and they navigated both with the aid of landmarks and through consultation with other pilots. They rarely left the pilothouse, which housed the wheel of up to 12 ft (4 m) across.

Steamboats were the rulers of the Mississippi for most of the 19th century, but progress took its toll. With the coming of first the train and then the car and lorry, their days were numbered.

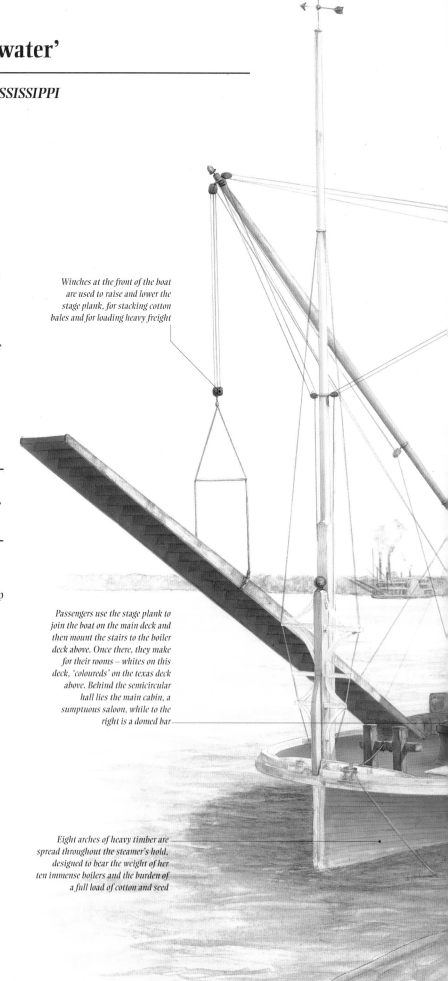

Winches at the front of the boat are used to raise and lower the stage plank, for stacking cotton bales and for loading heavy freight

Passengers use the stage plank to join the boat on the main deck and then mount the stairs to the boiler deck above. Once there, they make for their rooms – whites on this deck, 'coloureds' on the texas deck above. Behind the semicircular hall lies the main cabin, a sumptuous saloon, while to the right is a domed bar

Eight arches of heavy timber are spread throughout the steamer's hold, designed to bear the weight of her ten immense boilers and the burden of a full load of cotton and seed

The smokestacks of the White soar 80 ft (24 m) above the roof of the cabin and weigh up to 28 500 lb (12 900 kg). The moulded leaves at the top of the chimneys are each 7 ft (2 m) long, and like the chimneys themselves, are made from sheet-iron

The panelled and moulded pilothouse, reached from the texas deck below, holds the pilot wheel measuring 11 ft (3 m) across. One of the two pilots stands at the ship's helm, and outside the window of the pilothouse the ship's name is carved onto a nameboard and painted with gold leaf

Below the pilothouse lies the texas deck, 180 ft (55 m) long and 23 ft (7 m) wide. Here are the quarters for the captain and his crew, as well as 30 cabins for 'coloured' passengers. The two bridal chambers among these are slightly larger than the rest, but most measure a snug 5 ft 9 in (2 m) by 6 ft 2 in (1.9 m); all are furnished to the same high standards as those on the boiler deck below. Elegant walnut stairs connect the texas guard, the walkway round the deck, with the hurricane deck below

The ship's bell weighs 2880 lb (1300 kg). It was last put to use on December 14, 1886, when it was rung to warn the passengers of a rapidly spreading fire on board. Even so, some 20 people died in the fire, most of them women and children. After gracing the Mississippi for eight years, the White's working life was over

The main cabin is lit with seven gilded chandeliers, and has 24 doors on either side of its 233 ft (71 m) length. Two lead to bridal chambers for newlyweds, while 14 of the sleeping cabins hold double beds, eight with extra trundle beds stored underneath them; the remaining 34 have bunk beds. The doors and windows in the main cabin and other staterooms are filled with stained glass in blue, purple and gold, each panel a different design to the next

Just in front of the paddle wheel is the barber shop, and next to that a bathroom and water closet – probably the only ones on board. Below is a storeroom reached from the boiler deck

Close to the paddle wheels sit boilers made of steel, instead of the more usual iron. As a result, they are the safest on the river.

Black labourers use the stage planks at the front and sides of the J. M. White to load freight and coal, needed to stoke the boilers

SAILING THE MISSISSIPPI At each port of call, steamboats took aboard and unloaded tons of valuable freight including logs, iron, whiskey, butter and meat. They also conveyed mail along the river. The J.M. White. was built to carry more than 10 000 bales of cotton and seed, and its passengers included businessmen, politicians, planters, preachers and professional gamblers.

Higher than any other mortal

HOW AN INTREPID SCIENTIST SET A MOUNTAIN CLIMBING RECORD

WHEN A SURVEY PROVED THAT EVEREST WAS HIGHER than Mount Chimborazo, Friedrich von Humboldt was annoyed. 'All my life,' the Prussian scientist wrote, 'I have imagined that of all mortals I was the one who had risen highest in the world.' Although he never reached the summit, Humboldt's ascent of Chimborazo in the Ecuadorean Andes remained a mountaineering record for almost 30 years.

Climbing through the clouds

He began his attempt on June 23, 1802, as part of a survey of Latin America. He and his companions, Aimé Bonplaud and Carlos Montúfar, had only ponchos for warmth and relied on lungs acclimatised during practice ascents to help them breathe the rarefied air. They climbed through clouds and edged along a 10 in (25 cm) wide ridge. 'To our left,' wrote Humboldt, 'was a precipice of snow; to our right lay a terrifying abyss, 800 ft to 1000 ft (245-300 m) deep, with huge crags of naked rock sticking out of it.'

At 17 300 ft (5275 m), Humboldt's worst discomfort was his hands, cut on the rocks. After another hour, however, he was numb with cold and starting to feel nauseous and breathless. At 19 286 ft (5878 m), they reached an uncrossable ravine, 60 ft (18 m) across and 400 ft (120 m) deep. 'We delayed no longer than sufficed to collect fragments of rocks . . . as we foresaw that in Europe we should frequently be asked for a piece of Chimborazo,' wrote Humboldt.

HEAD FOR HEIGHTS *During his ascent of Mount Chimborazo, Friedrich von Humboldt deduced that altitude sickness – from which he suffered badly during the climb – was caused by a lack of oxygen.*

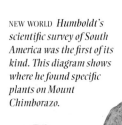

NEW WORLD *Humboldt's scientific survey of South America was the first of its kind. This diagram shows where he found specific plants on Mount Chimborazo.*

A great natural staircase

HOW THE MATTERHORN WAS FIRST SCALED

ON THE SUMMIT OF THE MATTERHORN, THE BRITISH artist Edward Whymper and the guides, Peter Taugwalder and his son, watched in horror as their four friends slid down the icy slope on their backs, spreading their arms as they tried to save themselves. 'They disappeared one by one,' wrote Whymper, 'and fell from precipice to precipice' – a drop of nearly 4000 ft (1200 m).

Triumph before the tragedy

Just before the climbers plummeted to their deaths, they had been jubilant. On July 14, 1865, the Matterhorn had finally been scaled. It was the 25-year-old Whymper's eighth attempt to scale this imposing 14 688 ft (4477 m) peak high in the Pennine Alps, but it was the first time anyone had seriously attempted to tackle the east slope. Until now, an optical illusion had deterred all comers: the upper section of the east face looked sheer and as Whymper wrote, 'repulsively smooth'. But shortly after leaving their camp at 11 000 ft (3350 m), Whymper noticed that it was in fact 'a huge natural staircase'. It turned out that the incline was just 40°. Only the last stage of the ascent was difficult and, ultimately, fatal – a slope of smooth, frozen snow a few hundred feet below the summit.

The mountaineers had completed the ascent in two days. An Italian party had set off by the south-west route on July 11, and the victors were anxious to defend their conquest. Spotting the Italians from the summit, but unable to make them hear their shouts, 'we drove our sticks in, and prised away the crags and soon a torrent of stones poured down the cliffs . . . The Italians turned and fled'.

GENTLEMAN CLIMBER *For his assault on the Matterhorn, Edward Whymper carried a spiked stick, known as an alpenstock, an ice axe and a length of rope.*

FATAL MOMENT *As they descended the mountain, one of Whymper's companions slipped, pulling down three more. The four men tumbled thousands of feet to their deaths, and Whymper and his guides were only saved because the rope that tied them together broke.*

The world's harshest mountain face

HOW THE NORTH FACE OF THE EIGER WAS CONQUERED

ON A CLEAR DAY, SIGHTSEERS COULD SPOT THE BODIES dangling from ropes against the north face of the Eiger. For a year they hung there – from the summer of 1935 to the summer of 1936 – before a further expedition cut them down. For more than 70 years the north face of the Eiger had been classed as too steep to climb, until two young climbers, Max Sedlmayer and Karl Mehringer, planned to claim the honour of conquering the most difficult climb in Europe for Germany. Two days later, they froze to death at the spot still called Death Bivouac.

Rivals crowd the slopes

That first attempt inspired emulators. When Heinrich Harrer and Fritz Kasparek, from Austria, started their climb on July 21, 1938, the mountain seemed crowded. Above them were the German master climbers Andreas Heckmair and Ludwig Vörg, using better equipment and 12-point crampons (boot spikes). After a while, the Germans turned back, distrusting the weather.

The first dangerous obstacle was the Hinterstoisser Traverse – a rope-assisted route to the first ice field. Kasparek led the way, fighting for balance on smooth ice. Beyond the Ice-Hose, a dripping, ice-lined groove, they paused for the night. The next day it took 5 hours to tackle the second ice field. As they reached its rim, Heckmair and Vörg suddenly reappeared. Starting afresh, they had caught up in a single day. Stifling rivalries, the climbers went on together.

Finally they reached the deadliest stretch of the ascent, the White Spider, a steep sheet of frozen snow, swept by avalanches. They arrived just as a storm broke. Kasparek ringed himself to a piton embedded in ice up to its hilt; Heckmair, clinging to Vörg's collar, buried his axe deep in the snow.

They bivouacked at 12 300 ft (3750 m), secured by rings to the wall of the Eiger. Despite worsening weather, the climbers went on, using all their strength to scramble up the icy vertical cracks out of the White Spider. When they emerged on the summit ice field it seemed flat by comparison. Even so, the last slope took hours to cross and when they reached the cornice they almost toppled over it in surprise onto the rocks of the south face. They reached the summit at 3.30 pm on July 24, 1938.

ENDURANCE AND COURAGE *The first to climb the north face of the Eiger – German for 'ogre' – were (from top to bottom) Andreas Heckmair, Ludwig Vörg, Heinrich Harrer and Fritz Kasparek. They returned, in Harrer's words, 'conscious of the privilege of having been allowed to live'.*

ABANDONING SHIP *In desperation, the crew quit Franklin's two ice-bound ships, at first in smaller boats and then by walking over the ice, leaving their comrades' corpses behind. They all died before reaching safety.*

Death in the ice

HOW SIR JOHN FRANKLIN LOOKED FOR THE NORTHWEST PASSAGE

WANTED *The government offered a generous sum to anyone who found the lost explorers.*

SIR JOHN FRANKLIN WAS GRIPPED BY AN OBSESSION THAT eventually killed him. He died in June 1847, his ship clasped in the ruthless embrace of Arctic ice, having devoted his life to the search for the elusive Northwest Passage – a long-sought-after short cut from the Atlantic to the Pacific around the Arctic coast of North America.

Franklin's first attempt to solve the riddle of the Northwest Passage was made as a young Royal Navy lieutenant in 1819, when he led an expedition that followed the mainland coast by land and inshore canoe. By the end, Franklin was eating lichen scraped from the rocks – but he had identified a substantial stretch of navigable coast. Although the sufferings he endured would have deterred most men, he remained determined to continue the survey.

Lessons of experience

In 1824, the British Admiralty authorised a new expedition for the following year. This time, Franklin hired a fishing expert to help to feed his crew, and took light boats that could be broken up for porterage and reassembled in 20 minutes. After surveying 1100 miles (1770 km) of coast, however, the results were disappointing: no navigable route hugged the mainland coast. Franklin pressed for a new attempt by sea, but it took 18 years, by which time he was nearly 60, before the Admiralty gave in. Franklin's two ships – stocked with three years' provisions preserved in cans, the latest innovation – sailed westwards from Disko Island, off Greenland, on July 13, 1845, and were never seen again.

Records later discovered under scattered cairns told a grisly story. In September 1846, the ships were beset by ice 30 miles (48 km) north-west of King William Island, and there they remained through two winters and two thaws. Franklin and 34 of his men perished on board of mysterious maladies before the 104 survivors abandoned ship in April 1848. They headed south across the ice, looking for a Hudson's Bay Company trading post situated on the Great Fish River, but they never reached it; some had died of starvation, some of scurvy, while others were slowly poisoned by the lead used to solder their tins of food. Their bodies were finally found on King William Island at the end of the 1850s.

Meanwhile, in England, Lady Franklin became the personification of a Victorian ideal of womanhood: first the patient wife, waiting for her hero to come home; then the stoical widow, sacrificing her fortune to search for his body. The relief parties she lobbied for came too late, but the search parties did find the route Franklin had sought, east of his course, around the south of King William Island. Lady Franklin died in 1875, just too early to see the monument in Westminster Abbey which was erected in her husband's honour.

BRAVE WIFE *Lady Franklin never lost hope.*

INTREPID HUSBAND *Sir John Franklin was hailed as a hero.*

OUTWARD BOUND *Confident and determined, Peary stands on the deck of the* Theodore Roosevelt. *He wears an outfit of Arctic fox and polar bear skin.*

MISSION ACCOMPLISHED *Back again at base camp, the strain of Peary's final race towards the Pole shows on the explorer's weather-beaten face.*

WARM WELCOME *Peary and the crew were happy for local families to make their homes on board ship. Inuit women supplied members of the expedition with warm clothing and strong shoes.*

TOUGH TERRAIN *Snow ridges up to 25 ft (7.6 m) high were 'heavy and tortuous going' with loaded sleds in tow, wrote Peary.*

Frozen pemmican, burning ambition

HOW ROBERT PEARY ATTEMPTED TO REACH THE NORTH POLE

FREEZING FOG BELCHED FROM BROAD CRACKS IN THE shifting ice. On March 11, 1909, after waiting huddled in an igloo for four days, watching the water freeze over, listening for the telltale grinding of the ice, the American explorer Robert Edwin Peary and his five companions picked their way over rafts of young ice that shuddered, tilted and crushed into piles. It was 'like crossing a river on a succession of gigantic shingles', Peary wrote.

'On the polar ice,' Peary explained, 'we gladly hail the extreme cold, as higher temperatures and light snow always mean open water, danger and delay. Of course, such minor incidents as frosted and bleeding cheeks and noses we reckon as part of the great game. Frosted heels and toes are far more serious, because they lessen a man's ability to travel, and travelling is what we are there for.'

Surviving the Arctic conditions

For nearly 20 years Peary had wanted to be the first man to reach the North Pole, and he had already spent nine winters in the far north, befriending the local Inuit. They taught him the secrets of Arctic survival, and he learnt how to hunt in the snowy wastes, how to build a snowhouse and how to handle a team of huskies. Now, at the age of 53, the tough US naval officer was playing what he called his 'last game on the Arctic chessboard'.

The expedition – numbering 22, including Matthew Henson, Peary's black servant and right-hand man – sailed in the *Theodore Roosevelt*, a 150 ft (46 m) schooner-rigged steamer. They travelled up Smith Sound to arrive at their base at Cape Sheridan in time for the winter of 1908-9. There they hired Inuit sledders and a short route across the ice to the Pole was decided on.

Peary relied on the muscle power of humans and huskies to reach his goal. The expedition's main source of food was 30000 lb (13 600 kg) of pemmican – a nutritious, if unappetising, mix of dried beef, fruit and suet – which was eked out with biscuit, tea and tobacco. Fresh meat was supplied mainly by hunting, although the Inuit sledders ate any dogs which died along the way.

Moving closer to the Pole

A small advance party on snowshoes led the way at every stage, supported by sledding-parties, each of which carried 500 lb (227 kg) of provisions. At each halt, superfluous mouths were sent back to keep the trail open and to build snow houses for the return journey. On April 3, beyond 86°N, with the weather

temperatures rising to 8°C below freezing (15°F), Peary decided to dash for the Pole with a single unsupported sledge and a picked crew of five men – four Inuits and Henson. The last 133 miles (214 km) were covered in five bursts before the party finally arrived at what they thought to be the North Pole at 10 am on April 6.

Later explorers have doubted the accuracy of Peary's calculations, and Peary himself had trouble believing that he had fulfilled his dream – 'The prize of three centuries,' he wrote in his journal. 'My goal for twenty years! I cannot bring myself to realise it. It seems all so simple.' Nonetheless, it was an epic trek and one that took Peary close to the North Pole.

SHIPS BUILT FOR THE ARCTIC

'THE ICE CAME … WITH SUCH SWIFTNESS THAT HAD THE Lord himself not been on our side we had surely perished; for sometimes the ship was hoisted aloft; and at other times she … would force great mighty pieces of ice to sink down on the one side of her and rise on the other.' Such hazards off Canada's Hudson's Bay, described by the Englishman William Baffin in 1615, became all too familiar to Arctic explorers.

Arctic challenges

There were various ways of dealing with Arctic conditions. Ships could be towed or winched free of clinging ice, or a way could be cut through with saws. Loose ice had to be fended off with boat hooks and poles, although allowing a wooden vessel to be squeezed up onto the surface of the ice could save the ship from certain disaster.

In the 1890s, when testing his hypothesis that ice is subject to currents, the Norwegian Fridtjof Nansen deliberately allowed his ship, the *Fram*, to freeze into pack ice near the New Siberian Islands. He drifted, embedded in the ice, for three years to emerge in the Greenland Sea.

Steam power and iron hulls transformed the possibilities for Arctic exploration, and early in the 20th century the first icebreakers cleaved the ice with their metal stems.

ICEBOUND *Nansen's ship, the* Fram, *was designed for the Arctic. Her sides sloped inward so she could slip out of the ice floes' grip and endure the pressure exerted by the ice by rising above it.*

Sport and sabotage in the saddle

HOW EARLY CYCLISTS RODE FOR PLEASURE – AND GLORY

THE TOUR DE FRANCE BICYCLE RACE WAS WELL UNDER way when things started to go mysteriously – and disastrously – wrong. One of the riders began to squirm and scratch as if his shirt had been filled with itching powder. Another cyclist could not sit still on the saddle, as though his shorts contained wire wool. Yet another competitor fell asleep over the handlebars and crashed. And at least one more cyclist bit the dust as his machine collapsed beneath him.

The passionate rivalry surrounding the second running of the Tour de France, in the summer of 1904, brought about a spate of sabotage. Soporific drugs were secreted in the water bottles of some competitors. The frames of several machines were damaged by filing. In addition, fanatical supporters threw nails in the path of 'enemy' cyclists, causing a string of punctures – and blocked their way with farm carts or chopped-down trees.

Out to win – by fair means or foul

The race favourite was the previous year's winner – the moustached Frenchman Maurice Garin. At one stage he was threatened by about 100 men brandishing sticks and stones and was struck on the head with a bottle. Garin went on to win; but was disqualified and banned for two years for allegedly travelling part of the way by car.

The world's toughest and most fiercely contested bicycle race had been founded in 1903 by a French cyclist and newspaper editor named Henri Desgrange. Divided into six stages – from Paris to Lyons, Marseille, Toulouse, Bordeaux, Nantes and back to Paris – it covered a distance of some 1509 miles (2428 km). Desgrange saw the race as a way of putting France on the sporting map of Europe – and of promoting national identity and pride.

Followers of fashion

When cycling first took to the roads and byways in the late 19th century, it had been regarded as no more than a pleasant and healthy pastime. Townspeople used bicycles to escape to the country at weekends, and in Britain their popularity increased when it was learned that Queen Victoria had ordered two of the latest machines for her household.

Many women cyclists replaced their skirts, which tended to get caught in the wheels, with knickerbockers – loose-fitting breeches gathered in at the knees – and long woollen stockings, such as those worn by men cyclists. Others, particularly Frenchwomen, favoured 'bloomers' – long, baggy trousers based on Turkish pantaloons which gathered modestly about the ankles. These had been made famous by an American campaigner for women's rights, Mrs Amelia Jenks Bloomer, who wore them in 1851 as part of a 'rational' costume for ladies.

DOGGED COMPETITOR
French racing cyclist Maurice Garin was known as the 'White Bulldog' from his distinctive white jersey.

THE FIRST BICYCLES

IN 1839, SCOTTISH BLACKSMITH Kirkpatrick Macmillan built his own personal transporter – a two-wheeled machine with pedals joined to rods which turned the back wheel and propelled the cycle forward. Manufacturers were soon making their own versions of the Scotsman's cycle.

In 1870, the penny-farthing appeared. Cyclists had to use steps to reach the high saddle, and dismounted by jumping off backwards while the cycle was still in motion. The 'safety' bicycle of 1885 was far easier to handle with its wheels of equal size, brakes, gears and sturdy diamond-shaped frame.

VELOCIPEDE
The 1861 Parisian cycle had a candle-lit lamp for night riding.

PENNY-FARTHING
The big front wheel of an 1870s cycle allowed more ground to be covered with each turn.

SAFETY BICYCLE
In 1895 the Elswick lady's bicycle had chain guards to protect skirts.

SIMPLE WEAR
A lady cyclist from the 1890s (left) dresses modestly in a long skirt, high-necked blouse, and leather boots.

SMART WEAR
Knee-length pantaloons (left) were popular in the 1860s, but by the 1890s, knickerbockers (right) were the preferred dress.

MEN'S WEAR
Knickerbockers, long stockings and a flat cap were fashionable cycling gear in the 1900s.

CHILD'S WEAR *A tricyclist shows off straw hat and button boots.*

Hazards of the 'horseless carriage'

HOW EARLY MOTORISTS DEALT WITH RIDICULE AND RISK

MOTORING FOR THE MASSES *In 1908 Henry Ford began to mass-produce the cheap and economical Model T, affectionately known as the 'Tin Lizzie'. From 1913 it was available only in black.*

MOTORING FOR THE FEW *Only the rich could afford the 1904 Daimler – a spacious and luxurious touring car ideal for country outings and seaside picnics. It became a favourite with Edward VII and the British royal family.*

THE HONOURABLE CHARLES ROLLS WAS PREPARED FOR trouble when he set out to drive solo through the night from Paris to the northern seaport of Le Havre in the harsh winter of 1900. But the 110 mile (177 km) journey was tougher than he had expected. More than a dozen mishaps, from burst tyres to oil leaks, troubled his journey. At one stage Rolls – who four years later was to team up with the engineer Henry Royce to form the luxury car-manufacturing firm of Rolls-Royce – spent 2 hours lying on his back under the radiator mending leaks. Freezing water trickled over him, and his leather motoring coat stiffened with ice. Even so, Rolls reached Le Havre by dawn – and went straight to bed to recover from his ordeal.

Barking dogs and startled horses

In his courage and perseverance Rolls was typical of the pioneer motorists who braved the elements – and much more – in their love of the motor car. There was the ever-present danger of skidding on wet, greasy and snowy roads. There was the risk of the starting handle slipping and hitting the driver in the face. There was the hazard of being chased by barking dogs, or troubled by startled horses. Garages were few and far between, so motorists carried their own tool kits, inner tubes, spark plugs, spare parts and petrol.

In cold weather, when the carburettor – which produces an explosive vapour of fuel and air in petrol engines – refused to work, it might be warmed by a hot iron. On hot days there was the risk of the engine overheating. In town, the cars' wheels got caught up in tram-lines, and in the country villagers jeered at the self-styled 'knights of the road'. To protect motorists' interests, automobile clubs were formed throughout Europe. The first of these was the Automobile Club of France, founded in Paris in 1895 and a keen promoter of annual car shows.

In the early days of the 'horseless carriage' in Britain a man carrying a red flag was legally obliged to walk in front of each machine to warn pedestrians of its approach. This absurdity was abolished by the Motor-car Act of 1896, and in 1903 a top speed of 20 mph (32 km/h) was set. Motoring was gradually taken up by professional men such as doctors, who abandoned horse carriages for their rounds and took to cars such as the German Opel 'Doktorwagen' – a two-seater with an adjustable hood – of 1909.

Brocade bed and mobile toilet

The rich spared no expense on their limousines. These large cars – with an enclosed passenger compartment – were named after the cloth from the French province of Limousin, which was used in the cars' interior. The limousine was associated with luxury. The Comte Boni de Castellane, for instance, had brocade armchairs which turned into a bed fitted in his *limousine de voyage* – as well as a wine cabinet, and a silver washbowl with hot and cold running water. One American millionaire even had a flush toilet built in the back of his 90-horsepower Charron.

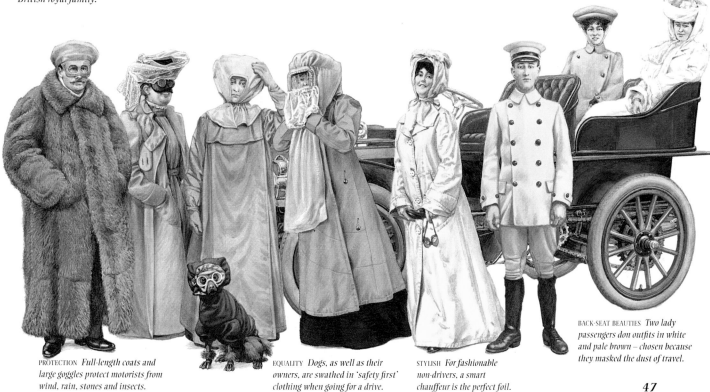

PROTECTION *Full-length coats and large goggles protect motorists from wind, rain, stones and insects.*

EQUALITY *Dogs, as well as their owners, are swathed in 'safety first' clothing when going for a drive.*

STYLISH *For fashionable non-drivers, a smart chauffeur is the perfect foil.*

BACK-SEAT BEAUTIES *Two lady passengers don outfits in white and pale brown – chosen because they masked the dust of travel.*

Those magnificent men in their flying balloons

HOW THE MONTGOLFIERS SENT THE FIRST AERONAUTS ALOFT

ONLOOKERS CHEERED WILDLY AS THE HOT-AIR BALLOON – a magnificent creation in blue, gold and red – rose into the air above the Bois de Boulogne and floated off over the rooftops of Paris. The two volunteer passengers – the historian Jean Pilâtre de Rozier and the Marquis d'Arlandes – raised their hats to acknowledge the acclaim. In the streets, people gasped and pointed in amazement at the balloon, the first ever to carry humans on a free flight.

Floating on hot air

The balloon, the *Montgolfière*, had been designed and built by two brothers, Joseph Michel and Jacques Étienne Montgolfier, owners of a papermaking factory near the small town of Annonay, near Lyons in south-east France. The airtight oval bag – measuring some 50 ft (15 m) across the middle and about 75 ft (23 m) high – consisted of paper-lined linen sections buttoned together. From the open neck of the bag hung a wire-mesh brazier. This burned straw and wool to produce the hot air that inflated the balloon and kept it airborne.

The flight took place on November 21, 1783 – about a year after the Montgolfier brothers had begun their experiments with hot-air flight. Noticing how the heat of a fire whisked charred bits of paper up the chimney, they tried holding small silk bags over a fire in their home, open ends downward. On release, the bags, filled with hot air, floated up to the ceiling. The brothers thought they had found a lifting force – which they called 'Montgolfier gas'. But they were simply taking advantage of the fact that air expands when heated, becoming less dense and, therefore, lighter than the cold air surrounding it. Fascinated by this phenomenon, the brothers

had staged a demonstration in Paris in September 1783 with a balloon carrying a wicker cage containing a sheep, a rooster and a duck. This flight lasted 8 minutes and the balloon and its passengers landed safely in the Forest of Vaucresson, 2 miles (3.2 km) from the starting point.

Fighting the flames

Two months later, the Montgolfiers achieved the first-ever manned flight. With its two passengers standing on either side of the balloon to keep balance, the craft rose to about 3000 ft (900 m). It drifted over the Seine. Suddenly, de Rozier and d'Arlandes noticed that the fire in the brazier was burning holes in the balloon's fabric and starting to sever the ropes connecting the gallery to the balloon. Desperately, the two men used wet sponges to put out the flames.

Finally, after a flight lasting 26 eventful minutes, the balloon touched down between two millhouses about 5 miles (8 km) from its take-off point. It was, declared the marquis, 'the most significant and exciting event of the century'.

UP THEY GO! *Two passengers – Jean Pilâtre de Rozier and the Marquis d'Arlandes – wave to the crowd as the Montgolfiers' balloon rises into the air. From now on the balloonists are at the mercy of the winds.*

UNMANNED FLIGHT Wilbur Wright (background) and an assistant hold onto guide lines as they make aerodynamic tests with a glider in September 1902.

SUCCESS STORY The Wrights wired news of their triumph to their father at home in Dayton, Ohio.

HAND-WRITTEN In his diary, Orville Wright wrote a detailed account of the brothers' four flights.

Man takes to the air in a network of wires

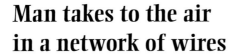

MANNED FLIGHT *Wilbur Wright pilots one of the 250 glider flights made by the brothers in two days in October 1902.*

HOW THE WRIGHT BROTHERS MADE THE WORLD'S FIRST POWERED FLIGHT

IT WAS A COLD AND WINDY MORNING ON THE BEACH near the village of Kitty Hawk, North Carolina, as – on December 17, 1903 – the American aviation pioneer Orville Wright attempted to become the first man to make a powered, controlled flight in a heavier-than-air machine. Lying face down on the lower wing of his biplane, the *Flyer*, he grasped a wooden strut for support with one hand and

than 40 ft (12 m). Its main features were the double wings, with ribs consisting of strips of wood glued together and tacked in place over supporting blocks. Upright wooden struts were attached to the blocks by metal hinges. The wings were crisscrossed with stay wire and covered with white muslin, sewn in place.

As the petrol engines then available were too heavy for flight, the Wrights built a lightweight version. Weighing 200 lb (91 kg), the four-cylinder, 12 horsepower engine was positioned slightly to the right of centre on the lower wing – so that it could not fall on the pilot's head in the event of a crash.

The pilot lay on his stomach to the left of the engine, and was held in place by a cradle-like harness around his hips. Wires running from the harness to the wingtips and to the rear wood-and-muslin double rudder – used for turning left or right – enabled him to adjust both the wings and the rudder merely by

FLYING FAN *Orville Wright was fascinated by the possibility of powered flight.*

CAUGHT IN THE ACT *Just as it lifted from the launch rail into the air, the* Flyer *– with Orville Wright at the controls, and Wilbur looking on – was photographed by John T. Daniels of the Kill Devil Hills Life Saving Station.*

prepared to work the controls with the other. Soon he shouted that he was ready to take off. The engine burst into life, the two rear-mounted propellers turned, and the plane's restraining rope was slipped. The biplane, mounted on a dolly, accelerated along a level, metal-edged wooden launch rail 60 ft (21 m) long beside a group of windswept sand dunes called Kill Devil Hills.

With Orville's older brother Wilbur running alongside to support the right wingtip, the *Flyer* gathered speed. Suddenly it became airborne. It flew low over the beach for around 120 ft (36 m). Twelve seconds after take off its skids crashed into the hard sand. Wilbur dashed up to the machine and shouted his congratulations. So the Wright brothers achieved their ambition of flying like the birds.

Copying hawks and eagles

Inspired by a series of glider flights made by Otto Lilienthal in Germany in the 1890s – and after carefully and repeatedly studying the way buzzards, hawks and eagles controlled their flight – the Wrights had begun to build the *Flyer* in their home town of Dayton, Ohio, early in 1903. It was just over 21 ft (6 m) long, with a wingspan of more

changing the position of his body. The wingtips could be 'warped' – or twisted up or down – to produce extra lift on one side, and less on the other, so that the plane could bank for a turn. The pilot's hands were free to hold on and to work the front rudder, or 'elevator', which raised or lowered the nose.

The Wrights also made the first practical aeroplane propellers, each 8 ft 6 in (2.6 m) long. These consisted of three thin layers of spruce, with the tips covered by canvas, and were coated in aluminium paint for protection. They were powered in opposite directions at about 330 revolutions a minute by chains attached to the engine.

In the autumn of 1903 the brothers transported the *Flyer* to Kitty Hawk, and assembled and tested it. Their first attempt to fly it, on December 14, ended in a crash. But on Thursday, December 17, the biplane was repaired and ready. Orville took his place on the lower wing, and a few minutes later was airborne. The Wrights made four flights in all on that day – in the longest, Wilbur flew for 59 seconds, covering 852 ft (259 m). They had confounded the scientists who dismissed heavier-than-air flight as impossible.

HANDY MAN *Wilbur Wright, with his brother Orville, ran a bicycle shop in Ohio.*

Circumfusa.sedet.digna
Talis.erat.quondam.patria
Mensa.fuit.lobo: sic.cum
Fac.Deus.ut.multos.haec.q
Germinet.ut.lobi.stirps.r
Fercula.praeclaro.donasti
Haec.habeant.longos.gai
Ao DNI.1567.

Alta 6

Alta

Aetatis svae

Everyday Life

THE WAY IN WHICH WE AND OTHER PEOPLE LIVED IS A SUBJECT OF ENDLESS FASCINATION. THE BASIC ELEMENTS – FINDING SHELTER AND CLOTHING, KEEPING CLEAN AND SATISFYING THE NEEDS OF THE FAMILY – HAVE ALWAYS BEEN THE SAME, BUT THE WAYS OF ACHIEVING THEM HAVE VARIED ACCORDING TO CUSTOM AND CULTURE

LORD COBHAM AND HIS FAMILY / HANS EWORTH

Powders, potions and scented head cones

HOW THE ANCIENT EGYPTIANS IMPROVED THEIR LOOKS

THE WEALTHY EGYPTIAN LAWYER TURNED TO HIS WIFE and asked her to check that his eye shadow was even. Although he was well practised in the art of applying make-up, he wanted to be absolutely confident that he looked his best for tonight's grand feast. It had taken him 3 hours to arrange his hair and make up his face, and still he was not satisfied.

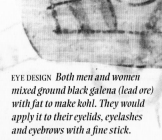

EYE DESIGN *Both men and women mixed ground black galena (lead ore) with fat to make kohl. They would apply it to their eyelids, eyelashes and eyebrows with a fine stick.*

Lotions and potions
Since both men and women wore elaborate make-up, one of the Egyptian household's most prized possessions was a cosmetics box. This contained a shiny copper disc used as a mirror, razors for removing body hair, combs and hair curlers, tweezers for plucking the eyebrows, thin eyeliner sticks, pots for eye shadow, rouge and lipsticks. Upper eyelids were shaded deep blue with powder shadow made from ground azurite; lower lids were painted bright green with eye colour made from ground malachite (copper ore). Both shades were mixed into a paste with water or gum. Women made rouge and lipstick from ground red ochre (iron-stained clay) mixed with fat, and used henna to colour the palms of their hands and their nails.

Crowning glory
Wealthy Egyptians were just as fussy about their hair, dying it with henna and applying plenty of perfume. Both men and women wore wigs made from human hair and vegetable fibre padding, piled high and styled with masses of curls and braids. Sometimes the wearers adorned the wig with semiprecious stones, or set it off with a miniature cone of scented animal fat, which slowly melted during a warm evening and dripped down their necks and faces.

CLOSE CUT *It was fashionable in ancient Egypt to shave a young boy's head, save for one side lock.*

In summertime, face creams – made of vegetable or animal oil, mixed with ground limestone – were worn more out of necessity than vanity; and to keep skin supple, Egyptians rubbed oil all over their skin. Heavy eye make-up had a dual purpose; it protected their eyes from the glare of the Sun and helped to ward off flies. The Egyptian preference for strong perfumes further helped to repel flies.

HAIR CARE *Women spent hours dressing each other's hair, brushing it, plaiting it elaborately, and adorning it with fancy combs.*

Preservation for an eventful afterlife

HOW THE EGYPTIANS EMBALMED AND MUMMIFIED THE DEAD

THE CHIEF EMBALMER HANDED A PRICE LIST TO THE bereaved family and, indicating a set of model mummies, asked them to make their choice. Several different types of treatment were on offer. The simplest and cheapest of these took only a couple of days to carry out and consisted of an internal rinse with a basic purgative, such as a mixture of salt, cassia and senna. More expensive was an injection of cedar oil into the belly: beforehand, the embalmer would block up the orifices of the body with padding, and after a few days he would remove it and drain out the corrosive oil together with the bowels and other liquefied organs.

Cleaning, packing and perfuming
The most elaborate and expensive mummification treatment took 70 days. First, the embalmer cut into the side of the deceased's abdomen and removed the liver, lungs, stomach and intestines, leaving the heart and kidneys in place. Next, the inside of the body was cleaned and packed with scented resins and natron – a salt compound similar to washing soda, which was used as a drying agent. The body was then buried in natron and left to dry out for 40 or so days. During this time the removed organs were cleaned, dried with natron and coated in molten resins, then placed in special jars, called canopic jars – painted wooden, stone or pottery bottles – the stoppers of which were

BREASTPLATE *The scarab beetle, symbol of rebirth, appears on many funerary items.*

SACRED ANIMAL *Cats, worshipped by the ancient Egyptians, were often mummified and buried in their own cemeteries.*

carved in the shapes of four Egyptian demigods, whose specific duty it was to safeguard the internal bodily organs.

Final preparations

When the body had dried out completely, the embalmer removed the packing materials from its insides and stuffed sand and clay under the skin to pad out the shrivelled flesh. He then packed the body with resin-soaked linen and bags of pure ground myrrh, cinnamon and sawdust, anointed it and covered it with molten resins.

Helped by two assistants, who passed him long strips of linen that they had dipped in sweet-smelling preservatives, the embalmer then bound the body. To safeguard the soul of the deceased, he placed magical amulets and jewels between each layer, then swathed the entire body in a shroud. The mummy was now ready for the journey to the underworld, so the embalmers carefully lifted the figure into an ornate coffin which in turn was placed on a couch in readiness for the funeral ceremony that would follow

when the body had been returned to its family. Before burial, ritual verses were inscribed on the coffin to safeguard further the spirit's journey, and food, clothes, jewellery and furniture were placed by the coffin for use by the deceased in the afterlife. Lastly, the canopic jars containing the bodily organs were placed in a box next to the coffin.

LIFE AFTER DEATH *It was customary for the chief embalmer to conduct the mummification ceremony wearing a jackal mask, symbol of the god Anubis, protector of the dead. Anubis and Osiris, ruler of the underworld, were the principal guardians of the ancient Egyptian death cult, which held that death was the moment of transition into another life.*

The Egyptians believed that two elements of a personality survived in the afterlife: the ba, *soul, and the* ka, *spirit. The* ka *returned periodically to Earth to eat and drink, but to do this it needed to reinhabit its earthly form. They therefore preserved the bodies of the dead and ensured that offerings of food and drink were placed by their tombs. The ultimate fate of the deceased was decided by Osiris, who weighed a dead person's heart on a scale, using a goddess, named Maat, meaning truth, as a counterweight.*

The heart of a wicked person would be heavier than the goddess: and he would be fed to a wild beast. But those who passed the test would live in luxury with the gods for all eternity, indulging in all the pleasures they had enjoyed in life.

SERENE IN DEATH *Wealthy Egyptians, such as Princess Thuya (right) and Prince Yuya (below right) could afford gold burial masks studded with semiprecious stones. For the less well-off, the common alternative was ornate masks made of cartonnage – pressed and glued linen or papyrus.*

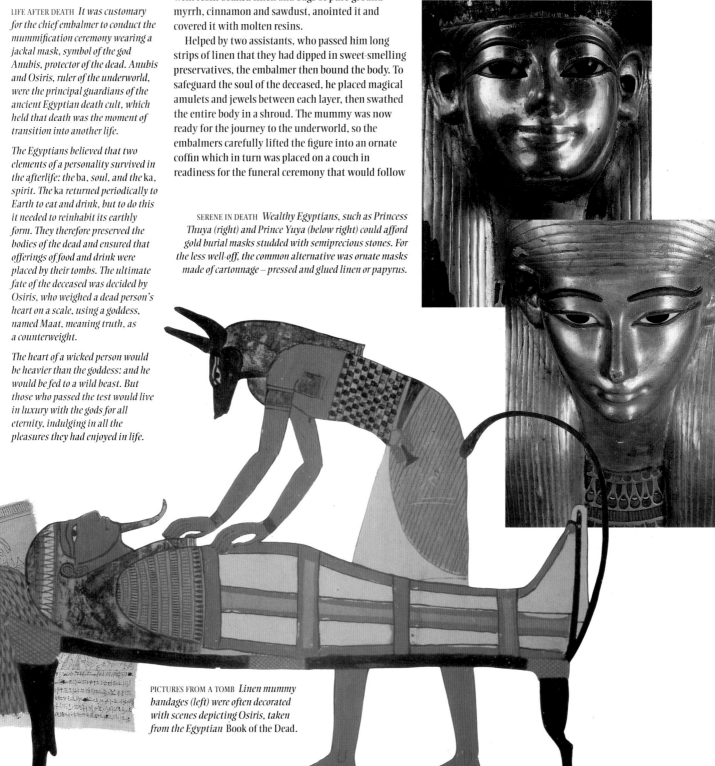

PICTURES FROM A TOMB *Linen mummy bandages (left) were often decorated with scenes depicting Osiris, taken from the Egyptian* Book of the Dead.

Bronze bones, wooden dolls and yoyos

HOW CHILDREN PLAYED IN CLASSICAL TIMES

CRIES OF DELIGHT RANG OUT AS THE SHELL, COATED ON one side with pitch, was tossed into the air. The eyes of the children were fixed on its descent. When it landed black side up, triumphant screams from the 'nights' (the black team) rang out as they chased the 'days' (the white team) around the courtyard in which they were playing. The game of *ostrakinda – ostra* meaning 'shell' – could keep children in ancient Rome amused for hours.

In the ancient world, much as today, parents considered play a vital part of preparation for later life. They made sure their children had plenty of toys – many of which, such as whipping tops, yoyos, hoops, kites, balls and swings, are still familiar.

Knucklebones, bronze fly and cooking pot

For older children there were all sorts of games. Knucklebones, or *astragaloi* – a game much like the present-day game of jacks – was particularly popular with Roman girls. The bones could be made from real bone, terracotta, glass, ivory, bronze, silver or brass.

Boys preferred a game – today called *morra* in Italy – in which two competitors each held up a hand at the same time, showing a number of fingers. The first to call out the correct total won the round. Scoring was done with a notched stick, which the boys held between them. Each time a player won he grasped the stick a notch higher. The first player to reach the middle of the stick won the game.

Bronze fly was like blindman's buff: a child was blindfolded and spun round and round. The other players stood around the blindfolded child and threw

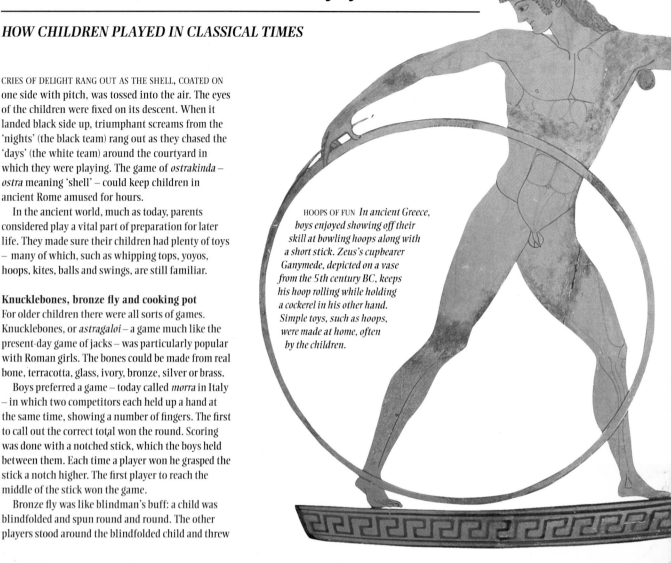

HOOPS OF FUN *In ancient Greece, boys enjoyed showing off their skill at bowling hoops along with a short stick. Zeus's cupbearer Ganymede, depicted on a vase from the 5th century BC, keeps his hoop rolling while holding a cockerel in his other hand. Simple toys, such as hoops, were made at home, often by the children.*

WOODEN HORSE *During the period of Roman rule (30 BC-AD 641) toddlers in Egypt could amuse themselves by pulling along small models of animals – such as this wooden horse.*

ROMAN FIGHTERS *Although Roman children, like any others, enjoyed playing games together, the competitive spirit could cause tempers to flare and fights, complete with hair-pulling, to break out.*

thin strips of papyrus at him until he caught one of them. The player who threw the strip that was caught had to take his turn in the centre. In the game of cooking pot, one child sat in the centre of a circle of others who prodded and jeered at him, while he tried to touch one of them with his foot. The child he touched had to take his place.

Board games for family fun

Whole families got together for long board game sessions. A popular Egyptian game was *senet*, played on a rectangular board, divided into three rows of 10 squares. The two players each lined up seven pieces, alternating them lengthways along the first 14 squares. A handful of sticks, flattened on one side, served as dice. Counting the flat surfaces to determine how many squares they could move, the players pushed their pieces up and down the three rows. The winner was the first to move all his pieces off the end of the board while blocking an opponent.

ROYAL GAME Senet *was played by ancient Egyptians from all walks of life. Even royalty enjoyed a game, as a wall painting from the 13th century BC of Queen Nefertari about to make a move shows.*

GOOSE POWER *Roman children shared their parents' passion for chariot racing. Young boys drove carts harnessed to dogs, sheep and goats. A mosaic floor in a Sicilian villa shows a cart drawn by geese with a young boy at the reins.*

GAMES FOR GROWN-UPS *At the end of a drinking party ancient Greeks sometimes played the simple game of kottabos. Players flicked the dregs of their wine from wide-brimmed cups, competing to hit a specified object, in this case a dummy duck.*

LITTLE WOMEN *Dolls were popular toys for children throughout the ancient world. Roman girls played with rag dolls, or carved wooden dolls with movable limbs (left). These could be very detailed, and were often dressed in fashionable clothes. In Greece, dolls were moulded from terracotta (above), and painted. Children in ancient Egypt played with wooden dolls, such as this black African baby doll (right) with beaded hair.*

STEADY HAND *In the game of knucklebones, players threw five pieces of bone into the air and caught as many as they could on the back of the hand. Any dropped pieces had to be picked up without dropping those already on the hand.*

Voyage to the afterlife

HOW THE CELTS BURIED THEIR WARRIOR DEAD

CART BURIAL *The skeleton of an ancient Briton lies beneath his iron sword in its bronze and iron scabbard. The cart beneath which he was buried has disintegrated. The five rings to the left of the body mark the position of the wooden yoke, and two bridle bits lie inside the wheel on the right-hand side.*

THE CELTIC WARRIOR WAS GIVEN A LAVISH burial. Laid on a bronze couch intricately decorated with dancing figures and horses, he rested on blankets of horsehair, wool and badger fur. He wore a gold band round his neck, his shoes were covered with strips of gold and an elaborate golden dagger lay at his side. His robes were made of the finest silk.

In the outer chamber of his two-room tomb, built about 550 BC on the edge of the Black Forest in southern Germany, stood a wooden chariot ready for the arduous journey to the underworld. Joints of meat and wine waited to sustain the warrior on his way, and iron nail clippers and a wooden comb were left so that he could look neatly groomed on arrival. The structure was finished with a layer of stones and covered with a mound of earth, making a tumulus which stood an impressive 24 ft (7 m) high. A carefully placed circle of stones marked the spot.

GRAVE GOODS *A Celtic warrior once wore this bronze helmet in battle. When he was buried, his armour was laid beside him.*

Long voyage to the Delightful Plain

Burial rituals varied from place to place and over time from early Celtic culture (*c.*700-500 BC) until the end of the Celts' domination of Europe in the first century BC. Cremation was another way to deal with the dead. The corpse was laid on a wooden pyre and surrounded with wood and straw. When the pyre was set on fire, the deceased's favourite servants would often show their devotion by hurling themselves into the flames, resolved to accompany their master into the next life. The ashes were transferred to a bronze or ceramic urn by friends and relatives and then buried alongside the warrior's wagon and other grave goods, possibly including a joint of pork and some bread.

BURIED TREASURE *A bronze torque – a neck ring – found in a grave in France. Torques were worn by both men and women. As well as being ornaments, they were status symbols.*

All Celts expected to be reunited with friends and relatives in the afterlife, which, according to a contemporary Roman writer, led to one curious and convenient custom. It was perfectly acceptable to borrow from your neighbour with the intention of repaying the debt only when you met again after death in the other world.

In Irish legends, the dead went to another world sometimes called the 'Land of the Young' or the 'Delightful Plain'. There, sickness, old age or death never disturbed the happiness of its lucky inhabitants, who were always young, and a hundred years passed as swiftly as a single day.

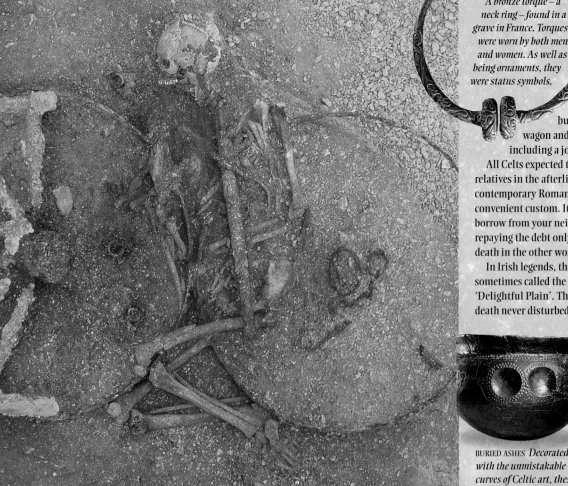

BURIED ASHES *Decorated with the unmistakable curves of Celtic art, these two ceramic funeral urns, unearthed from burial chambers in France, were used to hold the ashes of cremated dead.*

Going out in style

HOW A FUNERAL WAS CONDUCTED IN ROMAN TIMES

GAIUS JULIUS ALPINUS CLASSICIANUS WAS A POPULAR man in Britain, which was surprising considering he was the emperor Nero's representative. But as procurator of the province, in charge of financial affairs, he had tried to protect the British from the wrath of the governor, Suetonius Paullinus, in the aftermath of Boudicca's revolt in AD 60. Now, three years on, Classicianus was dead.

Call in the professionals
The undertaker arrived at the deceased's house to prepare the body for the funeral. Classicianus was dressed in his best toga and laid on a sofa, where he remained for three days.

The business of dying was not cheap. A funeral was the most important indicator of social status and so they tended to be as lavish as possible. As well as the undertaker's fee, professional mourners had to be hired to weep and wail throughout the proceedings. Musicians charged a fee for leading the procession to the burial place, and there were also fees for the cemetery staff, the costs of the funeral feast and sacrifice, and of the tomb itself. The poor often paid a monthly subscription to clubs to ensure a proper burial.

On the day of the funeral, a magnificent oration had to be delivered for someone as important as Classicianus, but even a local councillor would expect to have a speech delivered at his graveside. At the cemetery, the body was placed upon the funeral pyre, traditionally of oak and pine, and the pyre was lit. After the funeral the mourners washed their hands in wine or scented water to purify themselves from their contact with death, before sitting down to eat a funeral feast.

SOMBRE RITUALS *Members of the household carry the bier in a solemn procession to the cemetery (above), usually on the edge of town. Professional mourners and musicians lead the way, and family and closest friends follow behind. While the deceased embarks on his chariot journey to the Roman heaven (left), the family eats a funeral feast of meat, cakes and wine (right).*

Remembering the dead
Mourning lasted a further nine days, and was concluded with a sacrifice of a lamb or pig and another feast. The dead were honoured every February during the festival of *Parentalia*, when offerings of wreaths, corn, salt and bread were made at family tombs. During May, in the *Rosalia* festival, tombs were decorated with roses. Classicianus was certainly not forgotten, for his widow Julia Pacata Indiana bought him a massive – and very costly – tombstone.

ENSURING A WARM WELCOME *A sow is prepared for sacrifice to Ceres, the goddess of the earth. An animal was always killed to mark a bereavement, and to ensure the good wishes of the goddess and her help on the journey to the other world.*

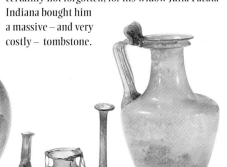

FUNERAL GLASS *After cremation the charred bones of the dead were washed in wine and put in a glass funeral urn. Then the urn would be set in a tomb or a burial chamber with niches for members of the family.*

Steaming off the grime

HOW ROMANS KEPT CLEAN AND RELAXED AT THE PUBLIC BATHS

AN ECHOING DIN OF EXCITED GIGGLES, CHEERS AND boisterous singing filled the marble halls of the great Baths of Caracalla in Rome. Masseurs pummelled their clients with slaps and thumps, and bathers splashed and shrieked in the pool.

For the Romans, taking a bath was not just a matter of keeping clean; it was one of their favourite leisure activities. The Baths of Caracalla, completed in AD 216 in honour of the Emperor Caracalla, had space for 1600 bathers at a time and were often filled to capacity. The complex covered 27 acres (11 ha)

and, in addition to the baths and exercise areas, included concert halls, libraries, shops and gardens. These bath houses first became popular in the 2nd century BC. The public baths were built by entrepreneurs to make money, or endowed by wealthy philanthropists.

When the baths first opened, men and women bathed together, but in later centuries this scandalised some citizens. The Emperor Hadrian, who ruled from AD 117 to 138, was the first to issue edicts in a bid to enforce separate bathing.

PALACE OF CLEANLINESS *Romans entering the massive Baths of Caracalla in the 3rd century AD could feel a justifiable sense of pride, for these were among the grandest baths in the whole empire. Two fountains play in front of the entrance, marble statues stand by the impressive arched doorways, the floor is made up of different patterned mosaics and the high vaulted ceiling is faced with copper. Bathers could enjoy a range of baths of different temperatures after paying one* quadrans, *the smallest Roman coin, or entering free if a benefactor was sponsoring a day or even a whole year. They could also make the business and political contacts that were so important in oiling the wheels of empire.*

GETING CHANGED *In the changing room, the* apodyterium, *bathers put their outdoor tunics on shelves. Some don short tunics and warm up with an exercise routine; others carry on to the exercise courts or gymnasium. Some go straight on to the* caldarium, *the first and the hottest room in the baths complex.*

KEEPING FIT *More active bathers in the gymnasium take part in weightlifting, wrestling, boxing or running. General exercises were also popular as were team sports and games with balls and hoops. Larger baths, such as those of Caracalla, sometimes also provided a colonnaded court yard, which was divided into areas for exercise or for just strolling or relaxing.*

GETTING CLEAN *Roman bathers wash in the hottest room, the* caldarium. *The man in the foreground is using a* strigil, *an implement made of wood, metal or bone, to scrape the cleansing olive oil off his skin. The oil has mixed with the dirt, and then been loosened by sweat in the steamy atmosphere. Another bather is receiving a massage, while others gather round a tub of hot water which provides clouds of steam.*

TAKING IT EASY *Women relax around the warm pool in the* tepidarium, *a cooler room after the hot blasts of the caldarium. Here customers could indulge in treatments which included massage, manicure and hair plucking.*

COOLING DOWN
Last of all, bathers at the Baths of Caracalla would dive into the cold plunge pool, large enough to accommodate dozens of people. Plunge pools were often left open to the sky, and were usually at the centre of the baths.

A PLACE TO MEET *Freshly scrubbed Romans gather around a large pool in the frigidarium, or cold room, at the Baths of Caracalla. Here gossip is being passed on, political intrigues set in motion and business plans sealed. The bathers play board games, place bets, buy snacks or wait to see jugglers and other entertainers. Bathing was a social activity which Romans found essential for a contented life.*

THE HAZARDS OF THE BARBER SHOP

THE ONE FACILITY THAT THE BATHS DID not normally provide was shaving; for that Romans had to go to a barber's shop, where an uncomfortable experience awaited them. The narrow bronze or iron bladed razors used by barbers were never sharp and with no soap to smooth the way, they left many a citizen wishing he had never reached puberty.

Plasters made of spider's webs
Martial (AD 40-104) said of one female barber he frequented, 'she does not so much shave you as skin you.' His contemporary, Pliny, prescribed a plaster made of spider's webs soaked in oil and vinegar to repair the damage done by heavy-handed barbers. Some men preferred to use a hair remover made of pitch and resin, which resulted in a smoother face but could be painful.

BEARDED RULER *The emperor Hadrian started a fashion for beards among Roman citizens when he sported one during his reign in the 2nd century AD.*

SHOPPING
Two women haggle over the price of a length of cloth at one of the many shops situated inside the baths. Wandering hawkers also offered goods to the baths' customers.

AL FRESCO *Visitors linger by the fountains outside the Baths of Caracalla. In warm weather, the social aspect of the baths would continue outside. This was especially the case in the 3rd and 4th centuries AD, when the baths built by the Romans grew ever larger and showed a distinct Greek influence in the design of the elaborate gardens and colonnades.*

PLAYING GAMES
A group of men gamble with dice and relax with a glass of wine. Gaming was a feature of baths' life.

WELL READ
Citizens read in the baths' extensive library: a good session at the baths would stimulate mind and body.

TURNING ON THE HEAT
A team of slaves load wood into the underground furnaces that provide hot water and power an underfloor central heating system during cold weather.

MUSICAL FINALE *Dancers and musicians provide entertainment in the large concert hall at the Baths of Caracalla.*

WHEN GETTING DRESSED MEANT WRAPPING UP

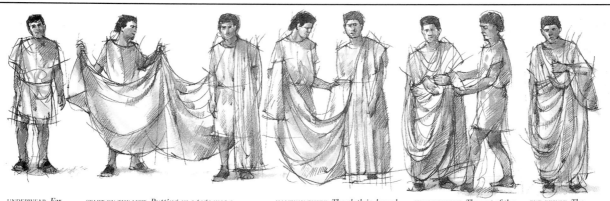

UNDERWEAR *For formal occasions, men wore the heavy white toga over a plain tunic.*

START ON THE LEFT *Putting on a toga was a complicated business. A slave takes the 18 ft (5.5 m) long and 7 ft (2 m) wide woollen garment and drapes it over his master's left shoulder. Then he gathers up the rest of the material.*

HALFWAY THERE *The cloth is draped around the back and under the right arm, and some of the cumbersome spare material is tucked into the tunic.*

FINAL TOUCHES *The rest of the cloth is placed carefully over the left shoulder and allowed to fall in graceful folds almost to the feet.*

END RESULT *The toga is elegantly pleated, and the wearer is ready for the day.*

A degree of suffering

HOW MEDIEVAL STUDENTS STRUGGLED IN THE SEARCH FOR KNOWLEDGE

SHARED KNOWLEDGE *Books were a precious commodity, so the texts for study – which were mainly classical – were read aloud by a master before discussion. The pupils would be lucky if they shared a single copy of the book.*

WITH SCARCELY A BRISTLE ON HIS CHIN, THE YOUNG student arrived, hungry, footsore and hundreds of miles from home, to take up his place at one of Europe's 14th-century universities. But his ordeal was far from over. His head was promptly shaved and older students raided his purse to throw a feast at the expense of the naïve newcomer.

In his own time he would inflict this ritual on others, but his student life, which stretched before him for up to 15 years, would involve ceaseless struggle, not only against the frozen fingers and empty stomach of poverty, but sometimes also against the state, the lecturers, other students and the town. On arrival the first priority was to find somewhere cheap to rent. Since there were few halls of residence, the students often took rooms in private homes. Sharing a room was common, and the poorest student might even share a bed for economy and warmth. The lot of the scholar at the old universities of Europe – from Bologna to Cambridge, from Paris to Salamanca – was not a happy one.

The growth of learning

The 12th century saw a revival of interest in learning throughout Europe – partly due to the rediscovery of works by the Greek philosopher Aristotle. Communities of teachers, later called universities, grew up in competition with the schools which were attached to monasteries and cathedrals. Entry was open to any free man who could pay the fees – there were no admission examinations. The idea of a woman studying was inconceivable, and would remain so until the 19th century.

Good teachers attracted students from far and wide. At each university, the scholars formed themselves into 'guilds', or 'nations' with common interests – usually their nationality, since many were studying in a foreign country. Scholars were known as 'clerks' since most were training for a career in the Church. In northern countries, they sported a monk's tonsure and a cassock, while in Italy students wore long fur-trimmed gowns called 'cappas'.

STUDENT LIFE *Studying at the college of* Ave Maria *in Paris entailed chores as well as learning. Each week, one student – known as the hebdomadary – had to keep track of library books (top) and feed the goldfinches (below right). The first student to wake was expected to ring the early morning bell and say the Ave Maria (below left). Feast days were celebrated with a hearty meal and a play depicting the founder being visited by the Virgin Mary, before lining up in traditional garb to be blessed (bottom).*

SEVEN YEARS TO MASTER LEARNING

BEFORE THE RISE OF UNIVERSITIES in the 13th century, traditional learning had been based on the *quadrivium* – geometry, arithmetic, astronomy and music – and the *trivium* – grammar, rhetoric and logic – from which the word trivial is derived. But these seven 'liberal arts' had little impact on the course of studies at universities.

Latin and philosophy were all a student needed to learn in order to gain a degree. A student studied for four years to become a Bachelor of Arts (BA) and another three to become Master of Arts (MA). Texts included Latin translations of Arabic and Greek mathematics and the works of Aristotle.

Proving your worth
Teaching and discussion was in Latin. Masters gave 'ordinary' lectures which focused on the more important subjects. 'Cursory' lectures were given by BAs and senior pupils.

Examination was by open debate. In an exercise called the 'disputation' a student had to defend ideas before his teachers in a demanding oral test which could last a week. If successful, he received a kiss of benediction from his teachers before taking part in a triumphant procession through the town.

SOLID CONCENTRATION
The professors of law at Bologna were men of international standing. Their splendidly carved tombs, such as this one depicting a class in progress, bear testament to their great wealth.

Rebelling against an austere regime
The student's day began at five o'clock in the morning with compulsory mass. Dinner was at ten in the morning and supper was at six; talking was forbidden at mealtimes when the Bible was read aloud. Discipline was strict – games and music were usually banned. However, some students baulked at being told what to do. An aggrieved father in Besançon, France, wrote to his son: 'I have recently discovered that you live dissolutely and slothfully, preferring licence to restraint and strumming a guitar whilst others are at their studies.'

Tension between the students and townspeople often ran high. In Oxford, England, in 1355, a street battle lasted three days and left 63 dead after an argument over the quality of wine in the Swyndlestock Tavern. The students were not always without blame and in Oxford the proctors, who were supposed to keep discipline, sometimes led them into battle. It was an international problem. The students and teachers of Paris regularly clashed with the citizens, and it was one such brawl, in 1200, that resulted in the university receiving a charter from the king to protect the students.

SEAL OF APPROVAL *In the 13th century the University of Paris was recognised by church and state and it was given its own wax seal.*

Mobile homes on the great American grasslands

HOW AMERICAN INDIANS CONSTRUCTED AND LIVED IN TEPEES

FAR ACROSS THE GREAT PLAIN, THE SIOUX HUNTERS could see their homes glowing like lanterns against the darkening grasslands. The hunt had been successful: two buffalo had been killed. There would be feasting that night, and more hides to add to the new tepee the women were making.

'There was only one way for young folks to get together. The girl had to stand outside the family's tepee with a big blanket. Her lover would come up and she would cover both of them.'
LAME DEER OF THE SIOUX NATION

The Sioux, like the Arapaho, Comanche and Blackfoot, were recent arrivals on the American western plains in the 18th century, forced west by the arrival of white people. These tribes were nomadic all year round, following the migration of the buffalo, their main source of food, shelter and clothing. Those of the eastern plains, such as the Pawnee and the Dakota, lived in permanent settlements most of the year, taking to their tepees for the summer and autumn hunts.

The cone-shaped tepee made an ideal portable home for these nomadic hunters. Consisting of a frame draped in buffalo hides, it could easily be taken down and reassembled. Making the tepees was mostly women's work, carried out during the spring buffalo hunt. As many as 30 skins were stitched together to make a typical large tent.

To erect a tepee, about 20 pinewood poles up to 30 ft (9 m) tall were stripped from young trees. Three or four poles were lashed together near the top to form the basic cone framework, then more poles were leant against these for extra strength. The buffalo hides were pulled over them, smoothed out and secured with wooden pegs knocked into the ground. The two ends were joined at the front with big bone or wood pins, and the precut opening at the bottom of this seam was covered over with a doorflap.

The Plains Indians called their tepees 'good mothers', because they sheltered and protected them. On hot summer days the door flap was left open and the bottom of the outside covering rolled up. For extra insulation in winter, another layer of buffalo skin was stretched over the poles inside the tent and grass was stuffed in the gap. Snow was banked up around the bottom of the tepee to prevent draughts.

Traditionally, one family occupied each tepee and the interior was crowded. Firewood, bedding, rugs, sewing materials, household equipment and willow-wood backrests all had to be stored. The inner lining was decked with leather bags filled with food and medicine, bows and arrows and ceremonial shields.

FIRESIDE TALES *Inside their tepee, a Cree family gathers around the warmth of a fire, sharing a pipe of tobacco, and waiting for dinner to cook in the heavy iron pot that has probably been bought from a white fur trader. Smoke escapes through a flap in the roof which can be adjusted with poles.*

A haven in a desert of snow and ice

HOW INUIT BUILT HOMES OUT OF FROZEN BLOCKS

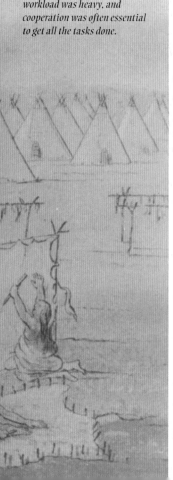

WOMEN'S WORK *During the spring buffalo hunt Comanche women on the Great Plains set about cleaning buffalo hides for a tepee. Stretched taut, the hides are scraped with a bone tool to remove hair and tissue. Later they will be stitched with thread made of buffalo tendons. The women's workload was heavy, and cooperation was often essential to get all the tasks done.*

ARCTIC LIFE *Sheltered from the harsh Arctic wind by a high snowdrift – which has been marked with a harpoon in case of a freak snowstorm – a group of Inuit igloos nestle together against the cold. Sledges, too large to fit inside, lean against the igloos ready to be used as ladders in case repairs are necessary to the roofs. A family stretch their legs, while two other people hurry home through the gathering dusk.*

WITH A LARGE KNIFE CARVED OUT OF A WALRUS TUSK, the Inuk cut blocks of hard-frozen snow, instructing his son to pay attention, for this was a skill that would ensure his survival in years to come. The father had already shown his son how to check that the snow was suitably hard packed and how to slice a hole, knee deep and about 10 ft (3 m) across, into it.

In the Arctic winter of northern Canada, biting winds, driving blizzards and temperatures as low as –46°C (–51°F), mean that death is certain without proper shelter. The Inuit were nomads, spending their lives in small, widely dispersed family groups. Their homes had to be easy to erect, as well as sturdy.

Using what comes to hand

Out on the frozen tundra during the long winter, the only available building material was snow and ice, and this the Inuit transformed into an igloo, a house built entirely of large blocks of hard-frozen snow. The blocks were used to erect a dome over a wide, shallow hole. Once the dome was finished and a doorway cut into it, a tunnel to the entrance was dug out.

Within 1 to 2 hours, even in dark and freezing conditions, a weatherproof structure big enough to house a whole family was ready. If the igloo was built near the sea, water was used to create a hard layer of ice on the outside. Inside, the igloo was about 6 ft (1.8 m) high in the centre, where a fire was lit. A platform of packed snow covered with caribou hides doubled as a divan in the day and a bed for the whole family at night.

LAYING THE FOUNDATIONS *An Inuk places blocks of frozen snow in a circle about 12 ft (4 m) across, leaving a gap for the doorway. The blocks are cut at an angle to create a dome shape. As each row is added, the blocks are positioned to slope inward until they reach the top, where a hole is left. This can be closed with ice or left open for smoke to escape.*

FINISHING TOUCHES *While one Inuk fills in cracks with snow, the other quickly smooths down the rough edges around the entrance hole – just wide enough for a grown man to crawl through – before adding the entrance tunnel.*

LONG EXIT *The tunnel around the entrance is the last section to be built. It protects those inside from blizzards, and on stormy nights gives shelter to their pack of dogs. A translucent pane of ice has been inserted to make a window. The slab of white ice positioned at right angles to the window reflects extra light into the igloo's interior.*

A marriage on trial

A WEDDING IN ANCIENT EGYPT

THE WEDDING PROCESSION OF THE 12-YEAR-OLD TAMIT moves joyfully through the streets of Thebes. Resplendent in a sparkling white dress and matching shawl, Tamit shyly acknowledges the good wishes of her smiling friends and relations. She is on her way from her father's home to that of her bridegroom, Amenhotep, to whom she now gives her allegiance. She takes with her all her worldly goods such as ornaments, clothing, mirrors and jewellery. There has been no formal marriage ceremony. No rings have been exchanged, Tamit has not changed her name to show her new status, and the dress she is wearing was not specially made for the wedding.

Like all marriages in Egypt in the 15th century BC, Tamit's is a trial marriage. It can be ended by either party at any time – especially if she fails to become pregnant and provide Amenhotep with a son. On the other hand, she can 'divorce' him on grounds such as cruelty or habitual drunkenness. All she has to do is return to her family home – taking her possessions and a third of everything that Amenhotep owns.

If all goes well Tamit will honour and obey her husband. And if one of them should die before the other, the bereaved person will be free to remarry. However, some devoted spouses prefer to wait alone to be reunited with their partners in the afterlife.

HIS AND HERS *A servant offers refreshments to an Egyptian bride and groom at their marriage feast. To show his goodwill, the groom has given his bride a token sum of money. In turn, her father provides the young couple with furniture – and enough grain to last for seven years.*

THE BLESSING *Using a jug of holy water, a priest blesses a newly married couple in Thebes. The bride is a true-born Egyptian, but the groom, named Sennufer, is from the southern realm of Nubia, which sporadically fell under Egyptian control.*

Toys for the gods

A WEDDING IN ANCIENT GREECE

ON THE EVENING BEFORE HER WEDDING, 13-YEAR-OLD Melissa takes her toys – including dolls, tops and a hobbyhorse – to a shrine devoted to Artemis, the ancient Greek goddess of childbirth and changing stages of life. She dedicates her childish playthings to the goddess, and offers up a clay figure of a grown woman making bread. This marks the end of Melissa's childhood, and the beginning of her new life as a loyal wife and devoted homemaker.

Red lips and white cheeks

The next morning, Melissa, the only daughter of a well-to-do Athenian merchant in the 5th century BC, bathes in water specially brought by slaves from a 'fountain of purification'. Attended by a team of slaves, she then spends hours being groomed and made up for the great occasion.

From an exquisitely carved toilet box – a wedding gift from her mother – she uses ochre to redden her lips, and chalk to lighten her complexion. She puts on leather sandals and an ankle-length yellow linen dress. She fastens the dress at the shoulder with a brooch, and gathers it with a belt so that it folds gracefully across her body.

Forsaking her youthful loose ringlets, Melissa piles her long hair on top of her head as befits a future wife and mother, and secures her new coiffure with pins and a headband. As a final touch, she puts on pendant earrings, a necklace and silver bracelets.

A dedication at the altar

As darkness descends, the bridegroom, a 31-year-old winegrower and dealer named Antinous, arrives at Melissa's home to 'claim' the bride. After a solemn and tearful ceremony in which her father hands her over to her husband, Melissa climbs into a mule-drawn cart to be driven to her new home.

On reaching Antinous's house, Melissa is greeted by her mother-in-law, who carries a blazing torch with which she ceremoniously lights the newlyweds over the threshold. The rest of the day is spent in feasting and drinking with close relatives in the homes of both sets of parents. Tomorrow, friends will arrive with presents for the newlyweds.

HOMECOMING
On arriving at her new home, a bride (on the right) is greeted by a musician playing a reed pipe. Once inside, she is led by her mother-in-law to the hearth, where they both kneel and are showered by close relatives with fruits, nuts and grains – symbols of prosperity and good fortune.

GIFT-BRINGING
On the day after the wedding, friends of the newly married couple arrive at their house bearing wedding presents such as a delicate, long-necked vase. By following tradition, the bride and groom ensure that their life together gets off to an auspicious start.

CARRIAGE ACROSS THE CITY *As is customary among the rich of Athens, an open carriage drawn by two handsome horses transports a bride across the city at night-time to the brand-new house she will share with her husband. Escorted by faithful servants and watched over by Eros, the Greek god of love, she will come to no harm.*

Jumping a broomstick

A WEDDING IN OLD VIRGINIA

SIDE BY SIDE, JOEL AND PHOEBE STAND NERVOUSLY IN front of their flimsy thatched cabin on a tobacco plantation in the British colony of Virginia. The couple are about to be married on a bright Sunday morning in the summer of 1742. Presently, they are approached by one of their fellow slaves, who places a broomstick on the ground. A signal is given and, hand-in-hand, Joel and Phoebe jump over the broom. The wedding guests applaud excitedly, calling out their best wishes. The marriage ceremony is over; the happy couple are now man and wife.

Banjos and bones

Laughing and shouting, the newlyweds and their guests hurry over to some trestle tables and benches set out beneath the shade of nearby trees. There, an entire hog is roasting on a spit. There are jugs of cider and bottles of corn whiskey for drinking toasts and for slaking thirsts. After the meal, banjos and bones – pieces of animal bone held between the fingers and rattled together – strike up music. The celebrations will last all night, the participants having been given the next day off: but at dawn on Tuesday they are due back in the fields for another 12 hours of muscle-aching work, picking tobacco.

Officially, marriages between slaves are forbidden, but the plantation owner has given Joel and Phoebe special permission. Married slaves are more settled and are less likely to run away, and children also help to keep married slaves 'at home'. The use of child labour spares the owner the expense of buying fresh slaves from the ships that bring cargoes of Africans from across the Atlantic.

A LIFE APART *Gathered outside their quarters on a tobacco plantation in the deep south of the USA, a group of slaves – guests at a wedding of two of their fellow workers – celebrate the happy event. To the accompaniment of a banjo, one of the men shows his skill at a traditional stick dance while the women onlookers sway and clap their hands. The slaves live in cramped and flimsy cabins, shared between two or more families. These are set well apart from the plantation owner's 'Big House' – an elegant mansion with pillared terraces and a sweeping, shrub-lined drive. In his own commercial interest, the owner will ensure that the slaves always have a roof over their heads, clothes to wear and food to eat. He encourages them to become Christians and abandon the traditional beliefs which crossed the ocean with the slaves to become widely established on the plantations of the southern United States.*

Dancing on egg yolks

A GYPSY WEDDING IN ANDALUSIA

SWEETMEATS OF ALL KINDS, SHAPES AND SIZES COVER a room in the bridegroom's house in the Andalusian city of Córdoba in southern Spain. There are fruit cakes, spice buns, cream puffs, date slices and, above all, *yémas* – the yolks of eggs coated with a crust of sugar – spread across the floor. Suddenly the gypsy bride and groom, Lucia and Antonio, burst into the room. A guitarist in the far corner starts strumming a flamenco; driven on by the music, the couple dance across the 'carpet' of confectionery.

Before long, the young newlyweds – who have just been wedded by the simple act of joining hands and swearing to be faithful to each other – are joined by their guests. With fierce cries the men spring high into the air, while the women clap their hands in encouragement. Then they fling themselves into the dance. Within a few minutes the sweetmeats have been trampled into a thick, yellowish paste, and the dancers are covered to their knees with egg yolks and squashed pieces of fruit.

Throwing money away
It is turning dark, but the traditional wedding celebrations will last until dawn and continue riotously for two more days after that. Antonio's doors are open to everyone. Every now and again, to express his joy, he throws handfuls of money – which he has earned as a travelling horse dealer – out of the window to passers-by. As the bridegroom, he pays for all the festivities and he will probably be penniless by the end of the week. However, relatives and friends will gladly provide enough for him and the 16-year-old Lucia to start their life together.

DANCING WITH DELIGHT *Clicking fingers loudly and insistently, like castanets, and performing a spirited flamenco – the traditional dance of Andalusia – Spanish gypsy women (below) join in the celebrations at a boisterous wedding.*

MUSICAL MERRYMAKING *Once the happy pair are united as man and wife, the wedding guests relax and enjoy themselves with food, wine, conversation and music. During the two years of the couple's betrothal they were closely watched over by their friends and relatives to ensure that they observed the proprieties. The wedding could go ahead only if the bride was 'pure'.*

Bride in a wooden box

A WEDDING IN MOROCCO

PREPARATIONS FOR THE WEDDING OF MALIKA AND HAMID begin in their village near Wazzan, in the Jabala foothills of the Rif, three days before the ceremony in the August of 1890.

At home, Malika receives her friends and relatives in her *haik* – an all-encompassing shawl-gown given to her by Hamid on the day of their betrothal. At the end of the day, the *haik* is returned to her betrothed, who will wear it during the marriage. Tired but happy, Malika eats with her female attendants before they paint her with henna to purify her for marriage.

In Hamid's home, the groom's relatives ritually clean and grind wheat to be used at the wedding feast as the men of the extended family gather for a special meal. This boisterous occasion, accompanied by music, song and the occasional gunshot – for good luck – culminates in painting the groom with henna.

A sweet and lucky life
The next morning, Malika is ritually bathed in oils before she visits relatives in the village. Towards the end of the day, both Malika and Hamid are decorated anew with henna. The following day, the bride is taken to the groom's house. Having been ritually bathed and dressed in clothes sent to her by Hamid, Malika awaits those from his family who will fetch

her, bringing with them the *ammariya*, a wooden box with doors which fits on a mule's saddle. Once Malika is settled in the box, the party sets off for her new home with a fanfare of drums, flutes and guns to scare off any harmful spirits. Resolutely, Malika stares straight ahead. To look back would mean the marriage will be a failure.

As she arrives at Hamid's door, dried fruit – a symbol of the marriage's future fruitfulness – is thrown over the *ammariya*. Her father carries Malika from the box to the nuptial bed, housed in a special tent in front of the house of her in-laws. Once inside, Hamid's mother presents her with a drink of milk to give her 'white', or good, luck. Malika then presides over a magnificent banquet in and around the tent.

Eventually, later that night, Hamid joins her in the bedchamber. The next day, the young couple visit his relatives in the village before returning to Hamid's home. Here, Malika will stay for 40 days before she is allowed to visit her parents for the first time as a married woman.

MARRIAGE DANCE *Ringed by wedding guests, a female entertainer performs a traditional dance to mark the marriage of a young couple in Marrakesh. From childhood, 19th-century Moroccan girls were brought up to become dutiful wives and mothers. Their husbands were chosen for them, and it was thought that providing the age-old customs were observed, the unions would be blessed with prosperity and healthy children.*

The Seven Blessings

A WEDDING IN LITHUANIA

IT IS SUNDAY, THE DAY AFTER THE JEWISH SABBATH, and a wedding is taking place in Kaunas, in Lithuania – one of the many *shtetls*, or small towns, scattered throughout eastern Europe in the 19th century. The bride and groom, Ruth and David, walk together through the cobbled streets to the wooden synagogue in the centre of the ramshackle town. There they stand side by side beneath the traditional *chuppah*, or canopy, which symbolises their future home, prosperity and happiness.

Gladness and good fellowship

The rabbi gives them a glass of wine to share, and thanks God for having brought the couple together. He then reads out the marriage contract which unites them as man and wife. David places a plain gold ring on the first finger of Ruth's left hand. He announces to the assembled relatives and friends that she is now his wife. Next he moves the ring to Ruth's third finger, and the rabbi intones the Seven Blessings – which praise God and thank Him for creating 'happiness and gladness, bridegroom and bride… love, brotherhood, peace and good fellowship'.

The rabbi hands David a small empty glass, which he breaks underfoot as a reminder of the destruction in Jerusalem of the First Temple in 586 BC by the Babylonian king, Nebuchadnezzar II, and of that of the Second Temple, by the Roman Emperor Titus in AD 70. This act demonstrates that there is no joy in the world without suffering. The congregation shouts '*Mazel tov!*' – 'Congratulations!' The newlyweds and the rest of the wedding party then proceed to David's house, where the celebrations last until nightfall.

BINDING BEAUTY *Ornate* ketuboth – *Jewish marriage contracts – record two marriages of the 18th century. They were written in Aramaic, the ancient language of the Jews, and were among married couples' most treasured possessions.*

JOINED TOGETHER *A young couple listen dutifully as a rabbi – his finger raised in emphasis – solemnly intones the duties and responsibilities expected of a Jewish man and wife. As the wedding is held outdoors, a prayer-shawl covers the couple's heads in place of the traditional canopy found in a synagogue. They are dressed according to tradition: a black suit, hat and white waistcoat for the groom; a white wedding dress with a blue velvet bodice, and a lace veil for the bride. Beside her, in another splendid dress, is her mother.*

FIRST LIGHT *A lighted wick in a stone bowl of animal fat gave cave-dwellers a lasting, if faint, light.*

The quest for a steady, bright flame

HOW PEOPLE LIT THEIR HOMES WITH LAMPS AND CANDLES

PORTABLE FLAMES *16th-century Swedes clasped blazing fir torches in their mouths to illuminate all manner of chores during the long months of daytime darkness.*

DRAWING A BURNING BRANCH FROM THE FIRE, A Stone Age hunter held it high above his head to light the path to his cave. As he stepped inside, the flames died and the welcoming sight of home vanished in the darkness. Even in Stone Age times, people sought better ways of lighting their homes than simply relying on firelight, such as dipping wooden splinters or rushes in resin, beeswax or pitch. When these were attached to the end of a stick and lit in a fire, they burned well but the light was soon gone. A steadier, although much smokier, flame could be kindled by putting a wick of moss, twine or grass in a bowl of animal fat. Similarly, people of the ancient world crafted stone or pottery lamps to hold animal or fish oil.

Around the 1st century AD, the Romans developed a candle made from refined beeswax. This, with its wick of twisted fibres, burned with a steady, clear and almost smokeless flame. Beeswax candles were found in Christian churches throughout the Middle Ages, too, although cheap tallow candles were in general use. Early candles were made by repeatedly dipping a flax or cotton strand into hot fat until the candle reached the required width. By the 17th century candlemakers were suspending the wick in a wooden or metal mould, full of molten wax or fat.

Trimming the wick

To keep candles burning – particularly tallow candles which burned at a low temperature and produced copious amounts of hot fat – it was crucial to trim or 'snuff' the wick so that it was long enough to allow excess fat to run down the sides, but not so long that it would bend over and cause pools of wax to drip over the candle's rim. Careless snuffing could extinguish the flame – hence the expression: 'to snuff out', meaning to extinguish. In 1820 in Paris, Jean Jacques Cambacérès designed a candle with a plaited, rather than twisted, wick which curled outwards and vaporised completely on burning.

FIERY SHELL *People living on the coast adapted shells to make lamps which burned fish oil.*

GOLDEN LIGHT *An early Roman oil lamp, made of gold, has two spouts for two wicks. Great numbers of such elaborate burners, fuelled with olive oil – a pint of which burned for around 40 hours – were used to fill the homes of the wealthy with light.*

STUDYING BY CANDLELIGHT *A 17th-century scholar (right) concentrates his reading light by positioning a candle behind a glass globe filled with water. In general use, however, the candle's naked flame shed a soft light that flattered rather than revealed (below).*

LAMPS FOR ALL OCCASIONS *Roman potteries mass-produced oil lamps in various designs; some had lids to keep the fuel pure and all had at least one spout to hold the wick.*

ROOM LIGHTS *Argand lamps, often arranged in a pair as a wall sconce, lit the homes of the wealthy in the late 18th century. A central oil reservoir fuelled both wicks, while air was drawn through the glass chimneys to sustain the flames. A mechanism to raise and lower the wick enabled users to vary the intensity of the lamp's light.*

LIGHT WORK *A street lamplighter is handed an oil lamp by his young assistant to light an open gas burner. By the early 19th century, gas lights were illuminating the streets of Britain's principal towns.*

Brighter and brighter

In 1784, Ami Argand, a Swiss chemist, patented a new lamp. Its woven wick was unlike any other, for it was rolled up into a small tube designed to draw air through its centre as well as from outside, making it burn brighter than earlier lamps. The wick, covered by a glass chimney, was gravity-fed by a reservoir above it that was filled with colza oil – the oil extracted from rape or cabbage seed. The Carcel lamp, introduced in 1798, took a different approach. Its clockwork pump drove the oil upwards from a reservoir located below the wick, with any excess oil simply flowing back down to the reservoir to be pumped upwards again.

The next major step forward came when paraffin lamps were introduced in the 1870s. The oil was cheap and clean, and burned with a clear, bright flame; it was also light enough to be drawn up into the wick by natural capillary action. Paraffin, in wax form, was also used to make cheap candles. But gas was already considered the lighting fuel of the future. In 1807 Frederick Winsor, a German businessman, lit London's Pall Mall with coal gas lamps. During the 1830s and 1840s, gas was piped to increasing numbers of town houses, where it was often used alongside candles and oil lamps. Early gas lamps were open flames shielded by glass lampshades. The first experiments in electric lighting took place in 1808, but it was not until the 1870s that

ELEGANT ILLUMINATION *In the 19th century, decorative oil lamps served as desk and bedside lamps. In this example from the 1870s, oil was gravity-fed from the brass reservoir at the side of the lamp (above) to the burning flame.*

RIVAL INVENTORS *Joseph Swan, a British chemist, and the American inventor Thomas Edison both produced incandescent electric light bulbs in the late 1870s. Swan's model (below) worked on the same principle as Edison's: a thin metal filament, suspended inside the airless glass globe, glowed when an electric current flowed through it.*

the incandescent light bulb was invented. The bulbs were low powered, and they became widespread only in the early 1900s. Women complained that electric light was harsh and less flattering than gas light, while many feared that electricity would leak from sockets. Later in the century, it became possible to flood a room with electric light equivalent to 100 candles.

DEFYING GRAVITY *The key at the base of the Carcel lamp needed regular winding in order to power the clockwork motor which kept the flame supplied with oil.*

Trying to predict the unpredictable

HOW EARLY FORECASTERS FOUND CLUES IN NATURE

AT THE GREAT EXHIBITION OF 1851, HELD IN LONDON'S glittering Crystal Palace, a certain Dr Merryweather unveiled his 'Tempest Prognosticator', an invention consisting of an apparatus 'by which one of at least 12 leeches confined in bottles of water rang a little bell when a tempest was expected'. He believed that the leeches would rise to the surface, where bells were placed, when a storm was approaching. Merryweather suggested the government should install 'leech-warning stations' along the coastline. But doubting officials rejected the offer, and Merryweather and his leeches returned to obscurity.

Watching nature for clues

Merryweather's work was part of a long-standing European tradition devoted to predicting the weather by observing the behaviour of animals, plants and other natural phenomena – a study which dated back to Aristotle's *Meteorologica*, a debate on the meaning of things such as rainbows, written in Greece in the 4th century BC. By the Middle Ages, Europe was awash with weather lore: traditional beliefs which had been condensed into rhymes and sayings. Many of these were based on fact, and remain valid today. The Moon's appearance and position was held to be of special significance. Rhymes such as 'Pale Moon doth rain/Red Moon doth blow/White Moon doth neither rain nor snow,' were in use long after scientists had found that the Moon did not affect the weather. 'The Moon and the weather/May change together/But a change of the Moon/Does not change the weather,' was more accurate.

ROYAL FORECAST *A Swedish forecaster uses nature's signs to predict the weather before his king.*

IN ALL WEATHERS *Lightning arcs in the sky, water spouts twist out at sea, waves crash against the coast, a gusty wind bows the trees and snow falls in this crowded round-up of almost every kind of weather, published in 1849. Other phenomena include the Aurora Borealis, or Northern Lights, a glacier in the mountains, dark storm clouds, the haloed Moon and just the suggestion of a sunny day.*

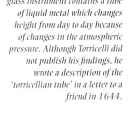

WATER PORTENTS *A medieval observer tries to interpret the meaning of rain and tides in a highly illuminated letter in a 13th-century manuscript.*

Country warnings of wind and rain

The most respected weather prophets, however, were those whose lives depended on it. Sailors and shepherds were known to be particularly expert. One of the most well known early texts of forecasts was first published in England in 1670 as *The Shepherd's Legacy* and then reprinted under the title *The Shepherd of Banbury's Rules*. These examples of weather lore arose from the study of winds, clouds, and mists which became part of tradition. Many of the Banbury shepherd's observations, such as 'If the sun rise red and fiery/wind and rain,' mirror sayings that are still familiar today, such as 'Red sky at night/Shepherd's delight/Red sky in the morning/ Shepherd's warning.'

'A red Evening and a grey Morning
Sets the Pilgrim a Walking.'
TRADITIONAL ENGLISH PROVERB

Some of his conclusions were incorrect, but modern meteorologists credit the unknown shepherd of Banbury with identifying one of the principal rules of weather – a tendency for a particular pattern to last several days once it has been established. Moreover, research conducted in the 1920s showed that on seven occasions out of ten, a red sunrise did indeed spell rain.

The behaviour of plants and animals also offers useful weather clues. The scarlet pimpernel was known as 'the ploughman's weather glass' in some areas, for instance, because of its prompt reaction to atmospheric changes. Fully extended flowers are a

sign of fine weather, but if an outbreak of rain is imminent, the plant reacts to the humidity and closes its petals in an attempt to keep its pollen dry.

Country sayings such as 'When ditch and pond offend the nose/Then look for rain and stormy blows,' show that these unpleasant effects of a change in atmospheric pressure had been accurately observed long before scientists understood why such smells escaped from mud and rotting vegetation on one day and not on the next.

DISCOVERY IN SILVER *The Italian physicist and mathematician Evangelista Torricelli (right) became renowned after accidentally discovering that the up-and-down movements of a silvery column of mercury can predict the weather. His slender glass instrument contains a tube of liquid metal which changes height from day to day because of changes in the atmospheric pressure. Although Torricelli did not publish his findings, he wrote a description of the 'torricellian tube' in a letter to a friend in 1644.*

A COLUMN OF MERCURY TO FORECAST THE WEATHER

EVANGELISTA TORRICELLI, AN ITALIAN SCIENTIST, WAS absorbed by the curious fluctuations of silvery mercury in the open-bottomed glass tube before him, its open end sitting in a bowl of the globular element. He did not realise it at the time, but the level of mercury's rise and fall reflected the changes in air pressure that were occurring and changing the weather. Torricelli had just invented what was to become known as the barometer. After its first appearance in 1643, it became clear that the column of mercury sank on the approach of bad weather and rose as the weather improved.

Early storm warning

Then, in 1660, a German physicist, Otto von Guericke, became the first person successfully to predict a bad storm using a barometer. It was soon realised, however, that weather is influenced by many factors other than air pressure, and that more accurate forecasts would be obtained by coordinating observations across a wide area. In the 18th century it was discovered that varying air pressures at different sites were related to wind speeds. This indicated that areas of high and low pressure moved around as weather systems.

In 1854 the British Meteorological Office was set up and began what became modern weather forecasts. Admiral Robert Fitzroy, the head of the Office, issued storm warnings, based on barometer readings, temperatures, rainfall rates and wind speeds from coastal weather stations.

WEATHERMAN *Barometer pioneer Otto von Guericke's work set up weather observation stations across Europe after he predicted a violent storm. It was later found that cyclones circle areas of low pressure, where the weather is bad; while anti-cyclones circle areas of high pressure, usually associated with fine weather.*

Willing victims of outrageous fashion

HOW THE WEARING OF WIGS WAS TAKEN TO EXTREMES

QUEEN OF ARTIFICE *By the 1770s, ornate hair structures had become high fashion for European women, who took their lead from Queen Marie Antoinette (left) and her followers at the French court of Louis XVI. Styles were diverse, the most extravagant decorated with flowers, imitation fruit and even model ships.*

CALLING ON HIS ELDERLY AUNT AT AN INOPPORTUNE moment, one elegant 18th-century gentleman found himself present as her hairdresser dismantled the intricate structure he had created the previous week. Relieved to have her hair loose again, she wondered out loud at an old woman's foolishness in putting up with such discomfort in the name of fashion.

Applying poison with a brush

Fashionable men and women powdered, primped and painted themselves almost beyond recognition throughout the 18th century, at times leading to a bizarre contrast between the dictates of high fashion and the commonsense demands of basic hygiene. Both sexes painted their faces white, although the dangers of using toxic white lead and mercury for this were already well known by the 1720s. The ugly pockmarks of smallpox were hidden by face patches made from black fabric. Conventional shapes such as moons and stars were replaced by fanciful lovebirds or tiny silhouettes of the wearer's friends.

Rouge, lipstick and even false eyebrows made of mouse skin were used by both men and women, and the loss of side teeth was disguised by little cork balls called 'plumpers', worn in the cheeks to fill out the hollows. Above them all sat the crowning glory – a wig or hairstyle of impressive proportions.

> '*All conditions of men were distinguished by the cut of the wig, merchants by the grave full bottom, tradesmen by the snug bob of natty scratch and coachmen with the curled hair of a water dog.*'
> JAMES STEWART, 1728

Periwigs, later shortened to wigs, had been readily available since the mid 17th century – Samuel Pepys bought his first in 1663 – and for much of the 18th century, no gentleman dared show himself in public without one. Once fixed in place – often over a scalp that had been cropped or shaven – wigs were dressed with pomade, a scented ointment extracted from beef bone marrow, and powdered with rice starch or wheatmeal. The powder was usually grey or white, although some of the more outrageous had their wigs powdered pink, blue or even black. The best wigs were made from human hair, although more affordable horsehair, goat's hair, wool and vegetable fibre were also used. One inventive wigmaker, advertising in 1750, even proposed 'a wig of copper wire which will resist all weathers and last forever'.

Going back to basics

Men began to abandon wigs in the mid 1700s, but still dressed their hair to look like wigs. Women used wire and padding to give height and body to their own hair. As the more extreme styles caused discomfort, even requiring women to sleep half-sitting, it was not long before simplicity returned. By the end of the century, hair was again worn long and free.

HEIGHT OF FASHION *Large, elaborate hairstyles such as those caricatured here required the costly services of a hairdresser. Society women would redo their hair every ten days or so.*

The rise and fall of the dandiest dandy of all

HOW BEAU BRUMMELL LEFT HIS MARK ON FASHIONS TO COME

IT WAS A BALMY MAY NIGHT AT THE START OF LONDON'S 1809 society 'season'. Almack's assembly rooms were buzzing with excited chatter, but towards 11 pm, a hush fell as a new arrival surveyed the room. Handsome 30-year-old Beau Brummell, arbiter of fashion, paused for a moment in the doorway. His sober, understated clothes reflected the impeccable style for which he was renowned.

Stylish dress and a ready wit

The first 30 years of the 19th century were a time when English gentlemen were the best-dressed and best-mannered in Europe. This reputation originated with the Dandies, a group of lighthearted, well-dressed and well-behaved young men who were Brummell's closest associates. Brummell and his followers felt that a true gentleman should never attract attention by his outward appearance.

Brummell himself was the central figure of the group, on intimate terms with the aristocracy and the royal family. His friendship with the Prince Regent, later to become George IV, sealed his position

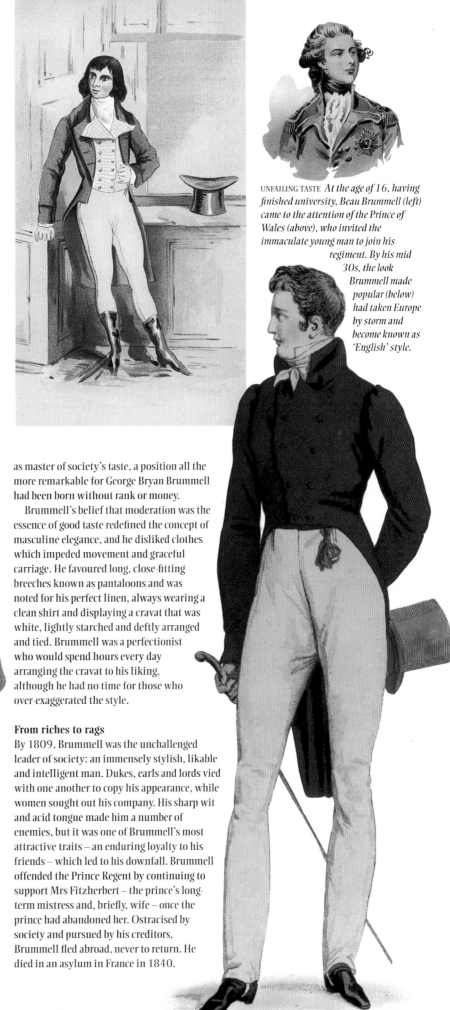

UNFAILING TASTE *At the age of 16, having finished university, Beau Brummell (left) came to the attention of the Prince of Wales (above), who invited the immaculate young man to join his regiment. By his mid 30s, the look Brummell made popular (below) had taken Europe by storm and become known as 'English' style.*

as master of society's taste, a position all the more remarkable for George Bryan Brummell had been born without rank or money.

Brummell's belief that moderation was the essence of good taste redefined the concept of masculine elegance, and he disliked clothes which impeded movement and graceful carriage. He favoured long, close-fitting breeches known as pantaloons and was noted for his perfect linen, always wearing a clean shirt and displaying a cravat that was white, lightly starched and deftly arranged and tied. Brummell was a perfectionist who would spend hours every day arranging the cravat to his liking, although he had no time for those who over-exaggerated the style.

From riches to rags

By 1809, Brummell was the unchallenged leader of society: an immensely stylish, likable and intelligent man. Dukes, earls and lords vied with one another to copy his appearance, while women sought out his company. His sharp wit and acid tongue made him a number of enemies, but it was one of Brummell's most attractive traits – an enduring loyalty to his friends – which led to his downfall. Brummell offended the Prince Regent by continuing to support Mrs Fitzherbert – the prince's long-term mistress and, briefly, wife – once the prince had abandoned her. Ostracised by society and pursued by his creditors, Brummell fled abroad, never to return. He died in an asylum in France in 1840.

CASUAL ELEGANCE *Brummell's look was a reworking of riding clothes, a style that was soon seen everywhere. His expertly cut and fitted outfits – knee breeches, pantaloons, trousers and woollen tailcoats – were the envy of all.*

A travesty of the female form

HOW VICTORIAN WOMEN SUFFERED PAIN AND ILL HEALTH IN THE NAME OF FASHION

PUSHING A PERAMBULATOR TOWARDS THE PARK FOR HER charge's constitutional one day in the 1860s, a 19-year-old nursery maid unexpectedly collapsed and died. According to the doctor who first attended the scene: 'Death was...accelerated by compression of the chest produced by tight-lacing'. In other words, the maid died because of her struggle to achieve the wasp-like waist which was the fashionable look of the day. It was one of many such incidents that led the medical profession to question the wisdom of women wearing such tight corsets.

Tiny waist, great discomfort

For much of the 19th century, the lace strings at the back of corsets were pulled ever tighter as women of all ranks and classes vied to achieve the neatest waistline. As the trim waist they sought was between 21-23 in (53-58 cm), for the vast majority the wearing of corsets (also known as stays) was the only way of achieving it.

Corsets laced so tight that their wearers could scarcely breathe was only one of the tortures women underwent in the name of fashion. Another was an uncomfortable rigid front panel known as a busk, which had originally been a separate piece of wood inserted into the front of the corset.

'HEALTHY' OPTION *This corset was claimed to cure a weak back while producing 'an elegant figure'.*

SPOILT FOR CHOICE *Corsets came in all shapes and sizes and at prices to suit everyone.*

CAGED BEAUTY *These 19th-century photographs parody the fuss necessary to get into a crinoline.*

HEIGHT OF FASHION *Large skirts supported by crinolines and tiny waists achieved by tight corsets were favoured by society women. The little girl wears a child-sized version of their costume.*

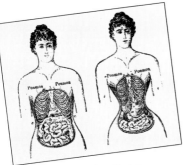

Health sacrificed for style

Some women claimed to find tight corsets pleasurable – one correspondent writing to the *Englishwoman's Domestic Magazine* described the 'delicious sensations, half pleasure, half pain' she experienced when she was being tight-laced into her stays – but this seems to have been the exception. Problems faced by corset wearers included giddiness and fainting, poor circulation and heart trouble.

'At Fashion's edict, stern and brief/ The waist must be compressed no more/ A suspiration of relief/ Goes up from shore to shore'
PUNCH MAGAZINE, APRIL 1909

The fashion was particularly damaging for pregnant women. They suffered far more than necessary in childbirth as a result of tight-lacing. It was claimed that many lost children because of the distortion of their natural figures by tight stays. 'Is it any wonder that persons so deformed should have bad health, or that so many young women should lament the loss of their first born?' asked Mrs Merrifield, a fashion writer, in her book *Dress As A Fine Art* in 1854.

Fashions changed in the late 1860s, when the tiny waist was rivalled as the focus of fashion by the bustle, an elegant projection to the rear of the dress. This required a wire contraption, jutting out at right angles. In the 1890s, flowing curves became the fashion, but tight-lacing survived until about 1910.

THE CHINESE 'LILIES' THAT FASHION FORCED TO WALK WITH TINY STEPS

THE OLD LADY SIGHED AS SHE LISTENED TO THE AGONISED sobbing coming from the courtyard. Slowly, with teetering steps she made her way towards her little granddaughter, who was sitting hunched over her newly bandaged feet. She knew from her own experience that the tight bandages would be searingly painful for the next two weeks, but she believed that the tiny 'lily' feet that would result from the treatment were the mark of a well-bred lady.

The practice of foot binding first appeared in the 10th century at the Chinese court. Some 200 years later it was well established among the upper classes, reaching its peak under the Ming dynasty of 1368–1644. The Manchus, who ruled China from 1644 until 1911, issued several edicts against the practice and never bound the feet of their own women. Nevertheless the custom persisted for many centuries.

Girls' feet were first bound between the ages of five and ten and they would wear the bandages all their lives. The intention was to prevent their feet from growing to more than four or five inches, about half the normal size. The four small toes of each foot were bent over and bound under the sole, leaving only the big toe pointing forward. Then the entire foot was swathed in bandages.

Crippled for life

As the foot grew, the arch was broken and all the bones became deformed. Although the woman got used to the pain, she was virtually crippled for life, and it would always be difficult to walk.

The binding was changed regularly, but the feet were never permanently unbound. After the fall of the old order in 1912, many women tried unbinding their feet. But those whose feet had been bound for ten years or more found that their feet splayed out and bled when they tried to walk.

BINDING FEET
A woman changes the bandages that bind her feet. They have been kept tightly swathed since childhood to produce the tiny feet that Chinese men admire.

FOOTWEAR *Shoes for women with bound feet were mainly ornamental, because the wearer was scarcely able to walk.*

CHANGING SHAPE *Bustles like these of the 1860s were achieved by wearing a boned or metal half cage or a petticoat with half hoops of metal.*

Coping with washday blues

HOW CLOTHES WERE CLEANED IN THE DAYS BEFORE WASHING MACHINES

WASHDAY HAS ALWAYS BEEN AT BEST A TRIAL, AND AT worst a curse, as the great Restoration diarist Samuel Pepys makes clear in his terse entry for April 4, 1666: 'Home, and being washing day, dined on cold meat.' But Pepys's disgruntlement at the dinner table was a minor cross to bear compared to the hardships which faced those confronted with the actual burden of having to do the washday chores.

For hundreds of years, the process barely changed, and until the first commercial washing machines began to appear in around 1860, nearly all washing was done laboriously by hand. An hour at the washtub was roughly equivalent to an hour of swimming breast stroke at an energetic pace, and the work was so physically exhausting that it is now thought to have been a leading cause of ill health.

The easiest way of dealing with washday was to put it off as long as possible. Until the 18th century, even the wealthiest households did their washing only once a month. People disguised the smell of unwashed clothes with perfume and deodorants.

One rule for the rich

Those who could afford it employed laundrymaids or sent their washing out. A formidable breed of professional washerwoman renowned for her muscular physique and lewd language dominated the communal washing-grounds, which held a central place in the life of every town and village. These public washing-grounds were situated near running water, which was collected in large tubs and heated on open fires. The women scrubbed the laundry with their bare hands, using soap made from animal fat

THE FAMILY WASH *All the womenfolk of a 19th-century German household help on washing day. Large houses had one room for washing and another for ironing. Because of the need for hot water these were usually near the kitchen.*

boiled with lye – an organic detergent derived from wood ash – and rubbing it against washboards. After the excess water was wrung out, washing was hung up to dry on the communal clothes lines.

Habits changed by the end of the 18th century. People could afford to change clothes more often as the price of cloth fell during the Industrial Revolution, so the quantity of laundry increased accordingly. Women wore at least three layers of underclothing alone and it was standard practice for these to be washed weekly. It became more economical for big households to do their washing at home in a laundry room, though commercial laundries continued to flourish in large cities.

Rituals of the laundry room

By the 19th century, the best private laundries had stone floors laid on brick piers which sloped gently towards a drainage gutter. Washing was done in a series of wooden tubs, the more sophisticated of which were fed by hot and cold taps. In winter and in cities without outside clothes lines, clothes were hung on wooden frames and left to air in drying rooms, heated by a furnace. Monday was for sorting the washing into piles of whites, coloureds and woollens. Maids removed ribbons, lace adornments and buttons that were too delicate to survive the wash, and stains and grease spots were rubbed with lye. Washing was left to soak in lukewarm water mixed with soda. On Tuesday

ALL IN ONE *Washing is cleaned on the ground floor and dried on the upper floors in a public laundry in Paris.*

WATER ON TAP *For the professional washerwoman, an ornamental fountain is just the place for doing the laundry.*

WHITER THAN WHITE *In the 19th century products with a hint of indigo were added to the white wash to give it a bluish tinge.*

the fires were lit. Whites were given at least three separate washes with soap in water as hot as the hand could bear and coloured clothes and woollens were washed in cooler water to stop them shrinking. Mrs Beeton in her *Book of Household Management* (1861) had a handy tip for preserving the colour of darker silks: she advised dabbing them with gin.

Mangling and ironing

The remainder of the week was devoted to mangling and ironing. The upright mangle was designed by George Jee in 1779. Turning its handle turned two rollers between which the cloth was passed. The action of the rollers squeezed out excess water and gave the sheets an initial pressing. By 1850 mangles were widely available. Once the clothes were nearly dry, they were ironed. A cloth was laid over a table so it could double as an ironing board, and heavy irons were heated by the fire.

Before metal alternatives were introduced in the mid 19th century, washboards, washtubs and other contraptions used to do the laundry were primarily made of wood. In order to make them water resistant, the wood had to be seasoned for up to 18 months before use. The newfangled washboards made of durable corrugated zinc, iron or glass were welcomed as valuable additions to the laundrymaid's armoury. Inventions to lighten the load had appeared as early as 1691, when the first washing machine was patented in England. These early machines mostly consisted of a tub with paddles inside. The tub was filled with clothes and the handle turned to make the paddles spin around. Few machines lived up to the extravagant claims of their inventors, however, and they frequently ruined the clothes they had been designed to wash.

The tedious task of washing changed very little until the introduction of electric-powered washing machines in 1906. Even then the proximity of water and electricity at first made these dangerous.

STREAMLINED STYLE *An early 20th-century mangle with an iron frame and wooden rollers has a screw to tighten the rollers' pressure.*

MANY HANDS *Those who can afford it revel in the novelty of a hand-cranked washing machine (above) in a mid-19th-century advertising poster. Poorer women were still tied to the drudgery of washing clothes manually (left), a job so time-consuming that the children had to help.*

Calling cards and kid gloves

HOW STRICT RULES OF ETIQUETTE GOVERNED VICTORIAN LIVES

FOLLOWING THE RULES *An unmarried Victorian belle was chaperoned at every social event she attended. At dances, for instance, etiquette dictated she could not dance more than twice with the same man. For his part, her partner would not dare to take her hand until it was offered.*

HE RECOGNISED HER AT ONCE, STROLLING THROUGH the crowded park, twirling a blue parasol on her shoulder. The young gentleman could hardly believe his luck, as he had not been able to stop thinking about her since they had met the evening before and shared two dances. His heart raced as she approached and he smiled, raising his hat.

She walked past him without even a flicker of recognition, her glance passing over him and moving swiftly on. His friend explained gently that although the couple had been introduced for the purpose of a dance, this was not the same as a social introduction. The man flushed, embarrassed by his error. A subtle code of conduct governed every aspect of Victorian social life. The 'cut', as it was known, was one of the deadliest weapons of the code. If a woman was approached by an unknown man she could completely ignore him – yet at the same time make him aware that the act was no accident. Upper-class women, ladies of ample leisure, were the keepers of the complex book of rules that governed their lives. The strict protocol of etiquette made sure that different social groups remained separate, and was useful for keeping poorer or socially undesirable people at bay.

The ritual of 'calls'

It began with what were called 'morning calls', although they were usually made in the afternoon. These were made to welcome a well-connected family to a new area, to announce one's return to the city from the country, to offer condolences or congratulations or to maintain a friendship.

Calling cards played an important role. Delivering the cards was a woman's responsibility. She could leave her card and two cards on behalf of her husband – one for the master of the house and one for the mistress – and depart, without asking to see the hostess. Alternatively, she could ask if the lady of the house were 'at home'. If she were, a meeting of not more than 20 minutes would follow. The visitor was not expected to remove her shawl or bonnet.

Subtle messages could be transmitted through such elaborate ritual. If a personal call was returned by a visit in which a lady simply left a card, it was clear the relationship was not expected to go further. A woman could determine whether her calls were welcome by the speed and enthusiasm of responses.

There were rules for all occasions. Handshakes were a potential minefield of social embarrassment. A married woman would extend her hand on being introduced to a gentleman but a young single woman would not. A man of junior status was expected to wait until his superior offered his hand. In a world where everyone was expected to know their place, etiquette ensured that people knew how to behave.

CALLING CARDS *Beautifully engraved cards, like the one below, were a vital part of the ritual of 'making calls'. Families would also send expensive invitations (middle) for their parties.*

DANCE CARD *Women used a pencil to enter the names of partners for each dance.*

AT DINNER *The woman being helped to her seat at this dinner party is wearing a pair of white kid gloves; any other material was considered to be vulgar.*

CHARM SCHOOL *The dancing master teaches young women the art of flirting with a fan.*

Eloquent gestures, private messages

HOW SOCIETY WOMEN USED FANS TO SIGNAL THEIR LOVE

THE YOUNG MAN LEANED AGAINST a pillar, idly scanning the crowded ballroom, where all London seemed to have gathered. He glanced at a dark-haired beauty carrying an ivory fan. He knew that she had seen him, because their eyes had briefly met. Now she held the fan fully open in her left hand in a gesture that he knew meant 'come and talk to me'. Heart pounding, he began to walk across the floor, but she started twirling the fan in her left hand. This signalled 'we are being watched'. He paused.

'Do not be impudent'

'I wish to speak to you'

'I love you'

'No!'

Romance in a gesture
Suddenly she rose from her seat, turned and walked towards the balcony, the closed fan in her right hand in front of her face a private invitation for him to follow.

Fans were essential items for any woman of fashion in the 18th and 19th centuries, but their true fascination lay in the way they could be used to send romantic messages. In the 18th century complex fan games were devised in England to spell out messages letter by letter. In the 19th century, a simpler scheme in which whole phrases (usually romantic) could be conveyed with a single gesture was devised in Spain.

'Follow me'

'Kiss me'

FLOWERS SAY 'I LOVE YOU'

MOTHER AND DAUGHTER GAZED ADMIRINGLY at the large bowl filled with beautiful tulips.
'A small thank you from young Mr Greville, Emily' said the mother, 'how perfectly sweet! I do believe you made rather an impression.'
'Really, Mama, I've never heard such nonsense…'
Emily was glad that her mother was unaware of the message conveyed by the flowers – a meaning she had checked in her much-thumbed copy of *The Language of Flowers* by Robert Tyas, a best seller of its day. The flowers were a declaration of love from her admirer.

Nineteenth-century gentlemen developed a great enthusiasm for using flowers to send romantic messages. Dictionaries listed the meanings of flowers: an azalea indicated temperance and ivy leaves meant friendship. The cult of the language of flowers flourished in Europe for around 50 years from about 1810. It was promoted by publishers, for whom it meant good business, but at length the craze fell out of fashion.

TRIBUTES *Roses, for beauty, and lily of the valley, for happiness, decorate Valentine cards.*

LOVE'S YOUNG DREAM *A girl sighs with delight as she reads a letter from her admirer. The pressed flowers enclosed in the note mean as much to her as the words.*

All in a day's work

HOW A VICTORIAN TOWN HOUSE WAS RUN

THE DAY BEGINS EARLY IN THE DETACHED VILLA IN Newcastle occupied by George Ridley, a director of a Tyneside shipyard, his family and their servants. It is still dark as Rose, the scullery-maid, rises from her attic bed on a chilly November morning in 1865 and drowsily goes downstairs to the kitchen. There she rakes the cinders out of the cast-iron stove, which she then coats with liquid black lead and polishes until it gleams. She brings coal up from the cellar, sets and lights the fire in the kitchen range, and boils the day's first kettle of water. Next she gathers up the family's boots and shoes ready for the footman to clean.

'Your newspaper is ironed, sir'

By now, the three housemaids are bustling from room to room on the ground floor. They sprinkle wet tea leaves on the carpets to catch the dust and make them easier to sweep with their stiff brooms. Before Mr and Mrs Ridley come down to breakfast, everything – from the gleaming mahogany furniture to the window frames – must be dusted and polished. At 8 o'clock Polly, the lady's maid, wakes Mrs Ridley with a cup of tea. At the same time Mr Ridley's valet takes him the morning newspaper – carefully ironed and folded to remove any creases – and a metal can full of hot water for shaving.

The Ridley's six children are awakened by their two nurses, who see that they get washed and dressed, and then send them into the first-floor nursery for their breakfast. Only baby Martin stays behind to be bathed and given his early morning bottle of barley water and milk. Presently one of the footmen rings a bell in the hall, and the entire household – including the butler, the cook, the children's governess, the two footmen, the coachman and groom – gathers in the ground-floor morning room, where Mr Ridley gives a short Bible reading and leads the assembled company in daily prayers. Master and mistress then breakfast in the dining room. There is tea and coffee; in addition to lashings of toast, rolls, butter, homemade jam and combs of honey, the Ridleys have a choice of freshly cooked cutlets, kidneys, fish, bacon, eggs and grilled chicken. Downstairs, in the kitchen, the staff eat their breakfast: hunks of bread, a cold game pie or slices of lamb from the previous evening's roast washed down with steaming cups of tea.

It is now 9 o'clock and time for Mr Ridley to be driven in his carriage to the nearby shipyard. His wife, Catherine, waves him goodbye from the pillared porch, and goes back into the house. The older children are in the upstairs schoolroom, receiving their first lesson of the day: arithmetic, from the governess, Miss Hardwick. The housemaids invade the now unoccupied bedrooms for their most thorough clean of the week, carrying an array of brooms, dusters and brushes. They throw open and clean the windows, sweep the carpets, polish the oak floor surrounds with beeswax, make the beds, shine the full-length mirrors, dust the ornaments and pictures, wipe down the wardrobes, shake the curtains and empty the chamberpots into the lidded slop pail.

By mid morning the upstairs work is done, and the housemaids launch into their next task. Armed with scrubbing brushes and buckets of hot water and soap suds, they get down on their hands and knees and clean the linoleum in the other rooms. They then polish the 'lino' until they can see their faces in it. Meanwhile, in the kitchen, the three senior servants – the butler, the cook and the housekeeper – are preparing a luncheon of cold meats for Mrs Ridley. The staff have their lunch, or 'dinner' as they call it, downstairs – cold meat or bread and cheese, washed down with water or beer. For the governess and children, some cottage pie is sent up to the nursery.

'Your bedroom fire is lit, ma'am'

At 2 o'clock, Mrs Ridley retires to her bedroom, where a maid has lit a fire that blazes cheerfully in the grate. As it grows dark, the footmen light the gas lamps and draw the maroon velvet curtains in the main rooms. At 6 o'clock Mr Ridley returns home, well in time to receive his dinner guests – businessmen like himself – and their bejewelled wives. Before the guests arrive, the butler, Rutherford, opens the fine hock and claret which he brought up from the wine cellar earlier to accompany the meal.

It is almost midnight when the guests finally leave. Their carriages trundle down the drive, past the stables and out through the wrought-iron gates. Indoors, the household at last prepares for sleep. Shivering, Rose climbs the stairs to her tiny bedroom as the wind off the North Sea whips against the side of the house. For her, it will be another early start in the morning.

THE SHINING HOUR *At first light, a hard-working maid scrubs the front doorstep of her mistress's spick-and-span town house until it shines.*

THERE'S A GOOD BOY *A nurse shows her affection for the latest member of the family to be put in her care.*

A GIRL'S BEST FRIEND *In a welcome off-duty moment, a housemaid sits with her pet cat cuddled on her lap.*

TIME TO BE SEEN AND HEARD *Well-off mothers in Victorian times set aside certain periods of the day to converse and play with their neatly dressed, well-disciplined children.*

WELL-ORDERED
The head cook of a household (centre) organises her staff and kitchen with military precision.

TIME FOR TEA An elegant tea, using the best silver, is nicely laid out in the garden for a well-to-do family.

SHORT BREAK Hands on hips, a smiling scullery-maid pauses during her scrubbing and cleaning duties.

WOMAN OF IRON
A stern-faced housekeeper stands with three of her young 'slaveys', whom she rules with an iron hand as they learn how a household is run.

HORSE PEOPLE As one of the household's senior servants, a coachman (above, left) takes a pride in his working attire. When it comes to an open carriage, however, the mistress of the house (above, right) drives it herself– with a little help at the start from a liveried groom.

SONGS AT TWILIGHT With the long, busy day behind them an Edwardian family gather in the comfort of their drawing room for an evening of musical entertainment. Everyone dresses for the occasion and everyone has something to offer, from playing the piano to rendering heart-tugging ballads, tender love romances and comic songs.

NANNYDOM Dressed in their Sunday best, five children and their nannies pose with their pet donkey.

Triumphs of Building and Engineering

ALTHOUGH OUR ANCESTORS HAD TO RELY ON TOOLS THAT WERE PRIMITIVE BY MODERN STANDARDS, THEIR AMBITION WAS BOUNDLESS. THE MIGHTY WORKS THEY LEFT BEHIND BEAR WITNESS TO THEIR SKILL AND INGENUITY, AND AS TECHNOLOGY DEVELOPED, THEIR ACHIEVEMENTS GREW EVER GREATER

A vast monument constructed by hand

HOW THE GREAT PYRAMID WAS BUILT TO HOUSE A DEAD PHARAOH

THIEVES BREAK INTO THE 'BURGLAR-PROOF' GREAT PYRAMID

FOR DAYS ON END, ABDULLAH AL MAMUN'S GANG OF WORKMEN scoured the face of the supposedly secure Great Pyramid searching for the secret entrance to the tomb of the Pharaoh Khufu. The Egyptian ruler had been buried there, along with a fortune in gold and other treasures, in the 26th century BC. Al Mamun – the Caliph of Egypt in the ninth century AD – ordered his blacksmiths to break in.

Bombs and battering-rams
Working on ramps, engineers lit bonfires beside the stone blocks in order to weaken them, then soaked them with vinegar, or sour wine, to 'explode' them like bombs. Next, they were smashed with battering-rams. Once inside, Al Mamun's men tunnelled their way with picks and hammers and chisels farther and farther into the interior. Finally they reached a flat-ceilinged burial room some 35 ft (10.5 m) long, 17 ft (5 m) wide and 20 ft (6 m) high. At last they stood, with torches held high, in the King's Chamber, which was dominated by a lidless box of glossy, dark-brown granite. But the pharaoh's body was not in it, nor was any of his fabulous treasure to be found. Al Mamun had been beaten to it by grave robbers who had plundered Khufu's riches more than 2000 years before, possibly using inside information obtained from workmen who had built the Great Pyramid.

INSIDE VIEW *Burglars clamber into the Grand Gallery, 30 ft (9 m) high and 153 ft (47 m) long (far left). This leads to the king's burial place. They also find the entrance to the so-called Queen's Chamber (left).*

ROBBERS' TOOL *A wooden mallet used by thieves 4500 years ago to wedge open the coffin lid in a prince's tomb at Maidum is still in place.*

WITH A BRUSH DIPPED IN RED OCHRE, A WORKMAN daubed 'the Craftsman Gang' on the side of the limestone block and, on a shouted command, the team began to pull. The block weighed two and a half tons, and there were only eight men to pull the sledge on which it lay. As long as the runners of the wooden sledge were splashed liberally with water, the heavy load could slide easily across the ground on wooden baulks or rollers. The block was to form part of the huge edifice which loomed over the men as they worked. Day by day, the pyramid rose by means of ramps made of earth and brick which were built up, layer by layer, as the pyramid grew in size.

The pyramids of Giza, near Cairo, were built as tombs for three Egyptian pharaohs. The largest, the Great Pyramid, originally stood 481 ft (147 m) tall and covered an area of more than 13 acres (about 5.3 hectares). It housed the remains of the Pharaoh Khufu, who died in about 2568 BC.

Guided by the stars
The pyramids were on foundations of solid rock. However, before building could begin, the site had to be levelled. This was achieved by cutting trenches into the rock, then filling the trenches with Nile water. Lines were cut in the sides of the trenches to mark the water level and the water was drained. Finally, the rock was cut down to the level of the marks and the trenches filled with rubble. The orientation of the pyramid – its four sides point north, south, east and west – was accurately fixed within fractions of a degree by making observations of the stars. The main stone used to build the pyramid was locally quarried limestone; the interior chambers were lined with granite quarried at Aswan, 600 miles (966 km) up the Nile; and fine white local limestone was used for the outer casing.

As the pyramid rose, the heavy blocks of stone had to be hauled up the ramps. In the absence of wheels or pulleys, everything had to be done by brute force. The final stage was to set in place the facing blocks of limestone. These were cut on the ground and polished with sand or small pieces of rock so that the vertical joints were a perfect fit. They were lowered onto a thin bed of mortar and levered into place. To complete the pyramid, a granite capstone was dragged to the top and manoeuvred into position. Only then could the ramps be demolished and the glory of the Great Pyramid be revealed at last.

TOWERING TOMB *The Great Pyramid rises majestically some 450 ft (137 m) above the desert sands; at its summit there is still part of the original white limestone facing. Three burial chambers, a network of sloping passages and corridors, and an escape tunnel were built into the pyramid.*

HARD AT WORK *An army of 10 000 specially recruited men laboured for 20 years to build the Great Pyramid. The work included jobs such as making mud bricks for the building ramps. Workmen shape the bricks in moulds from prepared clay (top, left). One workman braces his adze against a pile of bricks (bottom, left) while another bends double to carry the bricks in a yoke to the building site (bottom, right). Each team of workmen daubed their name on the blocks they worked on – probably as a means of keeping a tally of the number of stones they had put in place.*

Ropes, pits and mountains of timber

HOW STONEHENGE WAS BUILT

THE HUGE STONE ON THE GROUND WAS 20 FT (6 M) LONG and weighed 50 tons. At its foot lay a pit 8 ft (2.4 m) deep, three sides held up by wooden stakes and the fourth – nearest to the base of the stone – shaped as a ramp. A team of men had excavated the pit using deer antlers as picks and ox shoulder blades as shovels. Ropes of animal hide and plant fibre were lashed around the stone ready to haul it upright.

In 2000 BC Salisbury Plain, in the west of England, was the scene of backbreaking toil as the largest megalithic stones were erected at Stonehenge. Raising the head of the monster stone the first few feet off the ground was the hardest task, even with the help of shear legs – lifting tackle hung from two spars joined at the top and spread at the base. As the head of the stone rose, men stacked timber beneath it to stop it falling back again. The pulling became slightly easier as the angle increased, until the foot of the stone began to slip down the sloping side of the pit. The men levered the stone with long wooden poles, using the timber underneath as a fulcrum. Finally the base of the stone hit the bottom of the pit. Stakes along the pit's edge stopped the stone toppling forwards.

The building of Stonehenge took place in three phases over 1600 years. The first began in 3100 BC, more than 500 years before the Egyptians started work on the Great Pyramid. These Stone Age builders built a circle 380 ft (115 m) across, formed by a low outer bank surrounding a ditch, with another bank about 6 ft (1.8 m) high inside the ditch. At the start of the third phase, around 2100 BC, a double circle of 80 large bluish stones, known as bluestones, was added. These were dragged by men and oxen and floated on rafts from quarries 130 miles (210 km) away in the Preseli Mountains of south-west Wales.

Between 2000 and 1550 BC, early Bronze Age people removed the bluestones and erected a ring of 30 sandstone uprights, 16 ft (5 m) tall, linked at the top by horizontal lintels, each weighing 7 tons. Inside the ring they set five even taller 'triliths', grouped stones in which two uprights support a crossbeam; later, the bluestones were re-erected. The monument that these builders left behind is an astonishing tribute to their imagination and skill.

HEAVENLY MOTION *The Sun rises over the mysterious pillars of Stonehenge, on Salisbury Plain. The stones are aligned with the movements of the Sun and the Moon, and were probably used both in ceremonies celebrating the celestial movements and as a means of calculating a calendar of the seasons. For many years historians associated Stonehenge with the Druids, the Celtic priests, but in fact it was completed more than 1000 years before the Druids came to Britain, in 300 BC.*

MAGIC TOUCH *In the Middle Ages, scholars thought that Stonehenge was built by the 5th-century British king Vortigern and his wizard, Merlin. A 14th-century illustrator shows Merlin directing the work.*

LAST EFFORT *On one side of the great stone, men use a long pole as a lever, while on the other straining labourers give a final heave, and the vast stone slips down the sloping side of its specially prepared pit. Boulders and stones (left) are used to wedge the pillar in place. The boulders have been dragged onto the site on sledges that can be taken apart for easy storage.*

SHAPING *Using a rock as a hammer, a workman gives a smooth finish to one of the stones. As they were meant to be seen from within, the stones' inner faces and sides had a smoother finish.*

POSITIONING *The builders ease the lintel into place after raising it from the ground on a laboriously hand-built deck of logs. Holes on the side and bottom face of each lintel match projections on its neighbour and on the stone below. The stones are surrounded by three rings of holes, or pits, some of which were used for cremation burials.*

Shrine to a goddess of wisdom

HOW ATHENIANS BUILT THEIR PARTHENON IN JUST 15 YEARS

THE GODDESS'S IVORY SKIN GLOWED SOFTLY IN THE half-light of the temple's inner room. Her magnificent frame stood 33ft (10m) tall – the masterpiece of the Greek sculptor Pheidias. She was Athena, the patron of Athens and goddess of wisdom and the arts, and the building in which she stood was the heart of Athens' Parthenon temple. The building contained two chambers – one a treasury in which offerings were laid, the other housing the statue. It was a shrine to Athena Parthenos ('the Virgin') and gave its name to the whole temple.

The building of a shrine as a thank-offering to Athena had first been put forward after the Battle of Marathon in 490 BC, when the Greeks won a surprise victory over the invading Persians. But only the platform had been laid before the Persians attacked again, demolishing what there was of the new temple before they were driven back once more. In around 447 BC the project was resumed.

The temple was the brainchild of Pericles, ruler of Athens from 443 to 429, and was intended to symbolise the city's power. But as the building rose on the summit of the Acropolis, the hilltop that dominates Athens, it became a focus of scandal.

PRECIOUS CARGO *Little by little, the rounded block on its wooden sledge is lowered down the stone causeway. Ropes attached to the block and the sledge are tied to staves hammered into holes at the side of the slipway to prevent the block sliding out of control down the slope. It will form the capital of one of the 42 pillars in the Parthenon's stately colonnade.*

MOUNTAIN ROCK *Marble for the Parthenon is cut in a quarry high on Mount Pentelicon, 10 miles (16 km) from Athens. With mallets and chisels, workmen carve sockets for wedges to split stone blocks from the bedrock.*

SKILLED LABOUR *At the foot of the deep quarry, masons rough-hew the blocks to near their final shape. One block (right) has been hauled to the top using pulleys.*

FIGHTING GRAVITY *Workers pull ropes attached to giant winches, turning drums that hoist the block up the mountainside to the top of a causeway.*

READY FOR DELIVERY *At the mountain's foot, men transfer the block from its sledge onto wooden rollers, and ease it from the end of a specially built ramp onto a sturdy cart waiting below. Workers dismantle the sledge before returning it to the quarry.*

ANIMAL POWER *A straining team of mules pulls the massive block through the outskirts of Athens. The rising Parthenon is visible on the distant Acropolis.*

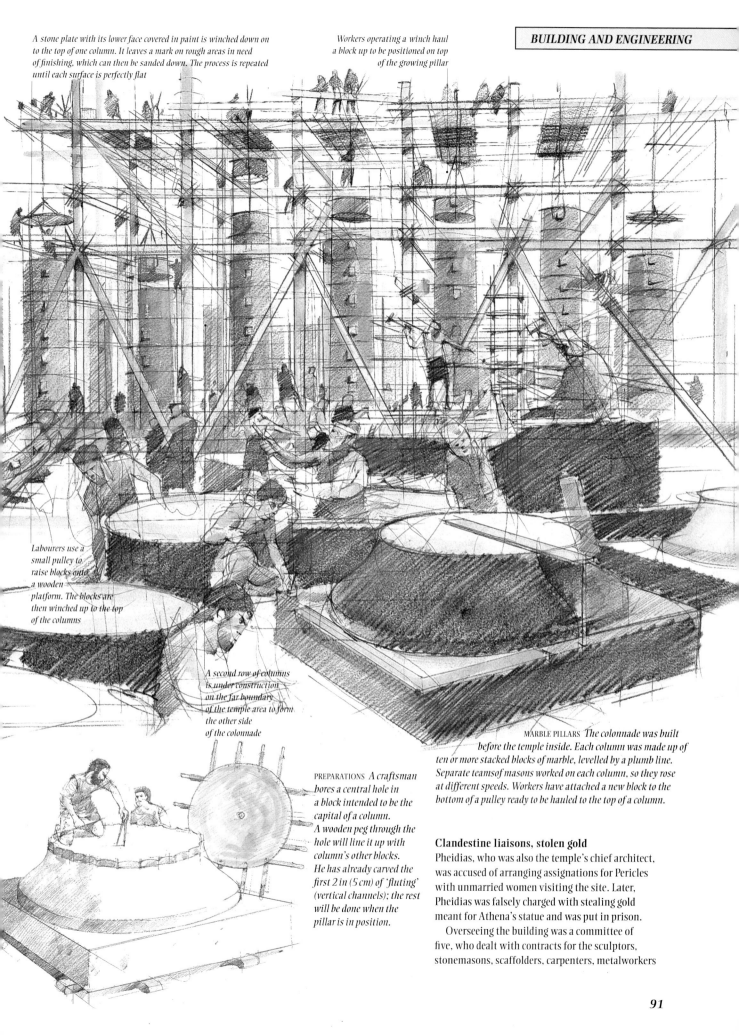

A stone plate with its lower face covered in paint is winched down on to the top of one column. It leaves a mark on rough areas in need of finishing, which can then be sanded down. The process is repeated until each surface is perfectly flat

Workers operating a winch haul a block up to be positioned on top of the growing pillar

Labourers use a small pulley to raise blocks onto a wooden platform. The blocks are then winched up to the top of the columns

A second row of columns is under construction on the far boundary of the temple area to form the other side of the colonnade

PREPARATIONS *A craftsman bores a central hole in a block intended to be the capital of a column. A wooden peg through the hole will line it up with column's other blocks. He has already carved the first 2 in (5 cm) of 'fluting' (vertical channels); the rest will be done when the pillar is in position.*

MARBLE PILLARS *The colonnade was built before the temple inside. Each column was made up of ten or more stacked blocks of marble, levelled by a plumb line. Separate teamsof masons worked on each column, so they rose at different speeds. Workers have attached a new block to the bottom of a pulley ready to be hauled to the top of a column.*

Clandestine liaisons, stolen gold

Pheidias, who was also the temple's chief architect, was accused of arranging assignations for Pericles with unmarried women visiting the site. Later, Pheidias was falsely charged with stealing gold meant for Athena's statue and was put in prison.

Overseeing the building was a committee of five, who dealt with contracts for the sculptors, stonemasons, scaffolders, carpenters, metalworkers

and other craftsmen, many of them travelling artisans. The team working on the building was fairly small, perhaps just 200 at any one time.

Carved from mountain rock

The temple was built almost entirely from white marble; only the doors and ceilings were made of timber. Some 22 000 tons of the stone were transported from nearby Mount Pentelicon. Once it was on site, the stonemasons dressed the blocks – already cut close to their finished shapes – so that they could be positioned. They used iron chisels, saws, drills, callipers, set squares and plumb lines. Because no mortar was used, the surfaces where two blocks met had to be a perfect fit. The masons flattened only the outer rim of the face of each block where it joined its neighbour, and left a rougher hollowed surface on the inner section – avoiding contact with the adjoining block. The blocks were clamped together with iron clamps set in lead.

The builders constructed a large flat base with three steps leading up to it, measuring 228 by 101 ft (69 by 31 m) on the top step. Next they raised a colonnade around the edge – 17 pillars on each side and eight at either end. In the centre they built the two-room inner temple, with six columns at each end.

Scenes of siege and battle

Before the roof was added, labourers put 92 carved panels depicting siege and battle scenes from Greek mythology above the colonnade, each panel measuring 4 ft 3 in (1.3 m) square. They were carved at ground level and then hoisted up on pulleys and positioned. The roof was of sturdy timber, covered with tiles hand-carved from white marble from the island of Paros were laid over it. By 432 BC the Parthenon was complete and it remained in good condition for 700 years. Its crumbling appearance today is the result of war rather than decay. In the 5th century AD Athena's statue was carried off to Constantinople, and in the 7th century the temple became a Christian church under the Byzantines. It was made into a mosque in 1460 under the Ottoman Turks. In 1687 the Turks, besieged by a Venetian army, used it as an ammunition store. Venetian shelling exploded a powder magazine, shattering columns and walls.

GILDED GODDESS *Athena, the temple deity, towers over workmen applying pure gold to her vast marble dress. Over a ton of gold was used, and the statue cost twice as much to create as the rest of the Parthenon.*

RELIGIOUS CARVING
Giant tongs lift the final section of marble for the great frieze on the Parthenon's inner temple depicting a procession in Athena's honour. Sculptors perched high on scaffolding are busily carving the frieze.

ROOFTOP STATUES *A section of one of the Parthenon's two marble pediments is lowered into position. The pediments will contain statues: those at the temple's eastern end represent Athena's birth; those at the western end show Poseidon, god of the sea, vying with Athena for the right to protect Attica, the region containing Athens.*

Men pulling on the arms of a revolving winch have hauled in the pulley's rope and lifted the marble pediment to the rooftop

Hauling on strong ropes, sweating workmen strain to help raise a fortified box containing an already carved piece of the pediment into its designated spot

Workmen standing beside wooden runners help to hoist a finely carved section of the pediment in its wooden box up to the front of the Parthenon. Many of the pediments were made on the ground, and then moved on the runners, ready to be hauled into place

FINISHING TOUCHES
A mason carves the fluting in one of the columns. He uses plumb lines hung from the top to ensure that the vertical lines match exactly with those already carved around the base.

CURVES THAT TRICK THE EYE

THE PARTHENON LOOKS PERFECTLY SYMMETRICAL, but there is hardly a vertical or horizontal line in the building. The columns lean very slightly inwards, and bulge slightly towards the top. These curves create the illusion of greater strength and beauty; straight lines would appear from a distance to bend inwards. A similar trick was used in the Ziggurat of Ur, a pyramid temple built to the moon god Nanna in Mesopotamia in about 2100 BC.

ILLUSION *The walls of the Ziggurat of Ur appear straight but in fact curve from top to bottom and corner to corner.*

Arena for an imperial bloodbath

HOW ROME'S COLOSSEUM WAS BUILT ON THE SITE OF A LAKE

AFTER SIX YEARS' TOIL BY AN ARMY OF ARTISANS, ALL was finally ready for the Emperor Titus to take his place in the imperial box. It was AD 80, and Rome was humming with visitors summoned from the corners of the empire to witness the opening of the Flavian Amphitheatre, named after the dynasty founded by Titus's father, Vespasian. For the next 100 days, a baying crowd enjoyed an orgy of blood as countless animals and men were slaughtered.

An imperial gift

Today the building is one of Rome's most majestic ruins. This huge amphitheatre, 620 ft by 513 ft (190 m by 155 m), has been known since the 8th century AD as the 'Colosseum' (from the Latin *colosseus*, 'gigantic') – but it got its name because it stood near a colossal statue of the Emperor Nero and not because of its size. It was planned by Vespasian, who rebuilt much of Rome in 69-79 after Nero's reign had ended in chaos. He chose a site in the centre of the city, where Nero had built a house beside an ornamental lake.

The first step was drainage. Stone sewers were built to drain the lake into the River Tiber. Foundations of concrete were sunk 18 ft (5.5 m) through the riverbed into gravel. Then 80 concrete walls were built from the perimeter of the oval building into the edge of the arena; these bore the weight of the tiers of marble seats, which were a total height of 160 ft (48 m). The perimeter wall was built of travertine stone, a white or light-coloured limestone quarried just outside Rome.

Silk cushions, sweet-smelling fountains

The Colosseum absorbed as much stone and mortar as three medieval cathedrals – 750 000 tons of dressed stone, 8000 tons of marble and 6000 tons of concrete. But detail and comfort were not neglected. Silk cushions and awnings were provided, and, according to the contemporary poet Calpurnius, fountains played jets of perfumed water.

The Colosseum's architect is unknown, for Vespasian was determined to claim the credit. Building, which began in AD 75 (according to some accounts, two years later or earlier), was carried out by teams of craftsmen belonging to more than a dozen workers' guilds, unaided by slaves.

Games were held in the arena for more than 400 years. Gladiators fought each other or wild animals, usually to the death. The majority of these men were slaves, criminals or prisoners of war, but some were professionals. The games attracted huge crowds, but their popularity began to fade during the reign of the Emperor Constantine (324-37): he was a Christian convert and disliked their violence. The last gladiatorial combats there were held in 404.

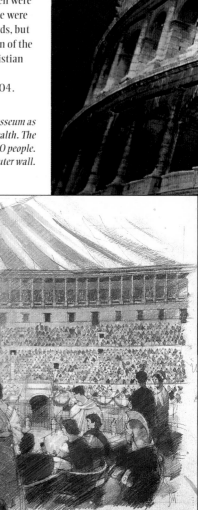

ROMAN PRIDE *The Emperor Vespasian saw the Colosseum as a way of giving citizens a share in Rome's wealth. The vast, richly decorated building could hold 60 000 people. Originally, statues filled the niches in the outer wall.*

ROUGH JUSTICE *A senator's wife gives the 'thumbs down' to the central pair of fighters, instantly condemning the fallen gladiator to death. The canopy was held up by ropes and taken down when weather permitted.*

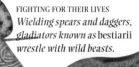

FIGHTING FOR THEIR LIVES *Wielding spears and daggers, gladiators known as* bestiarii *wrestle with wild beasts.*

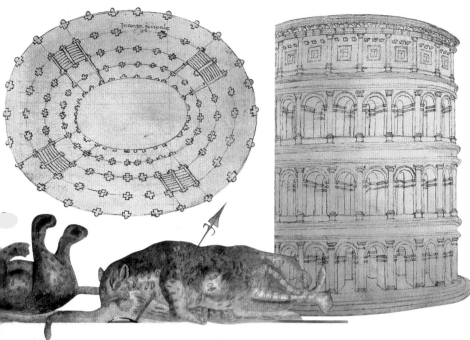

GRANDSTAND *The Colosseum provided three tiers of seating with standing room in a fourth tier at the top. These rather inaccurate views and plan were drawn by the 15th-century Italian architect, Francesco di Giorgio.*

Network of stone that linked an empire

HOW ROMAN ENGINEERS BUILT THE FIRST STRAIGHT ROADS

AT THE HEIGHT OF ITS POWER IN THE 2ND CENTURY AD, the imperial Roman army was a formidable force. It had conquered much of the known world, from northern Scotland to the banks of the Euphrates in Mesopotamia (modern Iraq). The army could build almost anything – bridges, tunnels, forts, canals, marketplaces and bath houses – but its greatest achievement was a network of roads over more than 55 000 miles (88 000 km) of the Roman Empire.

A smooth surface for marching feet

The first true Roman road was begun in 312 BC to link Rome with Capua in the south; it was later extended to Brindisi in the east. The route, known as the Appian Way, set new standards in road building by providing a dry, relatively smooth surface that was both firm underfoot and well drained. Moreover, in common with almost all Roman roads, the Way was remarkably straight, providing troops on foot or on horseback with the most direct route possible.

The course of the road was plotted by surveyors, using just two instruments: a portable sundial, to determine direction, and a *groma* – a horizontally held wooden cross with four vertical plumb lines – used for gauging right angles and marking out land. Once this was done, the roads were meticulously constructed, using layers of stones through which water could drain, and with a porous core that would solidify over years of continuous use.

Major roads started as a base of compacted earth at a depth of 3 ft (1 m), with a recommended width of 18 ft (5.5 m). This was followed by a shallow layer of small stones, and then a deeper one of broken stones, tile chips and concrete. This was topped with slag, lime, chalk or tiles and then finished with paving stones. Paving slabs ran along the sides of the roads which were occasionally divided by a central gutter.

The Roman Empire fell in the 5th century AD, but its remarkable roads survived in reasonable repair for another 1000 years or more; some remained the finest roads in Europe well into the 18th century.

ACROSS THE EMPIRE *Roman roads were built in the farthest corners of the empire, from Syria (top) to Algeria (far left). The stones of the Appian Way in Rome (left) endure to this day, a testament to the skill of its builders.*

THREE-IN-ONE *The Pont du Gard, built to carry fresh water across the River Gard in Provence, southern France, is the most glorious of all Roman aqueducts. The three-tiered structure, rising sturdily some 150 ft (45 m) above the river, was the tallest aqueduct in the empire.*

STONE CHANNEL *The top tier of the aqueduct held a canal, the incline of which simulated that which a natural stream between Uzès and Nîmes would have.*

A waterway stretching across the sky

HOW THE ROMANS BUILT THEIR MOST STRIKING AQUEDUCT

THE ROMAN GENERAL, AGRIPPA, KNEW HE WAS IN trouble as soon as he surveyed the area. In 19 BC the Emperor Augustus had ordered him to find a water supply for a fort at Nemausus (now Nîmes, in southern France), but the closest water source was at Uzès, 13 miles (21 km) to the north. As the source lay just 55 ft (17 m) above the level of the fort – close to the absolute minimum needed for gravity to make the water flow along an aqueduct – he had a problem.

Worse still, the only direct route between Uzès and Nîmes was through the mountainous Garrigues de Nîmes and across the River Gard. As the alternative – to circumvent the high ground – would more than double the length of the aqueduct, and would still mean crossing the river, the only answer was to cut across the mountains. After 35 years of toil, the result was the highest aqueduct ever built by Roman engineers, its most striking feature being the astonishing Pont du Gard, completed in AD 14.

The aqueduct was built as if it were three separate bridges, constructed one on top of the other. To achieve this, the Romans first built a wooden framework for the arches and then fitted blocks of local stone over the framework.

This was no easy task. Every arch was three blocks thick and each of the huge blocks – some of which weighed 6 tons – had to be lifted into position using vast wooden cranes operated by manpower. Labourers walking on the spot inside a timber wheel turned a winding drum, which was connected by rope to a large wooden arm. Each block was tied onto the wooden arm, and then – as the winding drum was turned and hauled in the rope – was raised to the top of the arm. There, workers waited to ease it into position over the timber framework. No mortar was needed because the blocks fitted together so precisely.

Built to last

The span of the biggest arch in the Pont du Gard was 80 ft (24 m), and the bridge – 885 ft (270 m) long – still stands today more than 150 ft (45 m) above the river below. Along the top of the aqueduct runs the cement-lined channel that carried the water, 4 ft (1.2 m) wide, 5 ft (1.5 m) deep. The quality of the construction was proved in the 16th century, when the piers in the second tier were halved in thickness to make room for a road, and the structure survived.

Memorials to the gods who were kings

HOW A CITY ROSE IN THE HEART OF THE CAMBODIAN JUNGLE

SACRED SYMBOL *The eastern front of Angkor Wat, the greatest temple in a city larger than ancient Rome, glows at sunset. Like all Khmer temples, it symbolises the cosmic mountain Mount Meru, home of the Hindu gods in the faraway Himalayas. Its terraces rise in three tiers to form a pyramid, on which stand five sanctuary towers. Inside the temple is a shrine dedicated by Suryavarman II to Vishnu, one of the principal gods in Khmer Hinduism.*

THE HEADS OF THE KHMER EMPIRE RULED AS THE deputies of the gods. Their kingdom – encompassing modern Cambodia and parts of South Vietnam, Laos and Thailand – flourished for 500 years, between the 9th and 15th centuries AD. As their capital, and a monument to their might, they created Angkor, a sprawling city of temples and shrines, covered in ornamental sandstone carvings embellished with gold and bright colours. Gilded bronze tridents crowned the towers and brocaded silk standards flew from brightly painted masts.

Of the 600 temples, the greatest was Angkor Wat, built by Suryavarman II in the 12th century. Today, all that is left is the sandstone, richly carved with animals, mythological and historical scenes. The Khmers built in wood as well as stone, but the timber buildings perished in the moist warmth of the tropical rain forest.

No written records of exactly how these great temples were built have survived, but it is clear that they were

LARGER THAN LIFE *Crumbling, moss-covered heads loom out of the dense rain forest, relics of an era when the Khmer empire stretched for thousands of miles.*

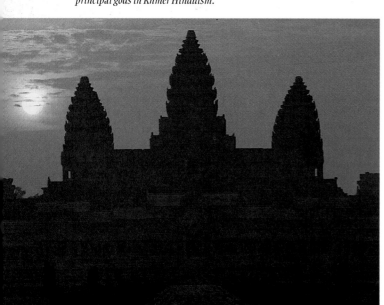

HIDDEN CITY *Angkor soon became covered by the jungle after its abandonment in the 15th century. It was first discovered by a European in 1850, when a French priest, Father Bouillevaux, came across the ruins in the dense growth.*

often piled together hurriedly by enormous armies of workmen, who set the sandstone blocks on top of one another without bothering to use any mortar.

Only when each section of the building was complete did the elaborate carving begin. A foreman traced the main design over the smooth sandstone, and then gangs of artisans chiselled out the reliefs. The scale of some of the carving was enormous – in the lower gallery of Angkor Wat, for example, stands a bas-relief 6 ft (1.8 m) high which stretches over a mile (1.6 km) in length.

The carvings include scenes both from history – stirring battle scenes and great pageants – and from myth. Powerful gods, serene goddesses and *apsarasas*, celestial dancers believed to be waiting in paradise as a reward for soldiers killed in war, are all evident, and many of the reliefs depict scenes from Hindu mythology, notably the *Mahabharata* and *Ramayana* epics. A superb surface patina is found on some of the carvings, created over the years by worshippers who pasted small pieces of gold leaf on to them as a way of building up favour with the gods.

Wealth gleaned from the rice harvest

The buildings were not the only glory of Angkor, for the city also boasted two huge reservoirs, numerous lesser ones and a network of irrigation canals. These reveal the Khmers' advanced skills in hydraulic engineering, and the water they provided enabled the people to harvest three or four crops of rice annually, the basis for the wealth that made the city possible in the first place. One moat, more than 2 miles (3 km) long, deviates from a straight course by less than half an inch (1.25 cm), while one of the canals runs straight for 30 miles (48 km).

The capital of the Khmer empire at Angkor was conquered by the Siamese in 1432, and the irrigation canals soon fell into disrepair. Only the fact that Angkor Wat, first built to honour Hindu gods, had by then become a Buddhist monastery kept it from destruction. Around the temple, hundreds of monuments succumbed to the encroaching jungle, although Angkor Wat itself survived for many years after 1432 as an important Buddhist pilgrimage site.

IMPORTED STYLE *A bas relief carved in Angkor in the 12th century shows the strong influence of Indian art.*

PRANCING INTO BATTLE *With their shields and spears, Khmer troops ostentatiously set off to war.*

RHYTHM IN STONE *Muscular, flexible figures dance out of the rock in a representation of court festivities.*

WAR ON WATER *A Khmer admiral commands his sailors during a 12th-century naval battle.*

BY THE WELL *Leafy trees surround this much-used well, and the water is carried away in sturdy yoked buckets.*

Mysterious ruins of a vanished nation

HOW THE CITY OF GREAT ZIMBABWE WAS BUILT

IN A LANDSCAPE OF GRANITE HILLS AND HEAVILY WOODED valleys, the dark outlines of Great Zimbabwe's stone structures come unexpectedly into view. In the local Shona dialect the word *zimbabwe* means 'stone houses' or 'venerated houses'. There are hundreds of other zimbabwes in this region, but this is the largest and by far the most impressive site.

The labyrinthine ruins are spread over 60 acres (24 ha) of land on a high plateau between the Zambezi and Limpopo rivers, some 250 miles (400km) inland from the Indian Ocean port of Beira. The most important ruins are the royal dwellings, located in the valley site and the earlier hilltop site,

some 150 ft (45 m) above the valley. Although the oldest parts date from the 8th century AD, the site may have been inhabited as early as AD 200.

A succession of peasant tribes settled here over the centuries, making their living by farming and trading. The huge amount of chippings lying about the site indicate that building stone was cut on the

VANTAGE POINT *The hilltop fortifications overlook the valley site (right) and the great enclosure in the far distance (below). Scattered between the two sites are the extensive foundations of many more circular enclosures, which once contained family dwellings and outbuildings.*

GREAT WALL *The most experienced masons would have been responsible for the final positioning of the stones in the enclosure wall. Sixteen thousand tons of stone were used – more than in the rest of the ruins combined.*

PRECISION CARVING *The graceful curve of one of the entrances to the enclosure wall (left) is characteristic of the finest architectural style at Great Zimbabwe.*

FIRM FOUNDATIONS *Enormous rounded granite boulders, worn smooth, provide a solid base for the early hilltop fortification walls.*

ENIGMA IN STONE *The solid masonry conical tower, 34 ft (10 m) high and 17 ft (5 m) in diameter at the base, may have had some religious significance in the past.*

spot, using iron tools. Huge blocks of granite would have been dragged overland – probably on sledges, for this was before knowledge of the wheel reached southern Africa – from quarries in the surrounding hills. The blocks were alternately heated and cooled until they cracked and rough stones could be chipped off. Early walls, dating from the 12th century, are rough and uneven, but by the time the city had reached the height of its prosperity in the 15th century, building techniques had improved immeasurably. Walls were made from carefully trimmed, even-sized stones rising in neat courses from a levelled trench, the stones so well cut that they could be assembled without mortar.

An enduring puzzle

The most striking structure in the valley is the great enclosure wall, which is 830 ft (253 m) long, 16-35 ft (5-11 m) high, and at least 4 ft (1.2 m) wide at the base. A workforce of 200-300 people would have been needed to quarry, haul and trim the stones and collect wood for the fires.

Within the enclosure stands Great Zimbabwe's most enigmatic structure – a conical tower. Fortune hunters in the 19th and early 20th centuries were convinced that it was an enormous treasury, perhaps containing some of the gold that was once mined in the district. But efforts to break in failed, for the tower is completely solid. A further puzzle is the maze of covered passageways in the hilltop palace that enabled the rulers to go about unseen. Why they wanted to conceal themselves from the sight of commoners is unknown, although legend has it that one 18th-century ruler wanted to become invisible and only ever appeared in public in heavy disguise. It was forbidden to look at him, and even courtiers had to approach on their bellies, with eyes cast down.

Decline and fall

In the mid 15th century, barely half a century after Great Zimbabwe had achieved religious, political and trading supremacy in the region, the society began to decline. The population had grown too large and moved on, and the surrounding natural resources were exhausted. The site continued to be occupied, but only a shadow of Great Zimbabwe's former glory remained. The Rozwi, the last inhabitants of the city, were conquered and either slaughtered or driven away in 1832 by the Zulu chieftain Shaka. The site was then finally abandoned, and the civilisation that created one of the greatest monuments in Africa was forgotten.

SOAPSTONE MESSENGER *A highly stylised carving (c.AD 1200-1400) may represent a mythical eagle that carried messages to the gods.*

PRECIOUS CURRENCY *Great Zimbabwe became wealthy in the 15th century by trading its locally mined gold – made into beads or beaten into thin flat sheets – with Portuguese merchants on the east coast.*

Masterpiece of the Gothic age

HOW THE GREAT CATHEDRAL OF CHARTRES WAS BUILT

THE CREATION OF CHARTRES CATHEDRAL, IN NORTH-WEST France, was a near miracle, achieved more by sheer willpower than by following any formal architectural plans. Built between the 12th and 13th centuries, the cathedral is a supreme example of Gothic architecture in its purest form, meriting its description as the finest cathedral ever built.

Out with the old
The new building was inspired by the Abbey of St Denis in Paris, Europe's first Gothic cathedral. While celebrating Mass there in June 1144, the Bishop of Chartres, Geoffrey de Lèves, was so impressed by its wide windows, soaring vaults and flying buttresses that he vowed he would rebuild his own cathedral, Chartres – with its thick Romanesque walls and massive vaulting – in the same style.

Funds were contributed by the townspeople of Chartres, rich and poor alike. Some gave generously from their own pockets, others through church collections. The first stage of the rebuilding schedule was the construction of three vast, heavily decorated

SPREADING THE LOAD
A flying buttress – an arch connecting an external wall to a free-standing support – transfers the downward pressure of the roof to the buttress and the foundations.

SOARING HEAVENWARDS *The sweeping verticals of the pillars draw the eye upwards to the vaulted roof, 120 ft (36 m) high at its uppermost point. The architect made no attempt to conceal the stone ribs or 'webs' of stone carried between them. On the contrary, he left the building's 'skeleton' in plain view, the vaults and columns contributing to the grandeur of the design.*

doorways in the western façade of the existing structure. This started in 1145, but the main building work did not begin for 50 years. The spur was a fire that destroyed much of the town and most of the original cathedral, leaving only the newly built façade intact. The fire was regarded not as a disaster, but as a sign from God to embark on a new masterpiece in his name.

Stonemasons who left their mark
The identity of the master mason who planned the overall design of Chartres Cathedral is unknown. Nor do any plans or drawings exist, probably because vellum parchment, the medium that preceded paper, was prohibitively expensive. Masons usually sketched out plans on the stone floors of their workshops for carpenters to copy onto wooden templates. These in turn were used by stonecutters

to fashion individual stones. The quarrymen marked each stone to show its place of origin, and the stonemasons used a similar marking system.

Every mason engraved his mark on each stone he cut so that his work could be easily identified. The paymaster could then see how many stones each man had cut and how well he had cut them, making payment a straightforward matter. The marks – simple triangles, arrows or intersecting Vs – were passed down from father to son, for the building of a cathedral took several generations. Some of the marks can still be seen on the pillars in the south transept. However, many of the best stones – cut by the most skilful masons – are unmarked because those men were contractors brought in to do that particular task; there was no need for them to mark their own handiwork.

The rebuilding that followed was a labour of love on the part of the whole community. Local people willingly hauled stone blocks 5 miles (8 km) from the quarry at Berchéres-l'Evêque to save paying exorbitant transport costs.

The most difficult aspect of a Gothic construction was the vaulted roof. It was composed of arches, and unlike those of earlier churches, it had no internal bracing. First the builders put up a false wooden roof, and placed on top of it precisely cut stones bonded with mortar. These were hoisted into position using a winching system and a huge treadmill, turned by men walking on the spot inside it. Once the keystones – the central 'locking' stones at the top of every arch – were in position, the vaults were stable

BELOVED EMPEROR *In one of the apse windows, the Holy Roman Emperor, Charlemagne – defender of the Christian faith – blesses the cathedral builders. Some 22 000 sq ft (2000 m²) of glass was used to make all the windows.*

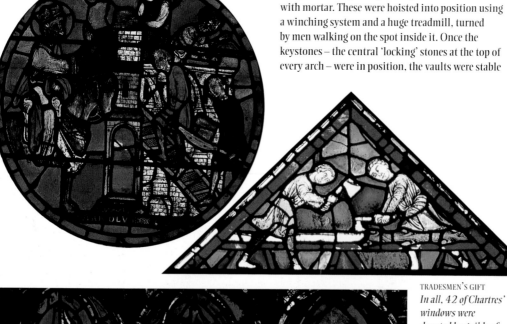

TRADESMEN'S GIFT
In all, 42 of Chartres' windows were donated by guilds of merchants or artisans. The carpenters' guild commissioned its own window (above).

GLASS MEMORIAL
The stonecutters who built the cathedral are depicted in their own window (left) carving a stone statue that was eventually positioned on the south portal.

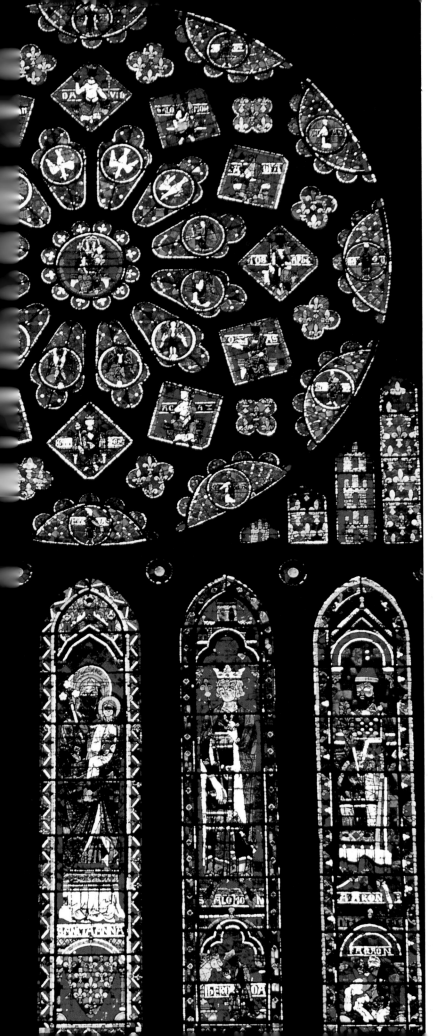

and the false roof was removed. But a vaulted roof with no internal supports placed enormous stress on the external supporting walls, so these had to be firmly buttressed to resist the outward thrust of the roof vaults.

Thin walls, wide windows

At Chartres, this reinforcement was provided by flying buttresses – arched masonry props projecting from a vertical support, rather like stone scaffolding. With vaulting and buttresses taking the strain, it became possible to build thinner walls than ever before, and to pierce them with wide windows. This had been impractical earlier, as chipping holes out of the walls would have weakened the structure.

By contrast with Romanesque churches, the interiors of Gothic cathedrals are flooded with myriad colours and light. This is nowhere more evident than in Chartres, whose windows in the nave are considered to be among the finest ever created. The three great rose windows are noted for their intricate designs, and each took a six-man team of glaziers almost three years to complete. The tradesmen who contributed to the building have also been immortalised in the cathedral's stained-glass windows.

MAJESTIC FACADE *A relief portraying Christ in majesty is part of a frieze telling the story of Christ's life set around the Royal Portal on Chartres' west face.*

KALEIDOSCOPE OF COLOUR *The north rose window of Chartres is a masterpiece of medieval craftsmanship. First, the glassmaker sketched the design on a table, then he cracked off pieces of glass in the required shade from huge coloured glass sheets. Next, he painted on details in dark enamel paint, fired the glass pieces and assembled them into panels. These were then set in the huge wall cavity.*

Temples that jumped up and down

HOW INCA STONEMASONS CONSTRUCTED THE WORLD'S FIRST EARTHQUAKE-PROOF BUILDINGS

DRESSING A FACE
Using a medium-sized hammer made of obsidian, or volcanic rock, a workman pounds a granite block at 15 to 20 degrees. He then twists the hammer to about 45 degrees to give the block a finer, smoother cut.

MAKING AN EDGE
With a smaller hammer weighing about 2 lb (1 kg), the workman gives the edge of the block a series of sharp, glancing blows. This ensures that the stone does not flake at the edges, creating a softer profile.

FITTING THE STONE
Helped by an overseer (in hat) the workman levers the stone free. The men have put the stone on top of a layer of dust on the stones already in place, and from the marks it made noted that it does not fit and needs more trimming.

FINAL DRESSING
Once the stone has been levered snugly back into position, the overseer carefully gives its outer face a final dressing with a medium-sized stone hammer. This provides a smooth and perfect fit.

HOMES, SHOPS AND OFFICES WENT CRASHING TO THE ground when an earthquake shattered the Peruvian city of Cuzco – former capital of the Inca empire – on May 21, 1950. Although the modern buildings were devastated, the ancient stone foundations on which they were built – including the walls of Inca temples – stood up to the worst of the shocks.

The Incas were master stonemasons who employed two main techniques for building their fortresses and temples. In one, irregularly shaped blocks of granite, some weighing around 100 tons, were fitted together without the use of mortar; in the event of an earthquake, the huge interlocking stones simply jumped up and then settled back into place. The other used smaller, rectangular blocks of stone laid horizontally and held together with sunken joints.

Working without knowledge of iron

The great age of Inca building was the 15th and early 16th centuries, when the empire was at its height. Granite blocks, which came from the quarries round Cuzco, were roughly hewn and transported to the capital on sledges dragged by vast teams of workmen. As the Incas did not have iron or wheeled transport, wooden rollers and levers were used when the sledges became stuck in ruts, or when the workmen had difficulty in pulling them up rocky inclines.

Once the blocks had been cut and dressed, huge teams of workmen hauled them on rollers to the site of the wall. As soon as the bottom course of stones had been laid, an earthen ramp was built, up which the next course of stones was dragged. This was done by a squad of workers at the front, harnessed to ropes tied to the stones; while a squad at the back used stout wooden levers to propel the stones along. On reaching the gap where it was to be placed, each stone was upended and levered into position.

The Incas' domestic buildings, usually rectangular in form, were built mainly with local stone. Typical features included trapezoid doors and windows, broader at the bottom than at the top which, together with the bevelled edges of the stone, made spectacular patterns of shadow as the sun blazed across the sky.

OLD AND NEW *This curved end wall of the Coricancha, the Inca Temple of the Sun in Cuzco, is a rare example of Inca masonry which has a smooth-fitted surface. It stands next to Santo Domingo, built by the Spanish invaders after 1532.*

CITY OF THE MOUNTAINS *Standing on a ridge high in the Peruvian Andes, Machu Picchu, the 'lost city' of the Incas, was built of white granite blocks, fitted precisely together without mortar. On three sides its rock walls plunge 1000 ft (305 m) into a gorge, and its now-ruined temples, palaces and houses were linked by more than 3000 steps. On the slopes below the city, which was only discovered in 1911, is a spectacular series of dry-stone terraces. The economy of the Inca Empire was based on farming, and the terraces – with their network of irrigation canals and connecting stairways – were used for growing staples such as maize and beans.*

HIGH REFUGE *The hilltop fort of Sacsayhuamán was built as a stronghold and a place of refuge overlooking Cuzco. A temple was later added to the complex of buildings within its walls.*

RUGGED RUINS *Resembling huge tree-trunks, the foundation stones of Sacsayhuamán were the result of three decades of back-breaking toil carried out by 20 000 workmen who hauled the granite blocks up the hill.*

A dome to rival the glories of Rome

HOW FILIPPO BRUNELLESCHI CROWNED FLORENCE CATHEDRAL

A NEWLY RETURNED EXILE FROM FLORENCE, WRITING IN the late 1420s, suggested that the dome of its new cathedral would be able to cover all the people of Tuscany with its shade. Even if this was a slight exaggeration, the spectacular dome has dominated the city ever since its completion a few years later.

Things might easily have turned out differently. Nothing like the dome had been attempted before, and there were many who thought the audacious attempt foolish. Such was the relief of all concerned when they saw the graceful structure completed that its chief designer, the artist Filippo Brunelleschi, received the highest honours and soon became idolised in Renaissance myth.

Ambitious plan for a new dome

The cathedral stands on the site of the old church of Santa Reparata. Rebuilding work began in 1294 and proceeded intermittently until 1366-7, when artists and master masons drew up a new plan featuring a long nave with shorter side arms and choir. Where the nave met the side arms and choir, they planned

a clear octagonal space, covering the full combined width of nave and aisles, and crowned by a high dome set upon a drum. They decided that drum and dome should be octagonal in plan, and that the dome should be constructed with the slightly pointed profile that had become usual in Italy in recent centuries.

This requirement presented the architect and builders with a difficult problem. A circular dome can be built without a supporting framework, provided there is always a strong circular ring at the top to act as a keystone, preventing the incomplete structure from falling inwards. But an octagonal dome under construction needs to be supported by a temporary timber framework – known as 'centering'. Centering had

WORK BEGINS *The Bishop of Florence listens carefully as the architect and the master mason outline their plans to build a dome over the octagonal space at the centre of the cathedral.*

GIANT CRANE *A hoist moves stones along the platform to where they are needed.*

The platform supports the builders at work and is used to store materials

A door leads to the access corridor that runs in the space between the inner and outer domes

The bricks are built up at an angle to the horizontal, sloping down towards the inner rim of the dome

THE PLATFORM *Before the dome could be built, a broad wooden working platform had to be constructed on top of the 13 ft (4 m) wide drum built over the crossing of the nave and side arms. The platform was supported by sturdy timbers anchored in holes in the masonry which are still visible today, hidden by the projecting cornice at the base of the dome.*

BRICKWORK *For the first 23 ft (7.1 m), where the dome had little inward inclination, construction was in stone. Higher up, bricks were laid in a pattern known as spinapesce, or herringbone. Between every three or four bricks laid circumferentially, others were set upright, linking the successive courses and dividing the brickwork into wedge-shaped blocks that could not fall down.*

Vertically laid bricks bond the different layers of the rising dome

WORK IN PROGRESS *All the building material is tied to strong cables and raised to the main working platform using an enormous wooden hoist powered by oxen. Labouring on the cathedral floor, the animals are yoked to horizontal levers, turning a vertical shaft on which two gear wheels are set. This shaft is raised and lowered using a screw at its base so that either the lower or the upper gear engages with the toothed wheel on the horizontal drum to its right. The wheel at the far end of this drum engages with a wheel on a second drum behind the first (far right). As these twin drums turn, they haul in or release ropes to raise objects to, and lower them from, the platform above. The gears make it possible to switch the direction in which the drums turn – and so change from lowering to raising material – without turning the oxen around.*

Three wooden working platforms were needed to build the dome. The two upper platforms, like the main working platform, were supported by timbers tied to metal rings. Often five or six hoists were in use at once, lifting material to the platforms from the floor far below

A block is hauled up to one of the working platforms

Small cranes lift material down from a delivery cart

Workers mix the mortar that will be used to bind the brickwork

Timbers are stacked ready for use

Labourers attach a heavy stone block to a pulley

Stonemasons dress the great blocks of stone that will be used in the dome

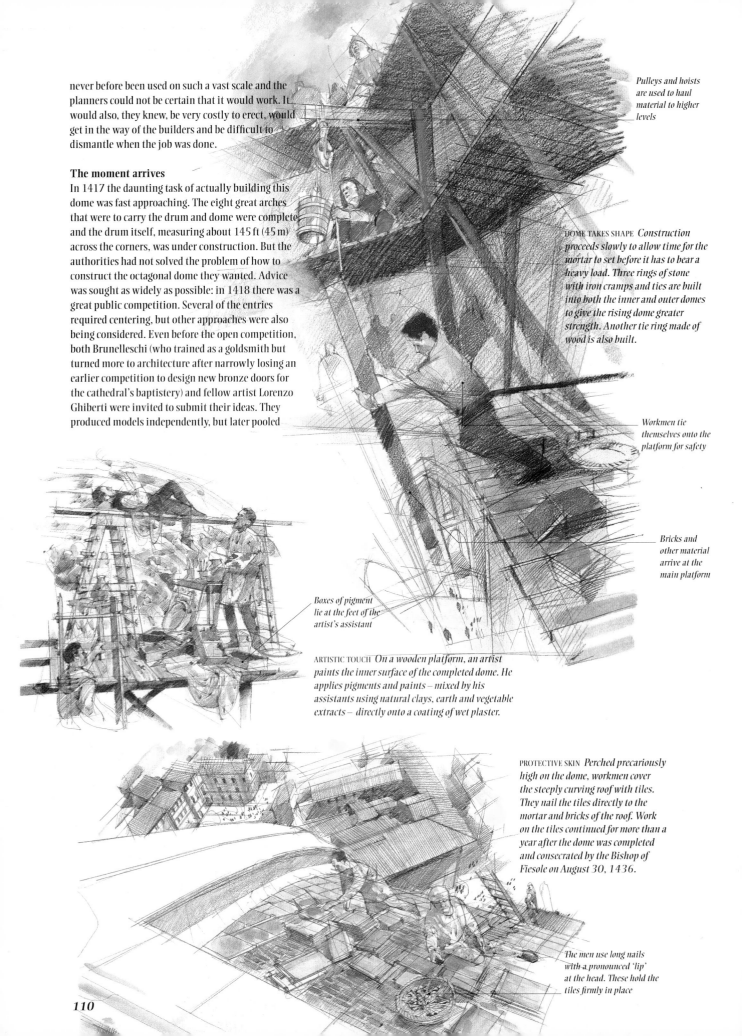

never before been used on such a vast scale and the planners could not be certain that it would work. It would also, they knew, be very costly to erect, would get in the way of the builders and be difficult to dismantle when the job was done.

The moment arrives

In 1417 the daunting task of actually building this dome was fast approaching. The eight great arches that were to carry the drum and dome were complete and the drum itself, measuring about 145 ft (45 m) across the corners, was under construction. But the authorities had not solved the problem of how to construct the octagonal dome they wanted. Advice was sought as widely as possible: in 1418 there was a great public competition. Several of the entries required centering, but other approaches were also being considered. Even before the open competition, both Brunelleschi (who trained as a goldsmith but turned more to architecture after narrowly losing an earlier competition to design new bronze doors for the cathedral's baptistery) and fellow artist Lorenzo Ghiberti were invited to submit their ideas. They produced models independently, but later pooled

Pulleys and hoists are used to haul material to higher levels

DOME TAKES SHAPE *Construction proceeds slowly to allow time for the mortar to set before it has to bear a heavy load. Three rings of stone with iron cramps and ties are built into both the inner and outer domes to give the rising dome greater strength. Another tie ring made of wood is also built.*

Workmen tie themselves onto the platform for safety

Bricks and other material arrive at the main platform

Boxes of pigment lie at the feet of the artist's assistant

ARTISTIC TOUCH *On a wooden platform, an artist paints the inner surface of the completed dome. He applies pigments and paints – mixed by his assistants using natural clays, earth and vegetable extracts – directly onto a coating of wet plaster.*

PROTECTIVE SKIN *Perched precariously high on the dome, workmen cover the steeply curving roof with tiles. They nail the tiles directly to the mortar and bricks of the roof. Work on the tiles continued for more than a year after the dome was completed and consecrated by the Bishop of Fiesole on August 30, 1436.*

The men use long nails with a pronounced 'lip' at the head. These hold the tiles firmly in place

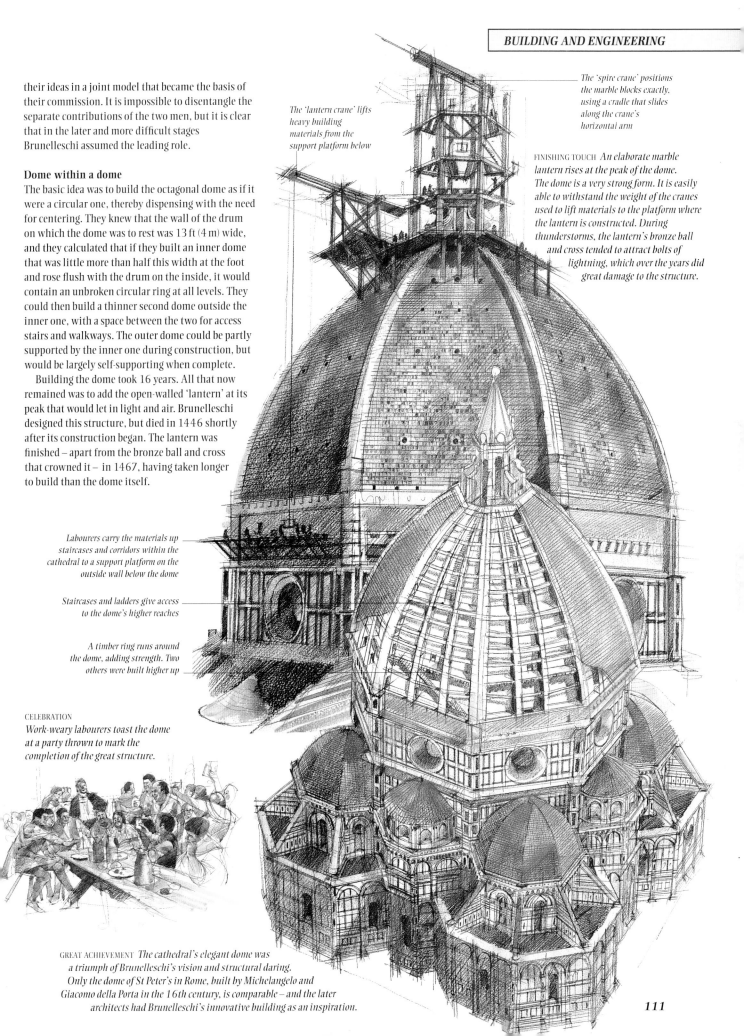

their ideas in a joint model that became the basis of their commission. It is impossible to disentangle the separate contributions of the two men, but it is clear that in the later and more difficult stages Brunelleschi assumed the leading role.

Dome within a dome

The basic idea was to build the octagonal dome as if it were a circular one, thereby dispensing with the need for centering. They knew that the wall of the drum on which the dome was to rest was 13 ft (4 m) wide, and they calculated that if they built an inner dome that was little more than half this width at the foot and rose flush with the drum on the inside, it would contain an unbroken circular ring at all levels. They could then build a thinner second dome outside the inner one, with a space between the two for access stairs and walkways. The outer dome could be partly supported by the inner one during construction, but would be largely self-supporting when complete.

Building the dome took 16 years. All that now remained was to add the open-walled 'lantern' at its peak that would let in light and air. Brunelleschi designed this structure, but died in 1446 shortly after its construction began. The lantern was finished – apart from the bronze ball and cross that crowned it – in 1467, having taken longer to build than the dome itself.

The 'lantern crane' lifts heavy building materials from the support platform below

The 'spire crane' positions the marble blocks exactly, using a cradle that slides along the crane's horizontal arm

FINISHING TOUCH *An elaborate marble lantern rises at the peak of the dome. The dome is a very strong form. It is easily able to withstand the weight of the cranes used to lift materials to the platform where the lantern is constructed. During thunderstorms, the lantern's bronze ball and cross tended to attract bolts of lightning, which over the years did great damage to the structure.*

Labourers carry the materials up staircases and corridors within the cathedral to a support platform on the outside wall below the dome

Staircases and ladders give access to the dome's higher reaches

A timber ring runs around the dome, adding strength. Two others were built higher up

CELEBRATION
Work-weary labourers toast the dome at a party thrown to mark the completion of the great structure.

GREAT ACHIEVEMENT *The cathedral's elegant dome was a triumph of Brunelleschi's vision and structural daring. Only the dome of St Peter's in Rome, built by Michelangelo and Giacomo della Porta in the 16th century, is comparable – and the later architects had Brunelleschi's innovative building as an inspiration.*

A framework built to last

HOW MEDIEVAL OAK HOUSES HAVE STOOD THE TEST OF TIME

THE TWO WORKMEN SPLATTERED TROWELFULS OF DAUB onto the basketweave partition between them. As the sticky mixture of wet mud, chopped straw, cow hair and cow dung clung to the closely woven wooden staves and twigs, the wall of the house took shape and the men lost view of each other.

The house the men were building was part of the great construction boom of the late 16th century, when many flimsy medieval structures were replaced with solid oak houses. Mud and wood walls were added to a sturdy load-bearing framework of oak beams, generally cut from oak that was still green so that it was easy to shape. As the timbers dried and seasoned, they often warped slightly, creating the uneven buildings seen today.

Constructing a home to last

One drawback to the wooden frame was rot could destroy the houses within decades because of damp. To counter this, the frame was always placed on a stone or brick foundation – a plinth wall – which was built either before the frame was in place, or once it

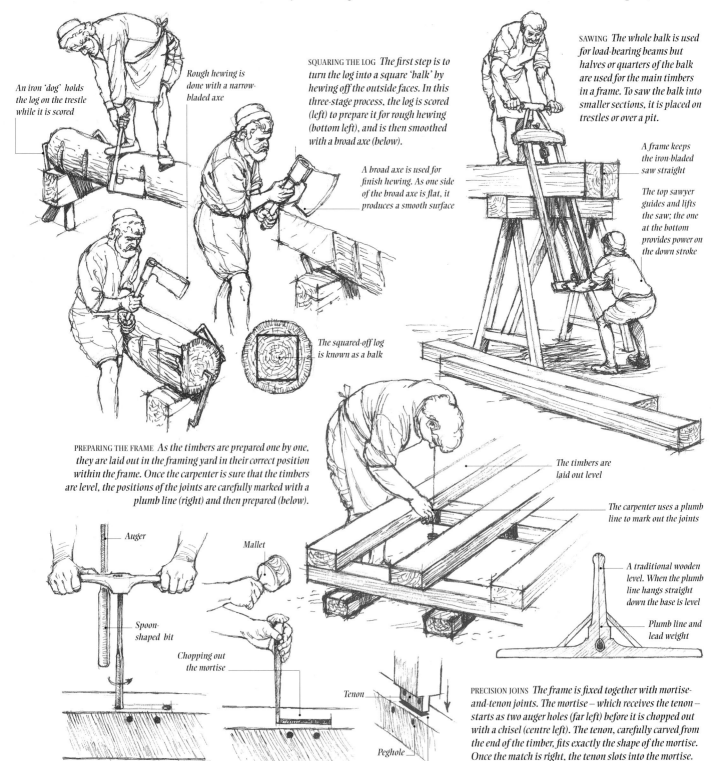

An iron 'dog' holds the log on the trestle while it is scored

Rough hewing is done with a narrow-bladed axe

SQUARING THE LOG *The first step is to turn the log into a square 'balk' by hewing off the outside faces. In this three-stage process, the log is scored (left) to prepare it for rough hewing (bottom left), and is then smoothed with a broad axe (below).*

A broad axe is used for finish hewing. As one side of the broad axe is flat, it produces a smooth surface

The squared-off log is known as a balk

SAWING *The whole balk is used for load-bearing beams but halves or quarters of the balk are used for the main timbers in a frame. To saw the balk into smaller sections, it is placed on trestles or over a pit.*

A frame keeps the iron-bladed saw straight

The top sawyer guides and lifts the saw; the one at the bottom provides power on the down stroke

PREPARING THE FRAME *As the timbers are prepared one by one, they are laid out in the framing yard in their correct position within the frame. Once the carpenter is sure that the timbers are level, the positions of the joints are carefully marked with a plumb line (right) and then prepared (below).*

The timbers are laid out level

The carpenter uses a plumb line to mark out the joints

Auger

Mallet

Spoon-shaped bit

Chopping out the mortise

A traditional wooden level. When the plumb line hangs straight down the base is level

Plumb line and lead weight

Tenon

Peghole

Mortise

PRECISION JOINS *The frame is fixed together with mortise-and-tenon joints. The mortise – which receives the tenon – starts as two auger holes (far left) before it is chopped out with a chisel (centre left). The tenon, carefully carved from the end of the timber, fits exactly the shape of the mortise. Once the match is right, the tenon slots into the mortise. Drilled pegholes (left) are used to secure the two sections.*

was erected onto temporary blocks. Box-framing, as seen here, was the most common form of construction, which used long, straight timbers linked by 'tie beams' to form a series of boxes in a complete frame.

The frames for a house – one for each side, partition, floor and roof slope – were usually prepared at a carpenter's workshop and then carted to the site. Once there, each of the frames was erected and pegged with wooden pins before the gaps between the timbers were filled with wattle and daub.

FINISHING TOUCHES *Once the frame is erected, the carpenter fixes the floorboards, doors and windows. Next, the masons construct the plinth wall and the chimneys before the tiler or thatcher covers the roof. In grander houses, joiners, plasterers and painters would finish the interior before the owner moved in.*

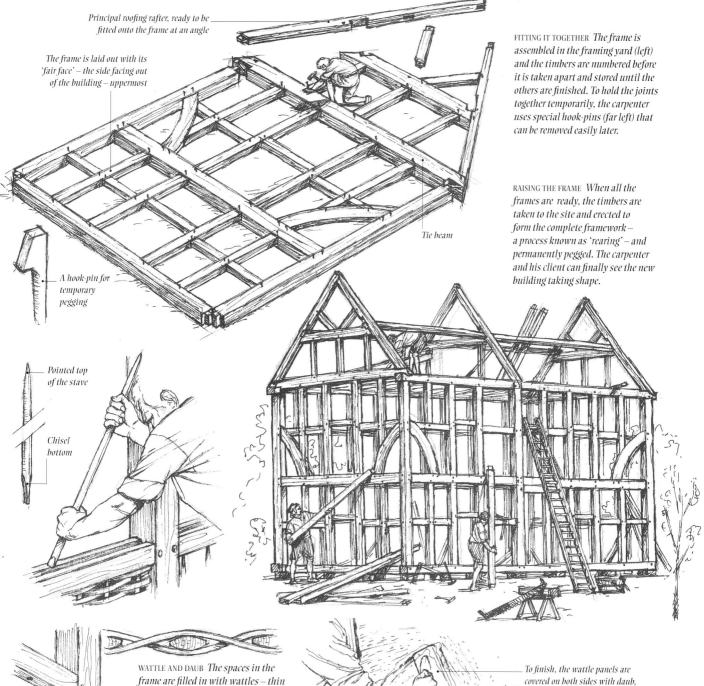

Principal roofing rafter, ready to be fitted onto the frame at an angle

The frame is laid out with its 'fair face' – the side facing out of the building – uppermost

A hook-pin for temporary pegging

FITTING IT TOGETHER *The frame is assembled in the framing yard (left) and the timbers are numbered before it is taken apart and stored until the others are finished. To hold the joints together temporarily, the carpenter uses special hook-pins (far left) that can be removed easily later.*

RAISING THE FRAME *When all the frames are ready, the timbers are taken to the site and erected to form the complete framework – a process known as 'rearing' – and permanently pegged. The carpenter and his client can finally see the new building taking shape.*

Tie beam

Pointed top of the stave

Chisel bottom

WATTLE AND DAUB *The spaces in the frame are filled in with wattles – thin twigs of hazel, ash or oak – woven around vertical staves of chestnut or oak (above). The pointed tops of the staves are fixed at the top of each panel (left) and the blunt end is then pushed into a groove on the bottom (top left).*

To finish, the wattle panels are covered on both sides with daub, an early type of plaster, and then painted with a lime wash

113

Memorial to the 'Chosen One'

HOW AN EMPEROR ERECTED A MONUMENT TO LOVE

IN 1666, AS SHAH JAHAN, THE ONE-TIME MOGUL emperor of India, lay dying in his apartments, he turned to the nearby mirror on the wall. In it he saw reflected the distant contours of the building upon which he had gazed unceasingly during his eight-year imprisonment. The building was the Taj Mahal, the mausoleum built by Shah Jahan 34 years before in honour of his favourite wife. Then, as today, it was accorded its place as one of history's most profound testimonies to marital love.

A PRISON FOR AN EMPEROR

THE TAJ MAHAL WAS NOT THE ONLY MONUMENT TO FINE craftsmanship at Agra. Across the river, within sight of the Shah's white marble palace, stood the Red Fort, its walls built out of red sandstone by local workmen renowned for their stone inlay work – still a speciality of Agra. They were employed by Shah Jahan in their hundreds to work on the fort from the late 16th century on, and by the time it was finished, palaces, gardens, barracks and other buildings were enclosed within its red walls. It also housed Shah Jahan after he was imprisoned there by his son.

MAN AND BEAST *A contemporary painting by the artist Tulsi Khurd shows building operations at the Red Fort. Wagonloads of the distinctive sandstone for the fort's walls were hauled from nearby quarries by elephants and bullocks.*

Wearing white for mourning

The Taj Mahal stands on the south bank of the Jumna River, outside Agra in northern India – capital of the Mogul Empire in the early 17th century. Its name was taken from the word 'Taj' – a corruption of the name 'Mumtaz' which belonged to Shah Jahan's favourite wife, Mumtaz Mahal. After 19 years of marriage, she died while giving birth to their fourteenth child in 1631. The emperor was inconsolable and locked himself away in his rooms, without eating, for nine days. When he emerged, pale and haggard, he ordered his entire kingdom to go into mourning. Music, perfume, public amusements and bright clothes were banned, and the emperor himself exchanged his finery for white robes.

During his confinement, the emperor had decided to build a monumental tomb as a lasting tribute to the memory of his wife. The world's greatest architects were summoned from India and beyond. Only the finest craftsmen were employed – specialist carvers, stonemasons, jewel-cutters and dome-builders – but the principal architect is unknown. The first stage was to excavate and level the site and lay a bed of sediment to prevent the Jumna River from seeping into the foundations. Tons of sandstone and marble were dragged to the site, and 43 different gemstones chosen for decoration included jade and crystal from China, lapis lazuli from Afghanistan and coral and mother-of-pearl from the Indian Ocean.

The shadowy interior of the Taj Mahal is arranged around a central octagonal chamber containing the tombs of Shah Jahan and his beloved. It is enclosed by a huge, pierced marble screen studded with precious stones, just one of many treasures found in one of the world's most magnificent buildings.

GRAND PLAN *The Taj Mahal's rectangular site measures 1902 ft × 1002 ft (580 m × 305 m). The entire mausoleum complex was planned as a whole, for later additions or amendments were not permitted in Mogul architecture.*

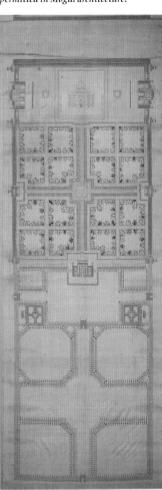

IMPRISONED ROMANTIC *Shah Jahan is said to have dreamt of the Taj Mahal, then encouraged one of his architects to take hallucinogenic drugs, so that he might experience the same vision. Shah Jahan ruled India from 1628 until 1658 when he was imprisoned in the Red Fort at Agra, by his son Aurangzeb, who declared himself ruler.*

SYMMETRY IN STONE *The perfect proportions of the Taj Mahal make a perfect silhouette. At the four corners of the 23 ft (7 m) high plinth stand four, three-tiered minarets, each 138 ft (42 m) tall. They lean outwards by a few inches – one slightly more so than the others, though no one knows why – so that in an earthquake they will not fall onto the mausoleum. The Taj Mahal lies in a bend of the Jumna River, which was diverted to flow past the very foot of the site to improve the view – the same view that both haunted and sustained Shah Jahan during his eight-year imprisonment. He planned to build his own tomb, a replica of the Taj Mahal in black marble, on the opposite bank of the river, but he never achieved his final ambition.*

POEM IN MARBLE *The main body of the mausoleum is composed of four identical façades, each with a towering archway, 108 ft (33 m) high. Each element of the design complements another. Plain surfaces are juxtaposed with intricately decorated marble, the numerous floral designs of inlaid stone reflecting Shah Jahan's passion for gardens.*

RICH PICKINGS *Visitors to the Taj Mahal have picked off and stolen many of the gemstone flowers.*

A masterpiece in the making

HOW THE MAGNIFICENT FOUNTAINS OF VERSAILLES WERE CREATED

IN 1661, THE YOUNG KING OF FRANCE, LOUIS XIV, visited Vaux, the superb new estate of the French finance minister Nicolas Fouquet. As the king strolled around the grounds admiring the elegant lawns and topiary, hundreds of fountains spurted jets of water high into the air. Louis was startled, and greatly impressed, but inwardly he was indignant. How dare a commoner flaunt such ostentation before the king? He, Louis, would put the upstart in his

CONCEIVED IN PIQUE *Building work on the immense palace of Versailles started in 1662 but the chateau was not completed until 1710, some 28 years after Louis XIV had made it his court.*

place by building the greatest palace and gardens on Earth at his favourite home – a small royal hunting lodge at Versailles, an insignificant village 11 miles (18 km) south-west of Paris.

Within three weeks Fouquet had been arrested on a charge of embezzlement and his advisers – the architect Louis Le Vau, painter Charles Le Brun, and garden designer André Le Nôtre – engaged by the king. At that time Versailles's only claim to fame was the 20-room lodge built by Louis' father in 1624. Now it was to host elaborate gardens similar to those of Vaux, but on a far grander scale. More than 1000 grandiose fountains were planned, with waterfalls and large reflecting pools, but there was one major obstacle: the estate was situated on high ground and the nearby stream was too small to provide a reliable supply of water.

Defying the laws of nature

Nevertheless, some 1400 fountains and waterfalls – of which 600 remain today – were built. The stream was dammed and the water raised to a reservoir near the lodge by horse-driven pumps and windmills. When two extra pumps installed to supplement the meagre supply failed to do the job, an engineer, Arnold de Ville, suggested piping water to Versailles from the king's estate at Marly, on the River Seine, 5 miles (8 km) to the east.

In 1678, a huge pumping plant, the *Machine de Marly*, was constructed across the river. Groups of pumps – 253 in all – were driven by 14 huge waterwheels 39 ft (12 m) across, and, over the distance of a mile, raised the water in three stages

REGAL FOLLY *Pumps built at Marly, on the Seine, were intended to supply the water features at Versailles, 5 miles (8 km) away. Even if they had worked as planned, the fountains would have flowed for just two hours a day.*

ARCHITECT *Louis Le Vau, appointed First Architect of the King in 1654, was responsible for much of the building work at Versailles.*

GARDENER *André Le Nôtre had full control over landscaping the gardens at Versailles from the early 1660s until his retirement in 1693.*

VISION OR VANITY? *Uneven terrain and a lack of local water did not deter Louis XIV from his dream of creating the most magnificent palace in the world at Versailles. The Sun King, as he became known, personally supervised every artistic detail of the work on the chateau and its gardens, which stretched as far as the eye could see.*

into a series of reservoirs. The highest of these lay 502 ft (153 m) above the Seine, and from here, the water flowed down to Versailles in cast iron pipes.

The *Machine de Marly* should have provided more than 1.3 million gallons (6 million litres) of water daily, but it never managed to raise more than half of that. By 1685 it was clear that Versailles would require more water, and a plan was made to divert the River Eure, some 50 miles (80 km) to the west. The French army was ordered to build a gigantic aqueduct, with arches twice as high as the towers of Notre Dame, but in 1688 France went to war and the 30 000 workers downed tools to take up arms. Work was never resumed.

THE ALLURE OF WATER *The fountains at Versailles were created by some of the greatest artists of the day. The Fountain of Saturn (left) was designed by François Girardon, while Jean Baptiste Tuby created the Apollo Basin (right). All designs, including those for the Fountain of Autumn (below), were passed on to sculptors, who then executed the designs after intensive discussion with the artist.*

ROUGH SKETCH *Painter Le Brun made delicate drawings of Versailles, including one for the Pyramid Fountain (right).*

ARTIST *Charles Le Brun was responsible for the painted interiors of the chateau and many of the fountains outside.*

A city built on a sunken forest

HOW VENICE ROSE FROM A LAGOON

AS THE ROMAN EMPIRE COLLAPSED IN THE 5TH CENTURY, marauding Goths and Vandals drove many inhabitants of the northern Italian province of Venetia to seek sanctuary on a group of mosquito-infested islands on a swampy lagoon in the north-east. Correctly, the refugees believed the islands to be so unappealing that the invaders would simply pass them by.

The early Venetians led a precarious existence, tethering their boats beside simple wooden huts, which were raised on stilts to protect them from high tides and flooding. On some of the higher islands, houses could be built directly on patches of gravel, but elsewhere the land had to be reclaimed little by little by digging drainage canals. The threat of invasion eventually diminished, and while some of the refugees returned home, others began to make a more permanent settlement in this isolated place.

Islands linked by 400 bridges

In AD 697, by which time building had begun in earnest, the settlements became an independent unit under an elected chief magistrate known as the doge. The most central settlement, Rivo Alto (later corrupted to Rialto), was to become the heart of Venice, linking 118 separate islands with some 400 bridges. To drain the land, more than 200 canals were dug branching off the Grand Canal, the principal waterway that traverses the islands in the shape of a backwards 'S' for 2 miles (3.2 km) and which would, in centuries to come, be lined with grand palaces.

On the soft ground of the island city, buildings required the firmest of foundations. The answer was to sink a forest of larch piles. They were driven into the clay subsoil by teams of builders using wooden hammers. Then further layers of larch and crushed birch were packed into the foundations. Locally produced birch was the most common building material, but workmen also used clay bricks, of a rich red-brown hue, from the mainland.

To hold back rising damp in the walls, stonemasons laid a layer of white limestone – brought in blocks from nearby Istria – at the high water mark of a building. This stone was both easy to carve and highly impervious to weathering, and from the Renaissance (15th century) onwards, masons faced grander buildings entirely in this dazzlingly bright material.

Minor shifts in the foundations of buildings were a constant hazard, and structures were designed to 'flex' with the ground beneath them. Roofs, for example, were rarely vaulted, and ceiling beams were closely spaced and topped by one or two layers of wooden planks.

CITY OF SPLENDOUR *The Grand Canal winds its serpentine way through Venice, with St Mark's Square centre left, in a 17th century view of the city. The Rialto bridge, built in the late 16th century, crosses the canal in a single 90 ft (27 m) span in the background. From unpromising beginnings in a marshy lagoon, the city that was born from the sea was to use the sea as a path to greatness. If they were to survive, the Venetians needed to be superb sailors; if they were to prosper, they needed to be enterprising traders. Venice built up a navy that was hired out at huge profit to the European powers; and Venetian merchants created a trading empire that brought the spices and silks of Asia to the rich but luxury-starved upper classes of Europe. Venetian families who grew rich on trade could afford to pay Italy's finest architects to design their beautiful palaces, and artists of the stature of Titian and Tintoretto to fill them with masterpieces.*

SHORING UP A CITY *Workmen hammer larch piles 12-15 ft (4-5 m) long deep into the soil, as foundations for a new building in Venice. The piles in the centre of the building will be topped by layers of crushed brick and by larch rafts, set in cement. Then further layers of larch planks will be laid, to form a base for the walls. Timber buried in clay resists decay because it is safe from attack by living organisms.*

A country wrested from the sea

HOW THE LOWLANDS OF HOLLAND WERE RECLAIMED

ALMOST TWO-FIFTHS OF THE NETHERLANDS IS MADE UP of land that once lay beneath the sea. It is guarded by dunes, dykes – and constant vigilance. Centuries ago, the first settlers in these low-lying areas built their houses on man-made mounds of clay, which gradually became larger as the settlements expanded. From about AD 1000, people began to build dykes and drainage ditches to claim yet more land from an ever-grudging, and sometimes revengeful, sea.

Building the dykes

First the builders constructed a sturdy clay embankment – a dyke – to defend against an inrush of sea water. They strengthened the seaward side with piles of branches, seaweed, straw or reeds – material that in later centuries was replaced by wooden piles with bundles of sticks to fill any gaps. The next phase was to drain the protected area. In places the land was high enough to drain at low tide, through sluices, or tide gates, that could then be closed to prevent the sea flowing back at high tide. But in areas at or below low water level, some means was needed to raise the water, so that it could flow away along canals, or channels cut along the tops of the dykes.

The solution was the windmill. The first pumps driven by wind power date from the 15th century, but the technique was not perfected for another century, by Jan Leeghwater, a self-taught hydraulics engineer and millwright, born in 1575.

The windmills used two types of pump to draw up water. The first was the scoopwheel, which had flat blades, covered with leather strips. As the wheel turned, the blades scooped up water, and raised it as near as possible to axle level. From there it flowed out through a channel. While the wheel was turning, a sluice opened to allow water to flow away; when it stopped, the sluice closed to prevent water flowing back. The drawback was that the lift was limited: a wheel 16 ft (5 m) across could raise water only about 4 ft (1.2 m).

The other type of pump was the Archimedean screw – a wooden axle with a ramp wound around it, like a giant corkscrew. As the screw rotated, water was drawn up the spiral ramp. The screw could lift water higher than a scoopwheel – 16 ft (5 m) was possible – but it was more difficult to make.

To drain deep areas required several windmills, working in stages. Between 1608 and 1612 Leeghwater drained the Beemster lake, in the north-west Netherlands, using a total of 26 windmills. The water was 10 ft (3 m) deep, and two sets of windmills were needed. The first raised water to a canal and the second set raised it still farther into the drainage outlets. Between 1615 and 1640 an area of nearly 50 000 acres (20 000 ha) was drained, creating the low-lying tracts of reclaimed land known as polders.

FURY OF THE SEA
On the night of March 5, 1651, the restless sea bursts through one of the dykes that protects Amsterdam.

DRY LAND
The fertile fields outside the city of Enkhuizen in 1610 show the richness of reclaimed land.

The capital that rose from nothing

HOW TSAR PETER THE GREAT BUILT ST PETERSBURG

AS THE ICY BALTIC WIND GUSTED ABOUT HIM, PETER the Great, tsar of Russia, surveyed the desolate landscape that he had just wrested from the occupying Swedes. Seizing a bayonet, he hacked four sods from the soil and, flinging them to the ground in the shape of a cross, roared: 'Here shall be a town!'

No one but he could have envisaged a town in this cold, damp and boggy flatland where man-eating wolves roamed in broad daylight. And when the freezing winds swept in from the Baltic Sea, the River Neva flowed back upstream, flooding the low-lying delta regularly and reducing it to an expanse of oozing mud.

The year was 1703, and this inhospitable area had been only sparsely populated for centuries. Yet now that Peter's army had captured the last Swedish-held fortress of Noteburg, he resolved to move his capital from Moscow to a newly erected city – to be named

St Petersburg – in this windswept wasteland. Peter the Great, who ruled from 1682 to 1725, abhorred Moscow, with its plots, deeply entrenched habits, superstitions and court intrigues.

It was this grim delta in the extreme north-west of Russia that aroused his enthusiasm. He regarded it as his 'window on the West', giving him at last the sea outlet to the Baltic he desired. In a base far from Moscow, he believed he could begin to mould a brave new Russia, remodelling it along the Western lines he so admired.

Man of the people

With visionary zeal Peter set about building a great city on the small island of Zayachy. But first the ground level had to be raised. To do this, workers brought tons of earth from neighbouring islands. They had no wheelbarrows, so they scraped up the soil with their bare hands and carried it in makeshift bags or inside their shirts.

Peter's far-reaching plans required a vast workforce. Peasants, convicts and soldiers were rounded up and forced to go to St Petersburg, some for a few months, others for life. Provincial governors were bound by law to provide some 40 000 workers a year to drive timber piles into the marshes for foundations, level the ground, and lay out the streets.

The workers had a miserable existence. They lived in squalid huts or slept in the open under thin blankets; their diet was poor and they were paid a pittance. Nonetheless, the first stage of construction was complete by 1710, and in 1712 the tsar made St Petersburg the capital of Russia.

A great city needed a great population, but Peter knew well that none of his subjects would move to this Baltic wilderness by choice, so he simply ordered his court and senior officials in Moscow to join him there. They did so reluctantly.

In the early years, life was hard. Almost every autumn the river flooded its banks, and harvests were scant or failed altogether. Without regular supplies from outside, St Petersburg would have starved, and the tsar's dream would have perished.

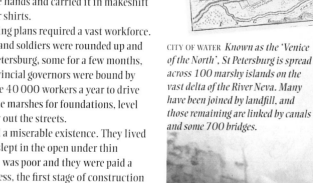

ROYAL PORT *Ships attracted by St Petersburg's low shipping tolls fire double broadsides in salute as they approach the new port and commercial depot at the mouth of the River Neva.*

CLASSIC CATHEDRAL *The elegant 19th-century cathedral of St Isaac is typical of the classical buildings that grace St Petersburg.*

CITY OF WATER *Known as the 'Venice of the North', St Petersburg is spread across 100 marshy islands on the vast delta of the River Neva. Many have been joined by landfill, and those remaining are linked by canals and some 700 bridges.*

PERFECT PLANNER *Peter the Great (left) hired some of Europe's finest architects, artisans and engineers to create a well-planned capital city which rivalled Rome and Paris in beauty and grandeur.*

City of words and music

The city had fewer than 35 000 inhabitants, most of whom wanted to leave. But Peter was content. He had escaped Moscow's stifling atmosphere and unsettled the nobility by uprooting them, thus strengthening his overall grip on the country.

As palaces started to rise along the Neva, a magnificent city began to emerge. Half-Russian, half-European in style, it was a city of golden spires and domes, of granite obelisks, palaces and galleries. The imperial court of St Petersburg became the home of Russian music, literature and poetry – the city of Pushkin, Dostoyevsky and Diaghilev. After the outbreak of war in 1914 the city was called Petrograd, and in 1924 it was renamed Leningrad. After Russian communism collapsed in 1991, however, it reverted to its original name – the only truly appropriate one for a city that owed its existence to the stubborn will and bold vision of one man.

BUILDING BLOCK
A severe shortage of building materials brought the tsar's grand building plans to a halt. He decreed that every cart, carriage or vessel of any kind entering the city must carry stones for building, along with its usual goods. Still there was a shortage, leading Peter to ban the use of stone for building anywhere outside St Petersburg, under pain of exile.

SURVEYING THE SCENE *Striding along a dam on the River Neva, Peter the Great leads a band of courtiers on an impromptu inspection of the work-in-progress of his new capital. A muscular figure some 6 ft 6 in (2 m) tall, the tsar thought nothing of exchanging his finery for a coarse jacket and, axe in hand, taking his place alongside a gang of workers.*

A masterpiece in metal

HOW THE WORLD'S FIRST CAST-IRON BRIDGE WAS BUILT

ON JULY 1, 1779, A TEAM OF WORKMEN RAISED A GREAT arch of cast iron over the River Severn in the English Midlands. The arch was the final part of an ambitious bid to erect a bridge made of cast iron, which would carry the road between Madeley and Broseley, near Coalbrookdale, Shropshire. Each rib of the Iron Bridge was lifted from a barge by ropes slung from wooden scaffolding, and then carefully placed on stone foundations. A well-known local ironmaster named Abraham Darby III was in charge of the work.

A dynasty of ironworkers

Iron was chosen for the bridge after months of heated discussion between local authorities and parliament, which had to sanction the construction. It was a daring decision, as until then bridges had been built of timber, brick or – more usually – stone. The strength and durability of masonry were well known, whereas iron – an expensive and relatively unproven material – had been used mainly in architecture to strengthen existing structures.

However, at their ironworks in Coalbrookdale, three generations of the Darby family had pioneered a cheaper method of making cast iron. Instead of charcoal, the traditional heating agent, they used coke – coal purified by being roasted to about 1000°C (1832°F) in an airless vessel.

The bridge's final span was 100 ft 6 in (30.6 m), and the entire structure required 378 tons of iron, cast into about 30 different patterns. Each of the five ribs of the bridge weighed 5¾ tons. Many details of the structure, including the way the castings were joined together, were borrowed from woodworking practice. Dovetail and mortise-and-tenon joints, as employed in medieval timber houses, were used. As cast iron cannot be welded, the parts had to slot together like a construction kit.

The bridge – the focal point of what is now called Ironbridge Gorge – was virtually completed in mid August 1779, and the accounts books record that £6 was spent on ale to celebrate the event. It was opened on New Year's Day, 1781. By then Abraham Darby had commissioned two artists to paint views of the bridge, and had advertised the beauty of the scene in a local newspaper, the *Shrewsbury Chronicle*.

In 1795 the River Severn flooded, destroying many of the bridges along its banks. But Darby's cast-iron masterpiece stood firm, as he had always said it would.

SYMBOL OF INDUSTRY *The fiery glow from the Bedlam Furnaces, near Abraham Darby's Iron Bridge, lights up the night sky. Working at full blast, the furnaces made iron for bridges, steam engines, kitchen ranges and even grave slabs. The area, in Shropshire in north-west England, was the birthplace of the Industrial Revolution in the second half of the 18th century.*

BRIDGE BEAUTIFUL *The elegant Iron Bridge (above), set against a background of wooded hills, was described by a visitor as a 'network wrought in iron'. As its fame spread, the bridge gave its name to the settlement that grew up around it.*

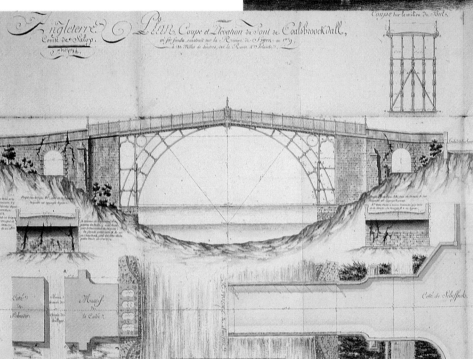

PLANS PERFECT *The plans for the Iron Bridge were drawn mainly by a respected Shrewsbury architect, Thomas Farnolls Pritchard, who died in 1777, aged 54, shortly before work on the site began. Four years earlier, Pritchard had put forward the then revolutionary idea of a cast-iron bridge over the Severn. His final design (left), which incorporated the details of several preliminary sketches, was greatly admired and was posthumously published throughout Europe.*

Gunpowder, shovels and picks

HOW TEAMS OF NAVVIES DUG THE THAMES TUNNEL BY HAND

SHELTERED BEHIND AN 80 TON CAST-IRON TUNNELLING shield, bare-chested labourers inched the device cautiously forward as work progressed on the world's first underwater tunnel, beneath the Thames between Wapping and Rotherhithe in east London. The shield – supporting the soft clay in the bed of the river – had been devised by the tunnel's chief engineer, Marc Isambard Brunel, to try to protect his men from flooding.

Hard and hazardous work

It was the shield that helped win Brunel – who was assisted by his son, Isambard Kingdom Brunel – the contract to build the tunnel. Work began on March 25, 1825, to the peal of nearby church bells, and with toasts and speeches from some of the 200 assembled guests. The tunnelling was done by

HARD LABOUR *Rubble is hoisted from the main shaft of the Kilsby Tunnel on the London and Birmingham Railway. Some 13 000 'navvies' – from the name given to 18th-century navigation canal diggers in England – slaved for almost three years in the late 1830s to build the tunnel.*

navvies – teams of men equipped with picks, shovels and, whenever necessary, gunpowder, used to blast through rock when building the railway tunnels that began to punctuate the countryside.

Meanwhile, news of the underwater tunnel – to be 1330 ft (405 m) long – spread throughout the world. It was one of the sights to be seen when foreign dignitaries visited London, and leaders of society arranged tours of inspection for themselves and their friends. The tunnel was opened on March 25, 1843, and some 50 000 people passed though it in the first 24 hours. Four months later Queen Victoria and Prince Albert paid a visit, when a stall-keeper selling souvenirs laid his entire stock of silk handkerchiefs on the ground for the royal pair to walk on.

TOPSIDE, BOTTOMSIDE *Ships sail by a short distance above the Thames Tunnel. The bricks lining the tunnel are taken to workers in the galleries (left). Used mainly by pedestrians, the tunnel was acquired by the East London Railway in 1869, and it still carries underground trains.*

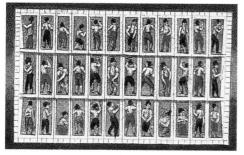

DEATH BY DROWNING *Water breaks through during excavations on the Thames Tunnel on January 12, 1828, killing six workers (right). As well as flooding, the navvies were prey to gases that stung their eyes, stinking black mud that gave them nausea and violent headaches, and foul air that knocked them senseless.*

HUMAN HONEYCOMB *Navvies squeeze into the working cells, each 3 ft (1 m) by 6 ft (2 m), of the tunnelling shield. It consisted of 12 parallel frames, divided into three cells.*

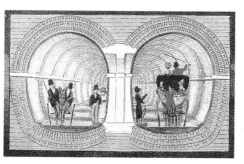

SCRUTINY *In his diary engineer Isambard Kingdom Brunel shows how he went by punt to inspect a serious water leak. A check was also made in a diving bell (right).*

HORSEDRAWN LUXURY *Artists imagined carriages using the tunnel; but the necessary ramp was never built and the tunnel was opened only to pedestrians.*

City of magnificent distances

HOW ONE MAN TURNED A SWAMP INTO A NATION'S NEW CAPITAL

IN 1791, MAJOR PIERRE-CHARLES L'ENFANT, A 37-YEAR-old French military engineer and artist, proudly presented the plans he had drawn up in response to the call for designs worthy of the new capital of the infant United States. Named after George Washington, the first president of the United States, L'Enfant's creation would be spacious, full of character and distinction – a city 'of magnificent distances'. But it would be some time before L'Enfant's vision became reality. Early statesmen and congressmen christened Washington 'wilderness city', while as late as 1842 the visiting British novelist Charles Dickens described it as a 'city of magnificent intentions, with spacious avenues that begin in nothing and lead nowhere'.

Visions of elegance

In 1785, the US Congress proposed the creation of a proper seat for government in a brand new 'Federal City'. The site, chosen by the newly elected president, was a diamond-shaped plot of malarial swampland spread over 100 sq miles (259 km²) at the confluence of the Potomac and Anacostia rivers. L'Enfant persuaded Washington, under whose command he had fought, to give him the job of designing 'a capital magnificent enough to grace a great nation'. He arrived in the tobacco port of Georgetown in March 1791 and drew up his plans at remarkable speed, with neither contract nor budget.

In his formative years L'Enfant had been greatly influenced by the baroque style of architecture, and he aimed to create a grand city like Paris or Christopher Wren's unrealised plans for London after the Great Fire. This city would incorporate grand avenues, or malls, that would terminate in imposing vistas and impressive parks. Building began in 1793 on the city's two most important government buildings and two main focal points, the Capitol (where the US Congress would sit) and the White House, the president's official residence. Impressively sited on natural rises at the end of long straight malls, both buildings incorporated elements of baroque design, which was at its peak in the late 18th century.

> '[Washington is a city with] streets a mile long that only want houses, roads and inhabitants.'
> CHARLES DICKENS, 1842

Two radiating patterns of broad, diagonal avenues extended from the buildings across the city, north-south from the White House and east-west from the Capitol. The Mall was inspired by the broad, tree-lined Champs-Élysées in Paris, but in reality it was more like a large lawn, and cattle were grazed there for much of the 19th century.

Too many streets, too few citizens

The architect had laid his plans with an eye to the future. He had envisaged a city inhabited by many hundreds of thousands of people, not just the few thousand that actually lived there, and at first the capital seemed too big. In 1842 Charles Dickens was merely echoing the general sentiment when he described Washington as a city with 'public buildings that need but a public to be complete; and ornaments of great thoroughfares, which need only great thoroughfares to ornament'.

L'Enfant never saw his idealistic plans reach maturity. His great new city had hardly begun to rise from the ground when he was summarily dismissed after a disagreement with a city commissioner. The excellence of his ideas, however, was duly acknowledged, and his plans were adopted by his successor.

After his dismissal, L'Enfant refused to accept the compensation offered – $2500 and a prime building site close to the planned White House. He spent the rest of his life petitioning Congress with rambling letters of complaint. Although he had drawn up one of the most brilliant town plans ever conceived, L'Enfant died nearly bankrupt in 1825.

SWEEPING VISTA By 1901, the 2 mile (3 km) long Mall, extending from the Lincoln Memorial (left) to Capitol Hill, had been transformed into the 'grand green corridor' of L'Enfant's imagination, with several imposing buildings, statues and monuments situated along it.

THE CITY TAKES SHAPE L'Enfant's blueprint for Washington DC (top), devised in 1791, was followed to the letter. The city survey (centre), taken in 1818, closely reflects L'Enfant's original plan. The 19th-century bird's eye view of Washington (bottom), however, reveals the city's main shortcoming: it was an under-populated wilderness. In 1808, the city had just 5000 inhabitants.

REPLANNING A NATION'S CAPITAL

CHOLERA AND MURDER WERE NO STRANGERS TO THE overcrowded, slum-ridden streets of mid-19th-century Paris. Exiled in England from 1838-40, the future Emperor Napoleon III delighted in London's spacious parks – the city's much-admired 'green lungs'. They made a striking contrast to France's congested capital. After Napoleon had declared himself Emperor in 1852, he appointed a civil servant – Georges Eugène Haussmann, the prefect of the Department of the Seine – to reinforce the prestige of Paris, and paid for the project by borrowing on an unprecedented scale.

Exuberance and evictions
Cutting a swathe through the city, Haussmann demolished 20000 houses and replaced them with wide boulevards and grassy squares where people could promenade and relax. He opened up the Bois de Boulogne and the Bois de Vincennes as public parks, and created several new ones. New public buildings were also built, such as the market at Les Halles, the Palais Garnier – new home of the Paris opera – and the Bibliothèque National. Haussmann aimed to create an exuberant metropolis filled with new shopping bazaars and lit at night by gas.

While the prosperous citizens of Paris enjoyed their reborn city, the poor suffered pitifully. They were speedily evicted from their homes without any compensation. A cartoon showed a man standing beside a heap of rubble, saying: 'But this is where I live and I can't even find my wife.' Nonetheless, rebuilding continued apace. With the emphasis on pleasure, Paris was turned into a sparkling showcase for the imperial regime. Within some 20 years, Napoleon and Haussmann created a new kind of city, with a spaciousness and vitality that no other could match.

FROM OUT OF THE ASHES *The Avenue de l'Opéra (below) was one of the elegant new thoroughfares that arose from the dust and rubble of the old Paris.*

BEWARE THE BEAVER *A satirical caricature depicted Haussmann as a beaver looming over the defenceless city he had been empowered to rebuild.*

NIGHT LIFE *An indoor horse-racing arena, gaslit shopping avenues and ornate theatres were all part of the new Paris – which came into its own after dark.*

MEETING PLACE *Dressed in the height of fashion, Parisians exchange gossip and admiring glances as they promenade the elegant Champs Elysées, or relax in style in the shade of the trees.*

An iron colossus by the Seine

HOW THE EIFFEL TOWER ROSE ABOVE THE PARIS SKYLINE

'FRANCE WILL BE THE ONLY COUNTRY IN THE WORLD with a 300m flagpole!' declared engineer Alexandre-Gustave Eiffel, of his plan to build the tallest tower in the world on the Left Bank of the river Seine. Among his critics were a group of artists and writers, which included Guy de Maupassant, who published a manifesto declaring that the tower would be 'a dishonour to Paris and a ridiculous dizzy tower like some gigantic and sombre factory chimney'.

Towering tribute
Eiffel's controversial design won a competition for a suitable monument to mark the 100th anniversary of the French Revolution as part of the Paris Exhibition of 1889. Among the rival ideas put forward was a proposal to build a mock guillotine 1000ft (305m) high.

Eiffel had become famous for his work in wrought iron – a material he was confident would be light and strong enough to build the record-breaking tower. However, the original design for the open lattice-work tower, rising from four semicircular arches, was not in fact Eiffel's at all: it was conceived by two subordinates – Maurice Koechlin and Emile Nougier – at Eiffel's engineering works. Under the master engineer's expert supervision, however, work commenced in January 1887.

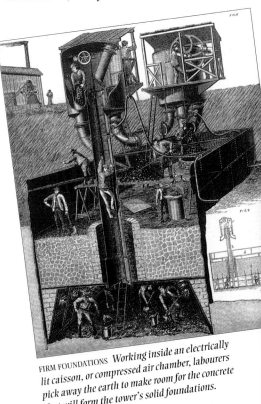

FIRM FOUNDATIONS *Working inside an electrically lit caisson, or compressed air chamber, labourers pick away the earth to make room for the concrete that will form the tower's solid foundations.*

Swaying in the wind

It took 50 engineers to produce the 3700 drawings required for Eiffel's masterpiece. Each piece of the tower was forged off-site, then lifted into place by two counterbalanced cranes, linked by a cable running over a pulley. As one crane lifted its load, the second descended, balancing the weight.

The first of the tower's three platforms was built 170 ft (52 m) from the ground and the second at 350 ft (106 m). By March 1889 the tower had reached its full height of 990 ft (302 m), twice the height of Cairo's Great Pyramid. Nothing like it had been built before and it remained the world's tallest building until the Chrysler Building was opened in New York in 1930.

Eiffel paid most of the tower's building costs of nearly 8 million francs himself, but he recovered his money within just a few years of opening it to the public. At first, it cost two francs for visitors to go up to the first platform, another franc to the second, and a further two francs to the third. Today the tower is as popular as ever. Thrill-seekers should pick a gusty day for their ascent. When a strong wind blows, the top sways by up to 4 in (10 cm).

HIGH-LEVEL CELEBRATIONS *Dignitaries made their way to the top of the Eiffel Tower to celebrate its completion in 1889. A workforce of only 230 men had taken two years to make and fit together its 18 000 components.*

SCENIC WONDERS *Paying visitors (right) flocked to marvel at the panoramic views of Paris and beyond seen from the tower's topmost platform.*

REACHING FOR THE SKY *Cranes at ground level lifted the pieces of the iron latticework into position for the tower's first 108 ft (33 m). As work progressed skywards, the cranes were moved up onto platforms, while workmen clustered about the structure.*

MAN OF VISION *Gustave Eiffel's dream of a landmark for Paris that would be known throughout the world became a reality – at a cost of £250 000.*

HEAVENLY JEWEL *Admirers of the Eiffel Tower were excited by its beauty when it was lit up at night. 'A shining jewel stretching to the heavens,' was one description.*

In pursuit of a legend

HOW AN AMATEUR ARCHAEOLOGIST SEARCHED FOR TROY

ON A BALMY AUGUST EVENING IN 1868, HEINRICH Schliemann, a former German businessman turned keen archaeologist, sat on a roof-top terrace in the town of Yenitsheri, in north-west Turkey, engrossed in Homer's epic poem, the *Iliad*. As the sun sank slowly into the Dardanelles, Schliemann shut his eyes and conjured up a vision of the passage he had just read – the vivid account of the sacking of Troy.

A burning ambition
Schliemann first read the *Iliad* as a boy, and as a result developed a fierce and lasting passion for Troy, the ancient Greek city besieged in the Trojan War. Now he was determined to find it, fulfilling a lifelong ambition. Although no accurate records of Troy's whereabouts existed, Schliemann was convinced that the ancient ruins lay buried under a small man-made mound near the Turkish town of Hissarlik.

Over the years, Heinrich Schliemann had become prosperous enough to finance his own archaeological digs. Although he was already highly regarded in the world of commerce, this roguish braggart yearned for the respect and admiration of the academic world. If he was successful in his attempt to excavate Troy, he felt confident that he would receive the recognition he so craved.

The classical ruins that lay on the surface of the site at Hissarlik showed that it had indeed been occupied in ancient times, and Schliemann was granted permission from the Turkish government to begin digging in April 1870. Over the next three years, a workforce of 100 men, equipped only with spades, dug vast trenches through a 30 ft (10 m) high mound, wantonly demolishing walls and other structures that failed to interest their employer.

Schliemann was convinced that at least four cities lay beneath the classical ruins, and that the Troy he sought was the second layer from the bottom: Troy II. But the ruins uncovered at that level occupied a site measuring only 300 ft (100 m) wide – far too modest to have contained the great walls and towers described in Homer's *Iliad*. Eventually the ruins of nine cities were discovered, and layer VIIa was the most

OBSESSED BY THE PAST
Schliemann retired from business at the age of 46, determined to make his special mark in the world of archaeology.

PLAN DE L'ACROPOLE
DE LA
DEUXIÈME CITÉ, LA VÉRITABLE TROIE.
MONTRANT L'ÉTAT DES EXCAVATIONS
à la fin de juillet 1892.
FAIT PAR W. DÖRPFELD ET J. HÖFLER.
Échelle en mètres.

LOST OPPORTUNITY *The western side of the mound at Hissarlik was Schliemann's objective, but it was the southern side – dug up by his assistant – that finally yielded the ruins of Troy VIIa.*

MISTAKEN IDENTITY *Heinrich Schliemann believed a gold death mask (right) was that of the Greek king, Agamemnon. But it belonged to a king who had lived some 300 years earlier.*

TROJAN GOLD *The solid gold vessels dug from the ruins of Troy II did not belong to Priam, but to an unknown king who ruled about 1000 years earlier.*

JUMBLED RUINS
Layer upon layer of crumbling stones at the Hissarlik site proved confusing for Schliemann and his workforce.

GRAND PLAN
Doors, walls and trenches uncovered in Troy II were mapped out in detail after Schliemann became convinced he had found Homer's Troy.

likely to have been Homer's Troy. Schliemann's extravagant claims might have been dismissed altogether, but at the end of his third season of digging, he discovered a treasure trove of silver and gold drinking vessels and jewellery. The hoard also included two golden headdresses.

Neither a scholar, nor a gentleman

His pledge to allow Turkish officials to examine any important finds he made forgotten, Schliemann swiftly smuggled the hoard out of Turkey to Athens. He proclaimed triumphantly that he had found King Priam's treasure and therefore Homer's Troy. But privately he still had his doubts. He had seen none of the palace decorations described in the *Iliad*, and the pottery shards that he had recovered from the site were too primitive to tally with the date at which they were supposed to have been made. Many scholars, too, still doubted the truth of his claims – with good reason. Schliemann's own later digs proved that the treasure in fact dated from around 2200 BC – 1000 years before King Priam's Troy.

Nevertheless, in 1878 Schliemann decided to continue digging at Hissarlik. He uncovered two further cities and established that one of them had had large walls and ramparts – tallying perfectly with Homer's description. In 1882 he boasted 'My work at Troy is now ended for ever'. Rumours that he was a fraud, however, forced him back to Turkey in 1889 in a final quest for Homer's Troy, this time with independent witnesses on hand. Although Schliemann uncovered a number of stirrup cups and pottery jars that once more he attributed to Homer's

Troy, it was clear by now that his whole chronology had been hopelessly adrift; all along he had been digging too deep. Schliemann resolved to return to Troy, but on December 26, 1890, he collapsed in Naples and died. Wilhelm Dörpfeld, his assistant, later discovered the ruins of a city that might have been Troy. Perhaps, after all, Homer had glorified the city beyond reality – a possibility that Schliemann would never have been able to consider.

ANCIENT RICHES
Contemporary-style silver earrings are among the treasures of Troy.

CROWNING GLORY
Academics were horrified to see Schliemann's wife, Sophia, wearing one of the golden headdresses unearthed by her husband.

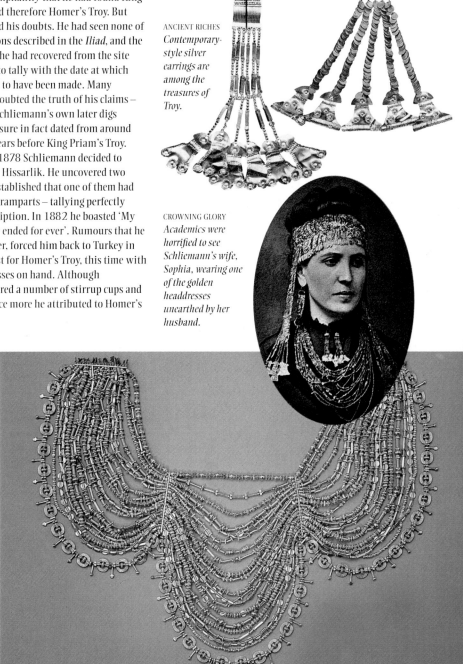

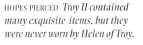
HOPES PIERCED *Troy II contained many exquisite items, but they were never worn by Helen of Troy.*

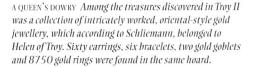
A QUEEN'S DOWRY *Among the treasures discovered in Troy II was a collection of intricately worked, oriental-style gold jewellery, which according to Schliemann, belonged to Helen of Troy. Sixty earrings, six bracelets, two gold goblets and 8750 gold rings were found in the same hoard.*

A steel seam forged across a vast wilderness

HOW THE CANADIAN PACIFIC RAILWAY LINKED EAST AND WEST

LAST STOP: THE EDGE OF THE WORLD

DESPERATE RUSSIAN CONVICTS AND poverty-stricken travellers were among the makeshift workforce that built the world's longest railway, in Russia between 1891 and 1904. The Trans-Siberian Railway ran 5787 miles (9313km), from Moscow to Vladivostok.

Primitive methods
The workmen used picks, wooden shovels and wheelbarrows to dig and transport tons of earth. They generally laid only the thinnest coating of ballast – the coarse gravel or crushed rock used to provide a bed for the rails – and sometimes used none at all. Foundation piles were driven into the earth by hauling a boulder to the top of a tall tripod using a pulley, then letting it fall and hammering in the piles when it crashed to the ground. The rails were thin, light, poor-quality steel.

For eight months of the year the ground was frozen solid, and digging was impossible. Then the workmen built stations and bridges along the line, collecting supplies by sledge. They lived in log cabins or in huts made from sleepers and covered with earth.

The railway was completed in time to rush thousands of Russian troops to the front in the 1904-5 war against Japan. It had taken 13 years and 4 months to build the line, at a cost of 500 million roubles (about £51000). But the line was still far from perfect: the first train to travel the length of the track was derailed ten times.

HAZARDOUS JOURNEY *The railway claimed many victims. In 1931 a locomotive crew froze to death at Nerčinsk, about 1400 miles (2250 km) from Vladivostok.*

ON JULY 28, 1883, WORKERS ON THE CANADIAN PACIFIC railway laid 2120 rails – covering nearly 6½ miles (10.5 km) – across the Canadian prairie in one day. It was a record-breaking feat of construction.

The previous year, Cornelius Van Horne, the pioneering general manager of the Canadian Pacific Railway, had declared his intention to complete the remaining sections of the most ambitious railway line in the world. Begun in 1873 by the Canadian government and contracted to the privately-owned Canadian Pacific Railway in 1880, the finished line would snake across prairies and wind over mountains from Montreal in the east to Vancouver in the west – a distance of 2920 miles (4699 km).

Starting from the centre
In 1882, vowing to lay 500 miles (800 km) of track in a single season, Van Horne hired 3000 men, supplied them with 4000 horses and began work in April when the weather had begun to improve. The starting point for the westward track was about half way along the route at Flat Creek (now called Oak Lake) near the western boundary of Manitoba. By November 1883, 962 miles (1548 km) of track had been laid, extending to Kicking Horse Pass in the Rocky Mountains.

Here the men ran into a wall of rock. For safety reasons, the company's contract specified a maximum gradient of 116 ft per mile (22 m per km). To abide by this, Van Horne would have to bore a 1400 ft (427 m) tunnel through the Pass, delaying the line's completion by at least a year. But he managed to obtain government permission to build a temporary, much steeper route through the Pass, where the gradient was twice the official limit. This great climb, dubbed the Big Hill, was one of the most perilous stretches of railway anywhere in the world.

Rounding one of the Great Lakes
In the railway's eastern section, engineers were faced with the problem of laying the line around the northern edge of Lake Superior. The only way to transport construction supplies was across the lake, over water in summer and ice in winter. During the winter months, 300 small teams of men working non-stop were needed to keep the 15000-strong construction team supplied with building materials.

Disaster threatened at the start of 1885, when, with the line almost complete, the funds ran out. But the Canadian Parliament loaned extra money, and the railway was saved. On June 28, 1886, the first passenger train left from Montreal. The railroad had been completed five and a half years ahead of schedule, with all government loans repaid, thanks mainly to the vision of one determined man.

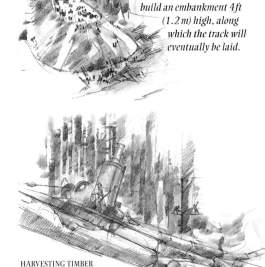

RAISED ROUTE *Workers cut through the Canadian wilderness to build an embankment 4ft (1.2 m) high, along which the track will eventually be laid.*

HARVESTING TIMBER *Men use steel cables to tie a felled tree to the steam engine that will haul it down to camp. Behind, lumbermen balance on a plank to tackle another tree trunk with a long saw.*

WORK CAMP *Sleepers are piled on a rail car (foreground). Washing hangs outside the tents of the crewmen (right); the cars beyond house the foremen and their families.*

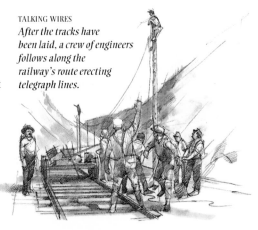

TALKING WIRES *After the tracks have been laid, a crew of engineers follows along the railway's route erecting telegraph lines.*

WEATHER WISE *On either side of the track, labourers dig ditches 18 ft (5.5 m) wide, into which snow will be thrown when the track is cleared during the long, harsh Canadian winter.*

LEVELLING *Horses hauling metal and wooden scrapers heavily weighed down with stones flatten the top of the railway embankment.*

LAYING THE TRACK *Men lower rails and sleepers into position. Each day, they use 65 wagon-loads of materials, delivered along the rails they laid on the previous day.*

BLASTING THE WAY *Dynamite crews watch as an explosion brings down part of the wall of mountain standing in the railway's path. The blasts continued day and night. General Manager Cornelius Van Horne set up three factories to make the necessary explosives.*

CROSSING THE RIVER *Engineers and labourers build a bridge using local wood. Steam engines haul the cut timbers to the edge of the river banks, and giant cranes lift the planks. As the railway became profitable, the wooden bridges were replaced by steel and stone constructions.*

CLIMBING THE MOUNTAIN *Three locomotives haul a passenger train up the steep slope to Kicking Horse Pass in the Rocky Mountains. Freight trains needed four engines to scale the incline.*

TAKING A BREAK *Workers pause for a cup of tea, served by one of their Chinese comrades.*

CHECKING THE TRACK *Men use long bars and shovels to ensure that the railway is level. Inspection crews walk along the line, checking that it is in good condition.*

Across a continent by ship

HOW THE PANAMA CANAL WAS CUT THROUGH CENTRAL AMERICA

A TORRENT OF MUD SWEPT DOWN ONE SIDE OF THE cut and up the other. Charles Gaillard, the engineer in charge of the Panama Canal, stood stock-still, in a state of shock. The mudslide had wiped out several weeks' progress on the great project to link the Pacific and the Caribbean through the narrow

FIRST ATTEMPT *A French firm tried to build a canal across Panama in the 1880s. At first, its surveyors plotted a sea-level route, through often swampy land. US leaders watched nervously; they wanted control of any canal in the region.*

MOUNTAIN ROUTE *The French, led by engineer Ferdinand de Lesseps, found the low land too soft and tried to blast a high canal with locks.*

FOILED MAN *De Lesseps (standing) faced court charges after his canal attempt was bankrupted.*

isthmus of land that joins North and South America – so cutting sea routes from Europe to the Pacific by as much as 7000 miles (11 300 km). Gaillard called the chief of operations, George Goethals, over to inspect the damage. 'What are we going to do now?' he asked. Goethals shrugged, and lit a cigarette. 'Hell, dig it out again', he said.

It was January 1913. Gaillard and Goethals were standing beside part of the Culebra Cut, an 8-mile (13 km) stretch of the canal that was being driven through the mountains of the continental divide between Bas Obispo and Pedro Miguel. The plan had been to blast a narrow defile through the rock, but the floor of the cut was soft and could not resist the mountain's weight pressing down on either side. As the cut was dug out, its walls collapsed and its floor swelled. Huge avalanches of mud swept away the workings – and the deeper the workmen got, the worse the falls became. In the end it would take around 28 000 tons of explosive to force the cut through the mountain.

A breeding ground for disease

Before the Americans began work, the Panama project had already defeated one great engineer: the Frenchman Ferdinand de Lesseps, who had been in charge of construction of the Suez Canal, in Egypt, between 1859 and 1869. His attempt to create a waterway across the Panamanian isthmus took eight years, cost $287 million and ended disastrously in 1889. Around 20 000 workers died, killed by yellow fever and malaria spread by mosquitoes. Panama was one of the least healthy places in the world.

In 1904, the leaders of the United States, long conscious of how a canal through Central America would reduce distances between east and west coast ports in the USA, decided to intervene. Their first step was to eradicate the political problems that had bedevilled the project. Panama was a province of Colombia, and the Colombian government would not agree to the canal's construction. President Theodore Roosevelt, who strongly supported the canal, encouraged a group of Panamanians to rise against Colombia and proclaim an independent republic; the price the new country had to pay was giving the US control of the area

DIGGING IN *Thousands of workmen laboured in the Panama Canal's construction. The volume of earth dug by hand and machine would have been enough to build 63 pyramids the size of the Great Pyramid in Cairo.*

LONG TASK *The 8 mile (13 km) Culebra Cut, shown partly excavated in 1909, took 6000 men seven years to dig.*

DEEP CUT *The canal's minimum depth is 37 ft (11 m). Some 9 million tons of earth were excavated to clear the way.*

through which the canal was to pass. Next, Colonel William Gorgas, a US army doctor, set about wiping out the mosquitoes that had sealed the fate of de Lesseps' project. Gorgas repeatedly fumigated houses with insecticide, and prevented the females from laying eggs by clearing marshland and draining pools to deny the insects access to open water.

After three years' preparation, construction of the canal began in 1907. It was a vast undertaking. At the Culebra Cut, work started at 7 am every day except Sunday, after dawn trains had transported workmen to the site. All morning, gangs drilled holes for the dynamite that was used to blast through the rock, using more than 300 drills. At lunchtime the drillers paused and the dynamite crews took over. In the afternoon crowds of workers, assisted by steam shovels, shifted the spoil into trains for removal. At 5 pm, the main force stopped work and the drilling and blasting continued. Every night, while the labourers slept in temporary townships, maintenance men checked the trains and shovels.

Lifting and lowering the ships

The Americans opted for two sets of locks, one to raise the ships to the level of Lake Gatún – lying about 85 ft (26 m) above sea level at the canal's Caribbean end – and the other to lower them again at the opposite end of the canal. These vast locks, cast on site, used 13.2 million cu ft (3.4 million m³) of concrete. Their floors were 13-20 ft (4-6 m) thick and their walls, honeycombed with conduits through which water flowed to fill or empty the locks, were as much as 50 ft (15 m) thick at the base.

The Panama Canal cost the American government $352 million – four times the price of the world's other great artificial waterway, the Suez Canal – and had claimed 5609 lives from accidents and disease. Even so, it finished $23 million under budget, and opened six months ahead of schedule in August 1914. It was the first great construction project of the 20th century.

OFFICIAL BACKING *The canal had an advocate in US President Theodore Roosevelt, here seen on the site.*

OPEN LOCKS *A tug called the* Gatún *is the first boat to pass through the Gatún Locks after their opening in 1913.*

OCEANS JOINED *The last wall of rock separating the Pacific and Atlantic is demolished on September 12, 1913.*

WATER GATES *The vast steel doors in the Gatún Locks, seen under construction in 1911, were 7 ft (2 m) thick.*

SUCCESS AT LAST *The SS Ancon (above) was the first passenger ship to sail the length of the canal after its opening in August 1914. George Goethals (left), the tough chief of operations, made sure the work was completed on time.*

Cables and caissons across the East River

HOW BROOKLYN WAS LINKED TO MANHATTAN ISLAND

JOHN ROEBLING, THE MAN WHO DESIGNED THE Brooklyn Bridge, never lived to see its completion. In July 1869 his foot was accidentally crushed while he was surveying the site of one of the bridge towers, and three weeks later he died of tetanus.

The bridge that Roebling had designed was to be the largest suspension bridge in the world, joining Manhattan Island to the mainland. Over a mile (1.6 km) long, it was to be suspended from four cables, each weighing over 750 tons. These were to be supported by two towers taller than any building in the United States at the time.

Hard labour in underwater chambers

After Roebling's death, his son, Washington, took over. Work started with the laying of the foundations of the two towers. Two huge caissons – wood and iron boxes, open at one end – were lowered, open end down, to the riverbed. Air was pumped in at high pressure to keep the water from flooding in. Workmen then clambered in to dig out the sand on the river bottom. As they did so, the towers were built on top of the caissons, their growing weight helping to push the caissons deeper into the sand. In the dim light of lamps and candles, and in fear of being crushed by the weight above their heads, the men shovelled mud and blasted boulders, carrying the debris to buckets on pulleys, operated from above, for lifting to the surface.

As the caissons sank, workers began to suffer from caisson disease, or the 'bends'. This occurred when excess air, absorbed into the blood in the caissons' compressed-air atmosphere, formed bubbles when the men returned to the surface, causing excruciating pain, paralysis or even – in three cases – death.

Once the caissons were in place, cement was poured into the working areas to provide a solid base for the towers. These were finally completed in 1876. Next the cables were hung. These had to be spun in position: wires were pulled backwards and forwards over the towers, being looped around anchorages on either side of the bridge. Each finished cable consisted of more than 5000 loops of wire. Once this had been completed, former sailors, used to climbing ships' rigging, were brought in to attach the suspending cables. The bridge deck was then hung from these, section by section. Finally, in 1883, 14 years after building had begun, the bridge was completed.

CABLE WEB *When the Brooklyn Bridge opened on May 24, 1883, it was hailed as the eighth wonder of the world. The first bridge in the world to use steel-wire cables – and given greater strength by a series of radiating cables – it is still one of New York's most impressive landmarks. Half as long again as any other bridge of the time, it was built at a cost of $16 million, and took the lives of at least 20 workmen. It also made an invalid of Washington Roebling, the man who supervised its construction. After twice having collapsed with caisson disease, he was confined to his sick room for ten years, but, though he struggled to read and write, he continued to direct the work from his apartment in Brooklyn Heights. He watched the building progress through binoculars, sending messages to the site and receiving reports from his crew via his wife, Emily.*

STRANDS OF STEEL *A total of 200 miles (320 km) of wire was bound with iron wire to form each of the 19 thick strands that made up the four main cables.*

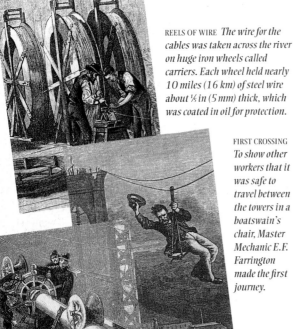

REELS OF WIRE *The wire for the cables was taken across the river on huge iron wheels called carriers. Each wheel held nearly 10 miles (16 km) of steel wire about ⅕ in (5 mm) thick, which was coated in oil for protection.*

FIRST CROSSING *To show other workers that it was safe to travel between the towers in a boatswain's chair, Master Mechanic E.F. Farrington made the first journey.*

CABLE WRAP *After the thick strands of steel wire had been clamped, workers suspended in a 'buggy' wrapped them in copper wire.*

BRIDGE ON BRIDGE *To allow labourers to reach their work, a wooden footbridge was strung across the gap between the two towers.*

WIRE SADDLE *Looped over a wheel attached to a rope, the cable wires were pulled one by one over the tops of the towers.*

DECK BUILDING *The bridge deck was hung from suspension cables piece by piece. More than 1100 cables were used to support the deck.*

ANCHORING THE CABLES *Strong anchorages were built at either end of the bridge, and the cable wires were wrapped around them.*

A massive arch across the water

HOW THE SYDNEY HARBOUR BRIDGE WAS BUILT

WHEN SYDNEY HARBOUR BRIDGE WAS FINISHED, in 1932, its huge steel arch carried four rail tracks and six lanes of traffic. Built at a cost of £9 600 000 to link the centre of Sydney with the suburbs on the harbour's northern shore, it was 3769 ft (1149 m) long, and 161 ft (49 m) wide.

A steel bridge with towers of stone

Construction began in 1923 with the erection of stone towers at each end of the bridge site to act as supports for temporary cables that would strengthen the structure during building. The bridge itself was made of steel plates riveted together. First, two half arches were built outward from each bank, until there was a 3 ft (1 m) gap between them. The supporting cables were then relaxed, allowing the half arches to lean against each other, so they could be fastened together with locking pins. Workers then moved the cranes back along the arch, hanging the steel deck-sections that would take the roads, footpaths and tram and railway lines as they went.

CRANE CONSTRUCTION *Creeping out from the shore, on top of the growing arch, cranes hoisted sections into place. The sections had been built in harbour workshops and floated into position on the water.*

STRONG AND SOUND *The successful tenderers for the construction of Sydney Harbour Bridge, the English company Dorman Long, built 50 ft (15 m) models to test for strength. Once the bridge itself was finished, 96 railway engines were driven onto it in February 1932 as a final test.*

A steel bridge over the Golden Gate

HOW THE ENTRANCE TO SAN FRANCISCO BAY WAS BRIDGED

THE DESIGNER OF THE GOLDEN GATE BRIDGE, CHARLES Ellis, never took part in its construction. Hired by Joseph Strauss – who had already designed a bridge to cross the 1.6 mile wide (2.6 km) strait between San Francisco Bay and the Pacific Ocean – Ellis drew up a new plan, involving the longest central span – 3 miles (4.8 km) – in the world, but Strauss, perhaps unwilling to share the credit, fired him.

Foundations in a 'bathtub'

Work began in 1933 with the building of two piers to act as foundations for the towers. The northern pier posed few problems, as a rock ledge just below the water surface provided a firm base. The southern pier's site, however, was 100 ft (30 m) under water, so a huge bathtub-shaped structure was built on the bedrock, its top projecting above the sea. The water was then pumped out to provide a dry area in which to build the pier for the tower. The towers themselves were built of steel cells hoisted into place by cranes.

Next, strands of cabling were pulled over the towers and bound together to form the main cables. Finally, the bridge deck was built with cranes which, starting at the towers, moved away from the shore laying the steelwork panel by panel in front of them.

HUNG DECK *Slowly moving towards each other, light cranes hang the deck of the bridge from lengths of cable. The finished bridge was 7 miles (11 km) long.*

FEARS FOR SAFETY *Security netting strung under the bridge to protect the workers was destroyed in 1936 by falling building materials. Ten men fell to their deaths at the same time.*

BRIDGE OF BEAUTY *San Francisco's notorious fogs were not the only problem that workmen on the Golden Gate Bridge had to contend with. They fought a constant battle against other elements, with treacherous tides and frequent storms making working conditions extremely difficult. Much work had to be done under water by deep-sea divers working for just four periods of 20 minutes every day, when the tide was turning and the currents were relatively slack. At any other time they would have been swept away by the powerful tidal surges. Even without these battles with currents and weather, the building of the bridge would have been a daunting task. When it was completed, in 1937, it had the world's highest bridge towers at 746 ft (227 m), and contained some 100 000 tons of steel and some 80 000 miles (129 000 km) of cable.*

CABLE WORK *Standing on a temporary deck, built to allow access, a group of workmen assembles the cabling from which the bridge deck will be suspended. The completed cables were just over 3 ft (1 m) in diameter.*

Castles in the sky

HOW NEW YORK CITY'S COLOSSAL SKYSCRAPERS WERE CONSTRUCTED

HIGH ABOVE THE STREETS OF MANHATTAN, THE TEAMS of riveters worked to a set routine. As an enormous crane hoisted each steel beam into place on the growing Empire State Building, a workman called the 'heater' warmed each rivet in a portable furnace until it glowed cherry-red, removed it with tongs and tossed it to a 'catcher', perched precariously on the very edge of nothing. Usually he caught it in his 'catching can', but sometimes he missed. Using tongs, the catcher knocked off the cinders and lodged the rivet in the prepared hole. Another workmate held it firmly with the aid of a heavy steel bar, while a third smashed the rivet into place with a compressed air hammer.

It took 60 000 tons of steel to build the Empire State. The beams and girders were cast in Pittsburgh, and within a day or two of being made, each numbered piece had been transported to Manhattan, hoisted into position and riveted into place. There was little storage space available on site, so elaborate charts and timetables were used to monitor progress and to ensure that deliveries kept precise pace with the erectors' and riveters' schedules.

Careful organisation builds a giant

The charts listed every lorry due to arrive, what it would carry, who would be responsible for it and where it ought to go. Each beam was hoisted by crane to the appropriate floor, then transported to wherever it was required on a miniature railway system. This methodical approach worked exceptionally well and on occasions the building rose by more than a storey in a single day.

The Empire State's 102 storeys were finished in record time. It took just six months to complete the 1250 ft (381 m) building instead of the anticipated 18 months, a feat that set new standards of efficiency for the construction industry. But it was 1931, in the early years of the Great Depression, and much of

the space remained unlet. The building was dubbed 'The Empty State Building'. It had cost $24 million to construct – which was cheap at the time – but for the first few years a major source of income, used by the developer to pay property taxes, was ticket sales for the observatories on the 86th and 102nd floors. From the top on a clear day, it is possible to see 50 miles (80 km) away.

A skyscraper made of steel

The steel skeleton of the riveted structure means that it is immensely strong – the building sways less than ¼ in (6 mm) on the 85th floor in a strong wind. In July 1945 an off-course US Air Force bomber, travelling at a speed of 250 mph (400 km/h) in fog and rain, crashed into the 78th and 79th floors. The three-man crew and 11 people in the building were killed, but the structure suffered no permanent damage. Survivors recall that the building simply rocked a couple of times.

Manhattan's distinctive skyline started to take shape when steel began to be used for tall buildings. Earlier buildings had been made from a variety of materials, including stone, brick, wood and cast iron. But a masonry building taller than about ten storeys would have required supporting walls so thick at the base that there would be hardly any floor space on the ground floor and, before lifts were invented, building height was limited to the number of steps people were prepared to climb.

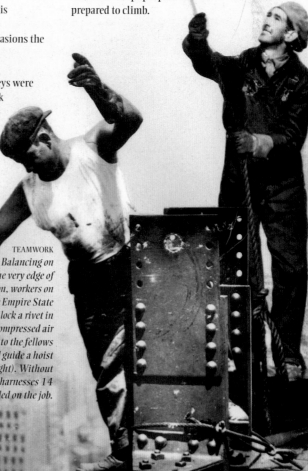

TEAMWORK
Balancing on the very edge of oblivion, workers on the Empire State Building lock a rivet in place with a compressed air hammer (left), signal to the fellows below (centre) and guide a hoist raising a beam (right). Without hard hats or safety harnesses 14 men were killed on the job.

HIGH CHIC *The Chrysler Building's steel façade epitomised Art Deco modernity when it was built in 1930.*

One of Manhattan's most striking early skyscrapers, built between 1901 and 1903, was the Flatiron building. It owes its unique shape to the narrow triangular site it occupies at the junction of Broadway and 5th Avenue at 23rd Street. Twenty storeys high, its riveted steel frame is clad in French Renaissance-style stonework.

Taller and taller

The Flatiron may also have been the first building to create strange aerodynamic effects in the surrounding streets. Even today, Manhattan's tall buildings create unusual wind currents, causing snowflakes to float upwards.

Before long, the Flatiron was dwarfed by other skyscrapers, including the Woolworth Building, completed in 1913. The architect, Cass Gilbert, chose the Gothic style for the 60-storey tower. The structure itself was made from steel and the exterior completely clad in terracotta. It could house 14 000 workers, serviced by 19 lifts and 2800 telephones – an astonishing number for the time.

When the building was finished it won immediate praise from the public – but some architectural purists were aghast, their sensibilities offended by Gilbert's use of Gothic detail purely for decorative effect, rather than for structural purposes. This was contrary to the modernist stricture that 'form should follow function'. In defence, Gilbert declared: 'The Gothic style gave us the possibility of expressing the greatest degree of aspiration . . . the ultimate note of the mass gradually gaining in spirituality the higher it mounts.'

WORLD BEATER *With its 102 storeys, the Empire State Building was for years the world's tallest building.*

OUTGROWN *The 1913 Woolworth Building is now dwarfed by the twin towers of the World Trade Center.*

HEAD FOR HEIGHTS *A labourer fits a steel beam into place (left). The working day began at 8.30 am and overtime started at 4.30 pm. The men stayed on site all day. For 40 cents, lunch was sent up from below.*

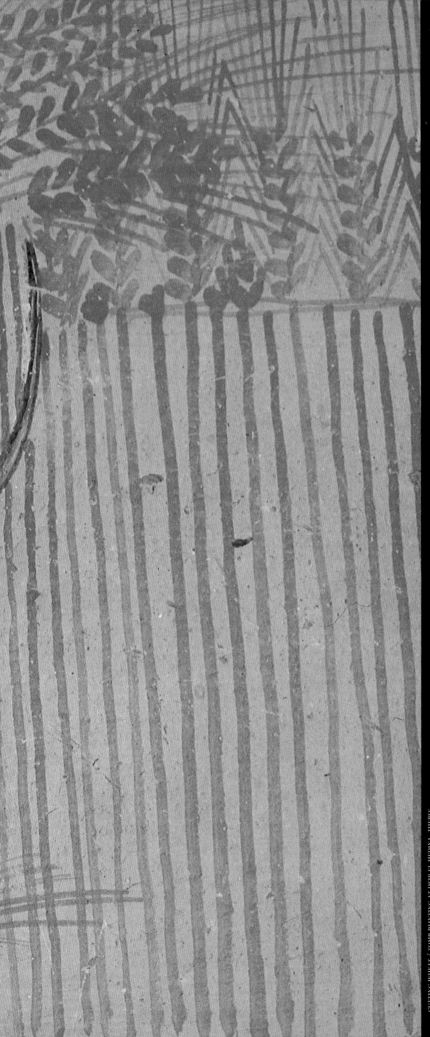

Food
and Drink

*AT FIRST, HUMANS
ROAMED THE EARTH
IN SEARCH OF FOOD.
THEN CAME THE
DECISION TO SETTLE
DOWN AND RAISE CROPS.
AS CIVILISATION
DEVELOPED, TASTES
GREW EVER MORE
SOPHISTICATED, AND
COOKING BECAME A
HIGHLY SKILLED ART*

Lying in wait for food

HOW PREHISTORIC MAN HUNTED

THE GROUP OF SILENT HUNTERS SQUATTED PATIENTLY in the bushes and tall grasses near a herd of grazing bison. At a signal, the men jumped to their feet, hurling stones, waving branches, shrieking and yelling. Heading for the only area that the men had left clear, the startled herd ran for their lives into apparent safety – but stampeded straight over the cliff edge to their deaths on the rocks below. The trick had paid off: there would be plenty of meat for the hunters and the other tribe members to eat.

Stalking the strong, preying on the weak
The earliest hunters preyed on weak animals – those that were injured, hungry or sick, stuck in a swamp or burdened by young offspring, for example. Alternatively, the herds were attacked with stone axes and spears at vulnerable moments, such as when the animals were crossing a deep river. Cunning hunters could mount ambushes at cliff edges. The development of the spear-thrower around 10 000 BC brought more quarry within range.

This was a short, notched baton in which the butt of a spear was wedged before throwing. In effect, it lengthened the thrower's arm and gave him greater leverage. By the time the bow and arrow had arrived, around 7500 BC, it was possible for small bands of hunters to slay large, fleet-footed animals.

Tracking skills were essential for successful hunting, although there were other ways to lure and ensnare prey. Hunters built enclosures near dry river beds and stampeded animals towards them. When the animals swerved at the last minute, they found themselves encircled. Early man also dug pits at the apex of a V-shape formed by two converging rock or earth walls, the widest side of which faced a stream or grazing pasture where the quarry gathered. The hunters drove the animals between the walls into the pit, from which escape was impossible. Some pits contained embedded stakes, sharpened at their tips, on which the animals were impaled.

The role of the gatherers
But even the most ingenious and cunning hunters could supply no more than a third of any group's annual food requirements. The meat contributed essential protein, but the bulk of prehistoric peoples' diet did not come from hunting.

Plants and smaller animals were a more reliable source of food, and gathering them was the role of women and children. They gathered berries, nuts, roots, grasses, grubs and insects, using flint axes and sticks to dig up shoots and roots. If they carried spears, it was not to kill, but mainly to protect themselves in case of attack by dangerous animals.

KEEPING TRACK *Early hunters knew the regular migration routes of large mammals such as bison and deer. They also made pictures of their prey; these cave paintings at Lascaux in France were made 17 000 years ago.*

EFFICIENT KILLING *The bow and arrow soon displaced stone axes and spears as the hunter's preferred weapon. In this cave painting of a wounded bison, the animal's woolly hide has been thoroughly pierced by sharp arrows.*

EXTENDED ARM *Two-legged man was always at a disadvantage against fleet-footed animals, so a bone spear-thrower – which allowed the hunter to attack from farther away – was a considerable asset.*

New weapons to exploit a different food source

HOW PREHISTORIC MAN FISHED

VERSATILE BONES *From the earliest days, people have carved and sharpened bone into weapons and tools. The angled barbs on the spearhead (below) and harpoon head (bottom) held fish in place. Both tools are around 30 000 years old. Later, more refined bone tools included early fish-hooks.*

AS MANY A PREHISTORIC HUNTER FOUND TO HIS COST, spear fishing required skill and patience. Crouched on a rock on a river bank, next to a dugout canoe, the fisherman scanned the water in front of him for the telltale flicker of a fish. As soon as he saw one, he plunged his bone or flint-tipped spear through the water and into the fish before hauling it onto land.

Some 10 000 years ago, as the last Ice Age came to an end, fish became an increasingly important source of food for many peoples of northern Europe. As the climate warmed up, the scrubby vegetation of the tundra was replaced first by light woodland and then by forests. The reindeer and mammoths – long an important food source – moved northwards with the receding ice and men learned to hunt smaller forest animals. Those who lived near the sea, rivers and lakes started to fish methodically. It was no longer sufficient to rely on picking up stranded or slow-moving fish with the bare hands. Specialised

spearheads were carved from antlers with barbs down the side and bound to wooden shafts – the barbs stopped the fish from sliding off after they had been speared. Three-pronged spears worked in the same way: the central prong speared the fish, while the others kept it from wriggling off.

Before fish-hooks were invented, the gorge – a baited sliver of bone attached to a line of leather or sinew – played an important role. As the fish swallowed the bait, the gorge stuck in its gullet and the fish could be hauled in.

Nets and funnels to trap the catch

A team of men working from the shore or in shallow waters could catch a large number of fish at once by using nets made of flax or hemp. These were weighted with stones and supported by wooden floats to keep them spread wide. More productive was the setting of permanent fish traps, however. Spawning salmon and other fish were caught in pool traps, where they could be easily speared or gathered by hand, or more sophisticated traps woven out of vegetable material. These long, funnel-shaped baskets had a wide entrance, which the fish could swim into, only to find themselves unable to escape.

HEADING HOME *Stone-carved salmon, on their long trek back to the spawning grounds, leap out of the water between the legs of a herd of deer. Many fish would have been trapped by hunters before reaching home.*

FAMILIAR FOOD *Just like their present-day counterparts, early peoples found fresh fish cooked over a fire irresistible. Stone carvings of a flat fish which closely resembles a sole (above), and the outline of a salmon scratched into the rock of a gorge in southern France (left), show that early peoples knew how to catch both sea and river fish.*

Mankind tames nature

HOW FARMERS RAISED THE FIRST HARVEST AND HERDS

STUMBLING ACROSS AN AREA OF WILD GRASSLAND, the band of hunter-gatherers stopped in their tracks, their migration in search of food postponed until this new-found supply of grass seed – used for grinding and baking into unleavened bread or adding to gruel-like stews – was finished. Then they moved on. When they returned to the same area the following year, they noticed a new growth of wild grass around the site of last year's grain store: spilt seed had rooted itself and germinated.

The first farms

Early nomadic peoples are first thought to have deliberately planted seeds to ensure a steady food supply, and then established settlements to wait for the crops to grow, c.8000 BC. The earliest 'farms' were mainly located in upland areas where wheat and barley grew naturally – the Zagros mountains of Iraq, the Taurus ranges of Turkey and the high valleys of Jordan.

The crops were tended with hoes until ready for harvesting, which was done with wooden or bone sickles fitted with a series of short flints; longer strips of metal, which gave a better cutting edge, replaced the flints by around 2000 BC. The crops were threshed by tossing the plant heads into the air – the wind separated the lighter outer chaff from the grain.

As the available farmland was outstripped by a growing population, farmers began clearing land for crops. Families moved away, carrying with them grains of wheat and barley to use as seed, and these strains adapted themselves to more hostile weather conditions – drought, damp and cold.

Herds of sheep and goats were already kept by nomadic tribes. The animals' small, blunt teeth and limited intelligence made them relatively easy to control, and they could survive on weeds and shrubs which were useless to people. As the expanding agricultural communities searched for ready supplies of food other than crops, big herds became the norm.

MAPPED OUT *The fertile area around Nippur, in Iraq, was extensively farmed, as this cuneiform clay tablet of the layout of local fields confirms.*

EARLY DAIRY *Domesticated cows descended from small wild cattle – the aurochs – have been kept for well over 8000 years. In this Sumerian frieze carved from around 2500 BC, milk from the cattle is being strained (far left), and the cream rocked back and forth in a large jug to turn it into butter. The cows (right) remain calm, since they are used to being milked by hand.*

SIGN OF WEALTH *Once the advantages of owning herds of livestock became clear – the process of domestication started in around 8500 BC – cattle quickly took on another role: as a symbol of a man's prosperity.*

TENDING THE FIELDS *Early farming tools were crude – staves of wood attached to heavy blades of stone made simple hoes – but the Egyptians improved on these. They used metal scythes for harvesting (above), and yoked oxen to thresh the wheat by trampling it (left). Winnowing, to remove the chaff, was done by hand (below).*

HAND-PICKED *The delights of the grape did not escape the Egyptians, who capitalised on the discovery of the early hunter-gatherers – that it was possible to plant what they liked where they liked – to establish vineyards, watered by the Nile, from one end of the kingdom to the other.*

SAFE-KEEPING *Large harvests produced quantities of grain to move and store. Donkeys carry baskets of grain to a granary (above), where porters are ready to add it to the store through the charge-hole.*

USEFUL FOWL *As early as 2500 BC, flocks of geese were being raised by Middle Eastern farmers. The birds were valued for their eggs, meat and attractive feathers.*

HARNESSING THE NILE

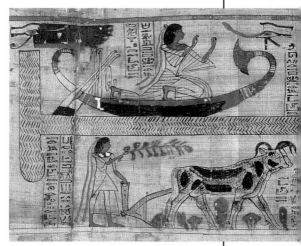

RICH RIVER *The rapid development of Egyptian farming was largely due to the River Nile. Its waters made fertile arid areas which would otherwise have lain fallow, and those who lived along its banks prospered as a result.*

ANCIENT EGYPT OWED NOT JUST ITS PROSPERITY but its very survival to one thing – the River Nile. Its waters and the rich, fertile soil deposited during its annual summer floods made it possible for the Egyptians to grow and harvest massive quantities of grain. Not for nothing was Egypt known in Roman times as the granary of the empire.

Source of life and death

Although the Nile could be relied on to flood every summer, the extent of the flooding was unpredictable. Sometimes the floods would swamp thousands of villages, while on other occasions they proved insufficient, with poor harvests and famine as a result.

To conserve as much of the Nile's precious floodwater as possible, farmers dug deep into the natural basins of the land and enclosed them with dykes. When the floods came, each village dug a network of canals and ditches on different levels, so that they could trap the water. Farmers moved the water from one area to another by making holes in the dykes and canals; to stop the flow, they simply filled the holes with mud.

To get the water from the canals and ditches to their fields, the Egyptians devised a number of simple ways of raising water. One was the *shaduf*, used for transferring water from a low ditch to a high one. It consisted of a long pivoting beam which rested on a wooden or clay pillar. At one end of the beam was a leather bucket suspended on a rope and at the other a mud weight to counterbalance the weight of the water. To lower the bucket into the water, a labourer would pull down on the rope; on the rope's release, the counterweight lifted the full bucket out of the water. The farmer could push the suspended bucket over the ditch and empty it where he wanted.

From gruel to ale

HOW EARLY MESOPOTAMIANS LEARNED THE ART OF BREWING

THE SUMERIAN HOUSEWIFE GATHERED UP THE DRIED, sprouted barley and crushed it into a powder. She mixed the grain with water, forming a dough, and shaped it into several loaves which she placed on hot stones around the edge of the fire. Half of the loaves she removed from the heat when they were only half-baked – these she would save for brewing.

Turning bread into beer

Mesopotamia (now Iraq) was a fertile land bordered by the Tigris and Euphrates rivers. Here, in about 3000 BC, the Sumerian people discovered – perhaps from observing the fermented leftovers of a serving of gruel – how to brew a crude form of beer. Uncooked loaves were broken up and left to soak in water for about a day, before the liquid was drained off and served as a weak ale.

Brewing was carried out by women, under the protection of the goddess Ninkasi. Ale was sold from private homes in great quantities, and by the time Babylonian civilisation emerged, about 1800 BC, the process had been refined. Unhusked barley was stewed in the purest available water and left to infuse.

Eight kinds of barley beer, eight wheat beers and three mixed varieties were commonplace, with herbs and honey added to enhance the aroma and aid preservation. But the popularity of the beer put tremendous pressure on Babylon's barley growers – up to 40 per cent of their crop was used for brewing. As Babylon's population grew, more and more of the barley was needed for food; beer was replaced by a wine made from dates.

HARD WORK *Grinding grain was the first step in brewing beer in the ancient Middle East. Holding a rubbing stone in their hands, women ground the grains to powder on a smooth stone. The brewing process itself was largely a matter of luck, although it was known that old jars were better for fermenting than new ones – probably because their cracks provided breeding grounds for the bacteria active in fermentation.*

LIQUID DIET *To avoid the mash floating in their cloudy beer, two men drink it through straws. As well as being a staple of the diet, beer was also essential for feasting or as a gift to the gods. Workmen were allocated 2 pints a day and senior officials received up to ten, although some of this may have been used as currency.*

Gift of the goddess

HOW THE GREEKS HARVESTED OLIVES AND MADE THEM INTO OIL

THE GREEKS RUBBED IT ALL OVER THEIR BODIES, lit lamps with it, beautified themselves with it, used it as a medicine, awarded it as a sports prize – and cooked with it. The fragrant oil extracted from the olive was a prized commodity.

Oil was extracted from olives in two main stages. First the olives were crushed to release the oil. The simplest method was to smash the olives under a stone roller, although from the 4th century BC special mills were occasionally used. The crushed mixture was ladled into sacks which were placed on a flat table with raised edges and pressed under a beam. After the first pressing, hot water was poured over the sacks and they were pressed again – perhaps three or more times. The mixture of oil, water and olive juice flowed through a spout at the edge of the table into a jar. As this filled up, the liquid was moved to large settling tanks, where the oil could float to the top, ready to be skimmed off.

HARVEST TIME *The first stage of making olive oil was gathering the olives. One harvester climbed into the olive tree and gave it a good shake, while others stood underneath and knocked its lower branches with sticks. The fruit fell to the ground, ready to be gathered up and put into baskets. The Athenians believed that the guardian goddess of their city, Athena, had herself given them the first olive tree, which grew on the Acropolis and was revered as sacred. Other sacred olive trees grew on private land all over Attica – the territory controlled by the Athenians – and it was regarded as a criminal offence to uproot one of them.*

OLIVE PRESS *A man uses the whole weight of his body to keep the beam of a large lever press in place while heavy sacks of stones are attached to keep steady pressure on the olives.*

The rise of an industry

HOW THE ROMANS BAKED BREAD

TRUDGING GRIMLY ROUND THE TINY ROOM, HER EYES fixed on nothing in particular, the slave pushed the beam that turned the heavy stone mill. As the two halves of the mill pulverised the gritty husks of grain, she daydreamed about how wonderful it would be to be given her freedom.

In around 30 BC there were more than 300 bakeries in Rome. Commercial bakeries were essential to provide city dwellers with their staple food, and also to meet the needs of garrisons, travellers, prisons and slave gangs. Many bakeries in Rome were owned by Greeks – masters of mixing doughs and creating different-shaped loaves – and comprised several rooms – a mill room, kneading room, oven and storeroom for the baked bread.

A woman's work

Milling was mostly done by female slaves, often prisoners of war. Once the grain was ground, some of the flour was simply mixed with water and used to make *maza* – a coarse unleavened bread, but bakers also mixed flour with a yeast sponge to produce a light, well-risen bread. The baker's assistants kneaded the dough by hand or, occasionally, in a large basin carved from a lump of lava, which had a central wooden shaft with two or three arms attached to the end. Wooden teeth were inserted into holes round the edge of the basin. As the baker turned the shaft, the arms pushed the dough forward and the wooden teeth caught it and held it back.

After kneading, the dough was shaped into loaves and left to rise. When the loaves had risen enough, they were passed through an opening in the wall into the oven. The charcoal or wood fire that warmed the oven was lit each morning. When the oven was hot enough, the baker raked out the ashes: the retained heat was enough to bake the bread.

Bakers catered to all tastes: apart from the standard loaves, speciality breads were also available. *Picenum* bread was made with dried fruits and cooked in earthenware moulds, which had to be broken to remove the loaf after baking. It was eaten soaked in honey-sweetened milk. Honey-and-oil bread, suet bread, cheese bread, and mushroom-shaped loaves covered in poppy seeds, were also on offer; alongside pancakes, sweet flaky pastries and *piada*, a flat *maza* bread topped with pickled fish and onions – an early form of pizza.

ANCIENT LOAF *Loaves were usually round, and marked with patterns of diamonds or flowers, or scored into wedges – like this one from the ashes of Pompeii.*

DONKEY WORK *In one of the bigger mills, a donkey turns the heavy stones that grind the grain to make flour. The donkey is chained to shafts projecting from the wooden frames of the mills, and a blindfold has been wrapped over its eyes to stop the animal becoming distracted.*

DAILY BREAD *Using a long-handled iron slice, a baker lifts a loaf of bread out of a large, dome-shaped brick oven. While the bread was being baked, the mouth of the oven would be covered over to keep in the heat.*

FLOURISHING TRADE *Surrounded by his produce, a baker sells loaves to eager customers. Once the day's bread had been taken out of the oven, bakers charged a small fee to cook meals that people brought from home.*

Dormice for dinner

HOW A WEALTHY ROMAN'S FEAST WAS PREPARED AND SERVED

AS REQUESTED, THE MALE GUESTS ARRIVED AT THE VILLA at 3 pm, in good time to have their feet washed – as was customary at a banquet in 3rd-century Rome, unless they had just come from the baths. The meal was being held in a vine-covered arbour, in which water from a fountain splashed into an ornamental pool, and lamps hung from the hands of the statues of gods and goddesses that lined the shady walk.

Best tableware put on display
Slaves had arranged three couches in a U-shape so that dishes could be easily brought in and taken away. On a side table were bronze basins containing rose-perfumed water, as well as fresh towels so that the diners could wash and dry their fingers between courses, as some foods were eaten by hand.

The female guests had arrived earlier, and had spent the time admiring the picturesque setting and the hostess's proudly displayed tableware. There were red ceramic dishes from Gaul, glassware from Egypt, special spoons for eating shellfish and, as the centrepiece of the table, a huge embossed silver dish filled with fruit for guests to cleanse their palates.

From pickled eggs to stuffed deer
The host then took the central place on the couch of honour, with his chief male guest seated on his right; it was time for the meal to be served. First came the appetisers: celery, lettuce, olives, pickled eggs, radishes, sea urchins, shallots, spicy sausages and a dish of mustard crushed into a paste with almonds and pine kernels.

Once the diners' appetites were stimulated, the main course was carried in. There was a choice between roasted dormice and braised cubed meat with onion and fennel; rabbits' and kids' livers, marinated in honey, eggs and milk, and cooked in a wine and fish sauce and served with a sweet custard sprinkled with pepper; and, for the hearty eaters, fallow deer stuffed with dates and damsons, and boar boiled in sea water. The whole meal was washed down with flagons of wine diluted with water and flavoured with honey.

The fruit course was then served, after which the host would shout 'Four!' – and the more courageous of the male guests might attempt to swallow four large cups of wine, each at a single gulp. At some eccentric and truly gargantuan feasts, the men then stumbled towards the *vomitorium* – where slaves helped them to perform the coarsest ritual of the evening. They stuck their fingers down their throats, brought up most of what they had consumed, and then returned to their places ready to start eating and drinking all over again.

DECORATIVE SILVER SERVING PLATE

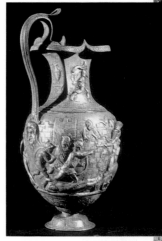
JUG WITH SCENES FROM HOMER'S *ILIAD*

DELICATE SILVER SPOON ON PLAIN PLATE

SILVER SERVICE *The Romans approached banquets with great ceremony, and regarded the quality of the food and tableware as a sign of status. Most of their bowls, jugs, spoons and large decorated dishes were made of silver; many were family heirlooms.*

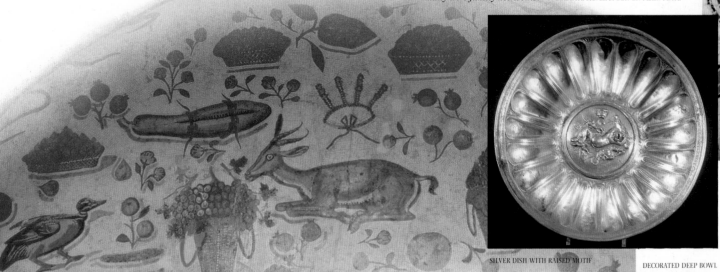

SILVER DISH WITH RAISED MOTIF

DECORATED DEEP BOWL

FLOOR SHOW *A stuffed gazelle, various fowl, vegetables and fruit make up the menu for a Roman feast depicted on this mosaic floor laid between the 2nd and 4th centuries AD in what is now Tunisia. The meat would be served with lots of spices and complicated sauces, or boiled and sugared.*

FOOD BEARERS
Two slaves help to prepare their master's table for a banquet. One carries flat bread, the other fish and fruit.

CHEERS! *Two diners toast each other while others prepare to listen to music and possibly a recitation of classical Greek verse from their host after a sumptuous banquet in the 1st or 2nd century AD. The diners are wearing headbands decorated with flowers or 'wreaths' as a symbol of festivity, and they have probably liberally anointed themselves with oil scented with fragrant perfumes.*

LEFTOVERS *A fish skeleton, a crab's leg, a wishbone and a snail's shell are intriguing details of a mosaic showing the debris of a good meal.*

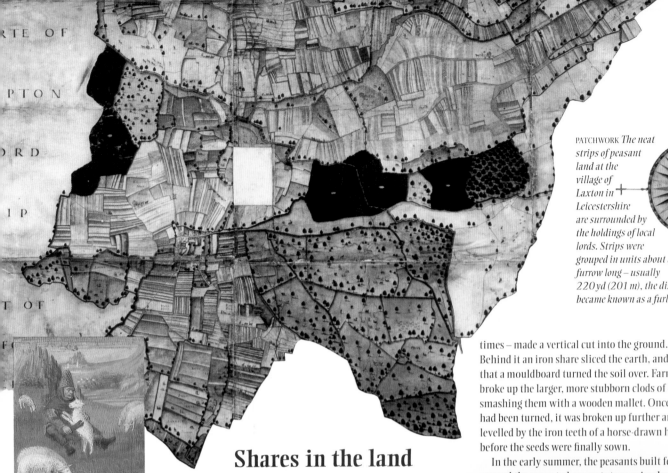

PATCHWORK *The neat strips of peasant land at the village of Laxton in Leicestershire are surrounded by the holdings of local lords. Strips were grouped in units about a furrow long – usually 220 yd (201 m), the distance that became known as a furlong.*

SHEARING
A shepherd shears a fleecy new year's lamb in April.

SOWING
While one man scatters grain, the other breaks up clods with a harrow.

WINEMAKING
After harvesting, the grapes are stamped to a pulp.

NUT HARVEST
A swineherd knocks down chestnuts to feed his hungry pigs.

Shares in the land

HOW MEDIEVAL PEASANTS MANAGED A STRIP FARM

AN ICY WIND BLEW HARD AGAINST THE TWO MEN working the heavy, wheeled plough. As one guided the plough through the frozen earth to make a straight furrow, the other cracked a whip and sang encouragement to the oxen. They were working a long, narrow strip of land, shaped to minimise the number of turns the team of oxen had to make. In return for working on his land two or three days a week, the local lord had allotted the peasants several strips of arable land to farm for their own benefit. Each strip took about a day to plough.

Letting the land lie fallow

The land was cultivated using a system of crop rotation. In the 'two-course' method, peasants sowed one half of their land with cereals, such as rye, oats, barley or wheat, and sometimes beans and peas, and left the other half fallow for grazing. The following year the process was reversed. In this way the land was alternately cultivated and left to rest and to be fertilised by grazing animals. In the 'three-course' system, farmers planted one third in autumn with wheat or rye, another in spring with oats and barley, and left the last third fallow until autumn.

To prepare strips of land for sowing, the peasants used wooden ploughs drawn by six or eight oxen to turn the soil. The blade at the front of the plough – the design of which had hardly changed since Roman

times – made a vertical cut into the ground. Behind it an iron share sliced the earth, and behind that a mouldboard turned the soil over. Farmers broke up the larger, more stubborn clods of earth by smashing them with a wooden mallet. Once the earth had been turned, it was broken up further and levelled by the iron teeth of a horse-drawn harrow before the seeds were finally sown.

In the early summer, the peasants built fences around the crops to keep out stray animals and fertilised the soil with animal manure, or seaweed, sand, straw and marl (a mixture of clay and limestone). Harvesting was back-breaking work. Sickles were used to reap the crops, which were tied into sheaves. The men threshed the sheaves of corn with flails – jointed sticks – while women scooped up the threshed corn with a shovel and hurled it up in the air, so that the lighter chaff was carried off on the wind, while the heavier grain fell to the ground. Grain was stored in sacks and would later be ground between stones into flour.

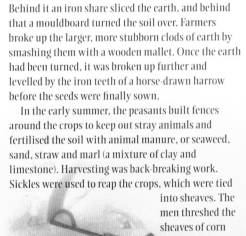

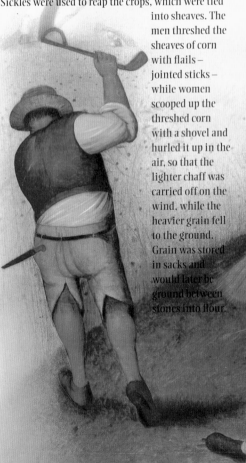

Living off the land

WHAT STRIP-FARMING PEASANTS HAD TO EAT

THE PIG NOSED AMONG THE ROOTS OF THE OAK TREE BY the village pond, gobbling up some of the last acorns. Its owner, a serf, eyed it from a distance; winter was beginning to draw in and it would soon be time to slaughter the animal. No part of it would be wasted. Most would be preserved by being smoked or salted. The blood and intestines would make puddings and sausages, and the fat would be used as a spread on bread or for cooking.

Boiled birds poached from his lordship

Apart from the pig, the peasant ate very little meat aside from the occasional boiled fowl or game poached from the master's land. Those who could afford it kept a cow, goat or sheep to provide milk, cheese, butter and wool. The animal was killed only when it had reached the end of its useful life. Beef and mutton were salted and smoked like pork, then roasted or pot-boiled.

Before the Black Death, few peasants were able to afford their own ovens for baking, but they could use a communal one, for which payment was required by their overlord. Most food was cooked on an open hearth in the main room of the house, although wealthier peasants might have a separate room or building. The main cooking pot was a cast-iron cauldron, suspended from an iron rod over the fire.

When housewives boiled meat in the cauldron, they made the most of limited supplies of fuel by wrapping cuts of mutton, beef or pork, puddings and sausages in separate bundles of cloth and cooking several items in the same pot. Flavourings or 'pot herbs', such as fennel, parsley, mustard and tarragon were used. Meat was forbidden during Lent,

on Fridays and, for most of the medieval period, on Wednesdays and Saturdays too. Fish made an excellent substitute. Peasants ate salted or pickled herrings and dried cod (stockfish), or fish caught in local rivers and streams. Fish could be wrapped in wet rushes and transported miles from their source, especially during cold weather spells.

But for most peasants the main meal was usually pottage, a kind of soup. The main ingredient was cereal — usually oats or barley — which often made the pottage thick enough to be cut into slices. Pease pottage was made from boiled dried peas; white porray included leeks, and green porray was a concoction of green vegetables, herbs and grain.

Green cheese from the dairy

Dairy food, such as buttermilk, whey, curds (often fed to pigs) and cheese were known as 'white meat'. The cheese, made from buttermilk and whey, was very hard, but kept well throughout winter. In summer peasants made a soft, moist curd cheese called *spermyse* or green cheese, which had only half the whey pressed out of it and was often flavoured with herbs. It was known as green cheese because it was new and ripe, not because of its colour.

The kitchen garden provided onions, leeks, garlic, cabbages, turnips, apples and pears, as well as herbs. Household waste made for a well-manured garden. Wild food — mushrooms, blackberries and nuts — added welcome variety to the monotonous diet.

WINTER RITE *The annual slaughter of the family pig that had been fattened since spring was a major event.*

PICNIC LUNCH *Tired peasants rest at midday during the harvest and enjoy a snack of bread and cheese – the staples in their diet – which has been provided by the lord of the manor. Ale – a weak brew of fermented barley and water – washes down the meal. After the harvest, the wheat will be ground and bulked out with rye to make the heavy country bread.*

ILLICIT AMBUSH *Two poachers at their secret work. The ferret (right) drives rabbits out of the warren into the waiting net. The penalty for poaching was a heavy fine.*

Feasts fit for a king

HOW SUMPTUOUS BANQUETS WERE HELD IN NOBLE HOUSEHOLDS

A FANFARE OF TRUMPETS HERALDED THE ENTRANCE OF an army of servants into the Great Hall of Leicester Castle in England. Each carried a huge platter, some of silver gilt and others of wood, brimming with food. On the first platter was a boar's head, its tusks decorated with flowers; swan, pheasant and capons followed. Then came a sturgeon garnished with pike sauce. The mouths of the 120 guests watered at the rich aromas of roasted heron, peacock, crane and partridge.

A sculpture on the table
As usual, no expense had been spared by Eleanor de Montfort, Countess of Leicester, and the sister of Henry III. Her monthly feasts were typical of the lavish dining of the wealthy classes in the 13th century. These feasts often included entertainment, such as jugglers, while diners waited for the next course.

The first three courses consisted mainly of fish and meat dishes, and were served with delicacies such as game-birds' eggs in jelly, and blancmange of white meat mixed with almonds and rice. Side dishes included beans cooked in milk with saffron. The fourth course consisted of fruits, nuts and sweetmeats. A gigantic baked custard with dried fruits in a golden pastry case was usually the main offering for dessert, but it was sometimes outdone by an elaborate showpiece dish called a subtlety. This might a be a swan served sitting upright in its

feathers with a gilded wax crown, or a sculpture of the host in wax and sugar. The countess and her chief guests sat at the high table, which was covered with an immaculate linen cloth and set with napkins, knives and spoons.

These sumptuous banquets held in England in the Middle Ages set a standard, in both cuisine and presentation, for all the feasts yet to come – particularly those staged by the country's leading trencherman, Henry VIII. The Tudor monarch thought nothing of eating until, wheezing and bloated, he collapsed into his specially strengthened bed. But even Henry's gluttony was put to the test at a banquet held at Hampton Court, his riverside palace near London, on New Year's Day, 1541.

Puddings for a royal union
To mark his marriage five months earlier to Catherine Howard – the fifth of his six wives – he ordered the cooks and kitchen staff to prepare a banquet of 60 different dishes. These ranged from tasty 'appetisers' such as Livering Puddinges (liver pâté with nutmeg), to main courses including Pudding in a Tench (whole fish stuffed with spinach, currants and spices, and baked in a wine sauce), a spread of side dishes such as Fritters of Spinnedge (spinach and date balls dipped in ale batter and then fried), and finally rich desserts like

WOODLAND FEAST *Members of a French 14th-century hunting party satisfy their hunger before starting out on the chase. The quarry were deer, whose meat – venison – featured as a centrepiece of many aristocratic banquets.*

KILLING SWOOP *Adult hawkers teach two youngsters by example in this 14th-century manuscript. The hounds were trained as retrievers, bringing back the birds the hawks drove from the sky.*

ANGLING FOR A MEAL *Servants use a scoop net and a small trawl net to catch 'farmed' fish for the enjoyment of the lord and his guests at the banqueting table. The fish are kept in a special pool in the grounds of the manor.*

A Dyschefull of Snowe (rosewater and sweetened heavy cream topped with egg whites and then garnished with sprigs of evergreen).

Over-indulgence in these banquets had a serious effect on Henry's health. The king was an imposing 6 ft 2 in (almost 2 m) tall, and measured 54 in (137 cm) around the waist. As well as being grossly fat, he suffered from leg ulcers and thrombosis of the lower limbs. On his doctors' advice, he tried to eat and drink less – but the temptation of rich, spicy dishes was too much for him.

To each according to his rank

No banquet, royal or otherwise, could begin until a scrap of every dish had been sampled by the official food tasters to make sure it was not poisoned. The top table was served first, and each dish was passed down to the other guests according to rank. Those at the lower tables got humbler meats, such as black pudding or venison stewed with beans. For special feasts these were flavoured with vinegar, honey or herbs and spices, such as parsley, sage, onions and fennel.

The meal was washed down with quantities of alcohol, usually mixed with water. The nobility drank fine wine from Bordeaux in south-west France; others were given weak ale.

Food was served on trenchers – thick slices of stale bread with shallow depressions in the middle that soaked up surplus juices. These were not eaten but collected after the feast and given away to the poor.

A DUKE'S DINING TABLE *The Duc de Berri (centre right) converses with a priestly guest. Knives are the only implements on the table. While one man feeds scraps to a dog, two puppies eat straight from the table. Banquets like this could not be held on Wednesdays, Fridays or Saturdays, which were fasting, or non-meat, days in Catholic Europe.*

GOD'S BOUNTY *A wealthy Flemish family says grace before tucking into a side of beef roasted with sprigs of herbs, while the maid brings in a roast fowl. Only the wealthy could afford white bread like the loaf on the table – so it became a status symbol.*

You are what you eat

HOW TUDOR DOCTORS DEFINED A HEALTHY DIET

THE DOCTOR TOOK ONE LOOK AT HIS PATIENT'S RUDDY face and knew what was needed: no red meat and plenty of fish. Any Tudor doctor worth his salt could identify the temperament of his patients simply by looking at their complexions, and prescribe the appropriate diet. This man, for instance, was clearly a sanguine character, lusty and fearless. Meat was thought to inflame the passions, whereas fish – being a cold-blooded water creature – cooled them. The three other character types – melancholic, choleric and phlegmatic – each had particular diets as well.

Fads for the rich
Only the rich could afford the specially prescribed diets or any of the various different dietary books that were available. Many of the recommendations in these books were dubious by modern standards. Those wealthy enough to buy sugar could try it as a

cold cure. The diplomat and writer Sir Thomas Elyot reported in his *Castel of Helth*: 'It is now in daily experience that sugar is a thing very temperate and nourishing.' In fact, sugar ruined the teeth of the well-to-do. Queen Elizabeth's teeth, for example, were completely black.

Different authorities had strong, sometimes contradictory, views on what was good or bad in food and drink. Wynkyn de Worde's *Boke of Kervinge* (1500), advised diners to 'beware of green salettes and raw fruytes' – possibly a sensible precaution if food was not properly washed. In 1542 Andrew Boorde in his *Compendyous Regyment or a Dyetary of Helth* stressed the danger of drinking water, which was often dirty and disease-ridden. If water had to be drunk, he stated, it should be either rainwater or a stream running 'from East into the West upon stones or pebbles'.

Sir Thomas Elyot, on the other hand, thought drinking water was excellent but should be avoided between meals as it hindered digestion. He also advised against drinking cider because he believed it gave people wrinkles.

SELF-HELP *Books of advice like Sir Thomas Elyot's* Castel of Helth *(above and right) of 1539, were among the best-sellers of their day.*

Stocking up for the winter

HOW MEAT AND FISH WERE PRESERVED IN THE MIDDLE AGES

THE WIND HOWLED OUTSIDE THE FARMHOUSE, BUT A fire burned cheerily in the hearth. The mistress of the house could not resist a feeling of satisfaction as she glanced about the snug kitchen. She was well prepared for winter. On the wall, salted herrings threaded onto thick sticks gleamed golden in the candlelight; beside them were strips of smoked eel and salmon. Above these, bulging nets held onions and pumpkins, and cored and dried apples were strung from the ceiling. She could smell the hams and sides of bacon hanging in the smoky recesses of the chimney, and she knew there were more provisions out of sight: dried cod and smoked herrings in the attic; waxed eggs stored in sawdust; dried meat packed in wheat straw.

Drying food by sun and wind

Sitting at the kitchen table, her husband ate his dish of stockfish (dried cod) with great relish. Before she had set to work on the fish, it had been as hard as board – as always – but she had soaked it in water for two hours before cooking it and then serving it drenched in butter.

Ancient techniques of preserving food by drying it in the sun or burying it in sand were popular in hot southern countries, but farther north it was discovered that the wind could also dry food. Fish preserved in this way kept almost indefinitely, and ships putting to sea on long voyages often carried whole fish and joints of bacon hung over the yardarm to be dried by the sea breezes.

Barrels of herring

Dried fish was a major part of the European diet in the Middle Ages as the Christian calendar forbade the eating of meat, eggs and dairy products on about half the days of the year. The dried or salted fish eaten in medieval Europe mainly came from commercial herring packers around the North Sea. Herring is a fatty fish whose oil turns rancid within 24 hours of being caught, so the salting process could not wait until the fishermen returned to port. Sailors gutted and salted fish on board and then packed them into barrels between layers of sea salt.

Smoking, unlike drying and salting, was very much a domestic occupation. Ham, bacon and red herrings were usually smoked simply by being hung inside the chimney. Pigs were prized for hams and bacon, but virtually every part of the animal was edible. The meat was cured by soaking it for several weeks in brine made out of salt, sugar, water and saltpetre. Some people preferred the dry-salt method, using salt only. Once cured, the hams were smoked for months on end. The kind of wood used on the fire gave the hams a distinctive flavour – oak was said to be the most delicious.

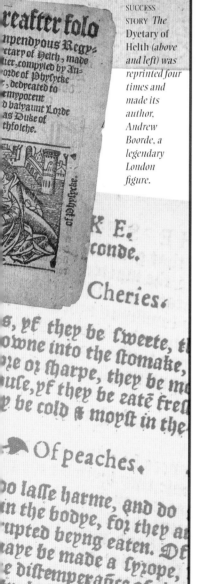

SUCCESS STORY *The* Dyetary of Helth *(above and left) was reprinted four times and made its author, Andrew Boorde, a legendary London figure.*

RAW INGREDIENTS *A kitchen maid unpacks her basket of vegetables in a kitchen already stocked with birds from the farmyard, a huge joint, and a large fillet of wild salmon. If they are not eaten at once the vegetables may be pickled in brine or vinegar to keep over the winter.*

SECRETS OF THE KITCHEN *Recipes for candying oranges and lemons are part of the lengthy repertoire of this 18th-century recipe book.*

The hand of friendship

HOW AMERICAN INDIANS SHOWED THE PILGRIM FATHERS WHAT TO EAT

WADING THROUGH THE WATER, SQUANTO FELT ABOUT with his feet and discovered eels in the mud beneath his toes. Scooping them out with his bare hands, the young brave tossed them onto the nearby riverbank. Spending the afternoon eel-catching was worth it for the thanks he would receive from his new European friends – who particularly enjoyed these fish.

The Pilgrims, a group of 102 English men and women, arrived at Cape Cod Bay on the west coast of north America aboard the *Mayflower* in 1620. But they were ill prepared to create the community they had envisaged. After a gruelling three-month voyage from southern England, they were malnourished and sickly, and they had not brought fresh provisions to see them through the winter. Disease and exposure killed off half their number in the first few months following their arrival and without the generosity of the local Wampanoag tribe, they might all have starved to death.

New home in a strange land

The settlers made their home near Cape Cod Bay, on land already cleared by the American Indians – whose own village had earlier been wiped out by an epidemic. They negotiated peace with Massasoit, the local chieftain, and without the threat of war were able to concentrate on building their settlement.

They regarded it as a miracle when they were befriended by Squanto, an American Indian who had once been a slave in England and had learnt English. He showed them which crops to sow and where to find wild food. He also introduced the Pilgrims to tribes with whom they could trade. With his help, the newcomers won the battle for survival.

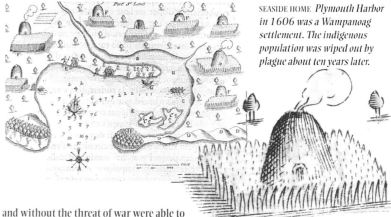

SEASIDE HOME *Plymouth Harbor in 1606 was a Wampanoag settlement. The indigenous population was wiped out by plague about ten years later.*

COTTAGE GARDEN *A New England summer wigwam stands in its own fenced garden. The Wampanoag cultivated kidney beans, squashes, Jerusalem artichokes and tobacco as well as other crops. Sweet corn, which was an important staple, was fertilised with dead herrings so that it would flourish in the poor soil.*

VILLAGE LIFE *New England natives lived a semi nomadic life. In spring and summer, the village would travel to good locations for roots and berries, or to rivers, marshes and the seaside in search of fish and shellfish. In the autumn they would harvest their crops before going to the forests to hunt meat to see them through the long, harsh winter.*

HOME FROM THE HILL *Four hunters sent out to catch fowl for the English Pilgrims' first Thanksgiving feast return heavily laden after a good night's work, and are welcomed back to the settlement.*

AN ANNUAL THANKSGIVING

IN NOVEMBER 1621, THE SURVIVORS OF THE GROUP of English settlers who had landed in north America the previous December called a period of Thanksgiving to celebrate their first harvest in the new land. They invited Massasoit, chief of the Wampanoag tribe that ruled the area where the immigrants had made their home, and 90 of his braves.

For three days, the Pilgrims and their guests played local and English games, practised target shooting and feasted on ducks, geese, venison, clams, lobsters and eels served with cornbread, leeks, watercress and other greens. Plums and wine made from wild grapes accompanied the food.

Feasting was rare in the first years of the colony, as harvests were generally poor – but settlers continued to hold modest harvest celebrations annually. In 1863 President Lincoln proclaimed Thanksgiving a national holiday, to fall on the last Thursday in November.

A new diet from a new world

HOW THE DISCOVERY OF THE AMERICAS TRANSFORMED EUROPEAN EATING HABITS

PEDRO CIECA SET OUT FROM PERU ON AN IMPORTANT mission. Francisco Pizarro, the *conquistador*, had asked for a small, edible tuber to be delivered safely to the botanists at the Spanish court. A drawing of the new tuber later immortalised Cieça's role in introducing the vegetable to Europe. Its Latin caption read: 'Little truffle received from Philippe de Sivry, at Vienna, January 26, 1588. *Papa* of the Peruvians, from Pedro Cieça.'

Although unprepossessing in appearance, this vegetable – the potato – was eventually to loosen the grip of famine that had hung over Europe for many centuries. The potato was native to the Andes, but grew easily in all types of climate and soil. It could be stored for months on end, ensuring a regular food supply through bitter winters. However, many were suspicious of the new vegetable – people in France thought it caused leprosy – and it was almost 200 years before the potato was accepted by Europeans as a good food source.

The conquistadores also brought maize back to Europe, where it had been cultivated at the time of the Roman Empire, but had not proved popular. In North, South and Central America it was a staple crop – roasted, ground into maize flour for bread or eaten whole when small. It was so significant in daily life that the Aztecs in Mexico, for example, made human sacrifices to a maize goddess, Xilonen.

Another food prized by the Aztecs was the *uexolotl* – a domesticated bird later known as the turkey. The Spanish first took it home in about 1520 and it became popular throughout Europe. In England, it was sold by Turkish merchants who sailed to English ports by way of southern Spain, and it was dubbed the 'turkie cock'.

TWO-WAY TRAFFIC *Plantain – brought by traders from Africa to southern Europe – was planted in the Caribbean islands by a Spanish missionary in 1516. Historians once believed that the fruit was a European export to the New World, but now think that it flourished independently in Peru and Mexico.*

Europeans enjoy exotic transplants

Throughout the Americas, beans of all types grew in abundance. Lima beans, scarlet runners, string and haricot beans provided an excellent source of food for the New World explorers, and could be dried and shipped back to Europe. Among the earliest to reach European tables were large kidney-shaped beans sent from the West Indies to Pope Clement VII, who gave them to Canon Piero Valeriano in 1528; the canon sowed the beans and, as *fagioli*, they became a popular element in Italian cuisine. Up until then, soya beans from the East and native broad beans were the only varieties available in Europe.

Some foods discovered by the travellers to the New World – such as peanuts and sweet potatoes – would not grow in Europe, but the hotter Mediterranean countries were able to cultivate other finds. Sweet peppers, hot chillies and tomatoes were adopted, and are now staple ingredients in European cooking.

SWEET TASTES *Preceded by the drink's rich aroma, an 18th-century maid (above) serves a cup of chocolate. The drink became popular among wealthy Europeans. Gardeners in Europe succeeded in planting and growing many of the exotic foods discovered in the Americas. Around 1640, the first pineapple grown in an English hot house was presented to Charles II of England (right).*

EDWARD & JOHN WHITE,
Late Edward & *William White,*
N.8. Greek *Street, Soho*

DEALERS IN CHOCOLATE, COFFEE & COCOA,

The only makers of Sir Hans Sloane's Milk Chocolate, greatly recommended by many eminent Physicians for its lightness on the Stomach, and good effects in consumptive cases

All kinds of Chocolate, Genuine Turkey Coffee, and unadulterated Cocoa

LONG JOURNEY *The cacao tree did not flourish in Europe. Its beans – from which chocolate is made – had to be imported by specialist dealers.*

A cold taste from the east

HOW THE FIRST ICE CREAMS WERE MADE

CURIOUS COURTIERS LOOKED IN THE DIRECTION OF THE royal table, the only one at which a rare new delicacy, 'ice cream', was being served. Charles II, king of England, watched as a manservant ladled out a generous portion of the ice-cold creamy substance and added some strawberries. Charles tasted it, then ate with obvious signs of enjoyment. The date was May, 1671; the banquet, a meeting of the Order of the Garter at Windsor – the first recorded occasion on which ice cream was served in England.

Ice cream already had a long history. It is widely believed, although it cannot be proved, that the recipe for the concoction was brought to Italy from China by the great traveller Marco Polo in the 13th century. The Chinese had discovered how to conserve ice for summer use as early as the 8th century BC. Their earliest ices

MAGICAL MIX *Fairies churn ice cream in a freezer made in Philadelphia, where vanilla seed was mixed with plentiful supplies of cream to make a tasty, speckled ice cream.*

A cold treat delights Europe

The first European ices appeared in the 1660s. Water-based ices could be bought in Paris and in the Italian cities of Florence and Naples. By the 1690s milk-based ice creams cast in elegant moulds were being made in southern Italy. The earliest English recipe for ice cream was published in 1718 in *Mrs Mary Eales' Receipts*, written by the late Queen Anne's official confectioner. This instructed the cook to fill a tin pot with cream and place it in a pail that had been packed with saltwater ice. The pail was covered with straw and placed in a dark place. The cream, mixed with flavoursome summer fruits such as cherries, apricots or raspberries, usually froze within 4 hours. Later chefs advised brisk stirring while the cream was freezing to reduce the tendency to form large ice crystals, and introduced eggs which gave the mixture a smoother, richer texture.

ICE ETIQUETTE *Ice-cream parlours were the exclusive domain of the wealthy until the mid 19th century. The icy treat was served in small portions in cone-shaped glass goblets. They were the precursor of the 'penny lick' – a small scoop of ice cream in a thick-bottomed glass cup, sold at seaside and summer fairs.*

HOKEY-POKEY *An Italian ice-cream vendor, popularly nicknamed the hokey-pokey man, waits for customers next to his wheeled stall. People fleeing from political turmoil in Italy in the 1860s brought with them their expertise in making ice creams and sorbets.*

TUTTI-FRUTTI *A 19th-century recipe book suggested mixing ice cream with dried fruit, spices and liqueur; spooning it into cornets, and dipping these into crushed pistachio nuts. Finally, the cornets were to be arranged into a pyramid and iced.*

were made with naturally occurring snow and ice, which was kept in special storehouses – often pits dug well below ground level – and mixed with fruit juices. Ice houses, packed with ice harvested in winter, were also built in Europe until a way to manufacture ice was discovered.

The technique for making artificial snow and ice from cold water was brought to Europe from the East by Arab scholars. Their breakthrough was the discovery that the addition of salt to ice could produce temperatures lower than freezing. Frozen salt water packed around a container of ice cream mix helped to freeze the mixture, whereas ice made from unsalted water would often melt before the ice cream could freeze properly.

Creating the 'king of wines'

HOW FRENCH WINEMAKERS FIRST MADE CHAMPAGNE

LAUGHTER RANG OUT AS ONE OF A SMALL GROUP OF ladies made a witty riposte to a gentleman's flattery. The gentleman in question proposed a gallant toast. Fashionably dressed courtiers were attending a supper party at the court of the French king, Louis XV, in 1745. A footman toured, replenishing the guests' glasses with sparkling white wine from the Champagne region of France.

In the 18th century, Champagne – an area of gently rolling chalk hills in north-eastern France – was gaining a new and unique reputation as the home of a delicate sparkling white wine. Some wines from this area had always sparkled, but this was viewed as a problem rather than an advantage. The bubbles were a natural phenomenon: in cold weather the yeast which turned the grapes' sugar into alcohol stopped fermenting, but began again when the weather was warmer. The carbon dioxide produced during this secondary fermentation created bubbles within the sealed bottles. This process could occur with any wine, but it happened particularly in Champagne because the region's chalky soil contained large amounts of carbon dioxide.

Of all the winemakers who were involved over many years in the development of a palatable sparkling wine, Dom Pérignon – a 17th-century monk – is credited with playing the most significant role in perfecting the drink.

Dom Pérignon was cellar master of the Abbey of Hautvilliers – in the Champagne district, close to Reims and east of Paris. The Champagne region was abundant in black-skinned Pinot grapes that yielded white juice; white wines made from these grapes were traditionally a yellow straw colour. Dom

TOAST OF SOCIETY *Madame de Pompadour said champagne 'is the only wine that lets a woman stay beautiful after she has drunk it' – perhaps inspiring an advertisement of the 1920s (left). A menu from the Ritz (below) shows cherubs pouring bottles of champagne.*

NOT A DROP TO DRINK *Dom Pérignon, the monk who helped to create one of the world's most celebrated drinks, is said to have been teetotal himself.*

Pérignon ran the abbey's estate, and oversaw all the tenants who grew grapes on its land. Over the years, by demanding the most exacting standards, he unlocked the secret of making a completely clear sparkling white wine, that did not quickly discolour, from black grapes.

The abbey imposed a tax on its tenants of one-eleventh of their produce. Dom Pérignon elected to take this in wine, and mixed wines from different vineyards and of different vintages to produce a blend known as *cuvée* (vat). Some historians of wine also credit Dom Pérignon with improving the bottling of sparkling wine by introducing high-quality Spanish corks.

Bubbles are big business

For decades sparkling wine was regarded as inferior to still wine, and in Champagne the worry was that the new drink would harm the region's winemaking reputation. But it slowly won admirers in cultured Parisian circles. The first firm to produce the new wine was formed in 1734 by Jacques Fourneaux in Reims, and is today owned by the Taittinger family. In 1743, Claude Moët, who owned vineyards, established the House of Moët in Epernay.

MENU

Caviar et Blinis
Huîtres côtes rouges
Tortue claire
Potage Germiny
Turbotin à l'amirale
Barquettes de laitances pimentées
Côtelettes de cailles à la Lucullus
Baron d'agneau de Pauillac à la grecque
Petits pois de Nice au beurre d'Isigny
Dindonneau de Noël truffé
Salade gauloise
Asperges vertes sauce crème
Parfait de foie gras à la gelée de Porto
Plum-pudding à l'anglaise
Mince-pies
Granité sicilienne
Friandises
Corbeilles de fruits

CHAMPAGNE MOËT & CHANDON

RESTAURANT RITZ

MOËT & CHANDON

The cup that cheers

HOW PIONEER TEA PLANTERS PRODUCED ONE OF THE WORLD'S FAVOURITE DRINKS

For more than three gruelling years Charles Bruce – a former British gunboat commander in the Far East – hacked his way through the dense rain forests of Assam, in north-east India. As a freelance adventurer, his latest mission was to locate a crude form of tea reported by previous explorers to grow wild in the region. Pioneer planters in several Indian provinces had already grown tea from seed imported from China. But the Governor-General of India, Lord William Bentinck, who had appointed a special Tea Committee in 1834, was keen to launch India's own version of the drink.

Burning and clearing the jungle

To help in his quest, Bruce – whose botanist brother, Major Robert Bruce, had already reported that tea was growing wild in Assam – recruited two Chinese tea experts. Together they located Assam's wild tea tracts, where tea trees grew up to 40 ft (12 m) high. Aided by native workers, they burned and cleared away the surrounding jungle to create a rough-and-ready 'plantation'. They withered the leaves in the sun, rolled them by hand and dried them over charcoal fires – turning them from green to blackish brown. Once the leaves were ready, they were carried by native bearers to the town of Nazira on the Brahmaputra river and then down river to Calcutta.

Excitement and enthusiasm

In 1838 Bruce's efforts were rewarded when the first-ever consignment of Indian tea – eight chests in all – left for England aboard the sailing ship *Calcutta*. The arrival of the tea in London the following January caused huge excitement. One of the first companies to market the 'new' tea was Twinings, which had

been trading in coffee, chocolate, cocoa and China tea since the beginning of the 18th century. The company realised that 'a nice cuppa' was – or soon would be – a basic part of the British way of life.

Equally enthusiastic, the British government ran a recruiting campaign for plantation managers with the newly founded Assam Company. In the summer of 1839, the first would-be planters set out on the three-month-long voyage to Calcutta. They then proceeded by steamship to the Brahmaputra and upriver to Nazira, which they reached some two to three weeks later. From Nazira, elephants ponderously bore the newcomers through the thick, damp jungle to the plantation sites. There the planters were housed in mosquito-ridden bamboo huts, with packing cases for furniture and candles or paraffin lamps for lighting. They suffered from extremes of heat and cold in the alternating dry and rainy seasons, while the spicy native food gave many of them chronic indigestion. In addition, they were exposed to the dangers of dysentery, malaria, yellow fever and cholera.

Meanwhile, armies of native workers cleared the ground and planted sackfuls

TEA PLANNER *As the first Governor-General of India, Lord William Bentinck (left) planned to set up a native tea industry. He envisaged tea estates, or gardens as they were also known, spread throughout the country. These, he hoped, would rival – if not outdo – those of neighbouring China. To achieve this, he proposed that commercial trading companies should 'resolutely undertake the cultivation of the tea plant on the Nepal hills and other suitable districts'.*

ONE LUMP OR TWO? *By the mid 18th century the drinking of China tea had become an accomplished art in high society drawing rooms. Ladies – and gentlemen – learned how to hold a cup correctly, with their fingers delicately poised. There was even a special 'language' of tea, in which, for instance, froth on the surface was supposedly an invitation to romance.*

COME FILL THE CUP *As tea-drinking became all the rage in 19th-century London, stores at which packets of tea were sold sprang up across the capital. This one in High Holborn advertised black, green, or mixed tea at specially reduced 'bargain' prices.*

THE GRAND OPENING OF THE OLD HOLBORN TEA ESTABLISHMENT, 74, HIGH HOLBORN. TAKES PLACE THIS DAY, AT TWO O'CLOCK.

HAWTHORN & COMPANY, Proprietors

of seed. The new plantations spread to Darjeeling, in the Himalayan foothills, where the 'Champagne of Teas' – delicately flavoured with a delicious bouquet – originated.

High life in the heat

The successful planters built homely, creeper-clad bungalows – so named after the Hindi word *bangla*, or 'house of Bengal'. They returned to Britain on leave in order to find wives, or *memsahibs*, to bring back with them. There were late nights and heavy drinking sessions at the whites-only clubs; adulterous affairs and illicit liaisons with Indian girls were commonplace. Shooting and fishing were the most fashionable outdoor pursuits, while golf and the Indian game of polo were also highly popular.

The planters worked hard and played hard. Their day began with an early breakfast, shortly after a misty dawn. First, they made sure that the workers were labouring in the sun-baked fields. During the morning they inspected the tea which had just been plucked for quality and quantity, and visited the primitive factories where the leaves were mechanically processed.

After a light lunch known as *tiffin*, the planters took a short siesta and then coped with their considerable daily quota of paperwork – such as

bills, accounts, employees' contracts and letters to the trading post at Nazira. They also dealt with any complaints from the field workers about conditions. In the late afternoon, they received reports from their managers and overseers and gave any instructions for the following day's work. The eventual success of the native tea plantations meant that Charles Alexander Bruce had fully justified the title which was later bestowed upon him: 'Superintendent of Tea Culture.'

THE CUP THAT CHEERS BUT NOT INEBRIATES!

SOBER DUTY *A waitress carefully pours a cup of tea from an urn at a teetotal music hall in south London. A night out there, including a penny for tea, came to just over a shilling in the 1880s.*

DRESSED TO DRINK *Heads turn and eyes swivel as a lady of elegance and style makes a sweeping entrance into a high-class London tearoom in 1909. It is five o'clock in the afternoon, and men and women of fashion have gathered to sip tea decorously – and to see and be seen.*

SIGNS OF APPROVAL *By the early 20th century, tea had become a favourite drink among royalty and their subjects. At least two of the big tea companies – Twinings (far left) and Lipton's (above) – enjoyed royal patronage. And Horniman's used 'fairies' to advertise their brew's purity.*

'Head 'em up, move 'em out!'

HOW A CATTLE DRIVE WAS ORGANISED IN THE WILD WEST

LIGHTNING REACTIONS *An electric storm provokes a frenzied stampede, captured on canvas by Frederic Remington. Every time cattle stampeded, they lost weight – and as a result were worth less when they reached market.*

ROUNDING UP *The spring roundup could take months to complete. Mounted men would form circles up to 20 miles (32 km) across, then move towards the centre, driving in any cattle they found. Calves were separated from the herd for branding.*

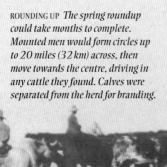

AT ANY TIME DURING THE DAY OR NIGHT, A SUDDEN noise might startle the herd of cattle, and within seconds every animal would be up and off – the thunder of flying hoofs and clashing horns as loud as gunfire. Moments later the cowboys would be on their horses and after them. Stampedes were just one of the daily hazards faced by Texan cattle drovers.

Seasoned cowboys knew that the slightest sound could panic a nervous herd, from a coyote's yelp to the clanking in the chuck-wagon as the cook prepared the evening meal of pinto beans, bacon and biscuits. Even attempts to soothe the

NIGHT WATCHMAN *After a long day on the trail, the cowboys are desperate for rest. But one of the men must always stay awake to guard the herd.*

CAMPFIRE TALES *Although some cowboys were drunken hell-raisers, most did little more than work hard all day, then gather for a quiet evening's storytelling and singing.*

animals by singing lullabies would be pointless if the stampede had been caused by marauding Indians, who would capture the runaway steers and demand a reward of cattle or money for their safe return. Rustlers were another problem, stealing steers and rebranding them with a new owner's mark.

Turning meat into money

The ranchers and their cowhands fully realised the risks involved in the 1000 mile (1600 km) trek from Texas to the Midwest railway termini, from where the livestock was transported to the butchers and packers of Chicago and Milwaukee. The dangers included hazardous river-crossings and armed bands of farmers who tried to stop Texan herds, possibly carrying tick-borne diseases to which their own stock was not immune, crossing their land. But the risks were well worth taking. The American Civil War had prevented the movement of any large herds, and by 1865 – when the Confederates surrendered – Texas was overrun with steers. A longhorn steer worth only about $3 in Texas could fetch $40 in Missouri, so a fortune could be made from a single successful drive of 2500 steers.

The 'Long Drive' north to Missouri began in the spring, when the cattle could gain weight on the hoof, grazing on the rich grasslands. Often, ranchers teamed up to hire an experienced trail boss to lead the expedition. Cowboys then rounded up the free-roaming cattle, identified by their brands, and set off.

The cowboys moved the cattle along slowly, while the trail boss rode ahead, scouting for resting places and water holes. The most experienced cowboys, known as point riders, led the herd. They were followed by the swing and flank riders, who supervised the middle of the herd, and the drag riders who brought up the rear and recovered any stragglers. The procession was completed by the ox or mule-drawn chuck-wagon and a herd of spare horses.

Payday celebrations

It was customary to cover up to 25 miles (40 km) on the first few days of the trail to prevent the cattle from turning for home, but after that the pace settled down to a leisurely 12-15 miles (19-24 km) a day. After an arduous trek, lasting 7-8 months on average, the cowboys were in high spirits as they reached their destinations, where they were paid. The inevitable celebrations might well account for the cowboys' reputation as whiskey-swilling, gun-toting, womanising reprobates. But in many cases this reputation was undeserved. According to President Theodore Roosevelt, cowboys were a unique and tough breed whose lined faces 'tell of dangers quietly fronted and hardships uncomplainingly endured'.

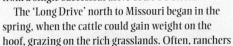

COWBOY OUTFIT *Cattle drivers wore practical clothes for the job: a big felt or leather hat for shelter, shade and to carry water in; a bandanna, used to prevent choking in dust storms or as a bandage; and heeled boots that did not slip in the stirrups.*

Creating the American breadbasket

HOW THE STEEL PLOUGH MADE THE GREAT PRAIRIES THE GRANARY OF THE USA

THE GREAT PRAIRIE LANDS – THE WEST'S FINAL FRONTIER – stretch for hundreds of miles between the Mississippi river and the Rockies. Early settlers, encouraged to move there in the 1830s by a government offer of 'free' homesteads, found not the promised land of their dreams but an impenetrable sea of grass. They were hard pushed to scratch even a subsistence living from the treeless plains.

Yet by the 1890s, the inhospitable plains had been transformed. The great prairies had become the nation's breadbasket. One thing that all the settlers quickly learned was the importance of making the most of limited resources, and the result was a breakthrough in prairie farming. Traditional cast-iron ploughs were unable to till the soil, which was extremely fertile but matted with dense grass roots up to 12 ft (4 m) deep. In 1836, however, John Deere, a blacksmith who had migrated from Vermont to Illinois, filed a patent for a revolutionary new plough which had a wrought-iron mouldboard and a steel cutting edge, or share. The highly polished steel cut through the thick, gummy soil of the plains so effectively that Deere's invention quickly earned the nickname of 'the singing plough'.

Deere's first workshop was in Grand Detour, Illinois, but such was the demand for his ploughs that he soon moved to the bigger town of Moline. Within 20 years, his firm (which became Deere & Company) was one of the world's largest manufacturers of farm

FARM HANDS *A farming family in Custer County, Nebraska, use horses and hand ploughs to work the land they were granted under a homesteaders act passed in 1862.*

STRONG TOOL *The steel blade of the plough invented by John Deere (above) was ideally suited to dealing with the tough grass and heavy soil of the prairies.*

TRADITIONAL HARVEST *A labourer uses a hand-held, multibladed scythe to cut crops while another gathers up the produce by hand. Labour-intensive, slow methods like these were ill suited to the vast, sprawling prairie lands and made it almost impossible for early homesteaders to scratch out a living.*

ENTER THE REAPER
Cyrus McCormick (right) revolutionised farming by mechanising the reaper. The demand for his machine was so great that his firm made more than 6 million of them.

REVOLUTIONARY CUT *A McCormick mechanical reaper makes short work of a field of wheat. The wheels rotate a series of blades fixed to a wooden structure; the blades slice through the crop and then expel the cut wheat to a side platform.*

equipment, producing more than 10000 ploughs a year, and was largely responsible for supporting the American steel industry in its infancy.

The 1830s also saw the development of a horse-drawn, mechanical, multibladed reaper which could cover up to 12 acres (5 ha) a day – far more than half a dozen men and their scythes could ever hope to achieve. Its inventor, Cyrus McCormick, grew up on a 1200 acre (485 ha) grain and livestock farm in Virginia, but, following the success of his reaper in the 1840s, he concentrated on building up his business, setting up a research department to improve his product. Later models of his reaper automatically bundled and tied the grain.

'I saw corn measured by the 40 bushel measure with as much ease as we measure an ounce of cheese.'
ANTHONY TROLLOPE, 1861

Other ideas for farm machinery came from across the plains. Among them was a thresher powered by horses driving a treadmill and, later, by a portable steam engine. This made one of the most laborious farm chores much easier, and less time-consuming.

Drought remained a perennial problem but, before long, improved windmill-driven pumps were capable of drawing vast quantities of water from underground springs. Another invention, barbed wire, helped to define land boundaries – a necessary development as thousands more immigrants flocked westwards – and kept out stray livestock. Among this new generation of settlers were 2000 or so Mennonites, a strict religious sect from the Crimea. They planted their nearly 100000 Kansas acres (40500 ha) with Turkey Red, an improved strain of wheat seed which they brought from Europe.

The demand for land reached extraordinary levels. In the early 1880s, officials in Garden City, Kansas, were even forced to climb through a back window to escape the rush of claimants storming their offices. But life remained hard for many of the new homesteaders, and thousands were ruined when their crops fell prey to the frequent prairie fires, tornadoes or even locust swarms, all of which could turn millions of acres into wasteland almost overnight. Small landholders also faced tough competition from land speculators, who bought up huge tracts of territory and experimented with intensive farming. This was to transform the prairie landscape: by 1890, speculators and factory farmers controlled half a billion acres (200 million ha) of cultivated land, compared to the 80 million acres (32 million ha) which were still in the hands of the smaller homesteaders.

TRACTOR POWER *One of the first petrol-powered tractors pulls a plough through the heavy prairie soils of Moro, Oregon. Continuing technical improvements and innovation in farm machinery made it possible for big business to tame the plains and make huge increases in the productivity of the land. Increasingly, the prairies were feeding the growing nation.*

HORSE POWER *A Nebraska farmer's family proudly stands on top of a reaper that also automatically ties bundles of grain. The machine, developed in the late 19th century, was so heavy that it required a team of at least 12 horses to haul it across the fields.*

Death in the pot

HOW MERCHANTS ADULTERATED FOOD AND DRINK IN THE 19TH CENTURY – AND HOW THEY WERE EXPOSED

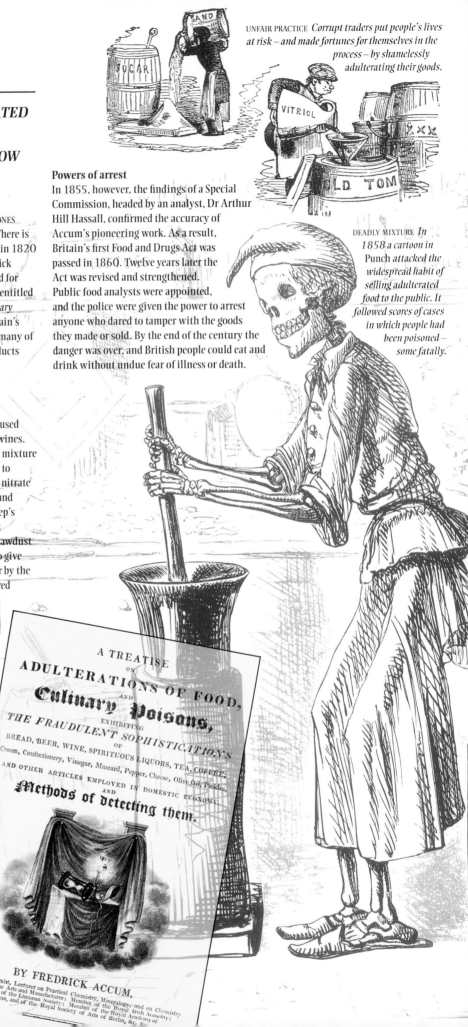

UNFAIR PRACTICE *Corrupt traders put people's lives at risk – and made fortunes for themselves in the process – by shamelessly adulterating their goods.*

DEADLY MIXTURE *In 1858 a cartoon in Punch attacked the widespread habit of selling adulterated food to the public. It followed scores of cases in which people had been poisoned – some fatally.*

THE BOOK HAD A MENACING SKULL AND CROSSBONES on the cover and a stark biblical quotation: 'There is DEATH in the Pot.' It was published in London in 1820 by a German analytical chemist named Fredrick Accum, who had lived and worked in England for almost 30 years. His sensational exposé was entitled *A Treatise on Adulterations of Food, and Culinary Poisons*. It revealed the criminal greed of Britain's food and drink manufacturers and retailers, many of whom enhanced their profits by mixing products with other cheaper, and sometimes dangerous, substances.

Sawdust and sweepings

Bitter almonds containing prussic acid were used to give a 'nutty' taste to many humble table wines. Bakers added alum – a potentially dangerous mixture of aluminium and potassium salts – to bread to give it extra whiteness. Tobacco dealers used nitrate of ammonia to give tobacco more pungency and aroma. Some dairy farmers put gum and sheep's brains in milk to thicken it, and increased its whiteness with chalk. Tea was padded with sawdust and sweepings, and roasted peas were used to give extra bulk to coffee. Brewers 'enhanced' beer by the addition of lethal drugs. And brightly coloured boiled sweets, so popular with Victorian children, were tinted with highly poisonous lead and copper salts – as were raspberry and gooseberry jam.

Many victims of the food fraudsters became violently ill with stomach cramps and pains, and some even died. In the Yorkshire mill town of Bradford, for instance, 15 people died after sucking lozenges coated with white arsenic. But the response to Accum's charges was a campaign to discredit him. In April 1821 he faced a trumped-up charge of 'mutilating' books in the library of the Royal Institution in London. Rather than endure trial and public disgrace, Accum fled back to Germany. With his departure the food poisoning scare died down.

Powers of arrest

In 1855, however, the findings of a Special Commission, headed by an analyst, Dr Arthur Hill Hassall, confirmed the accuracy of Accum's pioneering work. As a result, Britain's first Food and Drugs Act was passed in 1860. Twelve years later the Act was revised and strengthened. Public food analysts were appointed, and the police were given the power to arrest anyone who dared to tamper with the goods they made or sold. By the end of the century the danger was over, and British people could eat and drink without undue fear of illness or death.

A TREATISE ON ADULTERATIONS OF FOOD, AND Culinary Poisons, EXHIBITING THE FRAUDULENT SOPHISTICATIONS OF BREAD, BEER, WINE, SPIRITUOUS LIQUORS, TEA, COFFEE, Cream, Confectionery, Vinegar, Mustard, Pepper, Cheese, Olive Oil, Pickles, AND OTHER ARTICLES EMPLOYED IN DOMESTIC ECONOMY. AND Methods of detecting them.

BY FREDRICK ACCUM,
Operative Chemist, Lecturer on Practical Chemistry, Mineralogy and on Chemistry applied to the Arts and Manufactures; Member of the Royal Irish Academy; Fellow of the Linnæan Society; Member of the Royal Academy of Sciences, and of the Royal Society of Arts of Berlin, &c. &c.

HARD FACTS *Fredrick Accum pulled no punches in the frontispiece (right) of his outspoken book. On its publication, the press accused the government of turning a blind eye to the 'poisoning' of the nation's food and drink.*

Meals on wheels

HOW A MASTER CHEF FED IRELAND'S STARVING POOR

ALL MORNING LONG, HUNDREDS OF HUNGRY PEOPLE queued up before a tent erected on the esplanade outside the Royal Barracks in Dublin. From inside the tent came the appetising aroma of the beef-and-vegetable soup being made by one of Europe's finest cooks – Alexis Soyer, chef at the distinguished Reform Club in London. It was April 1848, and the total failure of the Irish potato crop for the past two years had caused widespread famine. Thousands of men, women and children were dying of hunger, and Soyer was so concerned about them that he launched a public appeal to fund his proposed soup kitchen.

Soon after arriving in Dublin, he set up a wood-and-canvas tent, which contained a coal-fired boiler holding up to 300 gallons (1360 litres) of soup. There were cutting tables for the vegetables, chopping blocks for the meat, and from the roof of the tent hung storage boxes for the condiments and thickeners. When ready, the first batch of soup was ladled into eight large saucepans.

KITCHEN KING *Alexis Soyer was proud of his enormous kitchen at London's Reform Club (above) – the partition walls of which are cut away in this drawing, allowing a clear view of the larders, sculleries and the staff at work. Soyer himself is seen (centre, right) explaining the layout to a lady visitor. Wearing his customary flowing cloak and hat set at a jaunty angle, Soyer – pictured (right) in 1855, when he was devising the British Army's diet in the Crimean War – was also known for his culinary inventions, such as 'improved' baking dishes. The most famous of these was his so-called 'magic stove', a lightweight, portable spirit stove, now known as a chafing dish. This was so adaptable that, according to a press report, 'a gentleman may cook his steak on the study table, or a lady may have it among her crochet or other work'.*

SOUP AND SAUCE *Even the rich and fashionable visited Soyer's soup kitchen for the poor in London's Leicester Square (below), to see how the other half ate. Later, the chef gained renown for his savoury Soyer's Sauce, sold in specially shaped bottles across the world.*

Drinking-up time

A bell rang to signal that the soup kitchen was open, and those in the queue were allowed in 100 at a time. After grace was said, each person was given 6 minutes in which to consume 2 pints (1.2 litres) of soup from an enamel bowl with a metal spoon chained to it, to clean the implements with a sponge soaked in hot water, and to leave the kitchen by the far exit. In this way, Soyer fed 1000 people an hour, 8 hours a day. For those too frail or too ill to attend the kitchen, he invented 'meals-on-wheels' wagons – horse-drawn metal carts in which the food was placed on top of small stoves to keep it hot on its way to the recipients' homes.

While in Dublin, Soyer published recipes for cheap but nourishing soups, stews, breads and puddings. He also gave unexpected advice about vegetable peelings. These, he declared, should not be thrown out, but used in soups to give extra body and flavour.

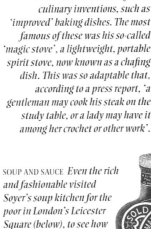

The king of chefs, the chef of kings

HOW AUGUSTE ESCOFFIER CREATED THE MODERN RESTAURANT

LUNCHEON AT THE SAVOY HOTEL IN LONDON WAS AT its height, and in the restaurant some 500 elegant diners were tucking into their lovingly prepared and mouth-watering meals. The food was the creation of the celebrated French master chef Auguste Escoffier, who sat in his glass-fronted office in the nearby kitchens, from where he kept an eye on his 80-strong army of chefs, under-chefs, pastry cooks, wine, soup, meat, fish, sauce and vegetable specialists – as well as dish-washers and apprentices.

When the 42-year-old Escoffier – who had perfected his culinary skills in restaurants and hotels in Nice, Paris and Monte Carlo – arrived at the newly opened Savoy in 1889, the preparation and serving of food still followed the dictates of the late Antonin Carême, chef to several crowned heads of Europe. He believed in serving meals *à la française* – that is, placing several different main and side dishes on the table at the same time. This gave an impression of a sumptuous banquet, and allowed the diners to sample a little of whatever took their fancy. It also blunted the palate, allowed the food to get cold, encouraged gluttony and caused indigestion.

A Russian revolution

Escoffier swept all this aside and introduced *service à la russe*: the 'Russian way', in which only one main dish and its accompaniment was served at a time – properly heated, correctly proportioned and perfectly balanced. During a meal of, say, sole in white wine sauce, pheasant with pâté de foie gras, and a crayfish soufflé, diners would be served a fruit or champagne sorbet after the strongly flavoured game course to refresh their palates. Each course was complemented by a carefully chosen wine. In addition, Escoffier did away with heavy, flour-based sauces in favour of lighter, easy-to-digest meat extracts.

As the Savoy's first-ever Director of Kitchens, Escoffier set fresh standards in obtaining the best possible supplies. He personally shopped for fish such as turbot, sole, trout and Scotch salmon in London's Billingsgate Market. Much of his meat, with the exception of Scotch beef, came from France, along with fruit and early vegetables from Les Halles, the great market in Paris. In addition, there were truffles from Périgord in south-west France, ducks from Rouen, butter from Brittany and Normandy, aubergines from the Italian Riviera, tomatoes from the Channel Islands, peaches from the Rhône valley, frogs' legs from the river Seine, and live turtles for making turtle soup – a favourite dish for Victorians – from the West Indies.

In the kitchens, Escoffier introduced working methods new to British hotels. He organised his staff, who worked in shifts from dawn until the small

hours, on military lines. His workers, under the overall command of a *chef de cuisine*, were split into several parties, or groups. Each of these was responsible for a particular commodity, and each was run by a *chef de partie*, or assistant chef.

As the twice-daily rush hours – lunchtime and dinnertime – built up, and waiters dashed in and out carrying trays of piping hot food, tempers often became frayed. To avoid this, Escoffier devised yet another series of innovations. First of all he replaced the traditional, loud-mouthed *aboyeur*, or 'barker', who bellowed out the orders, with a softer-voiced *annonceur*, or 'announcer'.

He forbade swearing and bullying, cracked down on gambling and smoking, and tackled the long-standing weakness of many chefs: alcohol. Instead of constantly swigging from bottles of wine, beer or spirits, kitchen staff at the Savoy slaked their thirst with draughts from a huge cauldron of refreshing cold barley water.

Keeping up appearances

A dapper dresser – who, because of his small stature, wore high-heeled shoes to raise him above the level of the stoves, with their intense heat – Escoffier encouraged hitherto slovenly workers to take a pride in their personal appearance. Their white hats and jackets were expected to be spotlessly clean, their shoes polished till they shone, and when off-duty they had to wear respectable suits.

Known as 'the king of chefs and the chef of kings', Escoffier later retired to Monte Carlo. He maintained his interest in cooking, and died there in 1935 aged 88.

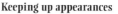

ROYAL OCCASION
Noble diners signed Escoffier's menu cards at a banquet celebrating the wedding of the Duke of Aosta to Princess Hélène of Orleans in 1895.

PERFECT SETTING *The glow from silk-shaded table lamps put diners in the mood to enjoy Escoffier's cuisine.*

HOT WORK, HOT EMOTIONS
Feelings sometimes ran high in the steamy heat and noisy bustle of hotel and restaurant kitchens. On one occasion Escoffier broke up a savage fight between two of his staff over the quality of a sauce. Overworked chefs had been known to go berserk in such conditions, attacking their fellow workers with whatever came to hand – such as meat cleavers, ladles and saucepans.

THE DISH NAMED AFTER A DIVA

ESCOFFIER'S MAGNIFICENT COOKING AT THE SAVOY HOTEL
attracted some of the world's most talented, powerful
and glamorous people, including political leaders,
actors, musicians and singers. Prominent among
these was the acclaimed Australian diva, or operatic
soprano, Nellie Melba. Escoffier was thrilled when she
sent him two tickets to hear her sing in Wagner's
grand opera *Lohengrin* at Covent Garden in the spring
of 1893. The next day Melba gave a supper party at the
hotel for her friend the Duke of Orleans. Escoffier, who
had been overwhelmed by Melba's performance,
decided to repay her by showing his own artistic skills.

The taste of paradise
In a burst of creation, he invented a new dessert which
he named *pêche Melba,* or 'peach Melba'. It consisted
of peeled peaches poached in a vanilla-flavoured
syrup and placed in a silver dish on a layer of vanilla
ice cream. He laid the dish between the wings of a
swan – his inspiration being the swan that draws
Lohengrin's boat in Wagner's opera – which he had
personally sculpted out of Norwegian ice and covered
with icing sugar. Melba adored the dessert, which she
said was like 'the taste of paradise'.

Escoffier, however, felt that the dessert still lacked a
'certain something'. He racked his brains for six years
– by which time he had moved to the newly opened
Carlton Hotel, in London's Haymarket – before
inspiration struck. He triumphantly added the finishing
touch to the peach Melba – a coating of raspberry
purée. 'Only then,' he declared, 'did I feel I had done
Melba justice.' On sampling the final version of the
dessert, Melba said she felt 'much as Eve did when
she tasted the first apple!'

SWEET THINGS *Melba, seen here in*
The Barber of Seville, *inspired*
Escoffier, who wrote his original
recipe for peach Melba in French.

LADIES' MAN *In the late*
1880s Auguste Escoffier
(above) introduced London
society to the delights of
'eating out' at the brand-new
Savoy Hotel. When asked for
the secret of his success, the
great chef said that he
always 'tried to please the
ladies' with his meals. 'If
they approve of the food,' he
stated, 'then it follows that
their menfolk will too!'

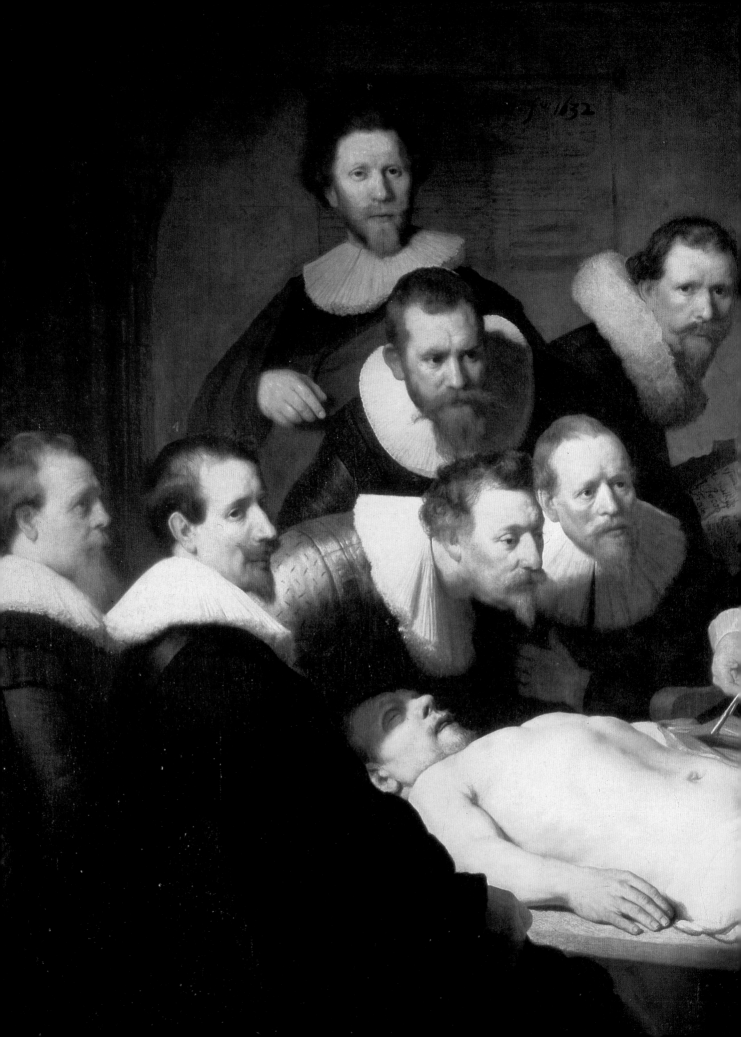

Milestones in Medicine

PAIN AND DISEASE ARE AN INEVITABLE PART OF LIFE. THE AGE-OLD STRUGGLE TO DEAL WITH THESE SCOURGES, AND TO DISCOVER THE SECRETS OF THE HUMAN BODY, HAS ENGAGED SOME OF THE FINEST MINDS IN SCIENCE

SAWING INTO THE SKULL

WHILE SEARCHING FOR REMAINS OF PREHISTORIC MAN IN the Peruvian mountains, Ephraim Squier, an American diplomat and anthropologist, made an astounding discovery. He came across the skull of a Stone Age man upon which surgery had apparently been performed. Two thin, parallel grooves had been cut in the skull with great precision. The grooves were bisected by two other grooves, as in a noughts-and-crosses figure, and the bone from the central square had been neatly removed to expose the patient's brain.

Instruments of flint

Squier's discovery, made in the 1860s, confirmed what had been suspected since finds in the 17th century: that what is now called trepanning – a procedure for removing part of the skull by means of a surgeon's cylindrical saw – had been performed by Stone Age man some 12000 years ago. The surgeons' instruments had included flint saws, polished stone knives and, for drilling holes, sharp-pointed flints set into a stick. A bowstring was looped around the stick, which rotated as the bow was sawed back and forth.

Following his discovery, Squier sent the Peruvian skull to a world-renowned French surgeon and anthropologist, Dr Paul Broca, who declared that it had been trepanned while the patient was alive. Signs of infection around the area showed that the patient had lived for about two weeks after the operation.

Broca's findings intrigued the medical world. As Stone Age skulls from France, Poland, Portugal, Alaska and Peru surfaced over the next 20 years, it became evident that trepanning was a technique used by Stone Age societies on several continents.

A creditable success rate

Surgeons appear to have operated on men, women and children not just in cases of physical head injury, but also to relieve fits, epilepsy, headaches, depression, mental disturbance and even lethargy. Such conditions were thought to have been caused by evil spirits trapped in the head, and by opening the skull to the air, the evil spirits were released. Primitive painkillers, such as leaves of the powerful coca plant from which cocaine is now derived, helped to lessen the pain.

Despite the crudeness of the Stone Age operations, many were successful. A study of trepanned skulls from eastern Europe revealed that more than 80 per cent of patients had survived.

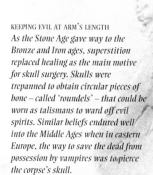

KEEPING EVIL AT ARM'S LENGTH *As the Stone Age gave way to the Bronze and Iron ages, superstition replaced healing as the main motive for skull surgery. Skulls were trepanned to obtain circular pieces of bone – called 'roundels' – that could be worn as talismans to ward off evil spirits. Similar beliefs endured well into the Middle Ages when in eastern Europe, the way to save the dead from possession by vampires was to pierce the corpse's skull.*

CONFIDENT CUT *A pottery statuette of an Inca surgeon in action shows how trepanning was carried out in Stone Age times. Sometimes patients underwent more than one operation – a skull found in Cuzco, Peru, has seven circular openings. Of 214 trepanning wounds found on skulls in the Americas, more than half appear to be completely healed.*

EARLY SURGERY *The skull Squier found in Peru clearly shows the marks of a delicate operation. Four straight cuts at right angles to each other outlined the large piece of bone that was removed.*

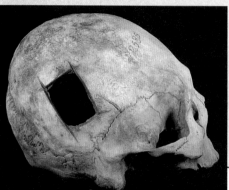

Spirits, gods and herbal cures

HOW SHAMANS AND MEDICINE MEN TRIED TO CURE DISEASE

AT MIDNIGHT IN THE DEPTHS OF THE SOUTH AMERICAN forest, a delirious woman lay on the rush matting sweating profusely. The medicine man moved around her, mumbling sacred chants before administering a potent brew of herbs and animal products. The effect was immediate. The patient raged and writhed, vomiting and wailing. For another 2 hours she was in torment, while the medicine man engaged the help of the spirits. Gradually, calm descended. As dawn approached the patient became more settled, and drifted into sleep. Two days later she was still weak, but on the path to recovery.

Keeping ills at bay

From the dawn of time, anyone able to offer a cure for illness played an important role in their society. In the past, as in many tribal groups today, disease was often interpreted as the effect of malignant forces – the work of spirits or gods who were displeased with the sufferer for wrongdoing or for breaking a taboo. Or perhaps it was the work of a sorcerer acting on behalf of another person who had a grudge to bear. In Melanesia, for example, sorcerers still point 'ghost shooters' – pieces of bamboo filled with the fragments of the bones of a dead man – at a victim in order to cause illness.

Folk medicine had an important religious element. Patients and doctors alike made no distinction between body, mind and soul: when all three were in harmony, it was believed, good health would prevail. The medicine man or woman was doctor, counsellor, psychiatrist and priest, and was entrusted with the spiritual and physical health of the tribe as a whole. Individual illnesses were seen as a blemish on the health of the tribe, and members had a responsibility to avoid damaging frictions and appease the gods, so as not to render themselves vulnerable to illness.

Curing with drugs and dancing

The position of medicine man could be handed down within families or earned by a patient recovering from a disease. The knowledge of how to treat various diseases was gathered over thousands of years, and common ailments were treated with herbal remedies which were administered as medicines, rubs or poultices. Australian Aboriginals, for example, ate clay for stomach ailments, just as clay-based kaolin is used in Western medicine to treat diarrhoea. Remedies used substances with symbolic associations, or worked on the assumption that 'like cures like'; the claws of a snake-hunting bird, for example, would be used to cure snakebite. Other treatments involved massage and manipulation, wearing an amulet, a change of diet or simply providing a listening ear, with the medicine man setting up a framework, or story, which made sense of frightening symptoms.

For more serious diseases, mental illness or problems such as infertility, it was sometimes necessary to commune with the spirit world. If an illness was ascribed to supernatural causes, healing could involve religious ritual, divination and trance dancing. If an illness was blamed on a spell inflicted by another person, the evil had to be unveiled and counteracted with similar magical techniques.

Medicine men still play an important role in societies around the world. Many people will still seek the advice of a traditional healer before consulting a doctor with more orthodox medical training.

CEREMONIAL MEDICINE *In 1832, Wun-New-Tow, medicine man of the Blackfoot tribe, attempts to save the life of his chief dressed from head to toe in the skin of a yellow bear.*

THE HEALING POWER OF SUPERSTITION *This portrait of an early medicine man was painted on the wall of a cave at Ariège, France, some 20-30 000 years ago. The influence of medicine men at this date may have had as much to do with their extravagant costume as their knowledge of medicines. By distracting patients from their suffering with a combination of disguise and ritual, Cro-Magnon medicine men may have been able to relieve pain, if nothing more.*

CERAMIC MEMORIAL *The way in which Inca shamans worked, mixing their medicines and making examinations of their patients to aid diagnosis, is commemorated in these figurines.*

Needles, pressure points and herbs

HOW DOCTORS IN ANCIENT CHINA DEALT WITH ILLNESS

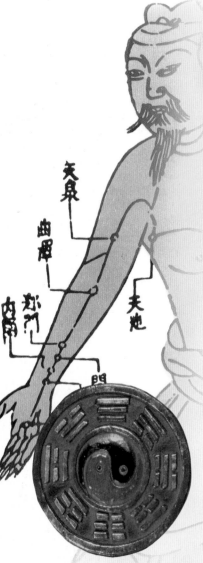

ACCORDING TO LEGEND, THE SAGES OF ANCIENT CHINA stumbled across something remarkable some 3500 years ago. They noticed that warriors who survived arrow wounds also made remarkable recoveries from long-term ailments once their wounds had healed. Was there some link between the two, they wondered? Accidentally, the sages had hit on the rudiments of a new form of medical treatment, and the science of acupuncture had been born.

Channelling the life force

Chinese faith in the medicinal value of acupuncture was bound up with their belief in a life force called *qi*. This, they held, flowed through the body along invisible energy channels known in English as meridians. Acupuncture was devised to correct any imbalance or stagnation of *qi*. To regulate the *qi*, sharp needles made of bone, bamboo or bits of ceramic were inserted into the appropriate points. In the West, acupuncture was first employed in the early 19th century to relieve pain and reduce fever. It has been used successfully in the treatment of arthritis, back pain and rheumatism, and recently it has also proved helpful in combating allergies, anxiety, sleeplessness and stress.

The healing power of herbs

As well as acupuncture, Chinese doctors used herbs to treat most ailments. Again, herbal doctors did not look for any single cause or symptom of illness in their patients; instead, they tried to find what they termed imbalances or patterns of disharmony. To do this, they not only determined the state of a patient's *qi*, they also assessed the balance of two other vital forces – the body's *yin* and *yang*. The former was held to be passive and female (white), while the latter was masculine and aggressive (black). As with *qi*, a balance between the two results in good health, while an imbalance is a cause of disease.

The first steps involved a careful examination of the tongue, checking its colour, texture, shape, size and coating, and of the pulse, among other things, to assess the state of the patient's *qi* and to identify the site of the disease. The appropriate remedy was then selected from more than 2000 substances available to doctors. Some ingredients, such as *gan cao* (liquorice), are still familiar today. The prescription sometimes changed weekly or even daily, depending on the patient's response. None of these beliefs has died – indeed, treatment with herbs and acupuncture has become a central strand of modern complementary medicine in the West.

SIGNS AND A SYMBOL *A traditional Chinese medical chart (top) signposts some of the 700 points in the body in which acupuncture needles can be inserted. The points lie along 14 commonly used energy channels called meridians, 12 of which are linked to individual body organs. To avoid illness, the flow of energy through the meridians must be kept in balance. An ancient lacquer ornament (above) shows the symbol of* yin *and* yang, *the two opposing forces in the body which dictate a person's state of health.*

A Book of Wounds

HOW DISEASE WAS DIAGNOSED AND TREATED IN ANCIENT EGYPT

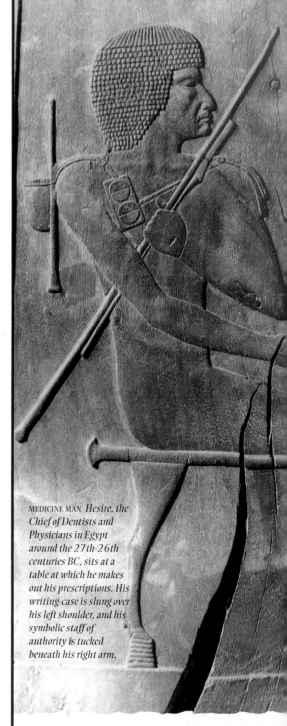

'INSTRUCTIONS CONCERNING A GAPING WOUND IN THE head, penetrating to the bone, smashing his skull and rending open the brain ...' So begins case number six in *The Book of Wounds*. This ancient Egyptian medical manual, written around 1600 BC, describes treatment for fractures, dislocations, wounds,

MEDICINE MAN *Hesire, the Chief of Dentists and Physicians in Egypt around the 27th-26th centuries BC, sits at a table at which he makes out his prescriptions. His writing-case is slung over his left shoulder, and his symbolic staff of authority is tucked beneath his right arm.*

tumours and other surgical disorders. The Greek historian Herodotus, writing in the 5th century BC, marvelled at the sheer number of doctors in Egypt, and at their degree of specialisation. 'Every physician is for one disease and not several', he wrote, 'and the whole country is full of physicians; for there are physicians of the eyes, others of the head, others of teeth, others of the belly, others of obscure diseases.'

The respect for medicine started at the very top in the royal household itself. One of the most powerful men in royal circles was the Superintendent of Physicians – and there was even a highly regarded specialist who regulated the pharaoh's bowels.

Physicians went to special medical schools, where they were taught anatomy and herbalism. They also learned to read and write, and produced a number of manuals that have survived to the present day.

Healthy minds in healthy bodies
The Book of Wounds, for instance, instructs physicians to make the most of their sense of touch, to feel their patients' injuries with their hands and to diagnose tumours by comparing them with the texture of fruit. The manuals also set out some modern principles of diagnosis, recognising the link between states of mind and physical well-being. One doctor, for example, reports that his patient is 'too oppressed to eat', while another, noting his patient's despondency, remarks that 'his face is as if he wept'.

Egyptian physicians modelled the internal workings of the body on the network of irrigation canals that sustained the countryside around them. They proposed that the bodily network consisted of a system of vessels, or *metu*, originating in the heart, which was believed to carry round all the bodily

liquids essential to good health. Just as crop failure followed when irrigation canals were blocked, a blockage in the *metu* would lead to ill health. The ancient Egyptians gained at least a rudimentary knowledge of human anatomy and physiology, not through dissection – which was forbidden – but by paying close attention to the removal of bodily organs before the work of embalming and mummification took place.

Brain of pig, blood of bat
An ailing Egyptian might confidently expect his doctor to arrive armed with a vast array of healing potions and remedies, among them perhaps burned hoof of ass, or the fat of a black snake. Many medicines were derived from animals, notably pigs' brains, the spleen or liver of an ox and hippopotamus fat – while others had more exotic origins. Blood of bat, gall of tortoise and dried swallow's liver were just a few of the remedies suggested for possible treatment of eye diseases, which were common in ancient Egypt.

Drugs were classified according to their desired effects, rather than by their ingredients. One jar might be labelled 'To improve the hearing', and another 'To expel disease in the belly'. All carried strict instructions on how they were to be used. Some had to be mixed with wine, others made into cakes, and all had to be measured out with the greatest accuracy. Polypharmacy, in which several remedies are prescribed in combination, was common in ancient Egypt – and many treatments specified ointments and suppositories in addition to drugs which were taken by mouth.

SEEKING A CURE *Supported by a stick, a young man with a withered leg offers a goblet to the gods in the hope that a surgeon can cure his lameness. In ancient Egypt the art of surgery was handed down from father to son.*

MEDICAL MANUAL *In 1862 an American Egyptologist, Edwin Smith, unearthed* The Book of Wounds *(left) in Luxor. It is one of dozens of ancient medical books which contain treatments for ailments from toothache to ulcers.*

TOOLS OF THE TRADE *A carved tablet (left) shows some of the instruments used by Egyptian surgeons, who routinely treated fractures and broken bones, as well as lancing abscesses and cutting out small tumours. They could also perform more complicated operations, such as amputations.*

'Nature acts without doctors!'

HOW HIPPOCRATES LAID THE FOUNDATIONS OF SCIENTIFIC MEDICINE

THE MEDICAL SCHOOL ON THE AEGEAN ISLAND OF COS was renowned throughout ancient Greece, for it was here that Hippocrates, the greatest doctor of his day, was a lecturer. Hippocrates, who was born on Cos in about 460 BC, the son of a doctor, taught the

students everything he knew about illnesses, their causes and their treatment. Instead of concentrating only on the affected part of the patient's body, he took account of the person's health as a whole, and was the first to forecast the likely outcome of a disease. He also asserted that the same illness might need different treatments depending on the sick person's build, temperament and age. His method of diagnosis paid as much attention to people's way of life, as it did to their symptoms: on one occasion, he reputedly cured King Perdiccas of Macedonia – after the king's own doctor had sought in vain for a physical cause of the illness – by treating him for mental stress, which he relieved with herbs.

BEDSIDE MANNER *Hippocrates taught his students that in order to inspire confidence in their patients they should look clean, healthy and well-fed, and adopt a kindly but grave demeanour.*

A PROFESSIONAL CODE OF CONDUCT

ONCE HIPPOCRATES' STUDENTS HAD QUALIFIED AS DOCTORS, they gathered under the plane tree which had provided shade for many of their lessons. There they took the so-called Hippocratic Oath, an ethical code attributed to the great physician, which, in modified form, is still used in graduation ceremonies in many of today's medical schools. It includes the following solemn pledges:

'I will use my skill to help the sick to the best of my ability and judgment. I will not harm or wrong any person by it.'

'I will not give a fatal draught to anyone if I am asked, neither will I suggest such a thing.'

'I will be chaste and religious in my life and in my practice.'

'Whenever I enter a house, I will go to help the sick and not with the intention of doing anyone injury or harm. I will not abuse my position to indulge in sexual relations with women or with men, be they masters or slaves.'

'Whatever I see or hear, whether professionally or privately, I will keep secret and tell to no one.'

'If I observe this sacred oath, and do not violate it, may I prosper in both my life and my profession, earning good repute among all men. If, however, I transgress and forswear this oath, may my lot be otherwise.'

'I will not cut, even for the stone, but I will leave such procedures to the practitioners of that craft.'

'FATHER OF MEDICINE' *Hippocrates' writings, and those of his contemporaries, were collated to form the* Corpus Hippocraticum, *or* Hippocratic Writings, *a 70-volume work containing their views on the practice of medicine.*

BATTLEFIELD FIRST AID *'War wounds,' Hippocrates stated, 'are best treated by those experienced in warfare.' Here, the legendary Greek hero Achilles bandages the wounds of his friend Patroclus.*

Nature's healing powers

Hippocrates emphasised the value of leading as orderly a life as possible, and gave classic descriptions of the symptoms of most common diseases. He advised on mental illness, for which bleeding and laxatives, as well as psychotherapy, were prescribed; and surgery, in which cauterisation, or burning with a hot iron, was sometimes used to heal wounds. He also wrote extensively on dislocations and fractures.

In many cases, Hippocrates blamed illness on a mixture of impure air and an unhealthy diet. To safeguard against these, he advised patients to eat more fresh fruit and vegetables. And he sometimes dismissed surgery or drugs altogether, instead advising patients to let the body cure itself. 'Nature,' he stated, 'acts without doctors!'

Hippocrates became noted for such sayings, 406 of which were later collected in his book of aphorisms, which was kept in the library of the Cos medical school. They included nuggets of advice that have since become household phrases, such as 'Desperate ills call for desperate remedies.'

The main weakness in Hippocrates' teaching lay in his lack of knowledge of human anatomy. This was because human dissection was taboo, as the ancient Greeks revered the dead, so animals were used.

Hippocrates died in Thessaly, in central Greece, in about 370 BC. However, his reputation continued to grow even after his death, until it was said that honey from the bees that made their hives near his grave was a miraculous cure for any ailment.

A lowly occupation

HOW ROMANS CAME TO TRUST THEIR DOCTORS

FOR THE FIRST SIX CENTURIES OF THEIR HISTORY THE Romans had no use for doctors, but they embraced the healing gods of other countries, notably Asclepius, whose cult arrived from Greece in 293 BC. When Greek doctors began emigrating to Rome in the 2nd and 1st centuries BC, the Romans suspected them of being frauds and fakers, and this was not far from the truth: quacks and opportunists abounded, for regulations were non-existent and anyone who chose to could call himself a doctor.

An exception was the Greek physician Archagathus, who arrived in Rome in 210 BC. His outstanding surgical skills quickly won over the sceptics, and he was welcomed into Roman society. But his reputation was not to last. After a succession of medical blunders and painful treatments he was dubbed *carnifex*, 'butcher'.

The atomic theory of healing

Asclepiades, who came to Rome in the 1st century BC, was more successful, and his moderate treatments improved the reputation of his profession. He opposed the accepted theories of Hippocrates by crediting doctors, and not nature, with the power to heal. Unusually for the time, he believed that disease should be treated 'safely, quickly and pleasantly', favouring such treatments as massage, diet, soothing medications – including wine – and singing. He based his methods on solidism, or the 'atomic' theory, originally advanced by Erasistratus in the 3rd century BC. This fanciful notion proposed that the body was made up of solid particles (atoms) which flowed through it continually. Sickness would result if their smooth movement was hindered due to the skin's pores being either too constricted or too relaxed.

Asclepiades' pupils Themison and Thessalus later worked these principles into a more formal healing system, which they christened Methodism. Thessalus, however, rejected formal medical training, claiming that anyone could learn all about medicine in just six months. As a result, many unsuitable people became doctors, blackening the good name that medicine had just begun to acquire.

Roman attitudes to doctors fluctuated widely over the centuries, but eventually suspicion and contempt softened into grudging respect. Some Romans even practised medicine themselves, though the majority of physicians remained foreigners – Greeks or Egyptians – and were mainly slaves or freed slaves.

CURE-ALLS *Roman dispensaries were piled high with medicinal plants and drugs: a typical remedy might call for hundreds of different ingredients.*

HEALING TOUCH *A Greek physician examines a youth. The oversized cupping vessel (right) is an ancient symbol of the medical craft.*

CULT FIGURE *The Romans built many temples to Asclepius, the Greek god of healing. Here, the god is shown bleeding a patient (left) who has been bitten on the arm.*

Dissecting apes for the sake of mankind

HOW GALEN EXPLAINED THE WORKINGS OF THE HUMAN BODY

MEDICAL GIANTS *The ideas of Galen (left) and the 5th-century BC Greek physician and surgeon Hippocrates (right) were influential for more than a thousand years, as the two men's presence in this 13th-century fresco shows. Though both men advanced medical knowledge, they believed that the humours governed people's well-being.*

THE SILENT PIG *In order to increase his standing in Rome, Galen held public dissections. One of his demonstrations was to dissect the nerves in the neck of a live pig, as shown in this woodcut from a 16th-century edition of his works. As he sliced through the nerves one by one, the pig squealed loudly. However, when he severed the nerve running from the brain to the larynx, the animal fell silent, proving that the brain – and not the heart, as Aristotle had claimed – governed the production of sound.*

GALEN, THE GREATEST PHYSICIAN OF THE ANCIENT world, was renowned as much for his arrogance as his talent. 'Never have I gone astray, whether in treatment or prognosis, as have so many other physicians of great repute,' he boasted. 'If anyone wishes to gain fame, all that he needs is to accept what I have been able to establish.'

Galen was born in AD 129 in Pergamon (present-day Bergama in Turkey). He studied medicine for 12 years, beginning at the age of 16. By the time he was 21, he had written six medical works and had begun to practise what was to become his speciality – the dissection of living animals.

In Galen's day, human dissection for medical research was considered unethical, so he had to gather his knowledge of anatomy from animals.

He also learned much about the human body – and became an expert in treating injuries working as a physician at a gladiatorial school in Pergamon.

Enemy of doctors, darling of patients

In 161, at the age of 32, Galen went to Rome, where his reputation as a brilliant practitioner had preceded him. Though he enhanced his reputation by successfully treating patients who were considered incurable, he antagonised his fellow physicians by accusing them of ignorance and greed. But his patients – many of them influential – revered him, and Galen soared to the top ranks of Roman society.

During his years in Rome, Galen made the majority of his important anatomical discoveries. He furthered his knowledge by dissecting more animals,

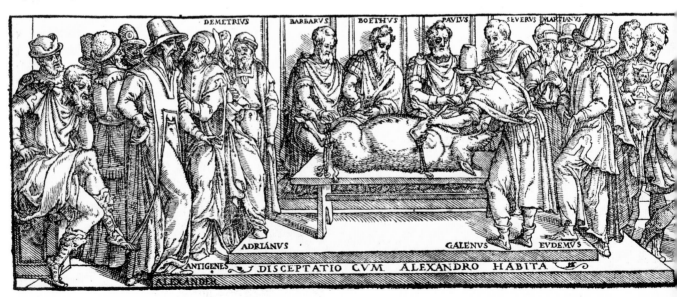

DEMETRIVS · BARBARVS · BOETHVS · PAVLVS · SEVERVS · MARTIANVS

ADRIANVS · GALENVS · EVDEMVS

ANTIGENES · DISCEPTATIO CVM · ALEXANDRO HABITA

ALEXANDER

in particular Barbary apes, which he believed to be built in much the same way as humans. One of his main achievements was to identify seven pairs of nerves originating in the brain, and he proved beyond doubt that the arteries contained blood, and not air, as had been thought for some 400 years.

'You can dissect an ape and learn each of the bones from it ... For this, you must choose apes that mostly resemble men ... in these apes which also run and walk on two legs, you will also find the other parts as in man.'
GALEN

Galen was not always correct in assuming that discoveries made by dissecting animals could be applied to humans, however. He attributed an important function to the *rete mirabilis*, a network of veins at the base of the brain. He did not know that this organ, present in cattle, did not exist in humans.

Galen also failed to understand the function of the heart, and did not realise that blood circulates around the body. Above all, his entire medical system remained based on the theory of humours, which had been conceived by ancient Greek physicians – even though his own discoveries about the workings of the human body sat uneasily with these traditional views. But many people were convinced by Galen's proud contention that he always knew the answers, and his medical theories – mistakes and all – were to go unchallenged until the 17th century.

THE FLUIDS THAT CAUSED DISEASE

THE PRACTICE OF 'READING' A PATIENT'S CHARACTER and treating him accordingly was laid down in the 'theory of humours'. This theory – believed by the ancient Greeks – stated that physical and mental health were dictated by four bodily fluids, or humours. These were: blood, phlegm or 'mucus', yellow bile and black bile. For good health, these should be evenly balanced: illness resulted when any one of the humours dominated the others.

Striking the right balance
Once a patient's illness had been diagnosed according to the humours, he had to try to restore the equilibrium. A patient whose disorder was ascribed to an over-sanguine constitution might be bled of his surplus blood, while those who were introspective and moody might be encouraged to sweat out the black bile thought to be causing this.

But a balance of the humours was not always easy to achieve. Much depended on a person's sex, age, diet, the weather and the time of year. For example, it was thought that phlegm tended to accumulate in the dark, chilly winter, causing sore throats and colds, while blood dominated in spring and summer causing sickness and vomiting. A balanced diet was regarded as important, and patients were advised not to eat or drink to excess.

In the 2nd century AD, Galen added his weight to the theory, claiming that each of the four humours caused certain temperaments in people. His belief in the theory of humours helped these ideas to spread throughout Europe and doctors adhered to them for centuries. Only in the 1800s was the idea that humours governed health discredited.

SANGUINE *Passion, courage and optimism were attributed to an excess of blood in the system.*

PHLEGMATIC *Calm, unexcitable or apathetic people were thought to be dominated by phlegm.*

CHOLERIC *Too much yellow bile was said to make you bad-tempered, yellow-faced, lean and hairy.*

MELANCHOLIC *Someone who was constantly gloomy or depressed was believed to have too much black bile.*

HEALING HANDS *Resting his hands lightly on his patient's abdomen, a doctor seeks to check the balance of the humours.*

179

Forming the ideal family

HOW MEN AND WOMEN OF DIFFERENT NATIONALITIES PRACTISED BIRTH CONTROL

THROUGHOUT THE AGES MARRIED COUPLES' DESIRE FOR children has sometimes come into conflict with an equally strong wish to limit the number of births. Fewer children generally meant a higher standard of living for a family. The property-owning ancient Greeks, for example, had strong views on what constituted the ideal family: parents who stayed together; one son and heir to maintain the family name and fortunes; and one daughter to make a profitable alliance with another, similar family.

The Greeks kept their families small in many ways. The most common was abortion, usually performed by midwives. Some women preferred drinks made from Cretan poplar or wild cucumber, which brought on violent vomiting and subsequent miscarriages. Repeated jumping was another means of terminating an unplanned or unwanted pregnancy.

Drinks and suppositories
There were also 'antifertility' drinks for women, made by dissolving a small piece of copper sulphate in water. This method was recommended by the Greek physician Hippocrates in the 4th century BC, who claimed that it guaranteed freedom from pregnancy for a year. Another drink made from the crushed seeds of parsley could – as modern tests have shown – block the production of progesterone, the hormone that readies the uterus for the implantation and growth of the fertilised egg. In addition, suppositories impregnated with honey or pepper were inserted in the womb, as were lead or frankincense ointments mixed with olive oil.

Barriers to pregnancy
Vaginal douches were used after intercourse to wash away semen; and many Grecian couples practised *coitus interruptus*, or interrupted intercourse. This method is related in the Bible, when Onan, ordered by his father to have intercourse with his brother's widow, withdrew from her and 'spilled his seed upon the ground'.

Unlike the Greeks, the ancient Romans favoured large families. Even so, some birth control was practised – especially by women who had married early. To guard against pregnancy, they used vaginal plugs containing alum, cedar gum and peppermint. The Romans also controlled the size of their families by killing or abandoning unwanted babies. This practice remained popular and was common in many European countries until the 18th century.

The first recorded use of contraceptives occurred in ancient Egypt. A papyrus dating from 1500 BC described how women who did not want to become pregnant used vaginal plugs. Those consisting of crocodile dung mixed with sour milk and honey were particularly effective – the acidity in the milk was a natural spermicide, while the honey acted as a barrier. Lint plugs soaked in honey and liquid made from acacia tips also proved useful.

The first condoms
Pessaries and other methods of birth control continued to be used over the centuries in north Africa and Europe. The 12th-century Arab physician Avicenna described many of them, including the age-old practice of jumping backwards and sneezing in order to bring about a miscarriage. The sheath or condom, described by the Italian anatomist Gabriel Fallopius in the 16th century, was initially used primarily to stop the spread of syphilis. It was sewn from medicated cloth and fitted over the penis. The protection offered by condoms made them popular with 18th-century rakes, such as Dr Johnson's friend and biographer James Boswell – and they were available in high-class brothels from London to St Petersburg. In England condoms were known as 'French letters', while in France they were called *les capotes anglaises* – 'English hoods'. Diaphragms – caps that fitted over the neck of the womb – were introduced in 1823; and with the introduction of vulcanised rubber in the mid 1840s, rubber condoms became available to the general public.

ROOT OF PASSION *The branched root of the purple-flowered mandrake plant – the shape of which resembles a man's body – was a 15th-century aphrodisiac. In keeping with the mysticism surrounding fertility drugs, the plant was said to scream in agony when it was pulled from the ground.*

COLD SHOULDER *The message implicit in this ancient Chinese silk painting dating from the late 4th century is simple: to avoid the risk of unwanted pregnancy, men and women should take care to keep well apart.*

'I am the Sky, you are the Earth'

HOW PEOPLE HAVE TRIED TO INCREASE THEIR FERTILITY

SINCE THE EARLIEST TIMES MEN AND WOMEN HAVE TRIED many means to aid reproduction. Some 30 000 years ago, for instance, people living in what is now Austria revered the so-called 'Venus of Willendorf', a carved limestone figurine 4 in (10 cm) high. Sexuality and pregnancy are displayed in the swollen abdomen, buttocks and breasts of the statuette, which may have acted as an erotic fertility charm.

Fertility rites and unusual aphrodisiacs

Early man believed that reproduction and growth, both of people and crops, was governed by such otherworldly beings. When the corn sprouted it was because Mother Earth had been impregnated by Father Sky, and an old Hindu marriage ritual contained the statement, spoken by the bridegroom: 'I am the Sky, you are the Earth.'

For many people, fertility was not a problem – their main concern was to limit the number of births. More children only meant more mouths to feed. But others needed help to conceive, and for these folklore was rich in remedies. It was a common belief that women were more likely to become pregnant if they had intercourse with their menfolk upon ploughed land. In England some couples made love on the surface of symbolic chalk giants cut into hillsides, including the nude male figure overlooking Cerne Abbas in Dorset. The Cerne Giant is thought to be a British version of the hero-god Hercules, whose cult was probably brought to Britain by the Romans.

The Greeks took a more scientific approach to fertility practices and rituals, producing a large body of medical writings on the subject. Good health, they believed, was the key to fecundity: a diet rich in nuts, pulses and cereals would make pregnancy more likely. They also advocated the use of aphrodisiacs, as did the Romans. Ingredients included elephant's trunk, vulture's lung and hairs from a she-mule's tail.

With the decline of the Roman Empire, the use of aphrodisiacs as an aid to fertility moved to north Africa. The Arabs published texts giving recipes for drugs said to increase sexual appetites and improve sexual techniques.

How to conceive boys every time

By the Middle Ages Europe had its own collection of fertility books, including *The Treasury of Health*, written in the 13th century by Petrus Hispanus, the future Pope John XXI, containing 56 prescriptions to aid fertility. Other books contained advice on curing sexual problems and on ways to guarantee the kind of children – boys – that most readers wanted. The manual *Tableau de l'Amour Conjugal* by Nicolas Venette, first published in France in 1686, recommended horse riding as a cure for impotent men; and the anonymous English authors of *Aristotle's Master Piece*, published in 1684, recommended that women wanting boys should have intercourse lying on their right sides, because males would grow only in the womb's right-hand side.

ELIXIR OF LOVE *A Roman magician prepares a love potion: it might contain nettles or thistles in wine.*

HOLY HELP *Even after conception, prayers were needed to guard against stillbirth.*

KITH AND KIN *Pierre de Moucheron (centre), a wealthy French businessman, enjoys a meal with his 19-strong family. High mortality rates in the mid 16th century made such large families a rarity.*

Spilling blood in pursuit of a cure

HOW BARBER-SURGEONS PRACTISED THEIR TRADE DURING THE MIDDLE AGES AND BEYOND

MEDICINE SHOW *Sitting on his horse, a 17th-century Dutch travelling surgeon holds his audience spellbound as he proclaims the marvels of his surgical prowess. Behind him, a huge billboard shows operations for hernia and removing stones and cataracts.*

DURING SPRINGTIME IN THE MIDDLE AGES, PEOPLE visited barbers' shops in towns and cities across Europe not necessarily to have a haircut or a shave – some of them went specifically to have a tooth pulled out or a pint of blood drained from an arm. People believed that an imbalance of the humours – blood, phlegm, yellow bile and black bile – was harmful, and that draining surplus blood each spring – the season of renewal – was the way to redress the balance. 'Let out the blood, let out the disease', was the principle behind the practice which, despite being potentially dangerous, was the most popular treatment for many serious illnesses for several centuries.

The job of letting blood had always been carried out by people with only the most basic surgical training, supervised by doctors. During the 13th century barbers began to take over this task,

and gradually they evolved into a professional body and widened their range of services to include minor medical treatments, such as lancing boils, dressing ulcers and extracting teeth. Some specialised in more serious surgery, treating cataracts or hernias. In smaller towns, the barber-surgeon might be called upon to perform relatively major operations, such as setting fractures or even trepanning – cutting away a portion of the cranium to relieve pressure on the brain. Initially, anyone who wanted to set himself up as a surgeon could do so without a licence, but by the 14th century in some parts of Europe, such as France and Italy, universities and guilds were beginning to regulate and license medical practice.

Shops for bloodshed

Most barber-surgeons conducted their business from a shop, often advertising their services with a sign graphically depicting a hand held aloft and blood dripping into a bowl. Bleeding bowls were made from tin-glazed earthenware, pewter or silver, and were usually marked on the inside to indicate how many ounces of blood had been drawn.

To let blood, the barber first soaked the patient's hand in warm water to swell the veins, making them easier to see. He then tied a tourniquet round the patient's upper arm and, with his bleeding bowl at the ready, decided which of five major veins – each believed to be associated with a separate vital organ – to cut. Gripping the patient's hand firmly in a napkin, the barber pierced the vein with a double-bladed lancet. When enough blood had dripped into the bowl, the barber-surgeon bound the wound loosely and sent the patient home.

Skills were passed down from generation to generation, so a trainee barber-surgeon would start off as an apprentice to a master, often his own father, or her own husband, and would learn all the necessary medical practices. In large cities and towns, such as London, Padua or Paris, however, an apprentice might be allowed to attend the same anatomy classes as student physicians.

A medical hierarchy takes shape

As well as the barber-surgeons, there were other types of medical practitioner in the Middle Ages, all of whom belonged to a company or guild, such as the Company of Barber-Surgeons, established in London in 1540. At the top were 'doctors of the long robe' – university-trained physicians, then master surgeons who studied Latin and performed serious operations; and at the bottom were barber-surgeons and wig makers. By the mid 18th century the gap between barber-surgeons skilled in surgery and those better suited to keeping barber shops was widening. In 1741, the London Company of Barber-Surgeons was dissolved, paving the way for a new breed of general practitioners.

PURGE OR CURE? *Grasping a pole to promote bleeding, a nervous patient watches blood from his right arm being let into a bowl by his barber. The amount of blood drained depended on the ailment.*

'DISCS FOR THE EYES'

THE 13TH-CENTURY ENGLISH SCIENTIST AND SCHOLAR Roger Bacon was the first person to suggest the use of pieces of curved glass to correct faulty vision. He did not develop the idea, but later that century, in about 1280, it was taken up independently in Italy by a Florentine monk called Alessandro di Spina. He produced a pair of spectacles, or eyeglasses, comprising two convex lenses joined together by a wire frame. The monk designed the spectacles as an aid for long-sightedness only, and the first written record of them was made in 1306 – a sermon preached in Florence that refers to the recent discovery of 'discs for the eyes'.

The first bifocals are invented

At first spectacles were extremely expensive, and were sometimes included in portraits to indicate the wealth of the sitters. The first such portrait was that of Hugh of Provence, painted in 1352 by the northern Italian artist Tommasa da Modena.

Spectacles with concave lenses, for aiding short-sightedness, did not appear until the middle of the 15th century, again in Italy. From then on the demand for reading glasses grew with the spread of printed books. In 1775, the American statesman and scientist Benjamin Franklin invented the first pair of 'bi-focals'. These combined both convex and concave lenses in one pair of glasses, and could be used for close or distant vision.

Contact lenses – small lenses which could be placed directly on the eye – were first suggested by the English physicist Sir John Herschel in 1827. However, they were not developed until 60 years later, when a Swiss scientist, Dr Eugen Frick of Zurich, produced lenses that were sufficiently delicate and precise.

STUDY OF A SCHOLARLY SAINT *A pair of eyeglasses dangles from St Jerome's desk in this imaginative portrait painted in 1480 by Domenico Ghirlandaio. The 4th-century scholar was adopted as the patron saint of the Guild of Spectacle-Makers.*

The art of anatomy

HOW RENAISSANCE ANATOMISTS LEARNED ABOUT THE HUMAN BODY

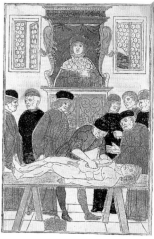

AFTER THE CRIMINAL WAS CUT DOWN FROM THE scaffold his body was put in a cart and sent to the local university in Padua. At the medical school, Andreas Vesalius, the 23-year-old anatomy lecturer, dissected the corpse in front of an audience of students and fellow teachers. Deftly, he revealed the fine network of nerves that ran from the brain down the spinal cord. The dissection, which was a public spectacle in 16th-century Padua, was believed to be the criminal's final punishment. Five years later, in 1543, Vesalius published the fruits of his research in *De humani corporis fabrica libri septem* (*The Seven Books on the Structure of the Human Body*), a huge, painstakingly illustrated work credited with bringing about the rebirth of anatomy as a science.

Vesalius – who had started dissecting animals as a child – relied on his own observations to draw a series of fascinating conclusions. His findings aroused great interest in medical and academic circles, because they disproved many long-held assumptions about the human body. Although his own teacher condemned Vesalius' work, its worth was recognised by younger scholars who followed his lead and pursued their own investigations, gradually building up a complete picture of the intricate workings of the human machine.

UNDER THE KNIFE *An assistant demonstrates the dissection of a corpse at a lesson directed from the chair by Mondino de Luzzi at Bologna University. In fact, unlike most of his peers, Mondino often did his own dissections. His* Anatomia, *published in 1316, helped to popularise the study of anatomy.*

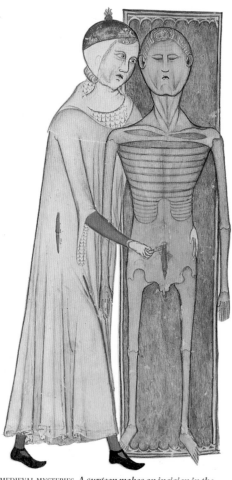

MEDIEVAL MYSTERIES *A surgeon makes an incision in the belly of a corpse. When this drawing was made in 1345, anatomy was a brand-new study in the West. Italian universities pioneered the subject despite a barrage of ethical and moral objections.*

LAID BARE *The internal organs of a woman have been delicately rendered by Leonardo da Vinci. He drew more dissections than any other artist of his time, but he wrote: 'Even if you have an interest in anatomy, you may perhaps be deterred by natural repugnance, or perhaps by the fear of passing the night hours in the company of these corpses, quartered and flayed, and horrible to behold.'*

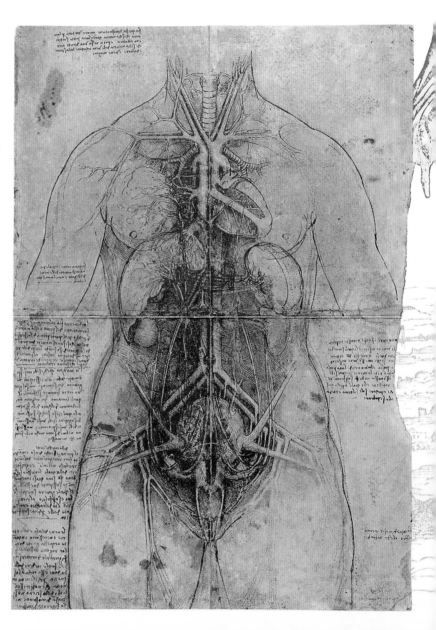

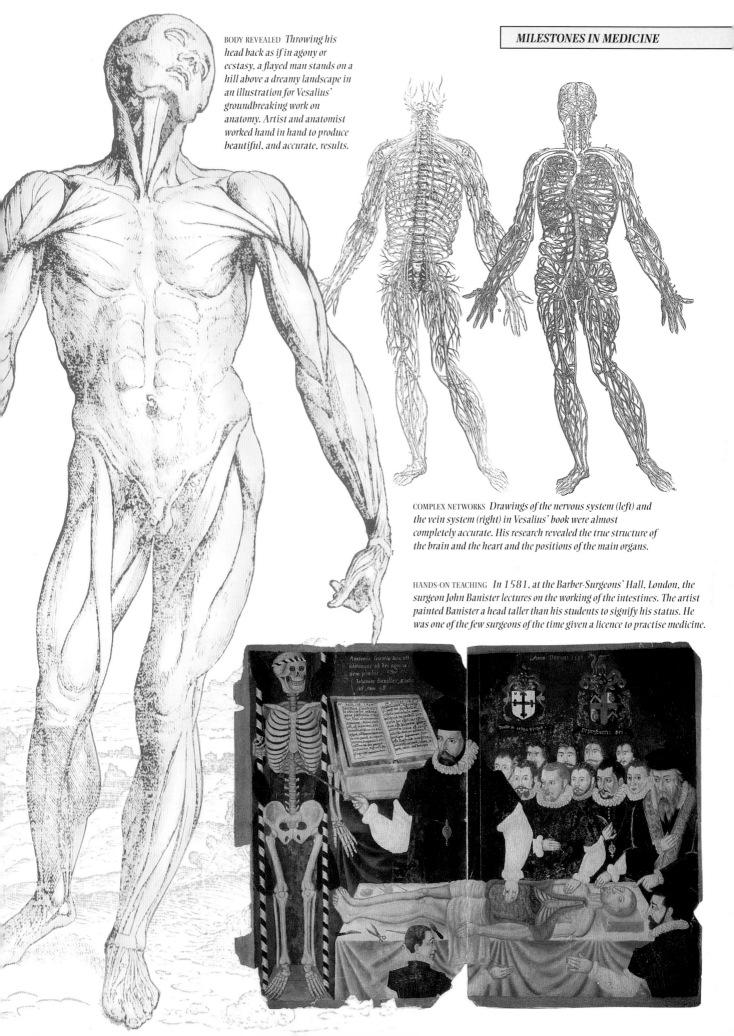

BODY REVEALED *Throwing his head back as if in agony or ecstasy, a flayed man stands on a hill above a dreamy landscape in an illustration for Vesalius' groundbreaking work on anatomy. Artist and anatomist worked hand in hand to produce beautiful, and accurate, results.*

COMPLEX NETWORKS *Drawings of the nervous system (left) and the vein system (right) in Vesalius' book were almost completely accurate. His research revealed the true structure of the brain and the heart and the positions of the main organs.*

HANDS-ON TEACHING *In 1581, at the Barber-Surgeons' Hall, London, the surgeon John Banister lectures on the working of the intestines. The artist painted Banister a head taller than his students to signify his status. He was one of the few surgeons of the time given a licence to practise medicine.*

Solving the mystery of the human heart

HOW WILLIAM HARVEY UNLOCKED THE SECRET OF BLOOD CIRCULATION

WHEN WILLIAM HARVEY PUBLISHED HIS REVOLUTIONARY book on the circulation of the blood in 1628, his reward was vicious criticism and abuse from fellow physicians. The reason was simple. In his *Exercitatio Anatomica de Motu Cordis et Sanguinis in Animalibus (An Anatomical Exercise Concerning the Motion of the Heart and Blood in Animals)*, Harvey challenged beliefs that had been accepted by the medical establishment since they were put forward by the Greek physician, Galen, 1400 years earlier, in spite of growing evidence against them.

Galen's explanation for one of the age-old mysteries – the role blood played in the human body – suggested that the body continuously made and then consumed large amounts of blood and thus the nutrition gained from food. He believed that blood from the arteries was used to 'top-up' the blood carried in the veins, passing from one to the other through tiny holes in the heart, from the right side to the left.

Harvey first heard this theory questioned as a medical student in around 1600. Having qualified as a doctor in 1602 and returned to England, he started a medical practice in London, later becoming a consultant physician at St Bartholomew's Hospital. His scepticism about Galen's theory led to 14 years of experiments and theorising before he arrived at a workable idea of how blood circulated. In 1616 he announced the results. Blood, he said, travels from the left of the heart through the arteries to the tissues, returns via the veins and enters the right side of the heart, travels on to the lungs and finally arrives back at the heart's left side.

CAREFUL DOCTOR
William Harvey backed up his radical theories with meticulous experimental proof.

STEMMING THE FLOW
The bulging veins of a tightly bound arm (above) can be deflated by the pressure of a single finger (below), proving that blood flows in only one direction.

MEDICAL DETECTIVE *Italian biologist Marcello Malpighi used a powerful microscope in 1661 to prove that Harvey's theory about tiny blood vessels was correct.*

Heartbeats provide the breakthrough

Harvey went on to conduct further experiments on live snakes, sheep and human corpses in an attempt to prove his hypothesis. Then he noticed when dissecting some of the animals that their hearts continued to beat after they had been removed from their bodies. This convinced him that the heart was a muscle which contracted to pump blood through the veins. 'The movement of blood occurs constantly in a circular manner,' he noted, 'and is the result of the beating of the heart.'

'Twas believed by the vulgar he [Harvey] was crack-brained, and all the physicians were against his opinion.'
JOHN AUBREY, 17TH-CENTURY ENGLISH WRITER

He also established that the left ventricle of the human heart contains 2 oz (60 g) of blood and that it beats between 60 and 200 times a minute, pumping out 4 lb (1.8 kg) of blood. This processed a colossal amount of blood – far more than the body could produce from eating food and drinking liquids. From this, he deduced that the flow of blood was contained in a closed system and was constantly being recycled rather than replaced. In an experiment on a live snake, Harvey showed that when a vein was tied with

A breath of understanding

HOW THE WAY LUNGS WORK WAS DISCOVERED

AS THE TWO MEN WATCHED, THE LARK IN THE BELL JAR began 'manifestly to droop and appear sick'. Within ten minutes, it was dead. The hands that had placed the bird in the jar were those of Robert Hooke, the 24-year-old assistant to the natural philosopher and chemist Robert Boyle. The men were conducting an experiment at Boyle's laboratory in Oxford to test the effect lack of air would have on living creatures. Using a pump, they slowly extracted all the air from the jar in which the lark was trapped. The bird's death, Boyle concluded, showed that 'there is some use of the Air, which we do not yet so well understand, that makes it so continually needful to the Life of Animals'.

Breathing 'to cool the heart'

Although the year was 1659, the accepted idea of the purpose of respiration had not changed since the 2nd century AD, when Galen had put forward the notion that the reason that animals breathed was simply to cool the heart with cold air.

Eight years after his work with Boyle, Hooke found by experiments on animals that the presence of air in the lungs was vital for life. But it was John Mayow, a Cornish physician, who came closest to explaining how lungs work. His research showed particles in the air were extracted by the lungs during breathing, and were absorbed by the blood for use by the body. Air exhaled by the lungs could not sustain life. A century later Joseph Priestley isolated the gas we now know supports animal life. He called it 'dephlogisticated air' – later known as oxygen.

AIR OF LIFE *Hooke and Boyle found that animals could not survive in the vacuum created by their air pump (above). Many years later (below) a lecturer repeated their experiment, demonstrating how a pet cockatoo dies when the air is pumped out of the globe where it is imprisoned.*

ANIMAL EVIDENCE *Malpighi was able to produce this detailed engraving of a frog's heart and blood vessels after observations based on Harvey's work. The lower part of the illustration is a microscopic cross-section of one 'cell' of the animal's lungs, showing the capillaries connecting arteries and veins which Harvey's theories said must exist.*

thread, the piece of vein between the thread and the heart emptied. But if an artery was tied, the section between the heart and the thread remained full of blood, while the section beyond it was empty. This proved that the blood flows in one direction; out from the heart through the arteries and back to the heart via the veins. It certainly did not flow from one side of the heart directly to the other.

Tiny vessels provide the proof

By 1628, Harvey had accurately established the workings of the heart and the movement of the blood. The only remaining puzzle was the apparent absence of any connection between the arteries and veins. Capillaries, minute blood vessels which supplied the missing link, were discovered by Marcello Malpighi four years after Harvey's death.

By the end of the 17th century Harvey's explanation of how blood circulated was an accepted medical fact. Galen's theory was seen as just another wrong turning in the progress of science.

Fighting off the plague

HOW ITALIAN CITIES TOOK STEPS TO KEEP DISEASE AND DEATH AT BAY

AS PLAGUE RAMPAGED THROUGH EUROPE IN THE 17TH century, leaving thousands of corpses in its wake, the cities of northern Italy – including Florence, Genoa, Milan and Venice – prepared to withstand its assault. The communities had been devastated by Black Death at intervals during the previous three centuries, and this time the Magistrates for Health were ready. Those who failed to cooperate faced severe punishment.

Health officials burst into people's homes to arrest anyone showing signs of plague, which took various forms. Bubonic plague was named from the buboes, or swellings, in the armpits and groin; septicaemic plague entered the blood and killed within 24 hours; and pneumonic plague attacked the lungs. All three were likely to end in death.

In search of a scapegoat

Some stricken families were locked in their houses along with their dogs, cats and livestock – which also fell prey to the disease. The doors were bolted and marked with a cross, and guards were posted to ensure that no one broke out. The isolated inmates obtained food by lowering baskets into the street. Later, when the risk of contagion had decreased, the houses were reopened and thoroughly disinfected with lime and burning sulphur.

Faced with a serious epidemic, doctors had little to offer victims except ineffective herbal medicines. One Italian physician and health officer blamed the spread of the disease on soldiers recently returned from abroad. They could be found by the 'unbearable odours of the rotting straw whereon they sleep and die' and were thrown into *lazzarettos*, or pest-homes, which one traveller called 'a replica of hell'. 'Here,' he lamented, 'you cannot walk but among corpses. Here you feel naught but the constant horror of death!'

Although the Italians did not know the exact means by which plague was spread – by fleas living on the rats which left ships calling at their ports – they made the connection between the disease and visiting ships. The authorities imposed a 5½-week period of quarantine – from *quaranta giorni*, the Italian for '40 days' – on the ships' cargoes and crews. At the end of this the vessels were thoroughly fumigated. Health officials also closed markets where there was a risk of contamination, burned suspect goods and instructed country justices how best to protect their communities from the plague.

It was all to no avail. Plague ravaged the cities, towns and countryside, and thousands of Italians died, as did people in other European countries at the time. As with the Black Death 300 years before, the epidemic of plague eventually petered out of its own accord – and life slowly struggled back to normal.

BEAUTIFUL ANTIDOTES *Elegantly crafted pomanders held spices or aromatic herbs intended to counteract the clouds of noxious air, emitted by rotting flesh or vegetation, thought to cause plague. The wealthy would wear them around the wrist or at the neck.*

UNTOLD HORRORS *As the plague spread throughout Europe, communities took desperate steps to try to avoid its consequences. Little could be done, however, and scenes of fearsome carnage were repeated from one end of the continent to the other. The plague struck Naples, in southern Italy, in 1656, and the market square soon became a mass mortuary for those who had succumbed to the dreaded disease, whose causes were still unknown.*

PLANETS, THE PLAGUE AND THE MISSING LINK

IN 17TH-CENTURY ITALY IT WAS SAID THAT the best remedy for the plague was 'run swiftly, go far and return slowly'. Doctors had no effective remedies against the disease. Some blamed 'miasmas' or polluted air, caused, they said, by a change in the weather combined with planetary influences, or from fumes given off by decaying corpses. In 1630, for example, a Dr Tadino of Milan asserted that, as Saturn and Jupiter were in conjunction, a fearful plague would occur – killing untold thousands of his fellow countrymen. Sure enough,

plague did break out – and it was swiftly transmitted from one person to another. At first, dogs and cats were blamed for the plague's progress and slaughtered in great numbers. No one realised that the dead rats littering the cities' streets were the 'missing link' they sought.

DOUBLE IMAGE *A Dr Zwinger of Basle, Switzerland, displayed this sign showing him in protective plague clothing (left). These included a beaked mask filled with spices, and a fumigating torch. He is also seen in his normal working attire (right).*

Battling for newborn life

HOW TRADITIONAL MIDWIVES DELIVERED BABIES SAFELY IN AN AGE OF SUPERSTITION

IN 1481, THE MIDWIFE AGNES MARSHALL STOOD BEFORE the Bishop's Court of York Minster, in the north of England, accused not just of lacking experience and skill, but also of the far more serious crime of using incantations during a birth. The court was preparing to hear her defence against the charge of witchcraft before deciding on her innocence or, if guilty, on the appropriate punishment to mete out.

This was a rare case – few midwives were ever accused of witchcraft – but until the 17th century, childbirth in Europe was an event steeped in superstition. The very act of giving birth was thought to defile the mother. For their part midwives were granted a licence to practise only after stringent examination by a church court.

In such examinations, maturity was considered an advantage. Aspiring midwives were required to show evidence of good character, and had to have borne a child themselves. Midwives were forbidden by the church to perform any act leading to a miscarriage or abortion, and were urged to do their best to make an unmarried woman name her child's father.

Hard labour and a stiff drink

The birth itself was always a hectic affair: friends and relatives who had themselves had children were summoned to witness the proceedings, as they could provide useful support and advice. Birth took place in a 'lying-in' chamber, usually a hastily prepared bedroom in which the curtains were drawn and the keyholes blocked to exclude light and air. Once the candles were lit the room became, symbolically, a different place. The midwife mixed a caudle – a hot, spicy drink based on ale or wine – for her charge. She also told the labouring mother-to-be when to sit or stand, rest or push, when to take a sip of caudle and when to use the 'birth stool' – a low seat on which the woman could squat.

After the baby was born, the mother underwent a month's 'lying-in', a period of transition during which the chamber was gradually dismantled. For the first week the mother remained in bed, visited only by other women, and the baby was taken to church by the midwife to be baptised. After this, the mother received fresh bedclothes and could carry the baby around the room. Eventually she was allowed access to the whole house and was able to meet male relatives. The 'lying-in' finally ended when the mother was led to church by the midwife. There, the ceremony of 'churching' purified her and enabled her to resume normal life.

MAGICAL PROTECTION *Before the 20th century, childbirth was a difficult and dangerous affair. This Egyptian magical amulet from the 13th century BC was supposed to help the wearer to give birth safely and easily.*

TAKING A BREAK *With the baby successfully delivered, relatives of the mother – known as gossips – who have gathered at this 17th-century Dutch home, chat among themselves, while a maid serves sweetmeats.*

Letting men into the secrets of childbirth

HOW FORCEPS AND BETTER HYGIENE MADE GIVING BIRTH SAFER

RESTRICTIVE CLOTHING
A plaque on a 15th-century hospital for children in Florence shows a child with swaddling clothes. These long strips of cloth were wrapped around a newborn baby before it was handed back to its mother. They would frequently stay on for months.

AS THE NIGHT WORE ON, IT BECAME APPARENT THAT THE delivery was going wrong; the baby was no longer alive. The midwife decided to break with the tradition that only women could attend a birth, and to send for a surgeon. Saving the mother's life was now paramount; all the doctor could do was extract the dead child.

A tool to help delivery

Until the 17th century, men were customarily barred from the delivery room except in emergencies. However, this started to change in Britain with the rise to prominence of the Chamberlen family.

Early in the century, Peter Chamberlen, the first of the Chamberlen 'man midwives', as they became known, invented a new instrument for extracting babies during difficult deliveries: the forceps. This instrument had two curved, hollow metal blades which could be inserted separately into a woman's pelvis, and fitted round the baby's head, allowing the doctor to pull the child out with far less risk of injury than had previously been possible. But the forceps remained a family secret until Hugh Chamberlen, grandson of Peter, revealed it to one Rogier van Roonhuysen in 1727, probably in return for having his debts written off. By the time the first illustrated account of the use of forceps was published in 1733, the demand for skilled practitioners – all of whom were male – was beginning to increase.

But childbirth remained a risky business, especially for those who opted for a hospital delivery. Puerperal fever killed up to 35 per cent of women who gave birth in hospital. After delivery, a woman would develop severe abdominal pain and a high fever, and death soon followed. The condition, caused by bacteria, could sweep through a ward, killing mother after mother. It was not until the 1860s that medical science started to solve the problem.

Please wash your hands

The great breakthrough was made by the Hungarian obstetrician Ignaz Semmelweis, working at Vienna's lying-in hospital. He noted three times as many deaths from puerperal fever in a ward visited by doctors and students than in one attended only by midwives. He was also interested when a colleague, who had dissected a corpse, died from a condition with symptoms identical to those of puerperal fever. Semmelweis reasoned that the problem was caused by the practice of leaving the dissecting rooms without washing the hands. Those who did this transferred something dangerous from the corpses to the mothers. Accordingly, he made everyone on his wards wash their hands in a chlorinated lime solution before tending patients. The results were dramatic: the death rate in maternity wards dropped from 16 per cent to 2 per cent. His work was savagely rejected by the medical establishment, however, in part because of personal rivalries, and it was only in the 1890s that modern germ theory vindicated him.

HALF MAN, HALF MIDWIFE *Although male midwives became fashionable in some parts of Europe during the 1700s, the idea was still alarming to many, as Isaac Cruikshank's cartoon 'A Man-Mid-Wife' demonstrates.*

The 'godlike delight' of saving lives

HOW A SOCIETY HOSTESS INTRODUCED SMALLPOX INOCULATION TO BRITAIN

EARLY IN 1715, LADY MARY WORTLEY MONTAGU, the 26-year-old daughter of the Duke of Kingston, was the toast of London society. Recently married, she was renowned for her brains as well as her raven-haired beauty. But by December her good looks were ruined, after a bout of smallpox left her face scarred with pockmarks and her long eyelashes gone.

The venom that stopped disease

Smallpox, a highly contagious and viral illness in which the victims become feverish and their skin covered with blisters, was one of Europe's most contagious diseases. In the summer of 1716 – when smallpox was rife in Turkey – Lady Mary travelled to Constantinople with her husband, Edward Wortley Montagu, who had been appointed British ambassador there, and three-year-old son, Edward.

During her stay, she became intrigued by the preventive measures being taken against smallpox. In a letter to England, she described how the Turks were inoculated against the disease 'by scratching open a vein in the patient and putting into it as much of smallpox venom as could lie on the head of a needle'. This process was known as variolation, from the Latin *variola*, 'pustule' or 'pox'.

'The smallpox, so fatal and so general among us, is here entirely harmless by the invention of ingrafting [inoculation] . . . the term they give it.'
LADY MARY WORTLEY MONTAGU, TURKEY, 1716

Lady Mary's observations were not quite accurate. The vaccine, a minute amount of fluid scraped from a smallpox pustule, was not injected into a vein, but smeared into a surface cut on the skin of a healthy person. If all went well the patient would develop a mild case of smallpox and recover without any scarring; from then on, he would be immune from the disease. But too large a dose would lead to death.

Nonetheless, Lady Mary had Edward inoculated in 1718, and he survived unharmed. On the Montagus' return to England in 1721, they found the country in the middle of a severe smallpox epidemic, and Lady Mary had her three-year-old daughter similarly inoculated by a British doctor to whom she taught the technique. Her example was followed by the Princess of Wales, who ran a successful controlled experiment on six 'volunteers' from Newgate Prison before her own daughters were inoculated.

Lady Mary's observation was to save thousands of British lives each year, but the technique was superseded by the end of the century by Dr Edward Jenner's vaccination, which was much safer.

LEADING FROM THE FRONT *By ensuring her own children were immunised against smallpox, Lady Mary set a clear example to her fellow Britons, encouraging the use of inoculation.*

LAUNCHING AN ALL-OUT CRUSADE AGAINST SMALLPOX

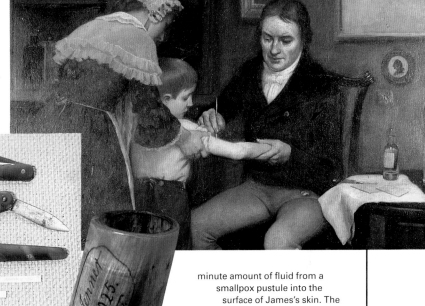

A SIMPLE SCRATCH *Jenner spent 25 years developing his theory that infection with cowpox made people immune to smallpox. He used lancets (below) to scratch a small amount of vaccine stored in a bone pot (centre) into the skin of his patients.*

AS A MEDICAL APPRENTICE IN THE west of England, Edward Jenner overheard a local milkmaid and a friend discussing smallpox, the commonest and most dreaded illness in 18th-century England. 'I cannot take that disease,' she declared, 'for I have had cowpox!' Years later, as the disease claimed more and more victims, Jenner became determined to find a means of combating it.

Cowpox caused cows' udders to blister, and similar blisters appeared on the hands of many of the girls who milked the animals. So Jenner took some fluid from blisters on the hand of an infected milkmaid, Sarah Nelmes, and scratched it with a surgical knife into the skin of an eight-year-old local boy, James Phipps. The youngster developed a single blister at the site of the inoculation. Three weeks later, on July 1, 1796, Jenner scratched a minute amount of fluid from a smallpox pustule into the surface of James's skin. The boy did not become infected. Over the next two years Jenner inoculated more than 20 people in a similar fashion. In 1798, he summarised his findings in a brief treatise, in which he coined the word 'vaccination' – from the Latin *vacca*, meaning 'cow'. It was 'intended to inspire the pleasing hope of its becoming essentially beneficial to mankind,' wrote Jenner.

Jenner's hope was soon reality. By the time of his death in 1823, vaccination against smallpox was a routine procedure and his war against the illness was on the way to eventual victory.

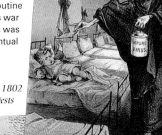

UNCERTAINTY *Jenner's linking of cowpox and smallpox was greeted with scepticism. This 1802 satirical cartoon* The Cow Pock, or the Wonderful Effects of the New Inoculation! *suggests that the new vaccine turns people into cows.*

FEAR *An 1880 lithograph sums up contemporary resistance to introducing foreign agents into the bloodstream. A spate of infections – including tuberculosis – inspired the slogan 'Better not vaccinate than vaccinate with impure virus'.*

Making use of magnetic attraction

HOW MESMER DISCOVERED THE POWER OF AUTOSUGGESTION

FRANZ MESMER GRAVELY CONSIDERED THE CASE OF a most puzzling patient, a young woman named Francisca Oesterlin. She suffered from what he called 'hysterical fever' causing convulsions, vomiting, fainting, temporary blindness, attacks of paralysis and 'other terrible symptoms'.

Creating 'tides' for the body

The symptoms, which came and went like the tides, were instrumental in giving Mesmer the idea of treating Francisca with magnets, which generated their own ebb and flow: at the time – 1774– the medical use of magnets to 'draw' out sickness was already being practised by some European doctors.

Mesmer placed three magnets on Francisca's body: one on the stomach and one on each leg. Her symptoms were greatly relieved, and by the following year she had made a complete recovery. Word spread, and soon people hastened to Mesmer's Vienna practice, eager to be treated by what he called 'magneto-therapy', or 'animal magnetism'. In some cases patients sat around a fountain, with their feet in the water. One patient held on to one pole of a magnetic bar, another to the opposite pole, and the rest of the circle joined hands to complete the circuit. As the magnetic power flowed through their bodies, Mesmer – wearing a magician's cloak – guided the proceedings. Some of the patients had seizures,

some went into trances, and others began to speak in strange tongues. Afterwards they all claimed to be cured of their nervous ailments.

In 1778 Mesmer moved to Paris, where he established the 'Magnetic Institute', specialising in the treatment of nervous disorders. Gradually he came to believe that he himself was a source of animal magnetism. Abandoning his magnets, Mesmer sometimes simply touched or stroked patients while placing them in a deep trance by monotonously repeating arcane words and gestures. His hypnotic suggestions apparently caused people to will themselves back to good health.

'The magnetic influence of the heavens affects all parts of the body, and has a direct effect on the nerves ...Consequently, an active magnetic force must exist in our bodies.'
FRANZ MESMER

Mesmer's popularity grew, but an ever increasing number of critics, including the king, Louis XVI, accused him of being a charlatan. Eventually Louis instructed the Academy of Science to set up a commission to investigate Mesmer's methods. In 1784 the commission, which included Benjamin Franklin, the American ambassador to France, reported that several of its members had undergone treatment at a clinic run by one of Mesmer's disciples, Dr Charles Deslon. None of them had experienced any kind of magnetism, and they concluded that mesmerism was a fraud. 'The doctrine of animal magnetism ... is pretty well laid to rest,' wrote Franklin afterwards to an enquirer. 'Reasonable men, if they ever paid any attention to such a hocus pocus theory, were thoroughly satisfied by the commissioners' report.'

Despite his detractors, Mesmer had demonstrated the value of self-suggestion in treating illness. And he had given a new word to the language: 'mesmerise', meaning 'to hypnotise'.

SALON OF HEALTH *Highly strung society ladies would flock to Mesmer's Paris clinic to press a bar, immersed in 'magnetised water', to the affected parts of their bodies.*

UNCHECKED MADNESS *Goya's disturbing depiction of crazed inmates at the insane asylum at Saragossa was a true record of the way the mentally ill were treated there and across Europe between the late 18th and early 19th centuries. A confidential report on the asylum at Charenton, commissioned by the French Minister of the Interior in 1812, complained that 'in an asylum where the passions that cause the disease should be damped down, there is no scruple about inflaming them by bringing the sexes together'. Depravity went hand in hand with degradation: lunatics were universally regarded as figures of fun. The director of Charenton yearned to belong to the Parisian smart set, and to this end mounted monthly theatrical entertainments at the asylum, declaring the shows to be part of a programme of 'moral treatment' for the patients. The director engaged the Marquis de Sade as writer, director and leading actor, and, as a result, the productions performed at the asylum drew the cream of Parisian society to its doors.*

Ice-cold showers and 'surprise baths'

HOW MENTAL ILLNESS WAS TREATED IN THE 19TH CENTURY

TIED TO A CHAIR, THE PATIENT SHUDDERED AS A JET OF icy water from a pipe above his head gushed out over him. The ensuing shock was intended to dislodge any fixed ideas or delusions that the inmate might have, but the harshness of the remedy was meant to be offset by kind words and jokes. But they did not come. Instead, the nurses openly jeered at the quivering wretch.

Torment masquerading as a cure

Hydrotherapy – cure by water – was the main form of treatment at Charenton, a 19th-century asylum near Paris. Besides ridding the patient of his false beliefs, the purifying powers of the water supposedly flushed unhealthy excesses out of his body. Inmates were also subjected to the 'surprise bath'. Two nurses would grab a patient, strip and blindfold him, then push him backwards along a corridor towards a 6 ft (1.8 m) deep pool; then, holding him by the hair, they would submerge him in cold water for several minutes.

There were approximately 350 patients, watched over by about 50 nurses. The inmates were not segregated according to their illnesses; as a result, the 'melancholics' or depressives, who wanted to be left alone, were constantly bothered by the aggressive and hyperactive 'maniacs'. Violent patients would be beaten and locked up, and anyone who annoyed the nurses would be locked in solitary confinement in a damp cellar.

However, conditions had improved since the French government took over the administration of the asylum in 1795. Before then, ventilation and sanitation were almost non-existent. Few nurses bothered to clean the walls, sweep the floors, or wash the chamber pots, glasses and crockery. Drinking water was in short supply. There were not enough sheets to go round, and blankets were dirty and torn.

The French author Donatien Alphonse François, Marquis de Sade, spent about 12 of his 27 years' imprisonment for sexual offences at Charenton. He lent prestige to the asylum, however, and his treatment, paid for by his family, was quite different from that of the other patients. The marquis had been transferred to Charenton from Bicêtre – an even more disreputable institution. By 1792, however, when Philippe Pinel was placed in charge of Bicêtre, attitudes towards the insane had already begun to improve. As well as unchaining the patients, the enlightened humanitarian gave them sunny rooms and let them exercise in the grounds.

PRIVILEGED INMATE *The Marquis de Sade was treated as an honoured guest at Charenton; his mistress was even allowed to live there with him.*

SET THEM FREE! *Philippe Pinel, director of the Bicêtre asylum, orders an old man's shackles to be removed. His approach to mental health was unorthodox, but patients were much improved for their new-found liberty, and in all only one man absconded.*

'Soothing, quieting and delightful beyond measure'

HOW CHLOROFORM BROUGHT ABOUT PAIN-FREE BIRTH

DOUBLE PIONEER *As well as ushering in chloroform for operations, James Simpson called for greater medical hygiene. Until this became standard, he warned, 'a man laid on the operating table of one of our surgical hospitals is exposed to more chances of death than the English soldier on the field of Waterloo'.*

WRONG CONCLUSION *Hearing Simpson and his friends crash to the floor after taking chloroform, a next-door neighbour assumed that they had collapsed from too much drink. In fact, a major medical experiment had taken place.*

ON THE EVENING OF NOVEMBER 4, 1847, JAMES YOUNG Simpson, professor of midwifery at the University of Edinburgh, invited two colleagues to his house to conduct an experiment. He poured a colourless liquid into three tumblers, and asked the men to inhale the vapour. After a brief period of lively conversation, the three men suddenly fell to the floor, unconscious. About 15 minutes later they came to. A triumphant Simpson exclaimed: 'This is far stronger and better than ether!'

The mystery liquid was chloroform, a substance Simpson hoped would end his search for the 'ideal' anaesthetic. Earlier in 1847 he had been the first doctor in Britain to use ether – a liquid similar to chloroform – as an anaesthetic in childbirth, even though ether's side effects included violent vomiting.

Simpson had conducted his experiment after a Liverpool chemist named David Waldie had pointed out that chloroform had been inhaled to relieve internal pains since the 1830s. Four days after the experiment, Simpson tried the new anaesthetic on a woman in painful labour in Edinburgh Infirmary. As she inhaled the vapours from a chloroform-soaked handkerchief, her pain diminished, and 25 minutes later, still conscious, she successfully gave birth.

Simpson's subsequent pamphlet, *Account of a New Anaesthetic Agent*, was hailed as a breakthrough. But despite its initial success, the use of chloroform was attacked by conservative obstetricians. One, Tyler Smith, argued that the more painful a woman's labour, the harder she would strive to give birth. Others thought chloroform was dangerous, as it slowed the workings of most of the body's organs.

Invoking the Bible

The church, too, raised objections. Theologians argued that women had endured labour pains since the time of Eve, quoting the book of Genesis: 'In sorrow thou shalt bring forth children.' Simpson, mounting a spirited defence, countered with another Biblical verse: 'Therefore to him that knoweth to do good and doeth it not, to him it is a Sin.' All women, he asserted, would soon demand pain-free childbirth.

As one of Queen Victoria's physicians, Simpson attended the birth of her eighth child, Leopold, in April 1853. The queen was given chloroform by Dr John Snow, Britain's first specialist anaesthetist. Her Majesty later described the effects as 'soothing, quieting and delightful beyond measure'.

ONE STEP FORWARD *Until the middle of the 19th century, amputations (top left) were performed with only alcohol as a crude anaesthetic. But unless the patient was stupefied, it was far from effective. The first public operation using ether, or laughing gas, was performed by Dr John Collins Warren in the USA in 1846 (bottom left). Despite this advance, surgeons regularly put patients' lives at risk by continuing to operate in their ordinary business suits, with bare hands and no masks (below).*

A revolution in the operating theatre

HOW A SCOTTISH SURGEON USED ANTISEPTICS TO SAVE THE LIVES OF HIS PATIENTS

JOINT EFFORT In 1867, Joseph Lister and his wife drafted an article (above) describing antiseptic surgery. The handwriting is Mrs Lister's, with corrections made by both of them.

THE LIFE OF 11-YEAR-OLD JAMES GREENLEES WAS IN danger. He had been run over by a cart and rushed to Glasgow Royal Infirmary with a broken leg. At the time, a compound fracture meant one of two things: death or amputation.

Surgery was a last resort in the 19th century. If the patients did not die of haemorrhaging or shock on the operating table, they were likely to succumb to postoperative infection. Fuelled by poor sanitation, overcrowding and chronic disease, infections spread rapidly round the wards. After an amputation a patient had only a 50 per cent chance of survival.

Luckily for James Greenlees, his surgeon decided to try a new technique. Joseph Lister first dabbed carbolic acid onto the wound, then he splinted the leg and left it undisturbed for a few days, after which he examined the injury. The wound looked healthy, there was no smell of decay and there was less inflammation than was usual. Lister re-dressed the wound and applied more carbolic. A healthy scar slowly formed, and after six weeks the bones had knitted themselves back together.

Safety through cleanliness

Lister described this and a further ten cases in the medical journal *The Lancet* in 1867. He had had one death and one amputation among his cases. The operation for compound fracture, regarded as frequently being fatal, was now a safe procedure. Only 1 in 50 of all the operations Lister later performed using carbolic acid ended in death.

At the time doctors believed that infection or sepsis – inflammation and pus – was caused by 'bad air'.

SAFE SURGERY Inspired by Lister's example, the 19th-century Austrian surgeon Theodor Billroth (with white beard) introduced antiseptic techniques at his clinic in Vienna. He performed operations that had previously been thought too dangerous because of the risk of infection.

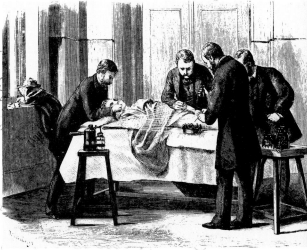

'PUFFING BILLY' In the mid 1860s Lister introduced an antiseptic spray by which a vapour of carbolic acid was played on patients during operations – especially when opening abscesses or stitching up wounds. The spray was nicknamed 'Puffing Billy' from the noise it made in action.

A few years earlier, the French scientist Louis Pasteur had discovered that airborne organisms caused food to rot. Lister reasoned that similar organisms must cause sepsis in wounds and if he could kill them he would stop infection. Meanwhile, a chemistry professor at Manchester, Frederick Crace Calvert, had found that carbolic acid delayed the decay of corpses. Lister deduced that the acid must be killing the micro-organisms, so in 1865 he began using it to protect wounds.

Steam cleaning

Over the next two years, Lister further promoted sterile conditions by ensuring that the operating theatre and staff were scrupulously clean. Throughout an operation the wound was sprayed with carbolic acid from a 'steam spray'. Lister took his work very seriously, and when a student shouted 'Let us spray!' prior to an operation, he was not amused.

FRESH FLUID In 1901 a French mouthwash was named after Lister, whose name is still associated with personal hygiene.

Some colleagues were sceptical of Lister's techniques, and the carbolic acid irritated surgeons' hands. Towards the end of the century other surgeons began advocating 'aseptic' rather than antiseptic methods, using a combination of heat and chemicals to sterilise the skin, the operating area and all instruments in contact with the wound. It remains the basis for all modern surgery.

The triumph of antiseptic surgery brought Lister fame, wealth and finally – after he lanced a boil in Queen Victoria's armpit – a baronetcy.

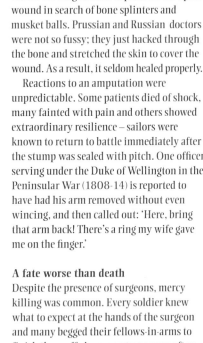

Brandy, bruises and bravado

HOW SURGEONS OPERATED ON THE BATTLEFIELDS OF NAPOLEON

FORTUNES OF WAR *A French soldier is treated during the 1812 battle for Moscow at Borodino. Medical staff operated an efficient ambulance service to back up Napoleon's army.*

THE SCREAMS OF THE WOUNDED WERE DROWNED OUT by the thump of falling cannonballs and the crackle of artillery fire. The stench of seared flesh mixed with the acrid odour of gunpowder as the doctor calmly started to saw through the captain's thighbone.

At the Battle of Borodino in 1812, Baron Dominique Jean Larrey sawed off more than 200 limbs in 24 hours. To ease his patients' agony, he would give them a swig of brandy and a napkin to bite on. One assistant held the wounded soldier down, while another firmly gripped the limb to be removed. Then Larrey used the French 'three-cut' technique, in which a flap of skin was pulled over the wound and sewn up. All wounds were disinfected with a preparation of marsh mallow and dressed with compresses of wine.

The operation could be over in minutes, unless the surgeon had to probe in the open wound in search of bone splinters and musket balls. Prussian and Russian doctors were not so fussy; they just hacked through the bone and stretched the skin to cover the wound. As a result, it seldom healed properly.

Reactions to an amputation were unpredictable. Some patients died of shock, many fainted with pain and others showed extraordinary resilience – sailors were known to return to battle immediately after the stump was sealed with pitch. One officer serving under the Duke of Wellington in the Peninsular War (1808-14) is reported to have had his arm removed without even wincing, and then called out: 'Here, bring that arm back! There's a ring my wife gave me on the finger.'

A fate worse than death

Despite the presence of surgeons, mercy killing was common. Every soldier knew what to expect at the hands of the surgeon and many begged their fellows-in-arms to finish them off. Army surgeons were often the least talented members of their profession, for the work was poorly paid and dangerous. As they had no official rank in the army, they were accorded little respect.

The Duke of Wellington, commander of the British army against Napoleon, was slow to acknowledge the work of his surgeons in hospitals behind the lines, and refused to accept ambulances. The wounded were removed on stretchers by regimental bandsmen or left to wait for the end of the battle. Transported on heavy wagons, they felt every bump of the war-rutted ground as their broken bones scraped together.

Meanwhile, Napoleon's army was using light, well-sprung horse-drawn ambulances attended by teams of medical staff. These

ambulances, built to Larrey's design, had horsehair mattresses which slid out on runners. The walls of the carriage were partly padded and lined with pockets for carrying medicines. If the conditions were suitable, the mattress was put on the ground and the operation performed on the spot. A four-wheel version of the carriage carried up to four men.

Troops during the Napoleonic wars were luckier than their predecessors. Up to that time, the ordinary soldier was considered expendable. Once wounded he was abandoned to the mercy of the locals or had to fend for himself. When the battle was over, the area was scoured by looters happy to cut a soldier's throat for the contents of his pockets, which were likely to contain the same valuables that had been plundered from the locals just a few days earlier.

THE QUICKER THE BETTER *The chief surgeon in Napoleon's army, Baron Larrey – at work (left) on the battlefield of Hanau – could amputate an arm or a leg in under 2 minutes by using the 'three-cut' technique (below). These drawings of Larrey's arm amputation date from 1829.*

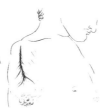

INCISION *The skin was cut at a special angle, leaving flaps to seal the wound.*

AFTER *The amputation finished, loose skin fell naturally over the hole.*

INSIDE *The surgeon tidied up ligaments and muscles underneath the skin flap.*

CLEAN DRESSING *Dressed and cleaned, the wound was left to heal.*

THE TWO HEROINES OF THE CRIMEAN WAR

WHEN BRITAIN WENT TO WAR WITH RUSSIA IN 1853, a young woman volunteered to take a task force of nurses to the front line to care for the sick and the wounded. In October 1854, Florence Nightingale set out for Scutari with 38 fellow nurses, arriving just in time to help care for over 10 000 war wounded from the battle of Inkerman.

Overworked and understaffed, Nightingale's group worked tirelessly to improve conditions for the wounded, and when the Crimean War ended abruptly in 1865, Nightingale returned home to find herself well known. She capitalised on both her experience and her contacts to start a training school for nurses in London and tried to improve on sanitary arrangements in the army.

Nightingale was not alone in her heroic work in the Crimea, however. Among those who wanted to work alongside her was Mary Seacole, a 48-year-old Jamaican who had considerable experience as a 'doctress' and in treating the illnesses that were wreaking havoc on the troops in the Crimea.

Seacole's offer of help was rejected, but she still left for the Crimea in 1854. Based outside Balaklava, she sold goods ranging from toothpowder to shoes to the soldiers. She used the profits to care for the sick, dispense medicine and provide healthy meals. Her work set a new standard for army suppliers.

HELPING HAND *Mary Seacole (above), a trained Jamaican 'doctress', volunteered for Florence Nightingale's task force. Disillusioned by her rejection – 'Did these ladies shrink from accepting my aid because my blood flowed beneath a duskier skin than theirs?' – she too left for the Crimea. There, she was just as admired as Nightingale (standing, left) but returned bankrupt to Britain after the war. Her debts were cleared by war veterans and well-wishers.*

Mad dogs versus a dedicated Frenchman

HOW LOUIS PASTEUR WAGED WAR ON RABIES

FOR MORE THAN THREE YEARS THE EMINENT FRENCH chemist Louis Pasteur had sought a safe cure for rabies. If successful, this would be the crowning achievement of his illustrious career. But the cure continued to elude him, and in the spring of 1884 he concentrated on his work in the forest of Meudon near Paris, where he kept 50 rabid dogs in cages. There he began a series of experiments that were to transform medicine.

A bite that ended in death

Each year, rabies, caused by a virus transmitted by the saliva of an infected animal, brought a slow and agonising death to hundreds of people throughout Europe. Those bitten by such animals, particularly by dogs and foxes, showed no symptoms for up to 12 months as the virus incubated in their bloodstream. Then delirium, convulsions, hallucinations and hydrophobia – an extreme fear of water – set in; within about four days the terrified victims were dead. Often, ironically, the victims were killed by the only treatment available: cauterising the bites with red-hot pokers or carbolic acid.

Pasteur had been inoculating rabbits with saliva and brain material from infected dogs – and humans – since he started research in 1881. Then, in the autumn of 1884, he dried strips of the spinal cord from infected rabbits in flasks containing caustic potash. He next injected experimental animals with an emulsion of the dried material, which greatly weakened the disease. He later discovered that this emulsified brain tissue was itself a vaccine which could save the lives of bite victims in whom the symptoms had not yet appeared. His chance to prove

FIGHTING DISEASE AND DECAY
Pasteur was 62 years old when his painstaking research finally prevailed against rabies. Some 20 years before that, in 1860, he had gained fame after a series of experiments in the pure, germ-free air of the Alps near Chamonix showed that germs in dirt and dust caused decay in wine, beer and milk.

VICTIMS IN SEARCH OF A CURE *As news of Pasteur's wonder vaccine spread, rabies victims, both children and adults, travelled from all over Europe to be treated at his clinic in Paris. By 1896 some 20000 patients – including four Russians (left and top) and six English children (above) – had been inoculated by Pasteur himself (far right) or colleagues at collaborating clinics. Fewer than 100 of these patients failed to survive.*

his theory in public came in July 1885, when a nine-year-old shepherd boy named Joseph Meister and his mother arrived at Pasteur's surgery in Paris. Joseph had been savaged by a rabid dog in his village in Alsace in north-east France, and his legs, thighs and hands were covered in deep bites. Appalled by his injuries, Pasteur had Joseph inoculated with the new vaccine made from the emulsified spinal material of a rabbit that had recently died of rabies.

Two weeks to live or die

Pasteur arranged lodgings for the Meisters, and started Joseph on a two-week course of daily injections, with the vaccine being strengthened each time. 'Each day,' said Pasteur later, 'I dreaded hearing that the worst had happened, and that little Joseph was dead!' But by the end of the fortnight Joseph had fully recovered – and Pasteur was hailed as a saviour.

In 1888 the Pasteur Institute was founded in Paris to undertake further research into the prevention and treatment of rabies and other diseases. Fittingly enough, one of its first concierges was Pasteur's grateful former patient, Joseph Meister.

Birth of a miracle drug

HOW A CANADIAN RESEARCH TEAM DISCOVERED INSULIN

LEONARD THOMPSON, A 14-YEAR-OLD schoolboy in Toronto, Canada, was dying. His weight had dropped to less than 70 lb (32 kg). He was existing on a starvation diet of only 450 Calories a day – less than a normal boy of his age usually eats in a single meal. His hair was beginning to fall out, his stomach was distended and, pale and listless, he lay in bed day after day in the city's General Hospital. He had diabetes, the wasting disease for which there was no known cure.

Diabetes is a chronic condition in which the body cannot properly process sugar because the pancreas, a gland near the stomach supplying digestive fluid, is not producing enough of a vital hormone called insulin. As a result, sugar gradually builds up in the blood and the tissues, with lethal effects.

Life-giving islets

In 1869 a German physician named Paul Langerhans had described the existence of clusters of tissue – later known as islets of Langerhans – in the pancreas. These islets, about 1 million to each pancreas, were later found to release insulin into the surrounding blood capillaries. Sporadic attempts were then made to extract and isolate insulin – from the Latin word *insula*, meaning 'island'.

In the autumn of 1920, Frederick Banting, a 29-year-old Canadian surgeon, devised an experiment to isolate insulin using the pancreases of dogs. He took his idea to the University of Toronto, where he discussed it with Professor John J.R. Macleod, an internationally renowned physiologist, who assigned a brilliant 21-year-old colleague, Charles Best, to help Banting to carry out his research work.

By the beginning of 1922 Banting and Best had produced their own supply of insulin. They injected it into a series of diabetic dogs with, in at least one case, encouraging results. After injecting each other to ensure there were no ill effects, Banting and Best were able to offer insulin to diabetic patients.

That January the doctors gave young Leonard Thompson a small amount of insulin. In May Leonard was released from hospital and resumed the more easily obtained dietary treatment of diabetes. By October, however, his condition had deteriorated, so he was readmitted to hospital and successfully put on regular injections of insulin. Leonard's case made front-page headlines throughout the world, giving new hope to millions of diabetics.

CRUSADERS THREE
Best (left) and Banting stand with one of the dogs which was injected with insulin in the crusade to combat diabetes.

A golden age for pill-pushers and charlatans

HOW QUACKS MADE FORTUNES ACROSS THE CENTURIES

IN THE LATE 19TH CENTURY, THERE WERE A GREAT MANY common ailments that doctors still could not cure – including worms, fever and rheumatic pains. For this reason the man or woman in the street was willing to try almost any remedy that promised relief. In such a climate, quack doctors flourished. For instance, pioneer families and townspeople across the American West rushed to see the famous Kickapoo Indian Medicine Company show whenever its wagon train came to town.

The proprietors, two patent medicine pedlars, hired native Americans to stage mock battles, war dances and wagon-train attacks. When the show was over, crowds would line up to buy bottles of 'World Famous Kickapoo Indian Sagwa', a concoction of herbs, buffalo fat, roots, bark, leaves, gum and alcohol advertised as a 'panacea for all ailments'. The owners made a fortune.

The Kickapoo Indian Medicine Company was the direct heir of the quack doctors of medieval Europe. These medical charlatans went from village to village with a jester who told jokes to the crowd while the stall was set up. After the entertainment, the quack related the amazing number of 'miracle cures' that he and his medicines effected.

The services of qualified doctors with proper medical training were expensive, and quack doctors competed fiercely for the custom of the lower classes. By the middle of the 18th century, well-regulated orthodox medicine was beginning to take hold in Europe, and quackery was frowned upon. Pedlars of patent drugs sought and found a new market where medical regulation was more or less non-existent – colonial America.

Dragon's blood and opium

The pedlars crossed the Atlantic, assumed the title of 'doctor' and journeyed from Maine to Georgia and from Virginia to the unexplored and unsophisticated West. Many claimed to be part Indian with knowledge of herbal cures or to come from Philadelphia, one of the centres of medical training. Their cures relied on a narrow range of ingredients, primarily 'dragon's blood' – dried Peruvian bark, containing quinine – and powders containing opium and ipecacuanha – a substance from a plant related to coffee. The opium caused constipation, and in large doses the ipecacuanha brought about vomiting.

The business of the quacks gained a huge boost with the spread of epidemics during the American Civil War. They grabbed this opportunity, and brought to the suffering public not only their remedies but also dog and pony circuses, Punch and Judy shows, magicians and minstrels. The Kickapoo Indian Medicine Company was the pinnacle of this type of quackery. With its demise in the early 20th century, a peculiar form of medical theatre was lost.

ZEILEIS'S MAGIC WAND

ONE OF THE MOST SUCCESSFUL QUACK DOCTORS OF THE 1920s was a peasant called Valentin Zeileis, who rose to fame and riches in Austria with a curious 'healing' machine. This was simply a glass vacuum tube that coloured fluorescent pink, blue and violet in turn when an electrical current was passed through it. Such devices were widely used by scientists to detect and measure electrical fields, but Zeileis used the tube to 'diagnose' and 'cure' illness. He passed it – like a magic wand – over a patient and the colours in the glass would mysteriously tell him which organs were diseased.

Sufferers treated half-naked

As Zeileis's renown spread, thousands of would-be patients flocked to the castle he had bought in the village of Gallspach. Dozens at a time would be herded, half-naked, into the treatment room. With a wave of Zeileis's hand, the room would be plunged into darkness, save for two red lights in the jaws of a gigantic sculpted snake. Then the electric current would be switched on. Zeileis would plunge into the crowd, working the tube over the sweating bodies of the sufferers.

The Austrian quack was finally undone when he attempted to legitimise himself and his work. For years he had successfully sued for libel anyone who denigrated his treatments. However, Zeileis made the mistake of publishing his 'findings' in pamphlets. In 1930 a Professor Lazarus of Berlin was able to demonstrate the uselessness of Zeileis's 'high-frequency' treatments, by refuting all of the extravagant claims made in these pamphlets.

VANISHING ACT *Exposed as a trickster, the long-bearded Zeileis fled from his Austrian castle, and was never heard of again.*

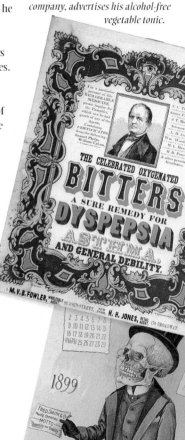

TWO IN ONE *Indian herbal pills for the bowels and a treatment for worms share a poster.*

BITTER MEDICINE *In 1846 George B. Green, proprietor of a popular drug company, advertises his alcohol-free vegetable tonic.*

DEATH'S CALLING CARD *An American cartoon gives a satirical warning about the dangers of travelling pills-salesmen, whose motto is 'Quantity not Quality', and whose goods are guaranteed to kill.*

BITTER-SWEET *By adding sugar and water, this otherwise bitter-tasting American tonic (left) is turned into the perfect pick-me-up.*

SURE CURE *For those with liver or kidney problems, Hops & Malt Bitters (below, left) promise speedy and lasting relief. The claim was that it purified the blood.*

THE REAL THING *A druggist (below) shows a customer that a box of liver pills has been signed as genuine by the inventor and the chemist.*

ROOT REMEDY *Sarsaparilla, a beverage flavoured with the roots of tropical plants, was taken to cure blood and liver disorders in the 19th-century USA. Among the most popular brands were Scovill's (top), Log Cabin (middle) and Ayer's (left).*

SHOCK TREATMENT *Blood and nervous disorders are said to be cured by means of this French-made electric rod known as 'Le Solitaire' (right). The young woman in the poster is preparing to use it to deal with a disease-causing, outsize microbe.*

GIFT OF THE GAB *An audience of rural Americans (above) listens avidly as a salesman extols the health-giving properties of a bottle of tonic, the ingredients of which are supposedly a secret. As the imposingly dressed quack emphasises each point, the American Indians beat their drums in approval.*

The man who investigated sex

HOW FREUD FORMED HIS THEORIES OF THE UNCONSCIOUS MIND

AN ATTRACTIVE AND INTELLIGENT YOUNG VIENNESE woman named Bertha Pappenheim suffered from a series of alarming fits. Her limbs became paralysed, parts of her body were numb, her sight and hearing were impaired and she even forgot how to speak her native German. Most disturbing of all, her personality changed to that of a mischievous child – and she did things which she could not recollect when her 'real' self took over again.

Cleansing the mind

In despair, the 21-year-old Bertha consulted Josef Breuer, an expert on diseases of the nervous system. In 1883 Breuer told his friend and colleague Sigmund Freud about Bertha, whose case, Freud claimed later, marked the start of psychoanalysis. Her symptoms had developed after she had nursed her father during a terminal illness. Breuer felt that it was this experience – and the unhappiness and guilt Bertha suffered – that had triggered her condition.

Freud agreed with Breuer's diagnosis of hysteria – the medical term for a condition with strong psychological and physical symptoms. He listened carefully as Breuer explained how, while in a trance, Bertha – who was given the case name of 'Anna O' – had been able to recall what she had said and done when her 'naughty' side took control. She also related the first experience of each of her symptoms. Breuer called this process 'catharsis', from the Greek word for 'purge' or 'cleanse'.

At the time, the 28-year-old Freud was the superintendent of the department for nervous diseases at Vienna's General Hospital. His subject was neurology, the study of the nervous system. But influenced by the work of the eminent French neurologist Jean-Martin Charcot, he increasingly became interested in the workings of the mind. He was also deeply impressed by Breuer's overall conclusion that although a patient's symptoms appeared in his or her normal conscious state, the cause of them remained locked in the unconscious mind. Hypnosis – the medically supervised return to the unconscious state – was the key to unlocking the cause and to curing the disease, asserted Breuer.

In 1886 Freud also started a private practice in Vienna specialising in the investigation and treatment of nervous diseases. To begin with, he treated his patients by traditional means such as electrotherapy, massage and cold baths. However, the treatments did not always succeed, and Freud turned to hypnosis and, later on, to Breuer's employment of catharsis.

Delving into the unconscious

After working closely together, Freud and Breuer jointly published *Studies in Hysteria* in 1895, a work generally regarded as the first written account of psychoanalysis – a method by which the cause of nervous diseases is traced to drives and emotions buried deep in a patient's unconscious mind. Such unorthodox views alienated Freud from his more conventionally minded colleagues in Vienna.

Working alone, he created what was to be regarded as his master work. This was *The Interpretation of Dreams* (1900), in which he propounded that, like neuroses, dreams were disguised manifestations of unconsciously repressed sexual wishes. From then on Freud published a string of books which confirmed his reputation as a bold and controversial thinker.

These included *Civilisation and Its Discontents* (1930), a pessimistic work coloured by the rise of fascism in Europe. In the book Freud contended that human existence was a continual battle between man's love of life and his morbid attraction towards death – with the death instinct proving to be the stronger.

The value of Freud's work has been challenged by many 'mind doctors', who feel it places too much emphasis on sex. However, his disciples regard him as someone who changed forever people's view and understanding of themselves and the world.

PATIENT LISTENER *When analysing a patient at his richly furnished private consulting room in Vienna, Freud encouraged her – or him – to try to relax by lying on a padded, rug-covered couch, while he himself sat on an armchair discreetly out of sight. The patient was then prompted to overcome any initial nervousness and to voice whatever thoughts came to mind. This type of therapy – known as 'free association' – was designed to cure mental ailments by bringing them out into the open, where they could be analysed and discussed.*

Among Freud's revolutionary theories was that of the so-called 'Oedipus complex', based on the tragic Greek myth about a man who unwittingly kills his father and marries his mother. Freud contended that all young boys unconsciously wanted to replace their fathers in their mothers' affections, while girls unknowingly sought to usurp their mothers and sleep with their fathers. By encouraging patients to talk about such forbidden desires, Freud enabled them to come to terms with their feelings of guilt, and to lead more emotionally stable lives.

Later, Freud was criticised for basing much of his work on the 'self-indulgent' ailments and sex problems of neurotic, middle-class Viennese women – who were not considered typical of the majority of hysterical patients.

Freud smoked throughout most of his consultations. He consumed 15-20 Dutch cigars every day, which probably caused the cancer of the jaw that killed him in 1939 – a year after he had fled from his native Austria to London to escape Hitler's persecution of the Jews.

DOODLES *Freud's own mental fears and fixations are shown in his strange scribbles.*

UNEXPECTED HONOUR *In the spring of 1935, to his surprise and delight, Freud was elected an honorary member of The Royal Society of Medicine in London.*

Law & Order

EVERY SOCIETY HAS HAD TO ANSWER THE BASIC QUESTIONS OF POLITICS: WHO WIELDS THE POWER? HOW DO RULERS GAIN OR KEEP THE CONSENT OF THOSE WHOM THEY GOVERN? HOW DO WE JUDGE AND PUNISH WRONGDOERS? AND ARE THERE ANY LIMITS TO AUTHORITY?

A citizen's duty

HOW AN ANCIENT ATHENIAN JURY DISPENSED JUSTICE

'ATHENIANS, I AM NOT GOING TO ARGUE FOR MY OWN sake, but for yours, that you may not sin against the gods by condemning me, who am his gift to you.' The orator paused as a terrible stillness fell in the *agora*, or marketplace. 'If you kill me, you will not easily find a successor, so I advise you to spare me.' There was a moment's silence while the 500-strong jury marvelled at Socrates' arrogance.

That year, 399 BC, the philosopher Socrates was on trial, accused of impiety and 'corrupting the youth of Athens' by questioning the traditional wisdom, values and gods of the Athenians. He had pleaded his case in front of a jury that was not exceptionally large for ancient Athens – juries might vary in size from 200 to 1000 members, depending on the severity of the charge.

Jury service was a right and privilege for every Athenian citizen, although the status of citizen did not include women, slaves or those born abroad. For administrative purposes, the citizens of Athens and the land of Attica that surrounded it were divided into ten tribes, each of which provided a tenth of the total number of jurors required. Anyone who wished to volunteer as a juror drew a token from among those allotted to his tribe, telling him in which court he could sit.

On the day of the trial, all the potential jurors presented themselves at the court and put the token with their name scratched on it into a *kleroterion* – an ingenious machine which randomly selected the tokens using a system of black and white balls, though exactly how it worked is unknown.

Athens relied on the initiative of individuals to bring a case, even for offences against the state, but only Athenian citizens could conduct suits in court. Women, outsiders and former slaves had to be represented by a citizen. Most cases were heard in local courts known as *dikasteria*, but murder and serious religious offences received special treatment.

Voting with tokens

Both plaintiff and defendant had to present their own cases. Although some hired professional speech-writers to help them, lawyers as such did not exist. The length of the litigants' speeches was strictly limited – water clocks were used to make sure they did not exceed the allotted time – but witnesses could be heard and documents read without restriction.

Each juror was given two circular bronze tokens inscribed with the words 'official ballot'. One had a solid hub, while the other was hollow in the centre. At the end of the case, jurors placed the token of their choice in an urn, the contents of which were then counted. The party with more votes won the case.

In Socrates' case, the plaintiff received 280 votes and Socrates 220. He was condemned to death, and 30 days later, the executioner arrived at his house with a drink poisoned with hemlock. The philosopher drank it and died a few minutes later.

THE POWER OF WORDS *The art of public speaking was rated highly in Roman society, and a skilled orator wielded considerable political power: in the absence of any evolved party system, a single speech could easily sway voting in the Senate. In this 19th-century painting, the consul Cicero (left) delivers one of the finest speeches of his career, in 63 BC. He is berating his bitter opponent, Catilinus (right) – who sits alone, having been deserted by his fellow senators. Catilinus, a former provincial governor, has gathered considerable support for a plot to overthrow the Republic, close the Senate and murder Cicero. But his scheme has been discovered, and Cicero now denounces him, asking 'How long will you try our patience?' Catilinus later fled Rome to lead a rebellion, in the course of which he was killed.*

LOTTERY OF JUSTICE *Potential jurors brought 4 in (10 cm) bronze tickets bearing their name to the court on the first day of a trial and waited to see if they were chosen.*

FOR OR AGAINST *Jurors received two tokens labelled 'official ballot'. The one with the solid core represented a vote for the defendant; the hollow centre meant a vote for the plaintiff.*

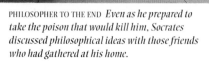

PHILOSOPHER TO THE END *Even as he prepared to take the poison that would kill him, Socrates discussed philosophical ideas with those friends who had gathered at his home.*

Ruling by vote

HOW THE ROMAN SENATE MADE ITS DECISIONS

HOWLS OF DISAPPROVAL FILLED THE PACKED SENATE house of 83 BC: Lucius Cornelius Sulla had proposed a bill to limit the annual rate of interest to 10 per cent. The senators, seeing their profits from moneylending under attack, shouted their objections, echoed by an angry mob outside. Sulla promptly ordered an 'executive session', closed the doors to the Senate and kept the mob at bay.

Before Sulla reformed the Senate, it consisted of 300 senators, citizens who were elected for life. But in 81 BC, soon after he was appointed dictator, Sulla increased the number of lifelong senators to 600. This use of his virtually unlimited powers – the first Roman dictator to hold them – spelt the beginning of the end for the Republic.

A position of power

Senators played an important role in Republican Rome, as the Senate controlled the Treasury and also advised the senior and junior consul – magistrates elected by the public to run the empire for one year – who were in charge of administering the city and the empire. In peacetime, the Senate usually met at three to six-week intervals, although it could be summoned by a consul up to three times a week.

Business began early in the day – the Senate had to finish sitting by sundown, as legislation passed after sunset was invalid – and started with the sacrifice of a live boar, goat, ram or bull. The shape and colour of the animal's liver was studied for omens, and providing the portents were favourable, the main business of the day was introduced. Senators could add any other business that concerned them and were then called on in order of seniority to speak on their subject before the matter was put to a vote. Speeches were often personal and vitriolic.

Enforcing the will of the Senate

The senior consul also acted as the presiding officer of the Senate, except when away from Rome. He had the right to set the agenda, to bring discussions to an end and to determine the order of voting. To vote, those in favour went to one side of the room, those against went to the other, and abstainers hovered in the middle.

After a vote was taken, a committee made up of the motion's supporters, chosen by the presiding officer, drafted the final form of the motion, known as a *Senatus Consultum*. It listed the time and place of the session, the issue in question, any decisions taken and the action that was to follow. The presiding officer deposited a copy of the statute in the State Treasury, housed in the Temple of Saturn, and made sure that the recommended action was carried out. If necessary, he would nominate people to execute the will of the Senate. He would also issue proclamations to be posted where those concerned could see them: legislation about the rafting of corn up the Tiber, for example, would most likely be posted at the docks.

But news of significant rulings usually reached the ears of the people long before it was read from a wall – in the case of Sulla's bill on interest rates, the mob carried word of his success faster than any poster.

ROMAN DICTATOR *Lucius Cornelius Sulla, appointed dictator in 81 BC after winning a war against King Mithradates VI of Pontus, was the first person in Rome's history to be appointed to this powerful position for an unlimited period of office.*

Laying down the law for a vast territory

HOW THE ROMAN EMPIRE WAS GOVERNED

THE EMPEROR TRAJAN WAS ANNOYED WITH GAIUS Plinius Caecilius Secundus, governor of the Roman province of Bithynia in northern Turkey. Why did he trouble him with with petty problems? Trajan had enough on his mind without being pestered to send an architect to the new theatre at Nicaea. The emperor scribbled a note to the governor, telling him bluntly: find your own architect.

Stretched to the limit

Trajan, emperor from AD 98 to 117, had only himself to blame if running the empire was getting too much for him. He had expanded it until it reached from the Persian Gulf in the east to Wales in the west, and from Scotland in the north to Egypt in the south. Governing an area peopled by over 50 million speaking dozens of languages was bound to be hard. Fortunately Augustus, Roman emperor from 4 BC-AD 14, foresaw some of the problems that would arise and devised partial solutions to them when he founded the empire a century before.

The key to running the empire, Augustus believed, was balance. An arch-pragmatist, he set out to find a way of maintaining imperial control while keeping down the cost of governing such a huge area. His solution was to devolve administrative tasks to local level, as far as possible. Accordingly, he divided the empire into 22 provinces, most based on previous native states, or groups of them, such as Lusitania (Portugal) and Cilicia (southern Anatolia). Each province was run by a governor, most of whom were chosen by the Roman Senate, although the emperor himself chose those posted to frontier provinces, such as Dacia, Judaea and Mauretania.

Augustus also tried to balance the provision of enough troops to defend the frontiers with the danger of giving too much power to ambitious generals. As a check on such aspirations, governors were kept in their posts for just three years, too short a time for them to build a personal power base among their command.

To prevent generals buying the loyalty of their troops, army pay and all other financial matters were dealt with by a separate official, the *procurator*. The governor and the procurator each had a small staff: a junior senator, a few clerks and scribes, a detachment of troops, and the *amici* – the governor's friends and relatives.

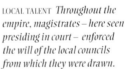

FORMIDABLE EFFICIENCY *The Romans kept detailed information on the territories they occupied. This unique plan found at Orange, France, marks roads and land boundaries and was used for tax collection.*

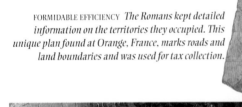

LOCAL TALENT *Throughout the empire, magistrates – here seen presiding in court – enforced the will of the local councils from which they were drawn.*

A STATE PROCESSION *A life-size marble sculpture on the Altar of Peace in Rome, shows magistrates, officials and members of the senate.*

Encouraging locals to govern themselves

In provinces like Gaul and Britain, former tribal states were allowed local self-government under the title of *civitas*, and cities were developed to act as their administrative centres. All over Britain, capital cities for tribal groups grew up, from Durovernum Cantiacorum (Canterbury) for the Cantiaci tribe in Kent to Venta Silurum (Caerwent) for the Silures in South Wales. Some existing towns were granted self-governing status, the title of *municipium* and a charter – Verulamium (St Albans) and Londinium (London), for example – and the Romans also created new towns, known as *colonia*, by settling retired legionaries at Glevum (Gloucester), Lindum (Lincoln) and Camulodunum (Colchester), among others.

All these self-governing communities had a similar form of administration. An all-male council, usually about 100 strong, was elected from the free-born citizens of the civitas, municipium or colonia. The council elected two senior magistrates who acted as judges, and two junior ones – the *quaestors* – who oversaw public works and managed financial affairs.

All roads lead to Rome

The most important function of the quaestors was the collection of taxes, which were assessed on the basis of a five-yearly census. The quaestors were primarily concerned with registering people for the *tributum capitis* (poll tax), and with registering property for the *tributum soli* (land tax).

Tax collection was generally contracted out to private agents known as 'tax farmers', a system which was sometimes subject to bitterly resented abuse. In Britain, the greed of procurator Catus Decianus was a major grievance of the Iceni tribe in East Anglia, leading to the tribe's bloody rebellion in AD 60 under their queen, Boudicca.

Different parts of the empire were linked by a sophisticated communications network. An imperial courier system made full use of the roads linking all the towns in a province to the provincial capital.

Stations and inns every 15-30 miles (25-50 km) provided the couriers with a change of horses and a place to stay overnight. This efficient system enabled the emperor in Rome to keep a close eye on all matters that interested or concerned him.

FROM PILLAR TO POST *A fast courier system – the* cursus publicus – *conveyed messages and officials alike around the empire on horseback or in horse-drawn carts.*

BALANCING THE BOOKS *Running an empire is expensive, and heavy taxes were levied in all provinces. Here, disgruntled Germans pay their annual dues to Roman 'tax farmers'.*

STAMP OF AUTHORITY *London was the largest centre of administration in Roman Britain and housed the offices of the procurator, who made his mark with this stamp.*

LEGAL COMPLEXITIES *The Romans introduced their complicated legal system to all provinces, thereby creating lucrative business for lawyers. Civil cases – which included theft and fraud – were tried by locally elected magistrates and were generally heard in the hall of the local government building.*

Bookish bureaucrats hold the reins of power

HOW CHINA'S SONG EMPERORS SELECTED THE FIRST CIVIL SERVICE 'MANDARINS'

THE CANDIDATES WAITED NERVOUSLY FOR THE examination to begin. They looked about at the imposing surroundings of the imperial palace in Lin'an (modern Hangzhou, in southern China). Then they looked at each other, aware that they were the cream of many thousands who had applied to sit the three-yearly examinations that controlled entrance to the Chinese civil service. After a first, local test and a second in their respective provincial capitals, they had reached the final hurdle. Those who passed the examination would be awarded the *Jinshi*, a coveted degree that marked a civil servant.

The Chinese civil service was established by the Qin emperors (221-206 BC). Staff were recruited on the basis of local officials' advice, a system that remained in place until the later Sui dynasty (AD 589-618) broadened the selection process by introducing central examinations for civil service candidates. These were expanded and greatly improved by the Song emperors (AD 960-1279), who developed the central and local school systems set up hundreds of years earlier during the Han dynasty (206 BC-AD 221) and established the *Jinshi* as the central test for prospective candidates.

No place for nepotism

The Song also developed an elaborate system to eliminate favouritism and corruption in the examinations. Papers were not marked with the student's name and three examiners read each one.

The changes introduced by the Song opened the way for a new type of bureaucrat to enter the service on the basis of talent alone – these elite scholars were later nicknamed 'mandarins' by westerners, referring to the Mandarin dialect of Chinese that they spoke and wrote. Periodically, there were complaints that the examinations did not test practical skills. But far from being simplified, the tests were made more formal. Candidates under the Ming dynasty (1368-1644) were asked to write the *baguwen* ('eight-legged essay'), which dealt with subjects in 700 characters or less. The essay had to follow a stylised structure of eight sections, including an introduction, a transition and a conclusion.

Model civil service

The dynasties that followed the Song made further practical alterations – the Ming, for instance, set up district recruitment quotas because they felt that too many government servants were drawn from the towns. But the Song reforms set the pattern for a service that lasted largely unchanged until the overthrow of imperial rule in China in 1912. It was also copied in western countries, notably in the renowned civil service that the British created to rule an empire in the 19th century.

REFORMER *Zhao Kuangyin, the first Song emperor, seized power in 960. A highly skilled diplomat, he bequeathed a stable government to his heirs.*

CITY LIFE *Business thrived under the Song, and large cities grew up. Such a vibrant empire needed efficient administration.*

IMPERIAL CELEBRATION *Soldiers gather to pay their respects to the emperor after army manoeuvres. The Song army lost prestige as the centralised civil service grew in importance.*

TENSE WAIT *Scholars form an expectant crowd ouside the imperial palace, where an official is due to post the names of those who have passed the civil service examination. For many, years of preparation have led to this moment. The price of success or failure is high, and the pressure of numbers is correspondingly great: only one candidate in 100 will be permitted to join the service. In the Song era, most successful applicants were in their mid thirties and many men tried again and again – some as many as 15 times – before they passed, or gave up. It was very difficult for poorer candidates to fund such a long period of education. Those who ultimately failed often became teachers, preparing others for the gruelling examination. Students were tested on the writings of Confucius, the great Chinese philosopher and teacher of the 5th century BC, and the work of his followers. A thorough knowledge of this literature was believed to be the mark of a good citizen.*

UNDENIABLE MAJESTY *Hauled by labouring elephants, the splendidly decorated carriage of the Ming emperor Wu Zong (1506-21) is the centrepiece of a grandiose procession. The civil service reforms resulted in a concentration of power in the emperor's hands, because the bureaucrats were dependent on his favour for their position and power.*

A king takes stock of a conquered nation

HOW ROYAL COMMISSIONERS COMPILED DOMESDAY BOOK

AT CHRISTMAS, 1085, WILLIAM OF NORMANDY, RULER by conquest of England, summoned the great barons and churchmen of the land to his court at Gloucester Abbey, in the west of England. He announced the greatest enquiry the kingdom had ever seen, to find out how much of the country's land and livestock was his as king, what was held by each of his barons and churchmen and what annual dues were lawfully his from each of the administrative units – the shires – into which the country was divided.

COUNTY SURVEY *A page of the Warwickshire section of Domesday Book shows what lands were held by William fitzAnsculf. The census for the book covered all but the most northern areas of England.*

ARRIVING *People from the surrounding area come to give evidence to the royal commission at Pickering Castle, one of many sites where evidence was taken.*

TESTIFYING *In the Great Hall, a landholder stands before the panel of commissioners to answer the questions put to him by the presiding bishop. The farmer has brought witnesses (seated, foreground) to confirm his claims. A monk makes careful notes of the evidence, while other local tenants await their turn to testify.*

WHAT THE DOMESDAY ENTRY SAYS

XXVII THE LAND OF WILLIAM FITZANSCULF

• WILLIAM fitzAnsculf holds of the king ASTON, and Godmund holds of him. There are 8 hides. There is land for 20 ploughs. In demesne is land for 6 ploughs, but the ploughs are not there. There are 30 villans with a priest and 1 slave and 12 bordars have 18 ploughs. There is a mill rendering 3s, and woodland 3 leagues long and half a league broad. It was worth £4; now 100s. Earl Edwin held it.

• From William, Stenkil holds 1 hide in WITTON. There is land for 4 ploughs. In demesne is 1 plough, and 2 slaves; and 1 villan and 2 bordars with 2 ploughs. It was worth 10s; now 20s. The same Stenkil held it freely.

• From William, Peter holds 3 hides in ERDINGTON. There is land for 6 ploughs. In demesne is 1 plough, and 2 slaves; and 9 villans and 3 bordars with 4 ploughs. There is a mill rendering 3s, and 5 acres of meadow. There is woodland 1 league long and a half broad, but it is in the king's preserve. It was worth 20s; now 30s. Earl Edwin held it.

• From William, Drogo holds 2 hides in EDGBASTON. There is land for 4 ploughs. In demesne are 1½ ploughs; and 3 villans and 7 bordars with 5 ploughs. There is woodland 3 furlongs broad and half a league long. It was worth 20s; now 30s. Aski and Alwig held it freely.

• From William, Richard holds 4 hides in BIRMINGHAM. There is land for 6 ploughs. In demesne is 1 plough; and 5 villans and 4 bordars with 2 ploughs. There is woodland half a league long and 2 furlongs broad. It was and is worth 20s. Wulfwine held it freely TRE.

WHAT THE DOMESDAY ENTRY MEANS

William fitzAnsculf – *William, son of Ansculf*

'William fitzAnsculf holds of the king . . . Godmund holds of him' – *Legally, all land was held by King William, who granted it to barons in return for their loyalty, taxes and military backing when needed. The king granted Aston to William fitzAnsculf, who held other lands elsewhere and in turn granted Aston to Godmund.*

A **hide** *was a unit for measuring land, often equivalent to 120 acres (48.6 hectares). It was used to calculate rates of tax and the levels of military service due to the king or other overlord. In 1084, King William imposed a land tax of six shillings per hide.*

'There is land for 20 ploughs' – *Enough land to support 20 teams of oxen. Teams normally consisted of eight oxen.*

'demesne' – *The lord's land, retained for his own use and profit rather than granted to a subtenant.*

'villans' – *The highest-ranking group of unfree farm workers, who were tied to working the lord's lands – although they often worked their own substantial holdings as well.*

'slaves' – *Unfree peasants who worked on the lord's land on his behalf.*

'bordars' – *Similar to villans, but with less status and smaller holdings of land – roughly equivalent to modern smallholders.*

'rendering 3s' – *This refers to the level of rent that the mill could attract: three shillings (15p) a year.*

'leagues' – *1 league was equivalent to about 3 miles (4.8 km). Woodland was a very valuable resource. It could be used for keeping pigs or the trees could be felled and sold for timber or firewood.*

'It was worth' – *In 1066, before the Norman invasion.*

'now 100s' – *Worth £5 in 1086, when the survey was conducted.*

'Earl Edwin held it' – *One of the inspectors' tasks was to find out who had held land before 1066.*

'Stenkil held it freely' – *Stenkil had held the lands in his own right and not from an overlord. The complex system of land grants was relatively new in 1086, having been extended by King William and his followers in the years after 1066.*

'in the king's preserve' – *This land remains under the king's control. It was probably used by King William and his nobles for hunting and hawking.*

'3 furlongs' – *A furlong is 220 yards, or 201 metres.*

' TRE' – *An abbreviation for the Latin* tempore regis Edwardi, *meaning 'in King Edward's time', that is before the 1066 Conquest. The commissioners had to compress a vast amount of information into a single book, and used many abbreviations.*

William's advisers divided England south of the River Tees into at least seven circuits, each of which was split into individual itineraries for three or four commissioners, recruited from the Church and aristocracy. They started to gather information in early 1086, holding a series of court hearings at which local representatives – including the sheriff, priests, public officials and ordinary villagers – gave sworn answers to a series of set questions.

Asking questions, checking answers

Everywhere they went, the commissioners asked the names of landholders in each county, the manors they held and their values, and the numbers and names of their immediate subtenants on each manor. Next, the king wanted to know about the resources of each area: how the land was farmed, how many plough teams, what sort of livestock was kept and how many mills there were, for example.

After this round of questions, William sent commissioners out again to crosscheck the information that their colleagues had been given, and to resolve any uncertainties. The whole enterprise was accomplished with breathtaking speed. By August 1086, many of the returns had been completed, perhaps helped by existing tax lists, and records of farm rents and land values. By September 1087 most of the information had been sifted and tabulated in Winchester, inscribed on parchment by one man, and checked and annotated by another.

William's men did their work well and a thorough survey resulted. The Anglo-Saxon Chronicle stated that 'not even one ox, not one cow, nor one pig escaped notice'. Its sheer scope reminded people of the Last Judgment, or Doomsday, and before long it was referred to simply as Domesday Book.

CHECKING *A second round of crosschecking followed the first, and the full penalty of the law would fall on anyone misleading the commissioners. Here a commissioner has discovered an undeclared horse.*

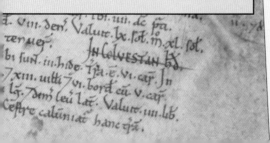

Law of the chieftains

HOW THE WORLD'S FIRST PARLIAMENT MADE DECISIONS

SINCE EARLY MORNING, CHIEFTAINS ON HORSEBACK from every corner of the land had been galloping down onto *Thingvellir*, a vast meeting plain surrounded by volcanic mountains in south-west Iceland. Each chief dismounted from his panting horses and joined the throng already gathered there to celebrate Iceland's midsummer holiday. The year was AD 930, and the arrival of high summer marked the onset of the *Althing* – the annual assembly of the world's first parliament.

While the chieftains, or *godar*, 36 of them in all, conversed with old friends and advisers, teams of men pitched tents across the plain and stallholders peddled ale to the crowd. Meanwhile young men hoping to find a wife at the *Althing* staged impressive mock battles on horseback to demonstrate their riding and fighting prowess.

Let the ceremony begin

Suddenly the crowd grew hushed and all eyes turned towards 'The Law Rock'. From this mound, the high priest, the *Allsherjargodi*, consecrated the plain and blessed the assembled crowd. They bade all present to lay down their arms and prepare to debate in peace. The *Althing* was in session.

Midsummer in Iceland is a period of uninterrupted daylight, and over the course of the next two weeks the business of the land was conducted in open-air public debates. In an age of rule by kings and tyrants, Iceland was a unique republic, and the duties that normally fell to one man were discharged instead by the *godar*. Ranking below the chieftains was the court of justice, a panel of judges made up of householders (elected by the *godar*) who helped to settle disputes. Every farmer was legally bound to belong to a chieftaincy, or *godord*, but he was permitted to switch allegiance from one *godar* to another as he pleased.

Much of the *Althing*'s business was the settling of disputes. The main event, however, was the *lögrétta*, or great council meeting, at which only the *godar* were permitted to debate and vote.

The main event

The *lögrétta* was conducted by the *lögsögumadur*, or lawspeaker, who was elected by the *godar* for a three-year term. The chieftains and their advisers all sat round him, while spectators watched the proceedings from the surrounding grassy slopes. It was the lawspeaker's duty to recite any relevant laws and settle knotty points of interpretation, but only the great council of chieftains was allowed to draft new laws and make important decisions of state, such as whether to make treaties with foreign powers.

When the fortnight of decision making, storytelling and merriment came to an end, the *Althing* was dissolved in a ceremony called *vapnatak*. Ritual demanded that all those present take up their arms once more, and with a clashing of swords that raised a great cheer from the crowd, confirm their intent to abide by the decisions that had been taken by the national parliament.

POETIC ORATOR *The great Icelandic poet Egill Skallagrímsson recited poems about his adventurous life to the annual* Althing *assembly.*

Tenants on the king's land

HOW SOCIETY WORKED IN MEDIEVAL EUROPE

ROYAL HOMAGE
*Richard the
Lionheart pays
homage to Philip II
of France. Although
Richard I ruled
England, and was in
command of a huge
army, he was obliged
to pay homage to
Philip for the French
lands he held.
Paying such homage
at public ceremonies
helped to cement the
bonds of
responsibility
between the lords
and their vassals.*

IT WAS A SIMPLE CEREMONY. IN THE GREAT HALL OF the manor a knight of the realm stood silent, accompanied only by one or two of his retainers. Facing him was one of the great men of the kingdom, perhaps an earl. The knight, in a symbolic gesture of submission, had ungirded his belt and removed his sword and spurs. His head was uncovered. He knelt down, placed his hands inside those of the man about to become his overlord, and in a few brief words swore his allegiance. The vassal then rose and kissed his lord on the cheek. The kiss was returned, on the mouth, by his lord.

The chain that bound free men

The bond between lord and vassal was the cement of society in medieval Europe, under a system that was dubbed 'feudalism' by 19th-century historians. This society has been likened to a vast pyramid, with the monarch at its apex. In theory, the sovereign owned every square inch of land in his domain. Even the greatest landowners were vassals of the monarch; they in turn parcelled out their estates to lesser tenants – knights or gentry. Such a land grant was sometimes called a 'fief' or, in Latin, *feudum*. The vassal agreed to support his lord, and the lord pledged himself to protect his vassal's life and property.

Martial horsemen

The knight's duty to ride into battle for his lord was the essential feudal obligation. But since maintaining a large force of knights was a very costly affair, the king gave estates to his closest and most trusted followers to support them in the enterprise. In time the ownership of these lands, together with the title of lord, became hereditary, and vast areas of Europe settled into rule by a territorial aristocracy. The theory of exchanging land for service was beautifully simple, but the reality was far more complex. Monarchs increasingly found it easier and cheaper to raise armies of mercenaries when they needed them, and personal service by the knights was supplanted by cash

payments. A system that had initially protected the peace of the realm now began to have the opposite effect, for it created powerful men with private armies. It spawned an almost endless series of conflicts between rival landowners – such as the 'Wars of the Roses', the contest waged by the houses of York and Lancaster in England in the 15th century.

ON THE LORD'S ESTATE *The humblest workers in the fields were serfs. They took orders from the local landowner and worked land that he kept for his own use and profit.*

FEUDAL MYTH *Kneeling before King Arthur, two knights pledge loyalty and faithful service in this illustration from a 15th-century French manuscript. Medieval customs were gradually incorporated into the legends of the past – such as that of this fearless 6th-century British king.*

The judgment of God

HOW PEOPLE FACED TRIAL BY ORDEAL AND BY BATTLE

SINK OR SWIM *A man undergoes trial by water. Attended by clergymen, he is stripped, tied up and lowered on a chain into the middle of a lake. If God allows the water to receive him he will be deemed innocent, but if, perhaps by wild struggling, he appears to stay on the surface, it will be taken as a sure sign of his guilt.*

BURNING BAR *A priest holds up the Bible as a boy assistant hands a red-hot iron bar to a man accused of assaulting a neighbour. He swears to his innocence then has to carry the bar in his bare hands for nine agonising steps, with the help of support from behind. A few days later, his hand will be inspected. If the burns appear to be healing well, he will be proclaimed guiltless and released.*

THE PRIEST AND THE ACCUSED MAN ENTERED THE church side by side and slowly approached the altar, before which a pot of boiling water containing a stone stood bubbling above a wood fire. The man was accused of stealing from and assaulting a neighbour. He denied both charges vigorously, and no one had been able to gather enough evidence to conclusively prove him either innocent or guilty.

As a final means of resolving the matter he had been condemned to be tried by ordeal using boiling water. The rules for this had been laid down in Anglo-Saxon England, perhaps as early as AD 700 during the reign of King Ina of Wessex.

God as the ultimate judge

Silently, the witnesses to the trial entered the church. Those whose loyalties lay with the accused man stood to the left of the boiling pot; those who supported the accuser gathered to the right. For the past three days the prisoner had attended mass, and had taken no nourishment save for bread, water, salt and herbs. His fate rested in the hands of God. The people believed that He would intervene to protect an innocent from injustice.

The priest sprinkled holy water over the witnesses, and gave them a crucifix to kiss. They remained quiet, and prayed to God that justice might be done. The accused then plunged his arm up to the elbow into the boiling water and hastily lifted the stone from it. For a less serious crime he would only have had to submerge his hand up to the wrist.

Three days later the bandage would be removed. If it was suppurating or discoloured, it was judged to be unclean – a sign that God considered him guilty. A specially convened court would then mete out punishment. But if the burns were considered to be healing well, he would be pronounced innocent. Other forms of ordeal included walking on burning ploughshares or grasping a red-hot iron bar.

The ordeal always had its critics and in 1215 the pope ordered priests not to take any further part in such trials. Without their support the custom soon vanished throughout Europe. But an alternative – trial by combat – remained valid as a means of settling legal disputes; from arguments over the ownership of land to murder charges.

The parties involved in the dispute sometimes chose champions to fight for them. Wearing red stockings to help hide any traces of blood, the professional champions would hack at each other with swords or axes until one of them gave in, so settling the matter.

TRIAL OF HONOUR *According to legend, the Holy Roman Empress Cunigunde opted for trial by fire when she was accused of adultery. Watched by her husband, Henry II, she walked barefoot over a glowing grate, aided by two bishops. She passed the test.*

Knights and commoners

Often, however, the principals preferred to fight in person. Knights battled it out on horseback, armed with swords or lances; commoners fought on foot with wooden staves tipped with iron.

After the Middle Ages, trial by combat became a forgotten right until 1817, when an Englishman named Thornton was accused of murdering Mary Ashford. Thornton challenged his accuser, the dead girl's brother, to a duel – and the law courts upheld his right to do so. But Mary Ashford's brother refused the challenge, and Thornton got off scot-free. The next year Parliament abolished trial by combat, stating that it was 'a mode of trial unfit to be used'.

RIGHTING WRONGS *People involved in a case could appeal against the verdict to the Court of King's Bench in London. This court had supreme jurisdiction over both criminal and civil cases. The appeals were heard before an imposing bench of red-cloaked judges.*

DOUBLE DEFEAT *A legal duel over some disputed land is fought in 13th-century England between Walter Blowberme and the nation's leading champion, Hamun le Stare. Blowberme wins the battle and le Stare, as the beaten champion, is sentenced to be hanged. This depiction, from a manuscript in the Tower of London, shows both the conflict and Hamun's ignominious end.*

Rope, rack, fire and water torture

HOW THE SPANISH INQUISITION TRIED HERETICS

A LONG LINE OF PRISONERS ACCUSED OF HERESY BY the Spanish Inquisition was herded towards the torture chamber. One by one the victims – men and women alike – were thrust into an underground cell where a team of black-hooded men waited to make them recant their beliefs and thereby save their souls. The Inquisition was headed by a fanatical Dominican monk named Tomás de Torquemada, who in 1483 persuaded Spain's joint rulers, Ferdinand and Isabella, to let him 'denounce all heretics'.

A glimpse of Hell

Torquemada's main targets were *conversos* – Jewish and Moorish converts to Catholicism who had returned to their original faiths. The heretics were arrested and those who refused to recant were, as a last resort, handed over to the torturers.

Although the Inquisition is today seen as a synonym for persecution and torture, in many cases defendants were sentenced to no more than fasting, prayer and the confiscation of their goods and property, as had been the case in earlier inquisitions in southern France and northern Italy during the 11th and 12th centuries. The Spanish Inquisition was, however, more brutal than its predecessors. Pope Sixtus IV who authorised it, in 1478, found the

inquisitors so severe that he tried to limit their powers. But the Spanish monarchs – who had only recently married to unite their kingdoms of Aragon and Castile – saw in the Inquisition a powerful weapon with which they could enforce political and religious unity, so Sixtus's efforts were in vain.

Often one glimpse of the dark, windowless torture chamber and its silent attendants was enough to make a prisoner recant. Those who did not were stripped naked and shown the instrument of torture to be used on them. If this failed to break their will, the torturers often set to work. In the late 15th century torture was authorised throughout most of Europe by the secular courts. Luckier victims would be held for questioning in cells in private houses.

TRIAL BY TERROR *Wearing a coroza, a pointed dunce's cap that symbolised the guilty, a victim of the Inquisition sits with head bowed as his accusers itemise the charges laid against him. To the side, other prisoners await their turn to be tried by the religious tribunal. If found guilty, they will be given every chance to recant. But if they refuse to do so, they will be sent to the torturers – who will make a final attempt to 'persuade' them.*

Torture in the name of God

One frequently used instrument of torture was the hoist, in which iron weights were tied to a victim's feet while his arms were fastened behind his back. A rope – attached to a pulley on the ceiling – slowly drew the sufferer upwards, his arms taking the weight of his body. Other ordeals included being hung upside down from a T-shaped frame, stretched on the rack, or tied to iron rings on the floor with fires burning beneath the bare feet.

According to the Inquisition's Holy Office, or rule book, no prisoner should be tortured more than twice – and torture sessions should last no longer than 15 minutes. But prisoners, some accused of offences as slight as smiling at the name of the Virgin, could be held for an indefinite period, as the inquisitors often added uncertainty and questioning at irregular intervals to produce a confession.

Despite the agony they endured, many heretics still refused to submit. They were then turned over to the civil authorities, who alone had the power to condemn them to death. With their heads shaven, and wearing tunics showing people burning in Hell, they were tied to stakes on the outskirts of town and burned alive in order to purify their souls and so allow them into heaven. Those who confessed their sins at the last minute were shown a degree of mercy: they were strangled before the fires were lit.

MERCILESS MONK *In Torquemada's 15 years as Grand Inquisitor in Spain, more than 2000 religious victims were burned at the stake. In 1494 complaints about his cruelty led Pope Alexander VI to appoint four deputy inquisitors to try to control him.*

FATAL FAITH *Parallels with the Crucifixion were drawn at the execution of many of the Inquisition's victims, who were tied to crosses before being burned at autos-da-fé, or 'acts of faith'. The ceremonies were regarded as warnings to the godless and those who abandoned Catholicism.*

ROAD TO CRUEL DEATH *Doomed heretics carry the crosses on which they will be crucified to their place of execution. They are stripped half-naked as a sign of their sins. Torquemada's persecution of the Jews culminated in a royal edict expelling 170 000 of them from Spain in 1492.*

The Sun King at play

HOW LOUIS XIV RULED OVER THE ROYAL COURT AT VERSAILLES

SETTING FOR A PAGEANT *The Basin of Apollo and Grand Canal at Versailles serve as a spectacular backdrop for Louis' endless, extravagant round of royal revels, picnics, plays and fêtes.*

'MY DOMINANT PASSION,' KING LOUIS XIV OF FRANCE confided in his memoirs, 'is certainly love of glory.' He lived his life as one long theatrical performance, with every action, every gesture, every word chosen and rehearsed for maximum effect.

A scene-stealing showman

High on the list of Louis' public appearances were his sumptuous outdoor royal fêtes, or carousels – celebrations held to mark great royal occasions, such as births and marriages. At these, as a young man, the king would sometimes take part in ballets created for him. He would always make a flamboyant entrance, on one occasion mounted on a bejewelled horse and wearing a silver-and-gold breastplate.

To commemorate one particularly splendid fête in the Tuileries gardens in Paris in the early 1660s, a medal was specially made for the 24-year-old king depicting the Sun rising over the Earth. From then on Louis was known as the 'Sun King' – a name entirely appropriate for a monarch of Louis' consuming vanity. As the Sun dazzled, so Louis would dazzle,

but even more brightly. As the planets revolved around the Sun, so every great nobleman in France danced attendance around Louis. And just as the rising and setting of the Sun framed the daily routines of every man and woman, so the two great ceremonies of the day at the court of Versailles were the rising and the setting of France's earthly star.

A divine right to rule

Louis, born in 1638, became king in 1643, but France was ruled by his mother and, until his death in 1661, the chief minister, Cardinal Mazarin, had encouraged the king to believe that he ruled by 'divine right', answerable only to God, and supremely powerful on Earth. Louis always remembered his minister's dying words: 'Never have a prime minister. Govern!' His officials were ordered to address all matters of state to the king himself. Though he almost certainly never spoke the famous words 'L'État c'est moi' ('I am the state'), they did express Louis' belief in absolute monarchy.

Life at Versailles, Louis' magnificent estate near Paris, where the court was established from 1682, followed two cardinal rules: nothing was to be done unless it pleased the king, and everything that was done should glorify him. Courtiers competed to flatter the king. When Louis asked the Duc d'Uzès when his wife expected to give birth, he received the reply, 'Ah, Sire, whenever Your Majesty pleases.'

NEVER ALONE *Louis, in his bath chair, is wheeled about the grounds of Versailles in one of his daily promenades. Observing the cluster of courtiers around the king, a visiting Italian nobleman likened the scene to that of 'a queen bee flying across the fields surrounded by her swarm'. The rules of the court meant French royalty never knew privacy.*

PUBLIC PIETY *The king's personal prayer book depicts him at prayer under the ever-watchful gaze of the lords and ladies of the court.*

Playing games with the king

To be excluded from Versailles was to be a social outcast, and great noblemen abandoned their estates and gave up political careers merely to become Louis' fawning courtiers. The king had little interest in literature or learning, and regarded new ideas as subversive, so the day was taken up with childish games and hunting; the evening in music and dancing. However, the numbers of people in the palace caused overcrowding, and a chronic shortage of toilets meant that courtiers sometimes had to disappear behind curtains to relieve themselves.

The king was never alone, day or night, and the tiniest gesture of approval meant everything to the recipient. To see the king eat was an honour indeed; to be addressed by him during dinner a supreme mark of royal favour; and actually to eat with him or serve him his food was a glimpse of Paradise on Earth. Even when Louis was not present in a room, everyone behaved as if he were. No one would dream of turning their back on the king's portrait, or of entering his dining room or bedchamber while wearing a hat or without bending a knee.

Louis ruled with an iron grip throughout his life, but his mismanagement of the country's finances and defeat in a succession of wars were to contribute to the downfall of the Bourbon dynasty within 80 years of his death. His great grandson (Louis XV) and the latter's grandson (Louis XVI) continued the royal tradition of high living, but neither could match Louis for charisma or decisiveness. The monarchy was finally toppled in the bloody revolution of 1789.

A KING'S DUTY *Louis rises from his silver, sunburst-decorated throne to receive the Doge of Genoa – who traditionally never leaves his city, but has come to Versailles at the Sun King's insistence. Also present at this audience, held in 1685, are the king's brother Philippe and his nephew, young Philippe (right), and his heir, the Dauphin (left).*

PAYING HOMAGE *Knights of the Order of St Louis kneel at the king's feet. The court seethed with plots as factions jostled for royal favour.*

L'ETAT C'EST MOI *Louis, seated, presides over a council meeting; the other members must stand. The king personally guided every aspect of government, and to disagree with him was tantamount to treason.*

FADED GLORY *A wax portrait shows Louis in 1705, aged 67. Missing most of his teeth, and plagued by gout and other painful ailments, the Sun King was beginning his final decline.*

A town in torment

HOW THE SALEM WITCHES FACED THEIR FATES

AS AN OLD WOMAN WAS LED BEFORE THE JUDGE, JURORS, and the packed court, the group of young women waiting in the front seats went into convulsions. Trembling, they threw themselves from side to side in fits and spasms, uttering frenzied screams as though they were being pinched and choked by invisible hands. The old woman obeyed the clerk of the court's order to touch each of the girls in turn, and as she did so they became calm. Surely, this was proof that the woman had Satanic powers over the girls – that she was, in fact, a witch?

An outbreak of voodoo

The events that unfolded in the quiet Massachusetts village of Salem – just west of the busy port of Salem Town, and 12 miles (20 km) north of Boston – between May and October 1692 resulted in 20 people, all of them innocent, going to their deaths having been convicted of witchcraft; nearly 200 more were arrested. The charge was one of 'afflicting' a number of girls and young women and being in league with the Devil.

The village of Salem was an offshoot of the first Puritan settlement at Plymouth, and along with Salem Town formed the nucleus of the new British colony of Massachusetts. Early in 1692 some of the village girls, aged between 9 and 19, began to suffer fits of sobbing and shrieking. They writhed, ran about, and flung themselves against walls.

All of them had spent hours listening to the fortune telling and voodoo folk tales of Tituba, a West Indian slave in the household of the Reverend Samuel Parris. Indeed it was Parris's own daughter, Elizabeth, and her cousin, Abigail Williams, who were the first to show hysterical symptoms. When questioned, they accused Tituba and two unpopular old women of witchcraft. Before long a dozen or so more young women and girls in Salem began to experience the same symptoms.

MAN OF GOD
The minister of Salem, Samuel Parris, was shocked when his nine-year-old daughter was afflicted by the hysteria. As a 'cure', he sent her away to live with relatives.

PLEADING HIS INNOCENCE *In a 19th-century painting, female accusers at one of the Salem witchcraft trials point at George Jacobs, who is charged with bewitching them. 'You may tax me for a wizard,' he shouts. 'You may as well tax me for a buzzard. I have done no harm!' Despite his spirited response, Jacobs was sentenced to be hanged.*

In flesh and in spirit

When questioned at the pre-trial examination they would 'cry out' the names of those whom, they claimed, tormented them through 'spectres' that took on their physical attributes. This 'spectral evidence' became central to the trial proceedings as well, for when the girls faced the defendant, they would go into fits, imitating the accused's gestures and words.

A special court was set up in Salem Town in May by the first royal governor of Massachusetts, Sir William Phips. He appointed his lieutenant governor, William Stoughton, known for his strictness, to preside over the trials. Stoughton and his fellow judges, including Samuel Sewall – who later repented – believed in the power of witchcraft and in the sincerity of the girls' afflictions.

Taking their word for it

Although the accused were frequently put under great pressure to confess, many did so willingly, since, if they did, they would be reprieved and set free; if they did not, they would be hanged.

The judges accepted unquestioningly any 'spectral evidence', and accusations spread from unpopular old women to others with spotless reputations. Soon, simply being a relative or associate of someone who had been accused was enough to bring a person under suspicion. Accusations then began to spread to residents of surrounding villages, and prominent members of the community. Among these were John Willard, a deputy constable who had suggested that the girls themselves be hanged, the Reverend George Burroughs, a former minister of Salem village, and even the royal governor's wife. By the end of the summer, however, people were beginning to turn against the girls, and distrust their testimonies.

When Sir William returned from a trip to the Canadian frontier in September 1692 and heard of all this, saw the overflowing jails, and registered the protests of responsible citizens, he stopped the special court sessions. The last hangings took place in that month; the remaining prisoners were eventually either acquitted or reprieved.

In all, 30 of the 200 people accused were sentenced to death. Of those, 19 were hanged, two died in prison, and Giles Corey, who refused to plead, was crushed to death by heavy stones. Of the other eight, two executions were delayed, one escaped and five confessed in order to escape the sentence.

Monday; Sept.r 19. 1692. aft. noon at Salem, Giles Corey was press'd to death for standing Mute

PRICE OF SILENCE *In his diary Judge Samuel Sewall recorded that Giles Corey was 'pressed to death' for refusing to make a plea.*

WRONGFUL DECISION *Judge Sewall played a major part in sentencing 30 people to death at Salem. Later, he was the only justice to admit that he had sinned.*

MARK OF THE DEVIL *A young woman accused of witchcraft is examined at her trial to see if she bears the so-called Devil's Mark. It was believed that Satan placed a piece of cold flesh like a wart on the body of anyone who entered into his service. He then sucked blood through what was known as the 'witch's tit'.*

ON PARADE *Ladies and gentlemen of fashion stroll with aristocratic prisoners through the courtyard garden of the Bastille. The inmates enjoyed playing 'host', and eagerly exchanged the gossip of the day with their social equals.*

Inmates of a luxury fortress

HOW PRISONERS WERE KEPT IN THE BASTILLE

SHORTLY AFTER THE STORMING OF THE BASTILLE ON July 14, 1789, fanciful engravings went on sale on the streets of Paris giving artists' impressions of conditions inside the notorious jail. Scores of prisoners apparently languished in chains in the rat-infested dungeons, next to skeletons still pinned upright in standing positions. Visitors could also tour the site: they were taken around the grisly dungeons and shown what were claimed to be the cruel instruments of torture the jailers had employed. In fact, these were nothing more than an old suit of armour and part of a printing press. The images in the engravings were the product of the artists' fevered imaginations, fuelled by the tumultuous events of the time and the accumulation of countless legends about the Bastille.

'The wine was not excellent, but was passable. No dessert: it was necessary to be deprived of something. On the whole I found that one dined very well in prison.'
JEAN FRANCOIS MARMONTEL, BASTILLE INMATE 1759-60

Built as a fortress in the 14th century, the Bastille, situated on the edge of the crowded and impoverished quarter of Faubourg Saint-Antoine, was a grim building with eight round towers and walls 5 ft (1.5 m) thick, surrounded by a dry moat. It had long served as a jail, gradually acquiring the reputation of a place where the innocent were incarcerated and condemned to oblivion. By the time Louis XVI came to the throne in 1774 the prisoners were not common criminals but people arrested at the express orders of the king or his ministers using a *lettre de cachet* (a special warrant bearing the king's seal) for offences such as conspiracy and subversion.

Among the most illustrious of the Bastille's inmates was the writer François-Marie Arouet, who took the pen name of Voltaire in 1718, while serving an 11-month sentence for lampooning the Duke of Orleans. During his imprisonment he rewrote his tragedy *Oedipe* (*Oedipus*), which was successfully performed in Paris a short while later. Another celebrated prisoner was the Marquis de Sade, imprisoned for gross indecency from 1784 until a week before the Bastille was stormed.

By the 1780s the Bastille was occupied by only a handful of inmates, guarded by a contingent of *invalides* – soldiers who had been invalided out of

regular service. Apart from the misery of incarceration, conditions for most prisoners were luxurious compared to those of inmates in other prisons. Treated in accordance to their rank, the prisoners had relaxed visiting hours and cosy, furnished lodgings which they were allowed to decorate with their own possessions.

The release of 'Julius Caesar'

To the people, however, the Bastille was a potent symbol of absolute royal rule. So, when the Parisian mob rose in violent revolution it was an obvious place to attack. After an assault lasting into the early evening, the Bastille fell at the cost of some 100 lives – including that of the governor, the Marquis de Launay, whose head was marched through Paris on the end of a pike. It was a high price to pay for the liberation of the prisoners. There were just seven of them: four forgers; the Comte de Solages, held for sexual misdemeanour; and two lunatics. One of these was a gaunt figure with a waist-length white beard. He was called Major Whyte, variously described as English and Irish, and being paraded triumphantly through the streets of Paris only served to confirm his delusion that he was Julius Caesar.

TAKING THE FORTRESS *The few soldiers guarding the Bastille were no match for the armed mob which stormed the prison intent on seizing the 250 barrels of gunpowder kept there.*

THE HIGH LIFE *Jean Fragonard sketched visiting day at the Bastille in 1785. The prison's well-bred inmates hold court in one of the fortress's spacious, high-ceilinged 'reception rooms'. The dank and infamous underground dungeons were no longer in use. Instead, the inmates were housed in airy rooms in the tall towers, where they led lives of comparative luxury. They were given a generous 'spending allowance', plenty of tobacco and alcohol, and even allowed to keep their own pet dogs and cats.*

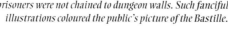

CREATING A MYTH *Despite this artist's impression, the prisoners were not chained to dungeon walls. Such fanciful illustrations coloured the public's picture of the Bastille.*

Cruel punishment for speaking out

HOW DUCKING STOOLS AND SCOLDS' BRIDLES MADE A MOCKERY OF JUSTICE

THE LANDLORD OF THE QUEEN'S HEAD ALEHOUSE IN Kingston upon Thames, near London, was tired of being nagged by his wife. So, in April 1745, he appealed to the local magistrates to settle their quarrel. To his wife's dismay she was found guilty of 'scolding', and was sentenced to be ducked in the River Thames under Kingston Bridge. Some 3000 people lined the banks of the river to watch the unfortunate woman being strapped into the ducking stool – hung from a pulley attached to a beam in the middle of the bridge – and then submerged.

At the time, nearly every village and town in England had a ducking stool kept ready for such use. It normally consisted of a tall vertical post planted in the ground with a long, pivoted crossbeam on top. A wooden stool or chair was suspended from one end of the beam, and a rope was attached to the other. The victim was tied in the stool, which was then lowered into the water.

Curbing an unruly tongue
An even crueller way of dealing with outspoken women in the Middle Ages was the scold's bridle. This was a helmet-shaped iron cage with a hinged collar complete with lock, which fitted tightly over the offender's head. It had eye-holes, a slit for the mouth, and a strip of iron which projected backwards and held down the tongue. Sometimes the iron strip was studded with spikes which cut agonisingly into the victim's tongue and mouth. Attached to the bridle was a long chain by which the luckless woman was led through the streets by the town jailer. The chain was tied to a pillory post or market cross and the victim left to be jeered at.

DEATH BY DROWNING *Women found guilty of witchcraft were sometimes drowned in the nearest pond – as happened to this victim in England in 1613. Village elders supervise the operation as the rope-bearers carry out their deadly task.*

UP AND DOWN *One ducking in cold, fast-flowing water was usually considered punishment enough for a 'sinful' woman. Persistent offenders, however, would be ducked several times, while their relatives, friends and neighbours looked on.*

Sharp blade gives a 'slight chill on the neck'

HOW THE GUILLOTINE DISPATCHED ITS VICTIMS

THE CLOCKS OF PARIS WERE STRIKING HALF-PAST THREE on the afternoon of April 25, 1792, when a red-shirted highwayman named Nicolas-Jacques Pelletier became the first victim of the French guillotine. He was led to the scaffold in the Place de Grève, where the public executioner, Charles-Henri Sanson, was waiting. Thousands of fascinated citizens jammed the square, among them the man responsible for introducing this new instrument of execution to France, Dr Joseph-Ignace Guillotin.

A quick death for all
The guillotine consisted of two upright posts, surmounted by a crossbeam, and grooved so that a heavy, diagonal-edged blade could descend between them. The condemned person was strapped to a board called a bascule which was then tipped forward, and the victim's neck was trapped between two pieces of specially shaped wood beneath the blade. The blade was suspended by a length of rope, the end of which was held by the executioner. When the executioner released the cord the blade fell swiftly and forcefully, slicing through the victim's neck. The severed head then dropped into one basket, and the body was pushed into another.

Until then only condemned aristocrats had been beheaded – by an executioner wielding an axe. Even so, some executions were clumsily performed, causing great suffering. 'Ordinary' criminals were sometimes broken on a wheel. Dr Guillotin, a professor at the Paris Faculty of Medicine, sought to make beheadings as humane as possible – and to extend the 'privilege' to all those sentenced to death, irrespective of rank or station. His reforms were incorporated as a humane amendment to the French Penal Code's previous decree that 'Every person condemned to the death penalty shall be beheaded.'

Practice makes perfect

To ensure that the instrument worked properly, Sanson practised on the corpses of criminals taken from the Hôpital de Bicêtre, near Paris. Finally, as a dress rehearsal for Pelletier's beheading, the heads of three corpses of 'Herculean dimensions' were cleanly and neatly severed by the blade. The highwayman's execution went without a hitch – with Sanson doing his duty with what an observer called 'the skill and love of an artist'. And Dr Guillotin stated that: 'The victim did not suffer at all. He was conscious of no more than a slight chill on the neck.'

GIBBET JUSTICE *A wooden gibbet set on a stone base was used to behead thieves in the Yorkshire woollen town of Halifax in the 17th century. In a communal execution the townsmen pulled on a rope which removed a pin that released the gibbet's 8 lb (3.6 kg) blade.*

DOCTOR DEATH *Guillotin traced the origins of 'beheading blades' back to 12th-century Naples, where they were used on criminals of noble birth.*

KING'S HEAD *The head of Louis XVI is shown to a jubilant Parisian crowd after he was guillotined in January 1793.*

Wet ordeal for those who challenged rules

HOW WATER WAS USED TO PUNISH AND TO EXTRACT CONFESSIONS

BY ORDER OF THE CAPTAIN, THE ENTIRE SHIP'S COMPANY gathered on deck to witness the keelhauling of a seaman who had been found guilty of 'gross insubordination'. A Dutch idea, keelhauling was a standard punishment in navies in the 16th and 17th centuries. First of all, the seaman had iron or lead weights attached to his legs. He was then tied to a rope passing crossways under the bottom of the ship and rigged between two yardarms – that is, the ends of the yard, or wooden crosspiece supporting the mainsail. He was dropped into the sea and hauled under the ship – to be badly cut by the jagged barnacles clinging to the keel – and hoisted up to the opposite yardarm. As soon as he regained his breath, the process was repeated.

While under water one of the ship's cannons would be fired to frighten the seaman even more. For lesser offences the culprit was merely ducked in the water from the yardarm and then pulled back up again.

Drip by drip

The ancient Chinese were probably the first to use water as a means of torture. The victims were chained to a cell wall and water from a narrow opening allowed to trickle down on their heads, drop by drop. The sheer repetition drove the prisoners insane. Water torture was later used to gain confessions in Europe and South America. A stream of water was directed from a height of about 6 ft (1.8 m) onto the forehead of a prisoner. Within minutes the pressure had the victim screaming for mercy and ready to confess to anything.

DEATH BY DEGREES *Bound hand and foot, a political prisoner in 16th-century France has large amounts of water poured into a cloth bag placed over his mouth. The bag is gradually forced down into the victim's throat until he chokes to death.*

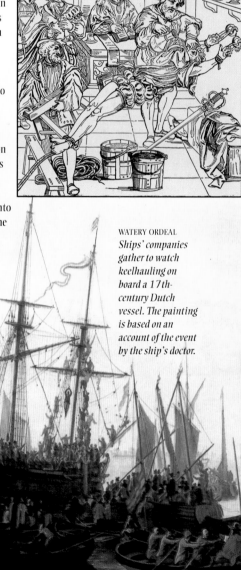

WATERY ORDEAL *Ships' companies gather to watch keelhauling on board a 17th-century Dutch vessel. The painting is based on an account of the event by the ship's doctor.*

Sailing under the skull and crossbones

HOW PIRATES BROUGHT TERROR TO THE HIGH SEAS

THE MERCHANT SHIP *CASSANDRA* SAILED PEACEFULLY into the harbour at Saint-Denis, on the French island of Bourbon (modern Réunion) in the Indian Ocean, on the afternoon of Sunday, April 26, 1721. She dropped anchor alongside a richly laden Portuguese East Indiaman, *Nossa Senhora do Cabo* (*Our Lady of the Cape*), while a second ship, the *Victory*, drew up on the other side of the vessel. All at once, the two newcomers lowered the English flags they were flying and hoisted the feared 'Jolly Roger' – a menacing black flag bearing a white skull and crossbones. Seconds later the cannons of the pirate ships opened fire, sending broadsides into the helpless East Indiaman.

No mercy shown

Driven on by beating drums and braying trumpets, the pirate crews screamed abuse at the Portuguese sailors, brandished swords, fired guns and hurled grenades onto the *Cabo*'s deck. Armed with short, broad-bladed cutlasses, they then swarmed onto the Portuguese ship and mercilessly butchered the crew, whom they outnumbered by almost two to one.

The *Cabo* was carrying diamonds, silks and porcelain worth a million pounds, most of which belonged to the Count of Ericeira, who was returning home on retiring as Viceroy of the Portuguese

FACE OF BRAVERY *The Count of Ericeira showed such courage when attacked by pirates that their leader spared his life on payment of a £400 ransom. The gallant nobleman was then rowed ashore to the accompaniment of a 21-gun salute.*

territory of Goa, on the west coast of India. It was the count, in a dashing scarlet coat, who made a gallant last stand on the *Cabo*'s quarterdeck. Sword in hand, he fought until the blade broke in two, and even then continued to lash out with the diamond-encrusted gold hilt as the pirates moved in for the kill.

Suddenly the pirate leader, Captain John Taylor of the *Cassandra* – which he had captured from British traders the previous year off Madagascar – shouted out the order to stop fighting, and the two pirate crews divided up the fabulous spoils. Rich beyond their highest hopes, most of the seamen – including Captain Taylor – gave up piracy and led less dangerous, more law-abiding lives.

Piracy was at its height in the early 18th century, when freebooters such as Taylor and the captain of the *Victory*, a notorious Frenchman named Oliver 'The Buzzard' La Buse, scoured the Indian Ocean and the Caribbean Sea. The crews consisted largely of ex-Navy men whose ships had been laid up at the end of hostilities such as the War of the Spanish Succession, and who could not settle down to everyday shore jobs. In addition, the crews of many privateers – privately owned vessels commissioned by governments to prey on enemy ships and to share in the spoils – turned to piracy in times of peace.

The buccaneers sailed mainly in fast, narrow-hulled schooners, and in single-masted sloops which could negotiate the sounds and channels in which the pirates hid. They also used sturdy, square-sailed brigantines carrying ten cannons – as well as three-masted ships with crews of up to 200, with 20 or more cannons and large cargo holds for storing booty.

Often the pirates' target – a heavily laden, poorly-armed merchant ship – would surrender on sighting the dreaded pirate flag without a shot being fired. Otherwise, a pirate captain would send a shot across his victim's bows, hoping this would be enough to frighten the merchantman into surrender. To avoid a prolonged fight, raiders would close in on smaller coastal vessels and throw grappling hooks onto their prey's uncluttered stern, and then overpower the crew on deck.

> **'Gentlemen, we want not your ship, but only your money. Money we want and money we shall have.'**
> A PIRATE CAPTAIN TO THE EAST INDIAMAN *DURRILL*, 1697

Different tactics were used with large ocean-going ships. The pirate vessels bore down on these until the bowsprit – festooned with heavily armed men – thrust over the victim's stern. The marauders threw smoke bombs made of rags soaked with tar and sulphur to confuse their opponents, before setting to with cutlasses and pistols.

Piracy gradually died out due to larger and better-armed merchant ships and regular naval patrols of the world's coasts and oceans. It was also condemned globally by governments as an international offence.

TURNING THE TABLES *Sword is met with sword and gun with gun as vengeful British sailors board a Barbary coast pirate ship which had preyed on merchant shipping in the Mediterranean in the early 19th century. After such an encounter the pirates were often shown more mercy than their sometimes unarmed and defenceless victims had been.*

BLACKBEARD, THE MAN WHO TERRIFIED THE AMERICAS

BEFORE GOING INTO ACTION THE INFAMOUS PIRATE KNOWN as Blackbeard – the bane of shipping off the east coast of America – stuck long, slow-burning fuses under the brim of his hat. His smoke-wreathed face, with its wild, staring eyes and tousled hair, made him seem like a creature from hell. His long, bushy black beard was described by the British writer Daniel Defoe as 'a . . . Meteor (which) covered his whole Face, and frightened America more than any Comet that has appeared there for a long time'.

Born Edward Teach, in Bristol, England, in the late 17th century, Blackbeard fought as a privateer before turning pirate in 1716. Sailing to America, he plundered vessels along the coasts of Carolina and Virginia. In May 1718 he audaciously blockaded the port of Charleston, South Carolina, and seized some eight or nine ships which tried to enter the harbour.

In November 1718 a British naval force under Lieutenant Robert Maynard cornered Blackbeard at his base at Ocracoke Inlet, South Carolina. During the battle Blackbeard was shot dead. Maynard cut off the pirate's head, hung it from the bowsprit of his sloop, *Jane*, and sailed home with his grisly trophy.

DEAD BUT DEFIANT *Blackbeard's severed head dangles from the bowsprit of the British sloop* Jane. *The pirate's body was pitched into the sea, where – according to legend – the decapitated corpse swam several times around the ship before sinking to the ocean bottom.*

An audience with the maharajah

HOW AN INDIAN PRINCE RULED HIS VAST TERRITORY

THE BEATING OF GIANT DRUMS RESOUNDED ACROSS THE city of Udaipur, in north-west India, summoning the 16 dukes, or *umraos*, of Mewar to attend the royal durbar at the court of Maharajah Bhim Singh. Although the maharajah's domain had shrunk considerably by the end of the 18th century, he remained an absolute monarch, so when he held a formal state reception, or durbar, the most powerful dignitaries in the land were expected to attend.

Procession to the palace

They arrived in style. Each duke approached the palace on horseback, escorted by lancers and followed by a procession of heralds, banners and drummers. When they reached the palace gates, set high above the city – its granite walls crowned with marble cupolas – the maharajah made them wait.

When at last Bhim Singh stepped out on his balcony, clad in fine white robes, his turban glittering with the most splendid jewels in the royal treasury, the nobles dismounted respectfully and their knight and squires bowed low. The commoners also prostrated themselves and, unaccustomed to such magnificence, shielded their eyes. The nobles

THE KING RIDES OUT *Bhim Singh and his attendants were exquisitely garbed, even when hunting. The Rajputs of Mewar were one of India's most highly respected royal families. In the 10th century, they ruled 12 000 villages.*

then filed into the Hall of Carpets – a large reception room with richly decorated pillars and walls, but containing very little furniture. After a pause, a royal herald announced the imminent arrival of the maharajah, proclaiming his titles, his ancestry and his achievements in battle. Preceding him came a procession of royal standard bearers carrying the royal symbols of authority – a golden sun, a yak's tail, ornamental fly whisks with solid silver handles, peacock feather fans set in gold, and battle trophies. Finally the maharajah himself entered and was greeted with loyal cheers of welcome as he lowered himself onto the only seat in the room – a cushion beneath a golden canopy. With a rattling of shields and swords, the assembled courtiers followed his lead and sat down on the carpeted floor. The durbar was about to begin.

Domestic disputes and dancers

The most important nobles, seated closest to the maharajah, stood up in turn. Each one stepped forward individually, offering him a gold coin on a handkerchief and pledging allegiance with the words: 'I am your child. My head and sword are at your command.' When they had finished, the lower orders offered their own tokens of allegiance.

With these formalities over, a band of musicians struck up and dancing girls twirled into the centre of the hall. Refreshments were served, including pungent betel nut, lime and leaf to chew, and diluted opium to sip. The maharajah turned to chat with those close by, and everyone relaxed. Then it was time for the varied business of the assembly to begin: a son might petition for the deeds to his late father's estate, a gift from the maharajah; two fathers might seek his blessing for a marriage between their children; or neighbours might ask him to settle a boundary dispute.

The end of an era

For all its pomp, the Rajput durbar served a real purpose. The ancient Hindu code of kingship stated that the maharajah is the 'father' of his people. And while his 'family' was expected to obey him and serve him unquestioningly, he in turn had to listen to their grievances, settle their disputes and offer them his protection. The durbar cemented the bond. In addition, a wise ruler took care to retain the loyalty of all his subjects by touring his kingdom during the milder winter months, holding informal durbars wherever his travelling court made camp.

Following the end of British rule in India in May 1948, however, Rajputana was absorbed into the new republic of India. The age of the traditional princely durbar had gone for ever.

BRITONS AND HINDUS *Bhim Singh's son, Jawan Singh, holding a sword and a shield, receives British officers in the Great Picture Hall at Udaipur in 1829. The two leading officers, in black coats and gold epaulettes, listen respectfully as the maharajah holds forth. Both Bhim and Jawan Singh, like many Indian rulers of the time, enjoyed warm relations with the British. In 1818, Bhim signed a treaty of alliance with the East India Company. The British protected him from the marauding of the warlike Marathas, and promised to defend his territory. In return, Bhim and his successors promised to submit to British overlordship, and to hand over to the Company three-eighths of Udaipur's revenues. Firm British leadership left the maharajahs free to enjoy the pursuit of pleasure and to exercise their power in the traditional ways – at magnificent durbars, held six times a year.*

Suffering and death in a floating cell

HOW CONVICTS WERE TREATED BEFORE AND DURING TRANSPORTATION

CONDEMNED TO THE HULKS *The relief guard rows out to a prison hulk at Deptford, on the Thames – one of several captured or decommissioned warships anchored permanently at six sites around Britain. The hulks were only supposed to serve as remand prisons housing convicts waiting to be assigned to a ship for transportation to Australia, but in fact many also served as long-term jails. These once proud, but now rotting ships lay still, their dank and foul-smelling interiors providing cramped conditions for the convicts. A three-deck ship might hold up to 600 men, the most desperate criminals housed on the lowest deck. But for all their grim appearance, each vessel had its own sick bay, workshops and chapel.*

EVEN THE MOST HARDENED OF BRITISH CRIMINALS shuddered to hear the solemn words 'transportation to the colonies' read out in court. For this meant spending possibly up to two years rotting in a prison hulk before being confined to the cramped hold of a convict ship for months, and transported to an unimaginable destination. The sentence was especially cruel for those who had committed only minor felonies such as poaching or pickpocketing.

Overcrowded, foul smelling and insanitary, conditions on board the prison hulks were grim, and petty crime and corruption were rampant. No sooner had a convict boarded the vessel than his clothes would be stripped from his back – if not by the warders by fellow prisoners – and his possessions taken by force to be sold.

Hard labour at the dockyard
While waiting to be assigned to a transportation ship, the prisoners toiled from dawn to dusk in the nearby dockyard. They were assigned to various forms of hard labour, including shipbuilding, painting, hauling timber and general cleaning. But if these conditions were bad, worse was to come.

Between 1787 and 1800, 42 convict ships, each carrying an average of 200 prisoners, set sail from British ports for Australia. After the Napoleonic Wars (1793-1815), the number of ships increased rapidly: in 1833 alone, 36 ships carried a total of 6776 convicts to the other side of the world.

In 1786, conditions on board the First Fleet, organised by the Royal Navy, had been reasonably humane. But after 1788 the Navy handed over the business of transportation to private contractors, whose only concern was to make a healthy profit. Of the 1006 prisoners who had sailed from Portsmouth on board the so-called Second Fleet's three transport ships, 267 perished at sea.

The disembarkation of that second consignment of British convicts at Botany Bay, New South Wales, in 1790, was a distressing sight for at least one eyewitness. 'Upon their being brought up to the open air,' wrote the Reverend Richard Johnson, chaplain to the infant colony, 'some fainted, some died upon deck, and others in the boat before they reached the shore. Many were not able to walk, to stand or to stir themselves in the least, hence they were led by others. Some creeped upon their hands and knees, and some were carried on the backs of others.'

Dying for a drink
For the duration of the four-month voyage, the felons were confined in an airless, vermin-infested hold, shackled together in pairs by short, rigid restraining bolts between their ankles. Hungry and parched with thirst, the listless prisoners – many of them feverish and close to death – lay cramped together on bedding sodden with vomit and excrement. Rations on board were so poor that prisoners sometimes resorted to eating the poultices covering the sores on their legs. When a prisoner died, fellow captives concealed the fact for as long as possible so that they could continue to receive the rations. Each one of them had one thought on their minds night and day – water. They were allowed just two, warm, foul-smelling

pints (1.15 litres) to drink a day. By 1815, a gathering outcry over such shipboard conditions forced the British government to draft stricter regulations. A surgeon had to accompany each voyage, and full payment for the convicts was deferred until they were safely delivered. These rules also kept in check the sometimes sadistic abuses by officers and crew. Conditions continued to improve over the next few decades: better food was provided, and the prisoners were assigned more open-air tasks that kept them fit and occupied.

By the time transportation came to an end, in 1868, a total of 162000 convicts had been delivered to Australia and the prisoners on a convict ship stood a better chance of surviving the three to four-month voyage than ordinary passengers travelling on standard merchant vessels.

TOKENS OF LOVE
While waiting in the hulks to be deported, some convicts passed the time by engraving pictures and farewell messages on copper pennies to give to the friends and relatives whom they would be forced to leave behind in Britain.

FLOATING PRISON *In the hulks, inmates are segregated by the severity of their crimes in galleries running the length of the deck, divided into 'wards' by iron bars.*

LET US PRAY *Prisoners were often whipped or put in solitary confinement, but each ship also had a chapel and a minister to cater for the spiritual needs of its inmates.*

The abode of misery

HOW THE PENAL COLONY AT PORT ARTHUR WAS RUN

FOR 50 YEARS MANY OF BRITAIN'S CRIMINALS WERE subjected to the 'lingering torment' of life in the Port Arthur penitentiary, on the south-eastern tip of Van Diemen's Land (later renamed Tasmania). In the 19th century, this island was little more than a vast penal colony. On arrival at Hobart, the main port, each convict was assigned to a settler as a servant. If he conformed to the regulations he might eventually gain his freedom, but for a serious offence, such as murder, he would be banished to Port Arthur. Between its establishment in 1830 and its closure in 1877, the penitentiary held a total of more than 12000 inmates.

Crimes and misdemeanours
Port Arthur and the penal colony of Van Diemen's Land was administered by the disciplinarian Lieutenant-Governor Sir George Arthur (after whom the port was named), who regarded it as his solemn duty to make Van Diemen's Land a 'Utopia of punishment and reform' and 'a mill for grinding rogues honest'. His aim was to break the convicts' spirit through hard, repetitive labour and strict discipline, inducing 'bovine acceptance', then to reform them through the teachings of Christianity.

A brutal flogging with a cat-o'-nine-tails was a frequent punishment – the leather 'tails', tied with 81 knots, would first be soaked in salt water and left to dry in the sun until sharp as wire. Above all, however, prisoners feared solitary confinement in a tiny, windowless cell. Long periods alone seriously affected a man's mental state, and many of the hardest convicts ended their days as gibbering wrecks in Port Arthur's lunatic asylum. For lesser crimes, offenders spent their days breaking piles of stones, or in a chain gang building roads and bridges across the island. Armed guards and vicious dogs barred the only land route out of Port Arthur, and rumours that the guards dumped blood and offal off the beaches to attract sharks ensured that only the most desperate would try to swim to freedom.

GEORGE WILLIS

JAMES MERCHANT

MICHAEL SULLIVAN

THOMAS JACKSON

CHARLES CLIFFORD

THOMAS HARRISON

WILLIAM BURLEY

ROGUES' GALLERY *The resigned expressions of some of the 'old crawlers' who were serving time at Port Arthur in 1874 are preserved in a photographic record at Hobart Library.*

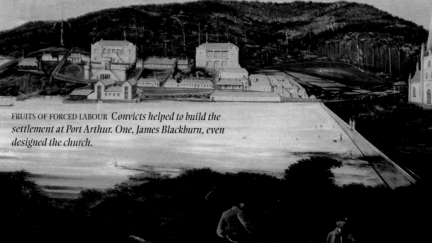

FRUITS OF FORCED LABOUR *Convicts helped to build the settlement at Port Arthur. One, James Blackburn, even designed the church.*

'Someone has stolen the *Mona Lisa*!'

HOW A SELF-PROCLAIMED ITALIAN PATRIOT TOOK LEONARDO'S MASTERPIECE

THE ASSISTANT CURATOR OF THE LOUVRE MUSEUM IN Paris stared in horror at the blank space on the wall where the world's most famous picture, *La Gioconda* – better known as the *Mona Lisa* – should have been. It was midday on Tuesday, August 22, 1911, when the official, Georges Benedite, recovering from his shock, telephoned the Paris police, exclaiming: 'Someone has stolen the *Mona Lisa*!'

Within 15 minutes more than 100 policemen descended on the Louvre. At first it was thought that the painting might still be in the building, and before long one of the officers came across the picture's heavy gold frame resting on a staircase. But the *Mona Lisa* itself – painted by Leonardo da Vinci on a wood panel measuring about 30 in (76 cm) by 21 in (52 cm) – had been skilfully removed.

Found in a cheap hotel

Two years went by without word of the painting's whereabouts. Then in November 1913 an art dealer in Florence named Alfredo Geri received a letter that began, 'I have stolen the *Mona Lisa*'. It was signed 'Leonard' – the alias of an Italian painter and decorator named Vincenzo Perrugia, who in 1910 had been one of four workmen in the Louvre who had reframed the picture. On December 11, 1913, Geri went to Perrugia's room in a cheap hotel in Florence. There the workman pulled a battered trunk from under his bed and released a false bottom. He took out an object wrapped in red silk, placed it on the bed, and slowly unwrapped the *Mona Lisa*.

The picture was duly returned to the Louvre, and Perrugia was arrested. In Florence in June 1914 he pleaded guilty to theft and was sentenced to a year and 15 days in prison, reduced on appeal to seven months. After serving in the First World War he returned to Paris – to open a decorators' paint shop.

PASSIONATE PATRIOT *Art thief Perrugia (right) said that he had stolen the* Mona Lisa *by posing as a maintenance worker on a day when the Louvre had been closed to the public. He had taken the picture, he claimed, 'for the glory of Italy'. The affair became headline news and gave the director of the Louvre nightmares.*

LEONARDO'S "LA GIOCONDA."

DISAPPEARANCE FROM THE LOUVRE.

(FROM OUR OWN CORRESPONDENT)

PARIS, AUG 22.

Paris has been startled this afternoon by the news of the disappearance of Leonardo da Vinci's masterpiece "La Gioconda" from the Louvre. The fact that what is, perhaps, the most famous picture in the Louvre should have...

"LA GIOCONDA" RECOVERED.

Famous Picture Stolen from Paris Louvre Now in Florence.

ITALIAN ARRESTED.

ROME, Dec. 12.—The Minister of Public Instruction, Professor L. Credaro, announced to-day that the famous picture, "La Gioconda," which was stolen from the Louvre, has been found at Florence.

The news of the finding of the Gioconda this evening was telegraphed directly to Professor...

Poison, bullets, beating and drowning

HOW RUSSIAN ARISTOCRATS ASSASSINATED RASPUTIN

THE SUPPER PARTY AT WHICH RASPUTIN WAS TO BE killed was arranged for the night of December 29, 1916. It was held at the St Petersburg palace of Prince Felix Yusupov, an ardent monarchist who considered the illiterate monk from Siberia to be 'the most evil and dangerous person in all Russia'. Rasputin was welcomed at the Imperial Court for his reputed mystical 'healing powers'. But his meddling in state affairs during the First World War angered loyal Russians, and by the winter of 1916 he had survived several attempts on his life.

A long time dying

Prince Yusupov decided to take a hand. He invited Rasputin to supper at his family home, where he offered the black-bearded monk a plate of chocolate cakes containing cyanide. Rasputin ate the cakes and drank some Madeira, also spiked with cyanide. He showed no signs of dying. So Yusupov produced a revolver and shot the monk in the back. With a roar, Rasputin fell to the floor. Yusupov's friends rushed in; one of them, a Dr Lazovert, declared that the bullet had penetrated the monk's heart, and that he was dead. As they started to drag the body from the room Rasputin suddenly came to and stumbled out into the snow-covered courtyard. Another conspirator fired two bullets from his revolver into the moving target, who still clung to life. Yusupov battered the body with a metal rod until it lay still.

Two guards, who had heard the shots in the courtyard, helped to tie the monk up with rope and wrap him in a curtain. He was driven to Petrovski Island, on the outskirts of the city. A hole was cut in the frozen River Neva and Rasputin, still breathing, was pushed into the icy water – where he drowned.

COURT CIRCLES *Prince Yusupov (right, in fancy dress) hated Rasputin's influence over Emperor Nicholas II, his wife Empress Alexandra and their family (far right). The empress was a mourner at Rasputin's funeral, which was held in secret.*

Revolution and the Red Menace

HOW A FAKED LETTER ABOUT A COMMUNIST PLOT LED TO THE FALL OF A BRITISH GOVERNMENT

AT BREAKFAST TIME ON OCTOBER 25, 1924, A POLITICAL bombshell burst on the British nation. The front page of the right-wing *Daily Mail* was dominated by the sensational story of a Russian plot to start an armed revolution in Britain. A letter allegedly written by the president of the Communist Third International in Moscow, Grigori Zinoviev, and intercepted by the Foreign Office, gave orders to 'paralyse' the British armed forces. With only four days to go before the general election, at which the Conservatives hoped to defeat Britain's first Labour government, the *Mail* proclaimed that a vote for Labour would be a vote for the so-called Red Menace: Communism.

Four conspirators and a single spy

People in the know, however, believed that the letter had been written by four White Russians who had fled from the Communist regime and taken shelter in Berlin. The emigrés were Alexander Gumansky and Alexis Bellegarde, former officers in the Imperial Russian Army, Edward Friede, a skilled forger, and one Druzhelovsky, a clerk at the Soviet Embassy in Berlin. They were bitterly opposed to the Anglo-Soviet treaty that Britain's prime minister, Ramsay MacDonald, had negotiated, which agreed that Britain would not only recognise the Communist regime, but would lend millions of pounds to the Russian government. If the Labour government fell, the treaty would fall with it. So, using notepaper stolen by Druzhelovsky from the Soviet Embassy, they wrote the letter and forged Zinoviev's signature on it.

The man who persuaded the four Russians to write the letter is thought to have been Captain Sidney Reilly, a shadowy figure known as the 'Ace of Spies'. Reilly did his job well, and the letter was 'leaked' to the *Daily Mail* in time for it to have a major effect upon the actual election result. On October 29 the Conservatives swept back to power, winning 419 seats to the Labour Party's 151.

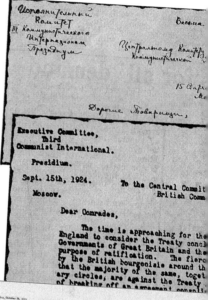

CALL TO ARMS *A letter written in Russian (below), and said to be the original Zinoviev letter, was found in the USA in 1970. It matched an English translation of the letter made by the Foreign Office in 1924.*

REVOLUTIONARY TALK *The picture paper* The Daily Mirror *also reported that Zinoviev (right) called for British Communists to enlist the army on their side, and to infiltrate munition factories and military barracks in a bid to overthrow the government.*

RASPUTIN'S SECRET BURIAL.

DISCOVERY OF THE BODY NEAR IMPERIAL PALACE.

(FROM OUR OWN CORRESPONDENT.)

PETROGRAD, MARCH 22.

Late last night an officer of the garrison at Tsarskoe Selo, who had long suspected that the body of Rasputin was secretly buried near the Imperial residence, found the remains of the impostor. They had been interred within the grounds under the altar of a small wooden chapel, recently built. The body, which was enclosed in a metallic coffin, had been re-interred outside the grounds in a neighbouring cemetery in the presence of a delegation from the municipality.

Gateway to a new life in a land of freedom

HOW THE RECEIVING STATION AT ELLIS ISLAND WELCOMED MILLIONS OF IMMIGRANTS TO THE USA

CHEERS BROKE OUT ON DECK AS THE SHIP PULLED INTO New York harbour and the crowd of immigrants got their first view of the Statue of Liberty. But their joy at escaping poverty and persecution was tinged with foreboding. Warned by friends and relatives who had already made the journey, the immigrants steeled themselves to face questioning by the notoriously tough US port authorities. One mistake, and they could be on the next boat back to Europe.

Immigration to the USA reached unprecedented levels during the early years of this century. In a peak year like 1907 ships delivered over a million people to New York. During the spring and summer months, immigration officials could be faced with thousands of hopefuls every day. Few spoke English. All were tired and uncertain of their fate.

Dealing with the deluge

Chaos was averted by the purpose-built immigration centre opened in 1892 on Ellis Island in New York harbour, which had previously served as a fort and an arsenal. After a fire in 1897 the site was expanded to take up 27 acres (11 ha), which included two hospitals, a restaurant, a customs house and a post office. There were also four enormous dormitories and a bathhouse which could accommodate 8000 people a day. Even so, the Ellis Island facilities were hard pressed to cope with the deluge of people.

The selection process began as soon as the ship docked. Officials made sure there was no epidemic on board and then supervised the transfer of passengers to the barges which took immigrants to Ellis Island.

SICKNESS AND SUFFERING AT SEA

FIRST-CLASS PASSENGERS HELD THEIR NOSES IN DISGUST AT the stench coming from the steerage quarters of the passenger ship carrying immigrants from Liverpool to New York. On breezy days the pungent smell of the 200 or so men, women and children huddled in cramped and filthy wooden bunks, or jammed among their own excrement and vomit, wafted throughout the entire ship.

Steerage – so called because steering equipment used to be housed there – cost about £3 per head for the arduous voyage to the New World. There was only one lavatory for every 100 steerage passengers, and rough seas often prevented these from being used at all. Water was doled out daily, enough being provided only for drinking and cooking. Most of the passengers slept in their clothes. They were surrounded by shabby bundles of possessions – and by what an observer in the 1850s called 'every sort of filth, broken biscuit, bones, rags and refuse of every description, putrefying with maggots'. 'Ship fever', or typhus, transmitted by lice, killed up to 1 in 10 of the migrants on each crossing.

Card playing and storytelling were the main ways of passing the time during the monotonous voyage – though, despite the lack of privacy, many babies were conceived at sea.

WAITING THEIR TURN *Immigrants on a steamer just arrived in New York wait as first and second-class passengers disembark. The poor are then taken to Ellis Island for vetting by immigration officials. Anyone considered a potential burden on the state will be rejected.*

FACING THE FUTURE *Their inspection ordeal over, and wearing their best clothes, immigrants of all ages wait to collect their baggage before leaving Ellis Island. They look forward to a life of freedom and happiness in the New World, where one person was as good as another – and anyone could become rich.*

CITY OF HOPES *New arrivals wait for the ferry from Ellis Island to New York City, whose distant skyline holds the promise of a new start in life.*

COLD WELCOME *Shouldering their worldly goods, a group of European peasant women trudge through the snow to the immigration centre.*

Two-minute medicals

The first medical examination was over so quickly that many people did not realise it had taken place. The immigrants checked their luggage into the baggage room and then filed up the main staircase while two doctors followed their progress. Physical deformities, lameness or potential heart problems were quickly spotted. Each immigrant was then given a cursory medical examination, often taking as little as two minutes. Those suspected of medical problems were labelled with single-letter chalk marks on their backs or chests. G, for example, meant suspected goitre, K hernia and X mental illness. These unfortunates were led into cage-like detention areas.

'Give me your tired, your poor,
Your huddled masses yearning to breathe free,
The wretched refuse of your teeming shore,
Send these, the homeless, tempest-tossed, to me.'
INSCRIPTION ON THE STATUE OF LIBERTY, NEW YORK CITY

The eye inspection was particularly frightening and painful – buttonhooks were used to turn eyelids inside out to check for signs of trachoma, an incurable infection which can eventually lead to blindness. After 1903, any immigrant suffering from this disease was deported immediately. The same fate awaited those suspected of having tuberculosis or an infectious disease of the scalp called favus.

Fears were heightened by the forbidding appearance of the immense registry room, or Great Hall, into which they now filed. A series of 22 iron-railed alleyways, like cattle pens, kept people in queues. At the end of each aisle sat an immigration officer at a raised desk like a judge in a courtroom. The bewildered immigrants, aided by interpreters, were now faced with a barrage of questions. A series of laws intended to protect American workers' rights

meant that immigrants had to avoid saying they had a job to go to. Those primed in advance spoke guardedly of 'good prospects' or 'a relative's promise'. People failing any of these tests were detained until the authorities were satisfied they were healthy enough and had no mental or legal problems. Any suspected criminals, polygamists, prostitutes, beggars or anarchists were detained or deported immediately.

Robbing the poor

Detention meant exposure to the multitude of conmen and dishonest lawyers who flocked to the island in search of victims. The authorities also fought a losing battle against corruption within their own ranks. It was common for immigrants changing money to US dollars to be short-changed. Many were also conned out of their life savings for bogus 'permits'. Finally free to leave, the immigrants either boarded ferries for New York or bought railway tickets to continue their journey into the USA. As a result of language difficulties, many people had their names 'Americanised' and duly left Ellis Island bearing a brand-new identity. Members of the same family often emerged each bearing a different name. It was their first faltering step towards a new life in the New World.

MELTING POT *Immigrants of all races flocked to the 'land of opportunity'. (From the top): a Finnish stowaway, a Syrian, an Albanian, an Armenian Jew and a Caribbean islander.*

The thirsty years of Prohibition

HOW THE USA'S NATIONWIDE DRINKING BAN WAS BROKEN

DOWN THE DRAIN *Axe in hand, a United States marshal breaks open barrels of wine and empties the contents into the gutter. A solemn crowd watches as gallon after gallon of confiscated alcohol is destroyed – by court order.*

LARGE HAUL *A river of contraband liquor runs down the chute straight into the sewers in Zion City, Illinois.*

SMALL HAUL *Agents empty bottles of whiskey down the drain to win another battle in their war against bootleggers.*

DRINKERS ACROSS THE UNITED STATES STARED GLOOMILY into their glasses as midnight on Friday, January 16, 1920, approached and the nation officially went dry. In New York, for instance, mourners gathered in the plush Hotel Vanderbilt to pay their last respects to alcohol by quaffing 100 cases of the best champagne. In nearby Maxim's, diners gazed mournfully at a bottle-shaped 'corpse' in a coffin in the middle of the dance floor. Then, as the clocks struck midnight, New York's crowded nightspots fell silent. People rose to their feet and stood with heads bowed in solemn contemplation of the moment – at 12.01 precisely – when the 18th Amendment to the US Constitution, banning the manufacture, sale and transport of alcohol, came into force.

The wets and the drys

This moment was the culmination of a drive to outlaw liquor that stretched back to the early 19th century. Maine went 'dry' in 1851, and by 1855 another 13 states had banned drink. In the early years of the 20th century the idea of prohibition gained more support and laws allowed communities to ban alcohol locally. In 1920 the 18th Amendment finally became law, enforcing national Prohibition.

It was a triumph for Minnesota Congressman Andrew J. Volstead, author of the Volstead Act, which enforced the 18th Amendment, and the temperance bodies. The amendment was passed by Congress in October 1917 and ratified by the required three-quarters of the 48 states on January 16, 1919 – to take effect a year later. Before the Act was passed the USA's 105 million citizens were already divided into two roughly equal camps: the wets and the drys.

Despite official confidence that the transition from wet to dry would be a smooth one, trouble broke out on the very first day of Prohibition. Trucks carrying liquor were seized in New York by law enforcement agents, and about a dozen thirsty citizens were arrested for breaking the new law. In vain they protested that they were drinking only cordial; the ban included beverages containing any more than 0.5 per cent of alcohol.

Smuggling and rum-running

Prohibition was barely two weeks old before truckloads of liquor were being smuggled over the long, largely unguarded borders with Canada and Mexico. Congress had provided fewer than 2000 agents to control the illicit trade. Another method of beating the law was rum-running by ships anchored out in the Atlantic, beyond American territorial waters. Fishing boats and small, fast craft were used to bring crates of alcohol from the West Indies to remote parts of the eastern seaboard.

LAWFUL CAPTURE *Agents seize huge stills disguised as part of a summer camp near Louisville, Kentucky.*

UNLAWFUL SEIZURE *Looting of contraband stores broke out across the USA as ordinary citizens became desperate for good-quality alcohol.*

CAUGHT IN THE ACT *Armed law enforcement officers sneak up on a group of distillers making 'rot-gut' whiskey in the backwoods. Sulphuric acid, antifreeze, rotten meat and dead rats were among ingredients used to produce the alcohol.*

SHOOTING THE FALLS *In a daring bid to smuggle liquor from Canada into the USA, a bootlegger – clutching onto a hollow raft filled with rum – is swept over Niagara Falls and drowned.*

HIDDEN IN THE HOLD *Prohibition agents – two of them disguised as dockers – examine the contents of 3000 sacks of liquor concealed in a coal steamer in New York harbour.*

Blind drunk, dead drunk

Some drinkers carried their own ingeniously concealed liquor supplies around with them. They hid flasks of whiskey in hip pockets, strapped bottles to the calves of their legs, and even poured the whiskey into hollow walking sticks. Others brewed their own crude alcohol at home. This consisted mainly of wine and bathtub gin – the latter made by distilling industrial alcohol and flavouring it with juniper berries. To give the gin an added 'kick', anything alcoholic from methylated spirits to cleaning fluid was added. But homemade alcohol, however rough and foul-tasting, was not as dangerous as the hooch produced by moonshiners – illicit whiskey distillers – and 'alky-cookers', brewing their potions in vermin-ridden tenement slums. Hundreds of people went blind or were paralysed by the brews, and others died.

Not surprisingly, the riches to be made from Prohibition attracted gangsters such as Johnny Torrio and Al 'Scarface' Capone – whose criminal activities in Chicago made them household names. Their suppliers were known as bootleggers from the bootleg, or upper part of a tall leather boot, in which old-time smugglers carried bottles of liquor. By 1927 Capone's gang was making an estimated $60 million a year from the liquor business – and a further $45 million from associated rackets such as prostitution and gambling. 'When I sell liquor,' complained Capone, 'it's called bootlegging. When my patrons serve it on a silver tray on Lake Shore Drive, it's called hospitality!'

Saloons and speakeasies

In Chicago, Capone also controlled many of the speakeasies – illegal drinking clubs – which sprang up all over the country. By 1929, in New York City alone, tens of thousands of speakeasies had taken the place of nearly 8000 saloons that had been forced to sell only alcohol-free drinks. The owners of the city's speakeasies paid an estimated $50 000 a year in so-called 'blind eye' money to policemen and low-salaried Prohibition agents – some of whom enjoyed a quiet drink 'on the side' themselves. But despite charges of corruption and inefficiency, the agents achieved remarkable results. The officers seized millions of gallons of illegal beer, spirits and wine. Some 50 000 liquor offenders were arrested each year, and prisons throughout the country were filled to bursting point.

By 1933 the USA was tired of being dry, and of the lawlessness that accompanied it. The Democratic Party and the newly elected President Franklin Delano Roosevelt endorsed the repeal of the law. Congress proposed the 21st Amendment, which repealed the generally derided, and more or less unenforceable, 18th Amendment. By the end of the year this had been ratified by three-quarters of the states. A few states in the south retained the ban until the 1950s and 60s. Mississippi became the last state to repeal Prohibition, in 1966.

CLOSING DOWN, OPENING UP *Prohibition officers (below right) close down a speakeasy fancifully called the Moulin Rouge. As one 'speak' closed, another opened. Straw-hatted tipplers (below left) raise their glasses in one of the USA's many illegal drinking clubs. The smartest of these were in New York, where numbered membership cards afforded entrance to those willing to pay high prices, and risk raids, for a drink.*

ARRESTING SMILE *New York speakeasy queen 'Texas' Guinan grins broadly as she is escorted into a police van after a raid on her club on West 45th Street.*

HIDDEN PLEASURES *One drinker (above) cheekily hides her whiskey flask inside a US Government directory. More stylish, and just as crafty, is the woman (left) with a dainty silver ankle-flask discreetly tucked away inside her calf-length Russian boot.*

The Great Escape

HOW SECOND WORLD WAR PRISONERS TUNNELLED THEIR WAY TO FREEDOM

AT THE TUNNEL FACE, A YOUNG RAF OFFICER DRENCHED in perspiration lay on his side, chipping at the sandy earth inches from his nose, while a second officer heaped the excavated earth into small trolleys that were hauled back to the entrance. To leave the tunnel, a man had to crawl along the excavated route and climb a wooden ladder up a shaft to a trap door that opened into the prisoners' barrack huts.

No detail too small

During the war, prisoners on both sides were duty-bound to try to escape and return to active service. For an escape bid to go smoothly, however, a great deal of preparation was necessary. Once outside the wire, the prisoners needed disguises, official papers and money which would aid their passage to a neutral country. A team of self-taught tailors set about transforming items of uniform into passable civilian clothing. Some prisoners proved adept at forging official German passes, while others made miniature compasses – all carefully hidden.

Every prisoner had to memorise a detailed cover story to satisfy the strict security checks he would face at railway stations and checkpoints. While some officers chose the disguise of a German soldier on leave, most opted to pose as a foreign worker from a country in Occupied Europe, since many such workers were employed in Germany at that time.

The plan for a mass escape from Stalag Luft III, a German Air Force camp for captured Allied airmen at Sagan, 80 miles (130 km) south-east of Berlin, was conceived in the spring of 1943. Three tunnels, codenamed Tom, Dick and Harry – the word 'tunnel' was never to be so much as whispered under the breath – were planned, so that if the guards discovered one of the tunnels, the other two might still escape notice. The depth for the tunnels was 30 ft (9 m), to avoid detection by the anti-tunnelling microphones that the Germans had sunk into the ground. The plan was put forward to the escape committee by Squadron Leader Roger Bushell, a fighter pilot who had been shot down over Dunkirk in 1940. The senior British officer in the camp, Group Captain Harry 'Wings' Day, and the engineering expert, Canadian Flight Lieutenant Wally Moody, were also involved.

Before digging could even start, however, the POWs had to solve a major problem – how to get rid of the 100 tons of bright yellow earth that would be excavated. They came up with a number of imaginative ideas. While strolling about with bags of earth concealed under their baggy 'penguin' trousers, the POWs discreetly let it run out of the bottoms, then scuffed it into the grey dust of the camp. They mixed sand with soil in the compound's gardens and, at every opportunity, staged boisterous open-air combat drills that raised clouds of dust, enabling them to trample yellow sand into the ground without attracting attention.

Ferrets on the scent

Attached to every POW camp were special guards, nicknamed 'ferrets' by the prisoners, whose sole duty it was to ferret out escape attempts. The POWs kept a constant lookout for them and devised an elaborate warning system to alert tunnellers if the ferrets came too near a tunnelling hut. Work was halted on Harry and Dick when word got out that a new compound was to be built over the very place where the tunnels were to have their exits, and all resources were immediately switched to Tom. The ferrets were becoming increasingly suspicious, however, and eventually they discovered Tom's entrance during a routine search and blew up the tunnel.

THE WORK SHOP

caving in piping

signs appro

'SNAPSHOTS' OF COURAGE *The artist Ley Kenyon (above) hid drawings of each stage of preparation and digging in 'Dick' until after the war.*

DUTY-BOUND TO BOLT *The POWs are quick-marched into Stalag Luft III. The commandant believed that they would be too content in this 'luxury' camp to make any escape bid.*

The final push

By early January 1944 the escape committee was confident that the Germans believed they had put an end to their escape attempts and work continued on Harry. By mid March the tunnel was complete, apart from the last few feet leading out into the fir trees that encircled the camp. The plan was to smuggle 200 officers into the tunnel in relays – if successful, this would be the biggest mass escape in history.

The operation began on March 24 – a moonless night – at 10.30. One by one the men scrambled down the shaft and wheeled themselves along the tunnel, switching trolleys in a shuttle system that enabled them to pass rapidly along the 336 ft (102 m) of single track. For a few hours all went smoothly and the officers began to pour out of the exit beyond the camp fence. But time was running out and more than 100 men had to turn back. Just before 5 o'clock in the morning disaster struck as a sentry discovered the tunnel's exit.

In all 76 prisoners had escaped. When news of the break-out reached Hitler he reacted furiously. His first order was to shoot all the escapees on recapture, but was eventually persuaded to reduce this total to 'more than half'. As a result, a nationwide hue and cry ensued: only three escapees reached Britain via Sweden. The rest were rounded up within two weeks of Hitler's orders, and 50 of them were shot while 'resisting arrest' or 'attempting further escape', among them Roger Bushell.

This vengeful act influenced the British military's decision in October 1944 that escape should no longer be a duty. But the fate of the Stalag Luft III POWs did not deter further attempts. Within just two months of the Great Escape, the daring officers of Stalag Luft III's north compound had already begun work on 'George', tunnel number four.

QUIET – INGENUITY AT WORK *Craftsmen in the tunnel workshop turned out a huge range of items to help with the excavations, from tools, tin piping and truck lines, to lamps and other electrical equipment.*

PRISON CAMP *Fences 9 ft (3 m) high strung with barbed wire surround the compound. The only sure way out is to tunnel into the dense surrounding woods.*

READY TO ROLL *In the stifling heat, the men often worked naked or stripped to their underwear. During the escape, they switched trolleys at two tiny rest stations, nicknamed Leicester Square and Piccadilly.*

LAST-MINUTE HITCH *The exit falls some feet short of the woods, so the first man out ties a rope to the ladder and crawls belly down over the snow into the trees behind the sentry box. Two tugs on the rope means 'all clear', and the next man snakes slowly across the snow into the fir trees.*

Feats of Science and Invention

CHANCE, PERSISTENCE AND GENIUS MEET IN THE INSPIRING STORY OF SCIENTIFIC DISCOVERY. THE SECRETS OF LIFE AND THE UNIVERSE HAVE BEEN REVEALED IN A PARADE OF BRILLIANT INVENTIONS AND VITAL BREAKTHROUGHS

A PHILOSOPHER GIVING A LECTURE ON THE ORRERY / JOSEPH WRIGHT OF DERBY

Getting the measure of the ancient world

HOW ANCIENT PEOPLES CALCULATED DISTANCE

AS A MARK OF RESPECT TO THE MOST POWERFUL GOD in the Egyptian pantheon, the identical stone sphinxes that were to line the great way up to the temple of Amun at Thebes had to be spaced precisely the same distance apart. The master builder was taking great care to measure the gaps with his palm-wood single cubit rod.

Traders follow Egyptian rules

By the time the New Kingdom section of the temple complex of Amun was being built – between 1524 and 1212 BC – the royal cubit had been the standard measure in Egypt for at least 2000 years. Most of the country's trading partners, from Nubia in the south to Babylonia in the north-east, had adopted it. Although the precise measurement varied locally, it was based on the dimensions of the human body.

A royal cubit corresponded to the distance from the elbow to the extended fingertips, around 21 in (53 cm). Each cubit was divided into 28 *djeba* (digits), each about one finger's breadth. Four digits equalled a *shesep*, the width of a palm. There were seven palms in a royal cubit and six palms in a short cubit –

around 17.7 in (45 cm). To measure longer distances the Egyptians used a length called the *khet* (rod) equal to 100 cubits (57 yd/52 m), and the *iteru* (river) that was equal to 20 000 cubits (6.5 miles/10 km).

'Man is the measure of all things.'
PROTAGORAS, 5TH-CENTURY GREEK PHILOSOPHER

By around 1000 BC, the Greeks were using a measurement system based on the Egyptian short cubit, calculating the distance from the elbow to outstretched fingers as about 18 in (46 cm). Each cubit was divided into 24 *daktyloi* (digits), and 16 *daktyloi* was equivalent to one 'foot' of about 12 in (30 cm) – corresponding to the length of the average man's foot.

The Romans later borrowed the Greek system, subdividing the foot into 12 *unciae* ('inches'). One pace was equal to five feet, and a thousand of these paces made a mile, which comes from the Latin for a thousand, *mille*.

BUILDER'S GUIDE *An ancient Greek stone relief from the island of Salamis shows standard lengths – from longest to shortest – for a cubit (top), a rule (centre), a foot and a hand span (bottom). The stone slab, covered in layers of whitewash and built into the side of a 15th-century chapel, was found in 1989. Builders and architects working on a big site, such as a temple, would have used a relief like this to keep their measurements to a standard.*

SAILOR'S DEPTH *Paying out lengths of sounding line over the sides of their boats gave 5th-century Greek sailors the measure for depth in the sea, a fathom – the length from fingertip to fingertip when both arms are at full stretch – as shown in this relief.*

A length to last

In Europe, the northern cubit – equivalent to about 18 Egyptian digits and also based on the forearm – was used to measure land. The Romans found their northern subjects so attached to their cubit that they gave up trying to impose Roman measurements and instead adopted local usage.

Variations of the northern cubit were used all over Europe from about 3000 BC up until the middle of the 19th century, although it had different names across the Continent. In the Low Countries, England and Germany, for example, it was called the *ell* – Saxon for forearm.

One problem with measurements based on parts of the body is the difficulty of setting a standard. The English solved this problem in the time of Henry I in the 12th century, by using the king's own arm as the length of a yard. Unfortunately, the measure continued to change because later monarchs believed their arms were closer to the true length of a yard. Finally, in the 16th century, Elizabeth I set the standard 36-inch yard.

ROYAL CUBIT *The bottom row of hieroglyphs on a reconstruction of an Egyptian cubit rod shows fractions; the top row catalogues the names of protecting gods. The middle row carries the different measures:* ⎸ *is one digit;* ▬ *a palm of four digits;* ▬ *a palm plus the thumb and* ⊏ *a fist of six digits.* ▬ ▬ *represents a double palm;* ⊿ ⌐ *a short span and* ▬ ⌐ *a long span. The symbol* ⌣ *represents 16 digits;* ⌐ *is a northern cubit of 18 digits;* ⊿ ⌐ *a short cubit;* ⌐ ≈ *a full royal cubit.*

PUBLIC RECORD *An Egyptian tax assessor measures a field of ripe grain using a long rope with the lengths of cubits marked by knots. The man wearing the linen shirt is a scribe, who will later write down the figures.*

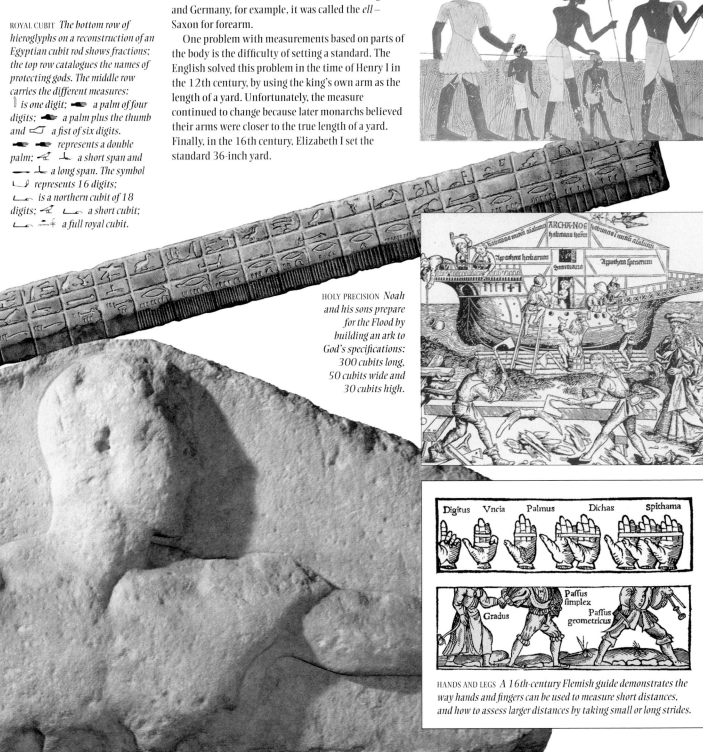

HOLY PRECISION *Noah and his sons prepare for the Flood by building an ark to God's specifications: 300 cubits long, 50 cubits wide and 30 cubits high.*

HANDS AND LEGS *A 16th-century Flemish guide demonstrates the way hands and fingers can be used to measure short distances, and how to assess larger distances by taking small or long strides.*

Choosing the right lever for the job

HOW A CHILDREN'S GAME REVEALED A LAW OF PHYSICS

STROLLING ALONG THE DOCKSIDE IN HIS NATIVE TOWN, Syracuse, the Greek mathematician Archimedes fell into conversation with one of the local boatbuilders, who complained that he had damaged his back. He had lifted a boat to apply some pitch, and something in his back had given way; he had been in constant pain ever since.

Archimedes, however, only gave the boatbuilder half his attention. He was distracted by two children playing seesaw, using a plank of wood balanced on a stone. A girl arrived and climbed behind one of her friends, but now the seesaw did not move: one end was too heavy. The children climbed off and moved the plank so one end was nearer the stone. With one child at the long end and two at the other, the plank balanced again and they continued with their game – one child easily lifting two others.

In a flash, Archimedes realised the implications of what he had just seen. He hurried the boatbuilder back to his yard to put this new notion to the test. He took a long spar, placed one end under a boat, and put a log under it close to the boat. Pulling the free end of the spar down with one hand, Archimedes easily raised the boat into the air.

He went on to write a simple formula to show how levers work: the weight that has to be exerted to lift a load, multiplied by its distance from the lever's fulcrum, equals the weight of the load multiplied by its distance from the fulcrum. In short, the longer the lever, the easier the job.

WORKERS' AID *Workmen in the Assyrian capital, Nineveh, 700 years before Christ, lever a huge statue of a bull-sphinx into place using a long plank. The longer the lever – the farther the force is from the fulcrum – the less pressure needs to be applied to move the sphinx. The fulcrum and the lever were used by builders throughout the ancient world, but only in the 3rd century BC did Archimedes find the mathematical formula to explain the relationship between the two.*

ULTIMATE CONCLUSION *After he had devised his mathematical formula, Archimedes boldly declared: 'Give me a place to stand, and a lever large enough, and I will lift the Earth.'*

The dream that solved a farmer's problem

HOW ARCHIMEDES MADE WATER FLOW UPHILL

ONE DAY IN THE 3RD CENTURY BC, ARCHIMEDES, the mathematician and inventor, watched an exhausted farmer making yet another journey up a steep bank to carry water from a stream to his fields. Surely, Archimedes thought, there must be an easier way of transporting the water?

Sketching in the sand

Sitting in the shade of an olive tree, Archimedes used a stick to sketch out ideas in the earth. His first thought was to build a ramp for the farmer to drag his water up to the field. Later, he fell asleep, and dreamed that he was walking up a ramp – but this one was wrapped in a spiral around a shaft. Waking with a start, he realised that this spiral ramp, or screw, could lift water from a stream to a higher level.

In his workshop, Archimedes made a wooden model of his idea: a screw, enclosed in a cylinder, with a handle at one end. Placing the end of the model in water, he turned the handle. With each turn, the liquid climbed a little higher inside the cylinder, until it poured from the top of the screw. Although similar devices had been used by the Egyptians for irrigation, and by the Romans to pump water, it was the observant mathematician who introduced the 'Archimedean screw' that is still used in many parts of the world to this day.

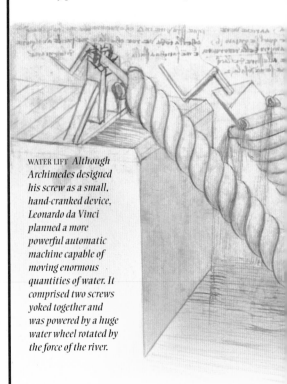

WATER LIFT *Although Archimedes designed his screw as a small, hand-cranked device, Leonardo da Vinci planned a more powerful automatic machine capable of moving enormous quantities of water. It comprised two screws yoked together and was powered by a huge water wheel rotated by the force of the river.*

'Eureka! Eureka!'

HOW ARCHIMEDES PROVED THAT A KING HAD BEEN CHEATED

THE GOLDSMITH TREMBLED WITH TERROR. EARLIER THAT day he had been dragged from his bed and taken to King Hieron's palace. It was something to do with the crown he had made for the king, one of the guards had warned him. Now he looked on in horror as his attempt to rob the king was exposed.

Tipped off by a sneak

Hieron was king of Syracuse, a Greek colony in Sicily. In celebration of a victory in battle in 212 BC, he vowed to thank the gods by placing a crown of gold in Syracuse's temple. To make sure no gold was stolen during the crown's manufacture, Hieron weighed out the right amount. When the crown was completed, he checked its weight and found it was the same. But next day an informer told the palace that some gold had been replaced by silver. Without delay the king set the philosopher Archimedes an apparently impossible task: to find out, without damaging the crown, whether the gold had been adulterated.

Over the following days, Archimedes could think of nothing but the king's puzzle. He stopped eating or washing. Eventually, his body odour became so overpowering that he was forced to go to the public baths. As he sank into the pool, water spilled over the sides and inspiration struck. Archimedes leapt out of the pool and ran naked down the street, shouting, '*Eureka! Eureka!*' ('I've found it! I've found it!')

What Archimedes had realised was that his body had displaced its own volume in water from the bath. If the crown was made of pure gold it would displace the same amount of water as a lump of unworked gold of the same weight. But if the gold was alloyed with another metal, the crown might weigh the same as the lump of gold, but it would be bigger, so displacing more water.

Archimedes wasted no time testing his hypothesis. Placing a jar in a dish, he filled it to the brim with water. As he lowered the lump of gold into the jar, water spilled into the dish, which he then weighed. Repeating the experiment with the crown he found more water was displaced. The crown was bigger than it should have been. Because the goldsmith had adulterated the gold with silver, he had made a larger crown to make up the weight. As Hieron thanked Archimedes, the palace guards removed the goldsmith and slit his throat.

BRUTAL END *Absorbed in complex calculations, Archimedes failed to notice the Roman sack of Syracuse in 212 BC. He was stabbed to death after refusing to obey a soldier's commands, insisting on finishing his sums first.*

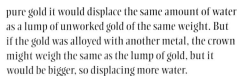

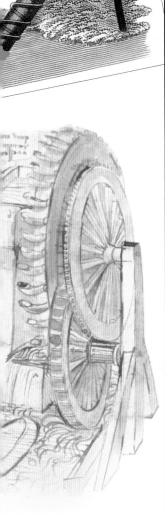

LABOUR SAVER *A variation on the Archimedean screw, a pipe wrapped around a tight cylinder and turned by a hand crank at the top, lifts water from a river or canal and into a trough.*

BATHTIME INSPIRATION *The greatest mathematician of the ancient world brought together mathematics and physics when he compared the weight of an object to its volume.*

Shadows at high noon

HOW THE CIRCUMFERENCE OF THE EARTH WAS ACCURATELY MEASURED IN ANCIENT TIMES

PEOPLE STARED AT THE MAN WHO STOOD BESIDE A TALL, vertical sundial in the Egyptian port of Alexandria, studying a shadow cast by the pillar. It was noon on June 21, the longest day of the year in 230 BC, and Eratosthenes was trying to measure the circumference of the Earth.

Simple geometry

Travellers had told Eratosthenes that at midday each June 21, the summer solstice, the Sun shone straight down a well in the town of Syene (present-day Aswan) – some 500 miles (800 km) to the south – without casting a shadow, and that its reflection could be seen on the surface of the water. This meant that, at that moment, the Sun was directly above it. By comparing the length of shadows in Alexandria and in Syene, Eratosthenes planned to use basic geometry to gauge the Earth's circumference.

Eratosthenes bent down and measured the shadow cast by the sundial. He then noted the height of the sundial itself. This gave him the lengths of two sides of a triangle. Using geometric theory, he was able to calculate the angle of the sun from the vertical. It was about seven degrees. Eratosthenes concluded that, with there being no shadow in Syene, the two cities were seven degrees apart, or one-fiftieth of a circle of 360 degrees – the circumference of the globe.

Eratosthenes had already reckoned how far Syene was from Alexandria. It took a camel 50 days to make the journey, at an average speed per day of 100 stadia – a Greek measure of distance equal to about one mile (1.6 km). This worked out to some 5000 stadia. Multiplied by 50 it meant that the circumference of the Earth was about 250 000 stadia, or 25 000 miles (40 230 km). The modern measurement of the Earth is 24 860 miles (40 007 km) for the circumference through the Poles.

THE LONGEST DAY *At midday on the longest day of the year the Sun shines directly into a well in Syene, a city in the far south of Egypt. The walls of the well cast no shadow.*

THE SHORTEST SHADOW *On the same day, at the same time, far to the north in Alexandria, a curious crowd gathers around the well-known citizen, the head of the great library, Eratosthenes, as he carefully measures the short length of the shadow cast by a tall pillar.*

Making the facts fit preconceived ideas

HOW PTOLEMY PUT THE EARTH AT THE CENTRE OF THE UNIVERSE

THE BUSY PORT OF ALEXANDRIA ON EGYPT'S NORTHERN coast was the intellectual hub of the Roman Empire. Scholars came from all around the Mediterranean to debate the meaning of life or the nature of the Universe, and to use the famous library. From around the middle of the 2nd century AD, a citizen of Greek extraction called Claudius Ptolemaeus, known to posterity as Ptolemy, was one of its leading lights.

A book to last a thousand years

This astronomer and mathematician presented everything known about astronomy at the time in a book of 13 volumes. The work included a catalogue of 1022 stars and an explanation, with Ptolemy's own mathematical proofs, of how the Universe was structured and how heavenly bodies moved.

According to the classical view, most famously expounded by Aristotle four centuries earlier, the heavens were perfect, and the cosmos was structured around the most perfect shape in geometry, the circle. A series of concentric transparent spheres rotated around the Earth, which was stationary. Ptolemy undertook a revision of this model, backing it up with logic, mathematics and his observations of the night sky. He argued that the Earth is stationary because, if it turned on its axis every 24 hours, as other earlier astronomers had suggested, an object thrown straight up into the air would not fall back to the exact spot from which it had been thrown.

Looking up at the clear Egyptian night, Ptolemy soon realised that the planets did not move in the way he expected. Rather than simply passing across

WISE MAN *The wisdom of Ptolemy, portrayed holding an armillary sphere, was revered for many centuries.*

ACUTE OBSERVER *Aided by Urania, the muse of astronomy, the great Ptolemy observes the stars.*

GREAT WORK *Ptolemy's work was held to be infallible in the Middle Ages. Its Arab translators called it* Almagest – *the Greatest.*

the sky from one side to the other, some of them appeared to move backwards at times, and others to stop. To fit the spherical model of the cosmos with what he had seen with his own eyes, Ptolemy tested and refined a system that had been proposed by a contemporary of Aristotle's, Apollonius of Perga,

who said that each planet moved in small circles, or epicycles, on the path of its own orbit of the Earth. Ptolemy proved the theory mathematically. After his work was translated into Latin by Gerard of Cremona in the 12th century, it became woven into the teachings of the Christian Church.

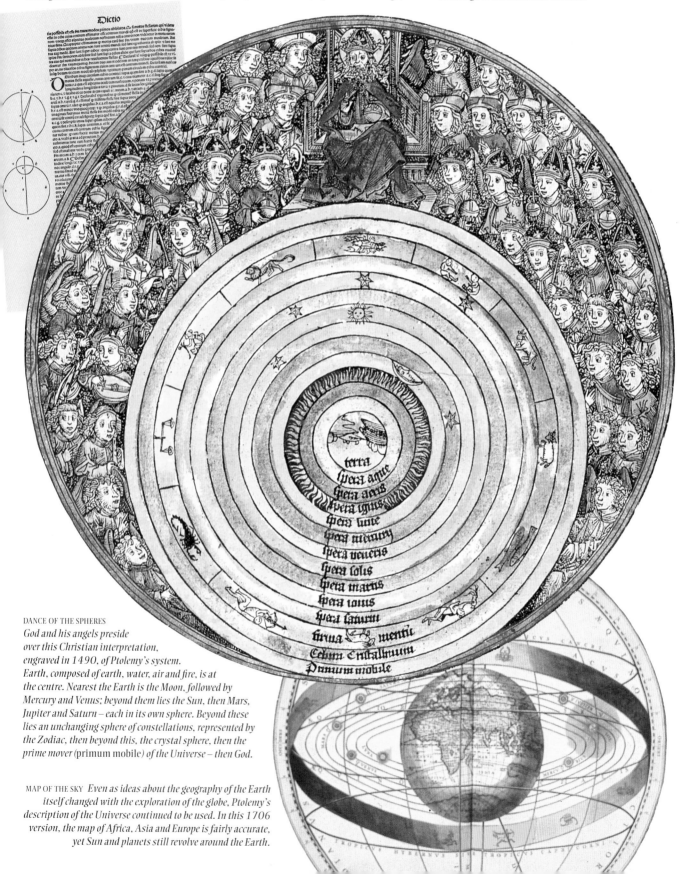

DANCE OF THE SPHERES
*God and his angels preside over this Christian interpretation, engraved in 1490, of Ptolemy's system. Earth, composed of earth, water, air and fire, is at the centre. Nearest the Earth is the Moon, followed by Mercury and Venus; beyond them lies the Sun, then Mars, Jupiter and Saturn – each in its own sphere. Beyond these lies an unchanging sphere of constellations, represented by the Zodiac, then beyond this, the crystal sphere, then the prime mover (*primum mobile*) of the Universe – then God.*

MAP OF THE SKY *Even as ideas about the geography of the Earth itself changed with the exploration of the globe, Ptolemy's description of the Universe continued to be used. In this 1706 version, the map of Africa, Asia and Europe is fairly accurate, yet Sun and planets still revolve around the Earth.*

Shadows and water as tellers of time

HOW PEOPLE MEASURED THE HOURS WITHOUT MECHANICAL CLOCKS

IN 1283, THE MONKS AT DUNSTABLE PRIORY IN Bedfordshire, England, became the first recorded users of a mechanical clock. They relied on it to fix the time of their chapel services. But for centuries before clockwork, people had marked the hours by watching the movement of shadows or of falling water.

The simplest – and oldest – means of gauging time was probably by looking at the Sun's position in the sky. In the third millennium BC the Babylonians studied the path of the Sun and divided daylight and darkness into 12 parts each, creating the 24-hour day. Each hour was divided into 60 minutes and each minute into 60 seconds, based on a number system devised by Babylonian mathematicians. The next step forward from merely tracking the Sun was to erect a column, or gnomon, by which time could be judged from the length and position of the shadow it cast; the Mesopotamians in South-east Asia were using gnomons in about 3500 BC. The sundial, with its calibrated scale, took a further 2000 years to perfect. Other devices used the fall of water or of grains of sand to mark time. A water clock was first used in Egypt in about 1350 BC.

ORNAMENTAL CLOCK *In this silver Chinese sundial of the 10th century BC (below), the 70 holes in the outer circle each represent 1 Kho (14 minutes) and together make up 16 hours and 33 minutes of maximum daylight. The central upright, used to cast the shadow, is missing.*

TIME DRAINS AWAY *The simplest water clock is a bowl which allows liquid to escape very slowly through a hole in its bottom (left). As the water level falls, marks representing the hours are uncovered on the sides. This Egyptian clock of the 14th century BC leaked water at the rate of about ten drops each second. The sides are angled to ensure that the water flows at a steady rate no matter how much is left inside.*

SHIFTING SHADOWS *Shadow boards were the forerunners of sundials. In this reconstruction (below), the shadow cast by the crosspiece indicates the hour. A water clock was used to ensure accurate marking of the wooden scale.*

TRAVELLING CLOCK *This portable two-disc Roman dial (above) told the time on any day, anywhere. A line on the upper disc can be set against the range of latitudes XXX–LX on the edge of the lower disc. The shaft used to cast a shadow moves on a separate scale to compensate for seasonal variations in the Sun's angle.*

MOVING PARTS *Operated by hand, this dismantled 5th to 6th-century Byzantine calendar (left) predicted the phases of the Moon. The large disc formed the front face of a circular box within which the gears were concealed. A lever or key was used to turn the device. An upright arm could be fitted to the front of the box, turning it into a sundial.*

THE FLOW OF TIME *Clocks operated by the natural movement of water could be very elaborate. This 13th-century illustration (left) shows the prophet Isaiah, standing, and King Hezekiah of Israel, who – as a sign that he would recover from illness – was granted a miracle and saw time go backwards. Here the water turns a wheel before filling a container. The clocks were not very accurate because of the difficulty of maintaining a regular flow of water.*

SCHOOL TIMEKEEPER *A tutor reminds his pupil that the lesson will not be over until the sand glass has run its course (right). Sand glasses were used in schools in Europe until the 17th century.*

MINUTES MEASURED IN GRAINS *This 16th-century English sand glass may have been used for timing workers paid by the hour. It is made from two glass bulbs. Between them, a brass diaphragm, with a hole about ten times bigger than each grain, is fixed with wax and string. Sand pours at a constant speed from the top bulb into the bottom.*

TIME IN HAND *Some sundials were small enough to fit in a traveller's palm or pocket. This portable 10th-century Saxon dial (above) has a different peg-hole for each pair of months corresponding to the Sun's changing position in the sky. The other six months of the year and their holes are on the reverse of the dial.*

WORKING HOURS *A skilled workman makes sand glasses in his workshop (above). Powdered eggshell could be used in place of grains of sand.*

TIME'S SWEET SMELL *The Chinese burned a trail of incense in a pierced case. Smoke emerged from different holes as time passed. This 15th-century case (right) comprises a box, two trays and two lids, each side of which is seen dismantled (above).*

A Mongol prince charts the heavens

HOW ISLAMIC ASTRONOMERS MAPPED THE NIGHT SKY

AT THE AGE OF 28, MOHAMMAD MARAGAY, ULUGH BEG or Great Prince and Governor of the central Asian states of Maveranakhr, Khorezm and Fergana, ordered a vast observatory to be built in his capital, Samarkand. The observatory was to complement the university he had founded four years earlier in 1420. Here the greatest minds of the Islamic world gathered to help the prince in his great project – to produce a new set of astronomical tables.

Instruments fit for a giant
An army of workers dug out the tons of rock needed to create a trench which housed the observatory's main feature: a massive sextant for measuring the heights of celestial bodies above the horizon. The rest of the instruments were almost as remarkable. Some were on a similarly large scale – a huge set of parallactic rulers and an armillary sphere which was 6 ft (2 m) across – which meant measurements could be made with greater precision than ever before. A variety of timekeeping instruments – astrolabes for measuring the elevation of the Sun and Moon, hourglasses and sundials – were used. These were especially useful for fixing the times of the five daily prayers required of all Muslims by the Koran.

The prince himself made many of the observations that went into the *Zij Guragoni*, a catalogue of 1018 stars, which was completed in 1437. Ulugh Beg determined the celestial year – the time taken for the Earth to go round the Sun – to be 365 days, 6 hours, 10 minutes and 8 seconds. Today's figure is 365 days, 6 hours, 9 minutes, 9.6 seconds. Ulugh Beg was less than a minute out.

The observatory's glories were short-lived. It was sacked by fanatics after the death of the prince in 1449 – murdered by assassins hired by his own son, Abdul-Latif – and was thought to be lost for ever until Russian archaeologists found its ruins in 1908 and excavated them.

INSIDE VIEW *The sextant, with a radius of 130 ft (40 m), stands in a building 100 ft (30 m) tall with a 30 ft (10 m) trench housing the lower half of the instrument.*

GRAND SCALE *Seen from above, an observer with two assistants uses the giant sextant. Light from the south-facing roof sight reaches the observer near the bottom of the steep marble steps. An inlaid brass strip, running down both sides of the sextant's steps, marks the degrees of the circle. The size of the instrument enables astronomers to measure the elevation of Sun, Moon, stars and planets extremely accurately.*

TEMPLE OF SCIENCE *Ulugh Beg's observatory stood about 100 ft (30 m) high. The small round aperture just below the dome is the observation window for the giant sextant housed inside. Outer rooms housed libraries, small observatories and laboratories.*

BOOK OF STARS *The great work includes a table of stars and an explanation of eclipses, planets and other phenomena.*

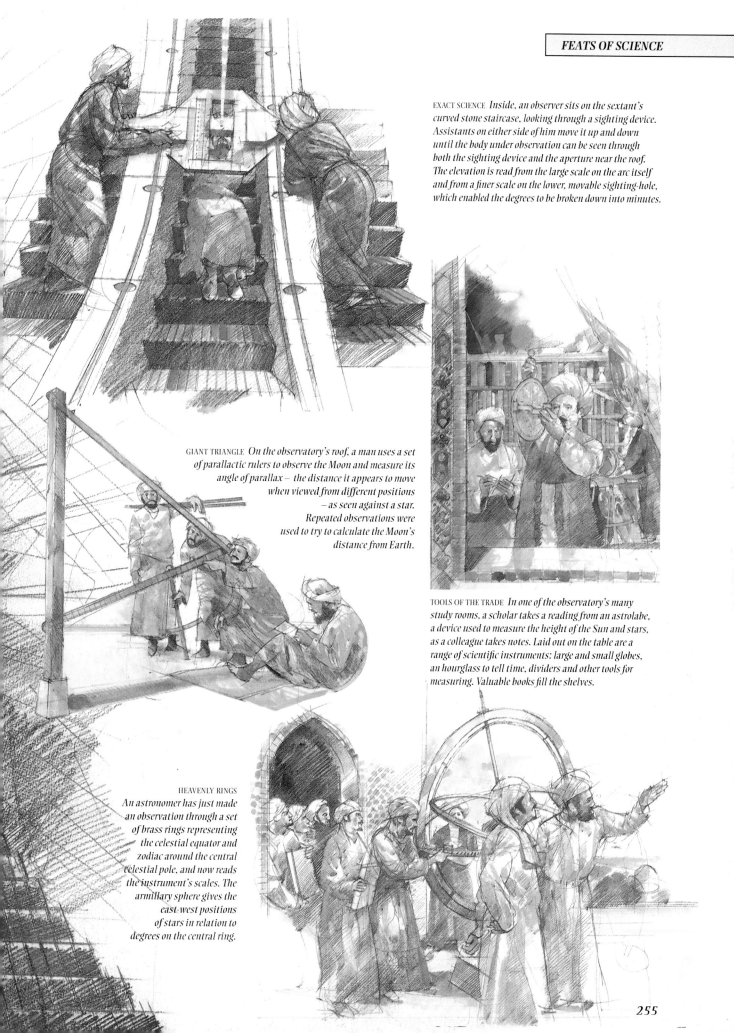

EXACT SCIENCE *Inside, an observer sits on the sextant's curved stone staircase, looking through a sighting device. Assistants on either side of him move it up and down until the body under observation can be seen through both the sighting device and the aperture near the roof. The elevation is read from the large scale on the arc itself and from a finer scale on the lower, movable sighting-hole, which enabled the degrees to be broken down into minutes.*

GIANT TRIANGLE *On the observatory's roof, a man uses a set of parallactic rulers to observe the Moon and measure its angle of parallax – the distance it appears to move when viewed from different positions – as seen against a star. Repeated observations were used to try to calculate the Moon's distance from Earth.*

TOOLS OF THE TRADE *In one of the observatory's many study rooms, a scholar takes a reading from an astrolabe, a device used to measure the height of the Sun and stars, as a colleague takes notes. Laid out on the table are a range of scientific instruments: large and small globes, an hourglass to tell time, dividers and other tools for measuring. Valuable books fill the shelves.*

HEAVENLY RINGS
An astronomer has just made an observation through a set of brass rings representing the celestial equator and zodiac around the central celestial pole, and now reads the instrument's scales. The armillary sphere gives the east-west positions of stars in relation to degrees on the central ring.

Worshippers of time

HOW THE AZTECS KEPT COMPLEX CALENDARS

AS DARKNESS FELL, A LINE OF PRIESTS CLIMBED THE sacred Hill of the Star, Uixachtlan, 7 miles (11 km) outside Tenochtitlan, the capital of the ancient kingdom of the Aztecs. On the summit they buried a bundle of 52 reeds. At that moment, they believed, time died. Then the priests climbed the temple steps and a crowd gathered below. The people of the town clambered onto the rooftops and waited for the ritual to begin. The sacrificial victim lay spread-eagled across the altar. When the star Alcyone reached its zenith, the senior priest struck, ripping out the victim's heart. Immediately, a fire was kindled in the chest cavity; the Aztec empire, so everyone believed, was safe for another 52 years.

Astronomer priests who told time
The Aztec priests used a complex dating system common all over Mexico since at least the first century AD. They were expert astronomers and mathematicians, who kept precise track of the progress of the Sun, Moon and planets. They used two calendars that ran concurrently. The first calendar of 260 days, the *tonalamatl*, kept track of religious events. It consisted of 20 name days combined with numbers from 1 to 13.

Alongside this calendar ran a solar year of 365 days, used to keep track of the agricultural year. This year was divided into 18 months of 20 days each. Days were counted only after they had passed, so the first day of each month was numbered 0. To make up the annual shortfall of five days, a period known as the hollow days was added on to the end of the year. This was a very inauspicious time, for evil forces were abroad. The population avoided conflict, fasted and did penance.

Every 52 solar years and 73 religious years, the two calendars coincided and began again on the same day. This marked the beginning of a new 'century', or 'bundle of years'. The end of each year was marked by

one of the priests setting aside a peeled reed, and it was these – made up into a bundle – which were buried at the ceremony marking the 'end of time'. Aztecs believed that the day of birth was very important in determining a person's future. Because of the two calendars, every day in a 52-year century had a unique combination of names. Each day was lucky or unlucky depending on which of the pantheon of gods was associated with it.

ROYAL RECORD *Mayan glyphs spell out the date* 1 Cimi 14 Muan *– the equivalent of January 16, AD 537. The inscription is part of a large stone carving recording the succession and relationships of the first ten rulers of the Yaxchilán dynasty in what is now Guatemala. The date highlights the importance of a particular ancestor.*

PERVADING INFLUENCE *The Sun and its god were worshipped by most pre-Columbian cultures in Central America. The sculpture (left) comes from Cotzumalhuapa in Guatemala and is carved in a local style. The Sun god, as provider of energy, needed sacrifices to keep up his strength. Failure to do this would mean the end of the world.*

The longest year of all

Aztec priests also followed the movements of Venus – the 'morning star' – which gave them a way to express long periods of time. They timed Venus' orbit of the Sun, which took 584 days. A cycle of 2920 days – a figure equal to eight solar years or five Venus cycles – linked the 365-day year and the Venusian year. The sacred calendar of 260 days, the Venusian year and the solar year coincided once every 104 solar years – creating a fourth cycle of 37 960 days.

The astronomical apparatus used to monitor the heavens and keep track of time was remarkably simple. The priests gazed at the stars from inside their temples which, being built at the tops of pyramids, had unobstructed views of the horizon.

A pair of crossed sticks was used to measure height above the horizon and from other stars. The spot where a heavenly body rose or set was measured in relation to a prominent landmark, such as a mountain or another pyramid.

On the first day of the new Aztec century the sacred flame, burning where the heart of a man had beaten only a few hours before, was used to ignite torches which teams of runners carried all over the empire, spreading the good news that the new century was born. The priests built a beacon on top of the Hill of the Stars that could be seen far and wide. Once temple fires were re-lit they were not allowed to die out for another 52 years, when again the nation would hold its breath during the 'end of time'.

LUCKY DAYS *Written on a strip of bark paper nearly 12 ft (3.6 m) long, the Dresden Codex, a Mayan almanac, shows which gods bring fortune to certain days of the year.*

SHARED BELIEFS *In Aztec times, the most potent sacrifices to the Sun god were human hearts. At the centre of the calendar stone (left) the god clutches two hearts in his hands. In the panels surrounding him are the legendary dates when the world temporarily ended, as previous suns were killed by whirlwinds (top left), a jaguar (top right), floods (bottom right) and fire (bottom left). The priest making the offering (right) is from an earlier culture.*

The search for the fabled philosophers' stone

HOW ALCHEMISTS TRIED TO CHANGE BASE METALS INTO GOLD

THE DEVIL AND DR DEE

IN LATE 16TH-CENTURY ENGLAND Dr John Dee, Astrologer Royal to Elizabeth I, devoted himself to searching for the philosophers' stone. He and his assistant Edward Kelly claimed to have found 'large quantities' of the stone among the ruins of Glastonbury Abbey. They used these to change a piece of copper cut from a warming-pan into gold, and then presented the pan and the gold to Queen Elizabeth to prove he could transmute metals.

Visions of Heaven and Hell
Despite his reputation as 'a companion of the Devil', the queen visited Dee at his home at Mortlake, near London. There she inspected his 'magic mirror' – a disc of highly polished black obsidian – and a crystal bowl, in both of which he claimed to see visions of hellhounds, an angel called Uriel, and the spirit of a girl named Madimi – who lectured him on the need to serve God.

After Elizabeth died in 1603 Dee petitioned her successor, James I, to clear him of the 'horrible slander' of being 'an invoker of imps and demons'. But the king refused to listen, and Dee recorded his last 'actions with spirits' in October 1607. He died, penniless, in December the following year at the age of 81.

MAGIC MIRROR *Dr Dee said that his round looking glass had been given to him by an angel, so that he could see spirits in it.*

ONE SNOWY DAY TOWARDS THE END OF DECEMBER 1666 a shabbily dressed stranger called on John Frederick Helvétius – physician to the Prince of Orange and one of Europe's leading authorities on the art of alchemy – at his home in The Hague. After introducing himself as Elias the Artist, the stranger produced a small ivory box, inside which were three glazed, sulphur-yellow objects the size of hazelnuts. According to Elias, the objects came from a legendary substance known as the philosophers' stone, which alchemists – named from the Arabic *al-kimiya*, meaning 'the art of transmutation', or the Greek *khemia*, meaning 'the melting and alloying of metals' – believed could change base metals into gold.

Glittering like gold
Helvétius begged Elias for a tiny piece of the stone. At first Elias refused to part with any of the 'magical' substance, which he said was enough to produce some 20 tons of gold. At a subsequent meeting, however, he grudgingly gave Helvétius a scrap of the stone. In the presence of his wife and son, the doctor put the scrap into a crucible which he heated until it was red hot, together with a small piece of lead. Helvétius later recorded in his journal: 'There was a hissing sound and a slight effervescence and the compound turned a vivid green. As soon as I poured it into the melting pot it assumed a hue like blood. When it began to cool it glittered like gold!'

Wild with excitement, Helvétius rushed with the still warm metal to a nearby goldsmith, who tested it and declared it was pure gold. News of the so-called 'miraculous change' spread through The Hague, and many distinguished

GOLD-MAKERS *In their quest alchemists used weird and wonderful laboratory apparatus. These devices included: a still in an oven (above), glass pelicans (top left) used to circulate vapours, a still (centre left), and a furnace (bottom left) for heating metals.*

visitors came to Helvétius's house to see this 'man-made' gold for themselves. These included Holland's General Inspector of the Mint, a man named Porelius, who tested the metal exhaustively and declared it to be genuine, high-quality gold.

Helvétius's claim is typical of tales told by alchemists throughout the centuries. Slaving over fiery furnaces, they concocted recipes of lead, sulphur, copper, tin and mercury in their search for gold. 'They give themselves diligently to their labours', wrote the 16th-century Swiss physician known as Paracelsus. 'They do not kill time with empty talk, but find their delight in the laboratory.'

In their hot and smoky laboratories, the alchemists also strived to produce the 'elixir of life', a legendary substance thought to prolong life indefinitely. According to Paracelsus, who studied alchemy at Basel University, it was relatively easy to make the elixir. All that was required was to dissolve the philosophers' stone in wine and drink the resulting solution. 'The elixir,' he wrote, 'cleanses the whole body of impurities by the introduction of new and more youthful forces which it joins to the nature of man.' Presumably Paracelsus did not succeed in his aim: he died in 1541 aged 47.

Shining like glass
However, the secret of the philosophers' stone did not die with him. In his book *Life Eternal*, a 17th-century Belgian physician and chemist named Johannes van Helmont claimed that he had often handled and used the stone. He said that it was heavy, the colour of saffron powder and shone like glass. His recipe to turn lead into gold was simply to add hot quicksilver, or mercury, to a sliver of the stone. That, he asserted, would produce 8 oz (225 g) of gold. And in the next century, a flamboyant Italian adventurer, the self-styled 'Count' Alessandro di Cagliostro, took an apartment in London's Leicester Square in which to practise alchemy. 'Gold-making,' he stated, 'is my hobby!'

Many people – especially women – fell under Cagliostro's spell, and gave him good money in exchange for what turned out to be bits of relatively worthless amber. Moving on to Paris, and then Rome, Cagliostro was later arrested by order of Pope Pius VI and imprisoned for life as a heretic. He died in the fortress prison of San Leone, in Italy, in 1795.

Alchemy was strongly disapproved of by the Roman Catholic Church, particularly as some alchemists maintained that the philosophers' stone symbolised Christ, and that their occult arts were of spiritual value to mankind. 'The only values the alchemists were interested in,' said a commentator later, 'were those of making money for themselves!'

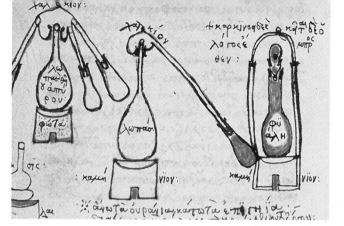

PUTTING IT ON PAPER *In the 4th century AD an Egyptian alchemist named Zosimos of Panopolis wrote and illustrated a treatise (right) on the art of alchemy. Zosimos believed that alchemy had been introduced by God's angels, who had also taught women how to enhance their natural beauty with precious gems and rare metals such as silver and gold.*

PUTTING IT INTO PRACTICE *A 16th-century alchemist (below) points a cautionary finger as his assistants strive diligently to turn base metals into gold. It is a painstaking matter of heating, stirring, pouring and distilling the ingredients, intended to mimic what they believed to be the processes within the Earth's crust that made the precious metal pure.*

A wooden helicopter

HOW LEONARDO DA VINCI INVENTED MACHINES AHEAD OF THEIR TIME

LEONARDO DIPPED HIS PEN IN THE INKWELL AND BEGAN to sketch rapidly. With each stroke of the pen his marvellous flying machine became more precise and more complex. When he had finished, he turned the page of his notebook and began to draw a giant catapult. Ludovico Sforza, the Duke of Milan and Leonardo's employer in 1486, would be impressed by the new weapons Leonardo was devising.

During his career, many of Leonardo's patrons – including several powerful northern Italian noble families, Pope Leo X and King François I – hired him for his inventive ability, as well as for his skill as a painter. They wanted him to create war machines.

Leonardo lived in turbulent times, when the Italian states were almost constantly at war. The Sforza family of Milan, like the Medici family of Florence, hoped that Leonardo would create a devastating weapon to give them victory on the battlefield.

Leonardo recorded and refined his inventions using the finest ink in his private notebooks, which were made from the best paper available. Very few of them, however, seem to have been made. Many of the sketches were perhaps little more than doodles made for his own amusement. Nevertheless, all Leonardo's drawings show a sound grasp of mechanics and meticulous observation of the world around him.

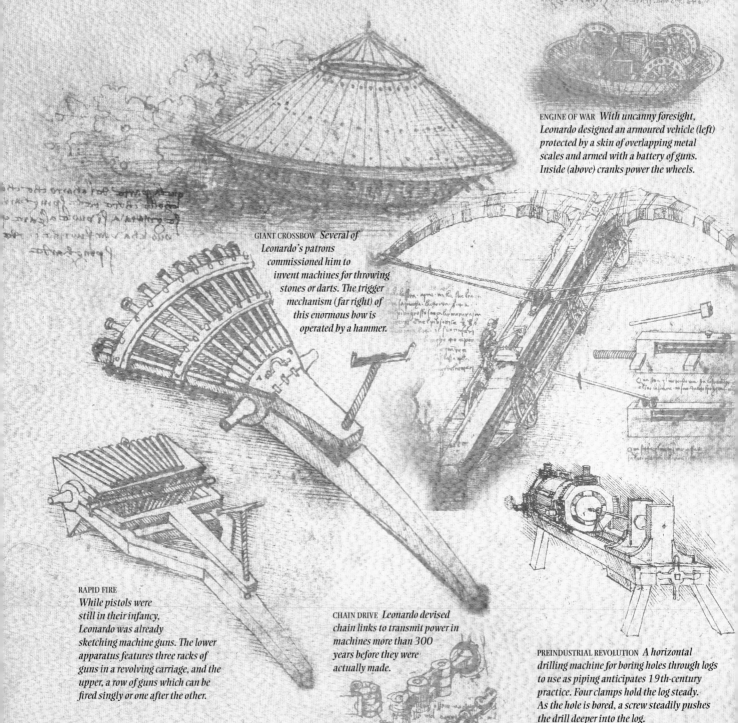

ENGINE OF WAR *With uncanny foresight, Leonardo designed an armoured vehicle (left) protected by a skin of overlapping metal scales and armed with a battery of guns. Inside (above) cranks power the wheels.*

GIANT CROSSBOW *Several of Leonardo's patrons commissioned him to invent machines for throwing stones or darts. The trigger mechanism (far right) of this enormous bow is operated by a hammer.*

RAPID FIRE
While pistols were still in their infancy, Leonardo was already sketching machine guns. The lower apparatus features three racks of guns in a revolving carriage, and the upper, a row of guns which can be fired singly or one after the other.

CHAIN DRIVE *Leonardo devised chain links to transmit power in machines more than 300 years before they were actually made.*

PREINDUSTRIAL REVOLUTION *A horizontal drilling machine for boring holes through logs to use as piping anticipates 19th-century practice. Four clamps hold the log steady. As the hole is bored, a screw steadily pushes the drill deeper into the log.*

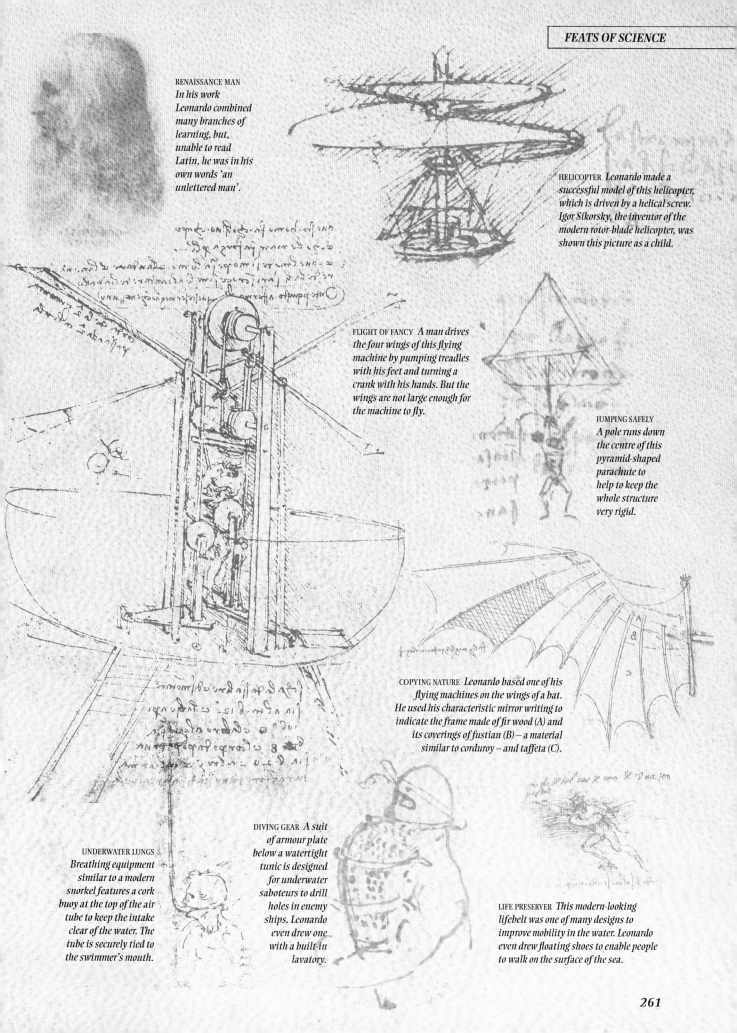

RENAISSANCE MAN
In his work Leonardo combined many branches of learning, but, unable to read Latin, he was in his own words 'an unlettered man'.

HELICOPTER *Leonardo made a successful model of this helicopter, which is driven by a helical screw. Igor Sikorsky, the inventor of the modern rotor-blade helicopter, was shown this picture as a child.*

FLIGHT OF FANCY *A man drives the four wings of this flying machine by pumping treadles with his feet and turning a crank with his hands. But the wings are not large enough for the machine to fly.*

JUMPING SAFELY
A pole runs down the centre of this pyramid-shaped parachute to help to keep the whole structure very rigid.

COPYING NATURE *Leonardo based one of his flying machines on the wings of a bat. He used his characteristic mirror writing to indicate the frame made of fir wood (A) and its coverings of fustian (B) – a material similar to corduroy – and taffeta (C).*

UNDERWATER LUNGS
Breathing equipment similar to a modern snorkel features a cork buoy at the top of the air tube to keep the intake clear of the water. The tube is securely tied to the swimmer's mouth.

DIVING GEAR *A suit of armour plate below a watertight tunic is designed for underwater saboteurs to drill holes in enemy ships. Leonardo even drew one with a built-in lavatory.*

LIFE PRESERVER *This modern-looking lifebelt was one of many designs to improve mobility in the water. Leonardo even drew floating shoes to enable people to walk on the surface of the sea.*

Turning the Universe around

HOW NICOLAUS COPERNICUS SHOWED THAT THE EARTH REVOLVES AROUND THE SUN

ONE EVENING IN 1543, THE DOOR OF A BEDROOM IN in a house in Frauenberg – a town in Poland – burst open and a figure carrying a book rushed across the room to an old, ill man in bed. The visitor placed the book on the coverlet. The invalid, the astronomer Nicolaus Copernicus, leafed through its pages. His life's work was done. A few days later, he was dead.

The book that the dying Copernicus examined, *De Revolutionibus Orbium Coelestium* (*Concerning the Revolutions of the Heavenly Bodies*), contained his theory that the Earth was not the centre of the Universe – a theory that the Roman Catholic Church was to ban for more than 200 years.

For over a thousand years, the accepted model of the cosmos was one proposed by the Greek scientist Ptolemy in the second century: the Earth was stationary at the centre of the Universe and the Sun, Moon, planets and stars moved around it. The Church embraced this model, because it supported the view that mankind and Earth were at the centre of God's Universe. But there were contradictions within the Ptolemaic system. Copernicus approached

the problem mathematically. He believed that the heavens were perfect and so the orbits of the planets must be circular – the most perfect geometrical shape. There was only one hypothesis which would allow for this. 'All the spheres,' he wrote, 'revolve around the Sun at midpoint and therefore the Sun is the centre of the Universe.'

Copernicus suggested that stars appeared to move across the night sky not because they were orbiting the Earth but because the Earth rotated on its axis every 24 hours. He devised the equations to make his model work, but he still had to make almost as many compromises as Ptolemy had done. This was because the planets actually move in ellipses, not circles, as Johannes Kepler showed 70 years later.

Dangerous displeasure of the Church

If he published his theory, Copernicus knew he risked the charge of heresy. His first book, *Commentariolus*, was written anonymously and circulated privately in manuscript form in 1514. Only after heavy pressure from his friends did Copernicus allow his ideas to be published in full. By that time he was 70 years old, and close to death.

The Catholic Church seemed at first untroubled by Copernicus's hypothesis. But the tacit acceptance of the book did not last. About 70 years later, in 1616, during the Counter Reformation, his writings were banned – a decision that was not revoked until 1835. This did not, however, stop his theory winning acceptance from scientists and laying the foundations of modern astronomy.

SUN CENTRED *The six known planets, including the Earth, and the 12 constellations of the zodiac, revolve around the Sun, which sits firmly at the centre of the Copernican Solar System.*

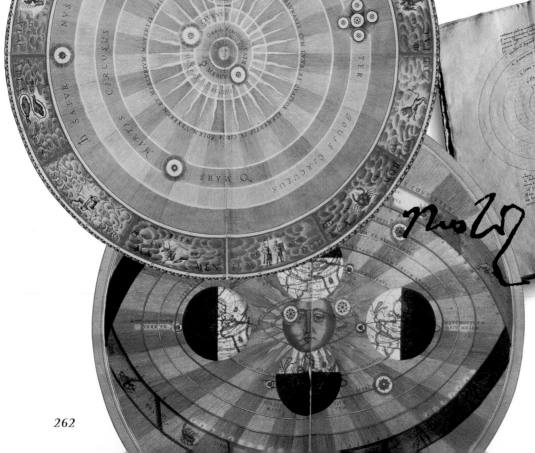

BRAVE BOOK *Copernicus's originality was as evident in his signature as in his books.* De Revolutionibus *caused little stir at first, but was later banned by the Church for 200 years.*

STAR MAP *The new place given to the planet Earth inspired a fresh look at the world. It was the first step of what became a revolution in science.*

Revelations from a castle of the stars

HOW TYCHO BRAHE OBSERVED THE HEAVENS WITHOUT A TELESCOPE

PERFECT CONDITIONS *Two observers use a giant sextant on a mobile stand above Tycho's secondary observatory, Stjerneborg – 'the castle of the stars' – built in 1584, near the main observatory. It was built so no breezes could disturb the delicate instruments. Beneath the far cupola is a giant quadrant – a quarter-circle scale with a radius of nearly 7 ft (2 m), fitted with sights that moved along its arc. With it, Tycho could measure angular distances of just five seconds (one-twelfth of a degree) between celestial objects. Beneath the other two cupolas are large armillary spheres, used to find the position of stars in relation to the celestial equator.*

WALKING HOME TO SUPPER FROM HIS LABORATORY on November 11, 1572, the Danish astronomer Tycho Brahe looked up at the sky and stopped in his tracks. The familiar constellation of Cassiopeia seemed to have gained a new star. Tycho asked his servants if they could see it. They could, but still disbelieving the evidence before his eyes, Tycho stopped a group of peasants on their way home from the fields. They, too, could see the new star.

This sudden change in the heavens was to shatter ideas about the Universe that had been taken for granted for more than 1500 years. Since the time of Aristotle, astronomers had believed that, beyond the Moon, the heavens were an unchanging series of spheres. Now it seemed that they were wrong.

Months of careful watching

Tycho rushed home and eagerly measured the new star's position in the constellation and checked its height with a cross-staff, a simple instrument which consists of one long stick held parallel with the horizon and a perpendicular bar that measures height above the horizon. Tycho watched the star for months and observed it becoming brighter and then slowly fading. He had read descriptions of comets and knew he was not seeing one of those, because it had no tail. The lack of parallax – the apparent change in position of an object when viewed from different places – proved that

the star was far beyond the Moon. What he was, in fact, observing was an erupting star – a supernova. The next year, Tycho wrote about his observations in *De Nova Stella* (*On the New Star*), a book which established his reputation as an astronomer.

The gift of an island

Impressed by Tycho, Frederick II of Denmark and Norway, bequeathed him an island in Copenhagen Sound in 1576. Here Tycho built an observatory, Uraniborg – Danish for 'the castle of the heavens' – where the best precision instruments were installed to his own designs. As a result, his star maps, though plotted without a telescope, were the most accurate to date. Tycho died in exile after falling out with the new king, Christian IV, but left his works to the astronomer Johannes Kepler, who used them to solve the Copernican puzzle and prove that planets moved in ellipses.

IMPOSING SKYLINE *Uraniborg, on the island of Hven, was Tycho's main observatory. Its striking exterior also housed a large library and a printing press.*

Planet Earth put in its place

HOW GALILEO PROVED THAT THE EARTH ORBITS THE SUN

IN 1600, A MAN WAS BURNED ALIVE BY ORDER OF THE Inquisition for the heretical suggestion that the Earth moved around the Sun. Thirty-four years later, charged with the same crime, 68-year-old Galileo Galilei knelt before ten cardinals, judges of the Inquisition. He looked down at his hands, twisted with arthritis. The choice was plain: recant or face torture and death. That afternoon he told his judges, 'I do not consider the view of Copernicus as true and have never considered it as true', and so saved himself from execution.

REVELATION IN THE CATHEDRAL

IN 1581, THE 18-YEAR-OLD GALILEO GALILEI KNELT TO PRAY in the cathedral at Pisa. His attention was distracted by an attendant lighting a lamp that dangled by a long chain from the ceiling. He pulled the lamp towards himself and after lighting it, let it go. The lamp began to swing backwards and forwards.

The pulse and the pendulum
Galileo watched with fascination. Here was something he had never noticed before. While the length of each swing was decreasing, the time taken to complete a backward and forward movement appeared to stay the same. Galileo timed the swings against his own pulse. He discovered that whatever the length of the swing, the number of pulse beats for one oscillation remained the same. Galileo realised that he could make an instrument – a simple pendulum – for measuring the frequency of the pulse. He called it the *pulsilogia* and it was used by a generation of doctors.

SCIENTIFIC HERO *Despite the disapproval of the Vatican, Galileo was much admired by generations of scientists and philosophers that came after him. This painting of him in Pisa Cathedral was executed more than 150 years after his death.*

The course of events that was to lead to Galileo's confrontation with the awesome powers of the Inquisition began in 1609 on a visit to Venice. Friends told him about the invention of a Dutchman, Hans Lippershey, which could make distant objects seem close – a telescope. At once, Galileo set about grinding glass into lenses. After many tries he made a telescope that magnified 30 times. In January 1610, Galileo pointed his telescope at the night sky and began to look at the heavens.

His startling observations seemed to demolish objections to the theories of the Polish astronomer Copernicus that the Earth and planets moved around the Sun. The Church held the view that the Earth could not move in space because it would leave its Moon behind. Yet Galileo discovered no fewer than four moons around Jupiter, which clearly moved across the sky.

'Philosophy is written in this grand book, the Universe. But the book cannot be read unless one first learns to comprehend the language and read the alphabet in which it is composed.'
GALILEO GALILEI

Another argument against Copernicus's theory was that Venus did not appear to have phases, like the Moon, and that if Venus were orbiting the Sun, its reflection of the Sun's light would have appeared in phases as it moved. Galileo found that, seen through a telescope, Venus did have phases.

In 1624, Galileo asked Pope Urban VIII for permission to publish support for Copernicus's view of the structure of the Universe. Urban agreed to allow Galileo to write a book that put forward the Copernican view, provided the Church's official view, based on the work of the 4th-century astronomer Ptolemy, was given equal prominence. Galileo's book, *Dialogue Concerning the Two Chief World Systems* –

BANNED BOOK *The frontispiece of Galileo's book about the structure of the Universe shows its three characters deep in debate. Salviati ('saved') argues in favour of Copernicus. Ptolemy's view is defended by Simplicius ('simple') while an impartial observer remains open to persuasion by either side.*

AWESOME POWER *The trial of Italy's most famous scientist, Galileo Galilei, attracts a large audience of clergy to the monastery of Santa Maria sopra Minerva. Although Galileo was forced to agree with the judges that the Earth was stationary, it is said that he muttered the words, 'And yet it moves' under his breath.*

Ptolemaic and Copernican,
was published in February 1632. By
August the Pope realised that the argument
in the *Dialogue* was biased against the Church. The
Copernican used clever reasoning, while the follower
of Ptolemy presented feeble arguments.

Forced to recant

In October 1633, Galileo was summoned to Rome to
face the Inquisition. There, the cardinals cross-
examined him over a period of ten months, but they
did not resort to torture. Initially, he put forward a
strong defence of his views, but eventually he said
his book was wrong. In 1634, the *Dialogue* was
banned and Galileo returned to his villa in the hills
above Florence, where he spent the last eight years of
his life under house arrest.

NEW MOON *Galileo's drawings of the
phases of the Moon show that its
surface is scattered with mountains
and craters. Earlier observers had
believed it was a perfectly smooth
crystal sphere.*

NEW INVENTION *The
telescope with which Galileo
saw the Moon had a fixed concave
lens at one end and a movable convex
lens as an eyepiece at the other.*

A new world – in a drop of water

HOW MICRO-ORGANISMS WERE FIRST DISCOVERED

PEERING THROUGH HIS HOMEMADE MICROSCOPE AT A drop of stagnant water, Anton van Leeuwenhoek was sure he could see tiny creatures on the move. He looked again. They were still there – and in liquid that, to the naked eye, showed no signs of life. What van Leeuwenhoek saw that day in 1674 were the single-celled organisms that are now called protists.

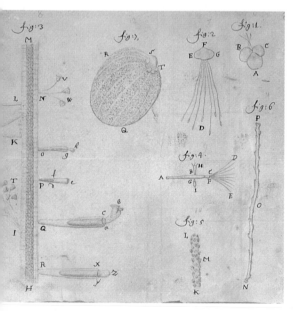

STRANGE LIVES
Van Leeuwenhoek carefully observed rotifers, water creatures that use tiny propellers to move through the water, and had them engraved by an illustrator.

HIDDEN WORLDS
Hundreds of tiny animals, never seen before by humans, swim across the pages of van Leeuwenhoek's notebooks.

SWIMMING CELL
In 1677, van Leeuwenhoek was the first to discern spermatozoa in seminal fluid.

He called them animalcules, 'small animals', and marvelled at the fact they were many times smaller than any living thing he had observed before – far too small to be seen with the naked eye. To gauge the size of these little animals, he compared them to a hair plucked from his beard.

The compound microscope, which used two lenses to magnify an object, had been developed in Europe at the beginning of the 17th century. It produced a blurred image with coloured edges, because the glass was of poor quality and the distortions could not be corrected. In his work as a draper, however, van Leeuwenhoek had obtained sharp magnification from the simple, single lens used by merchants to inspect material for closeness of weave.

Hand-crafted technology

Van Leeuwenhoek started grinding his own small, spherical lenses. These could magnify up to 200 times or more, and produced much clearer images. To make his microscope, van Leeuwenhoek drilled a hole through two brass plates and clamped them together, with the bead-shaped lens gripped in the aperture between the holes. Then he attached a bracket to one end of the metal plates and fixed a specimen holder – a long screw which twisted up and down behind the brass plates until its tip was positioned just below the lens. Van Leeuwenhoek carefully placed his chosen specimen on the point of the specimen holder and holding the microscope in front of a light source, such as a candle or a window, he observed the specimen.

In 1674 and 1676 van Leeuwenhoek reported his sightings of little animals to the leading scientific institution of the day, the Royal Society in London. His letters caused a sensation, and many fellows of the Society refused to believe van Leeuwenhoek had seen any tiny creatures. But in subsequent years, the 190 or so letters that van Leeuwenhoek sent to the Royal Society detailing his researches established his reputation. He made many pioneering discoveries. In 1683 he found tiny organisms in his tooth scrapings; the first bacteria ever seen with the human eye.

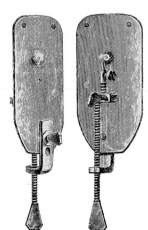

SINGLE LENS *Van Leeuwenhoek's simple microscope required a steady hand. It had to be held upright in front of the eye, while the specimen to be magnified was precariously balanced on the point of the screw.*

A sweaty shirt and some grains of wheat

HOW SCIENTISTS DEBATED THE EXISTENCE OF A 'VITAL PRINCIPLE'

A TINY, WHISKERED SNOUT PEEPED OUT OF THE SHIRT pocket. Johannes van Helmont beamed. His recipe for producing mice had worked. Take one sweaty shirt, add some grains of wheat, and leave for 21 days. The result: a nest of baby mice. The opinions of van Helmont, a 17th-century Flemish physician, did not appear as bizarre to his contemporaries as they do to us. It was then almost universally believed that life could arise from non-living material.

Conjuring maggots from thin air

Aristotle, the Greek philosopher, came up with the theory of spontaneous generation in the 4th century BC. He argued that all inanimate matter contained a 'vital principle' that could produce living things. Everyday observations seemed to verify this idea. Provided a free flow of air was present, rotting meat produced maggots, grain produced weevils.

By the 17th century, however, the concept of spontaneous generation was beginning to be questioned. In 1668 Francisco Redi, physician to the Tuscan court, devised an experiment to find out if, as he suspected, maggots came from eggs laid by flies. He took four large, wide-mouthed flasks. In one he put some fish, in the second a dead snake, in the third a dead eel from the River Arno, and in the fourth a slice of veal. He did exactly the same with four other flasks, but sealed them with fine Naples gauze, which kept out flies but let in air. After a few days, all the flasks stank, but only in the uncovered flasks did maggots appear. It seemed that air alone was not enough to produce life.

Gravy grows tiny animals

Despite the evidence put forward by Redi and other scientists, supporters of spontaneous generation did not give in easily. The debate raged throughout the 18th century. In 1748, for instance, an English priest, John Needham, set out to prove that spontaneous generation existed. Needham sealed hot mutton gravy in a flask and examined a sample of it under his microscope a few days later. It was swarming with small organisms. Needham concluded that a vital principle within the gravy had created these living forms.

However, an Italian scientist named Lazzaro Spallanzani heavily criticised Needham's research.

In 1767 he repeated Needham's experiment, but this time the flask was sealed after the gravy had been heated for some time. Two weeks later, no life had appeared in the liquid. Spallanzani argued that Needham had not heated the gravy sufficiently to kill the organisms already in the flask, and that these simply multiplied after the gravy was sealed. Needham counter-argued that heating at high temperature had destroyed the vital principle. Neither experiment was conclusive and the debate rumbled on for nearly 100 years until the French scientist Louis Pasteur proved that the organisms in Needham's gravy were airborne bacteria.

BEAUTIES OF NATURE *Francisco Redi (bottom), a noted poet as well as a scientist, wrote several fabulously illustrated treatises on biology, including (right) his experiments with spontaneous generation and (below) a ground-breaking study of insects.*

FERTILE EARTH *In this 17th-century woodcut, the fruits of the tree of life filled with the 'vital principle' fall into water to create fish and onto the ground to create birds.*

SIMPLE SOLUTION
In 1862, Louis Pasteur finally disproved the theory of spontaneous generation with the help of a swan-necked flask, which allows air to enter but excludes micro-organisms and dust. He put wine into the sterilised flask and the liquid remained clear. Then he broke the neck off and soon the wine fermented, proving that fermentation was caused by airborne bacteria.

A rainbow in the drawing room

HOW ISAAC NEWTON PROVED THAT SUNLIGHT WAS MADE UP OF MANY COLOURS

QUIET GENIUS *Isaac Newton laid the foundations for modern mathematics and physics before he was 25 years old, but he did not publish his findings until he became President of the Royal Society in London many years later.*

IN 1665, AS THE GREAT PLAGUE SWEPT THROUGH England, 22-year-old Isaac Newton fled from the unhealthy air of Cambridge, where he was a student, to take up residence with his mother in the safety of the Lincolnshire countryside. In the following year Newton laid the foundations for all his major discoveries, including his trailblazing exploration of the nature of light.

Disproving Descartes
The young Newton was fascinated by the French scientist René Descartes' account of his experiments with light, published in 1637. Descartes had observed how prisms could split white light into a rainbow-like spectrum of colours, and concluded that these blocks of glass or crystal transformed light permanently. The varying thickness of the prism, he thought, was responsible for the different colours.

Newton decided to test Descartes' hypothesis by designing a simple experiment of his own. He darkened one of the manor's south-facing rooms, allowing only a thin beam of light to shine through a hole in one of the shutters. On his workbench he arranged an assortment of items including a convex lens to make light rays converge, a triangular prism, and a board on which to project light. When the

PARLOUR GAMES *The young Isaac Newton experiments with sunbeams in the parlour of his mother's house, Woolsthorpe Manor.*

FRESH APPROACH *Newton ponders the rainbow from a ray of sunlight shining through a prism. It was his ability to observe with a fresh eye that led him to his discoveries.*

MASTER PLAN
Newton's own rough drawing of his apparatus shows the simplicity and elegance of his experiment.

beam of white sunlight shone through the prism, it produced the expected effect – a spectrum of red, orange, yellow, green, blue, indigo and violet and a change in the angle of the light.

Newton projected this spectrum onto a vertical board and drilled a hole where the red light hit it, allowing a beam of red light to shine through. When this was passed through a second prism, it changed angle – but failed to produce another spectrum. From this, Newton concluded that white light was composed of different colours, which the prism revealed rather than produced.

Bending rays of light

The phenomenon behind this effect, known as refraction, occurs when light passes through different substances that change its speed. Light travels faster through air, for example, than through glass. Because each of the constituent colours of white light has a slightly different wavelength, the prism bends it to a different angle, separating it from the others to make the spectrum.

To test his hypothesis, Newton passed all the coloured rays through a convex lens that focused them on a second prism. The rays were refracted in the opposite direction, emerging as a single beam of white light which, when passed through a third prism, split once again into seven colours. By separating, reassembling and separating it again, Newton had conclusively demonstrated that white light was made up of many different colours.

WHY THE MOON DOES NOT FALL TO EARTH

ONE AFTERNOON IN 1665, ISAAC NEWTON WAS sitting in the orchard of his mother's house in Lincolnshire. Suddenly, an apple fell from one of the trees to the ground. Newton pondered this event, reasoning that the apple had to fall to Earth because there was nothing to support it. But then he looked at the Moon. As far as he could see it had no support, yet it did not fall down. Then he had a flash of inspiration.

The mysterious pull of the Sun

What stopped the Moon falling down was its forward movement, which balanced the pull of gravity. The tug between the two is what caused the Moon to orbit continuously. The same forces – the gravitational pull of the Sun and forward motion – kept the Earth and planets orbiting the Sun. If gravity or the forward movement weakened, the Earth would fly out into space or fall into the Sun.

Although the Moon was attracted to the Earth, because it was farther away, the pull of gravity was less powerful than that exerted on the apple. Newton reasoned that if he calculated the force of bodies falling near the Earth's surface – like the apple – and compared this with the speed and size of the Moon's orbit, he would come up with a law for gravity.

Various theories had been put forward to explain why bodies fell to Earth, and why the planets remained in orbit around the Sun. Earlier in the same century, the French scientist René Descartes had proposed that a vortex of matter drew smaller bodies towards larger ones, and the Italian Giovanni Borelli had suggested that the planets were moved by invisible spokes that emanated from the Sun. None of them could be shown to work mathematically.

Newton worked out that the force of gravity was equal to the masses of the two objects times each other, in inverse proportion to the square of the distance between the two. In other words, if the Moon were twice as near the Earth, the force of gravity between the two would be four times greater.

Laws of motion

His work on gravity and the Moon led Newton to formulate three laws of motion. The first was that a body in motion or at rest will stay that way unless force is applied. This force, the second law states, equals the mass of the body multiplied by the acceleration it produces in the body. The third law is that if the first body exerts a force on the second, there is an equal and opposite reaction on the first body.

To prove his theories mathematically, Newton had to invent calculus, a form of advanced mathematics still used today. He published his findings in a book, *Principia Mathematica*, more than 20 years after he saw the apple fall in his mother's garden. No one knows why he left it so long, although he may have underestimated the size of the Earth initially, which would have made all his calculations inaccurate.

ACTION AND REACTION *A simple machine demonstrates the third law of motion. Steam blowing out in one direction pushes the engine in the opposite direction.*

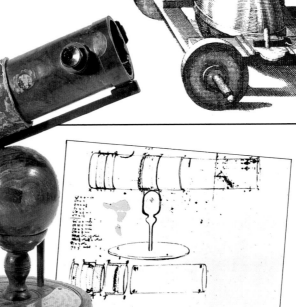

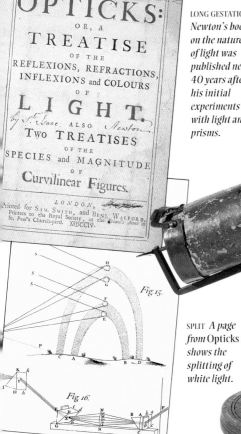

LONG GESTATION *Newton's book on the nature of light was published nearly 40 years after his initial experiments with light and prisms.*

SPLIT *A page from* Opticks *shows the splitting of white light.*

REVOLUTIONARY FOCUS *Newton's sketch of his telescope, built in 1671, shows how it uses a lens and a curved mirror to focus, instead of two lenses. Many modern telescopes are based on Newton's design.*

The elements of air

HOW THE GAS OXYGEN WAS DISCOVERED AND NAMED

REVOLUTIONARY TRACT *Lavoisier's radical new method of naming elements and compounds was published as a treatise in 1789.*

MEETING OF MINDS *Lavoisier and his wife, Marie Anne, pose over a table laden with laboratory equipment. She assisted with experiments and edited his work after his death.*

AS ANTOINE LAVOISIER ADJUSTED A GLASS JAR, HIS WIFE, Marie Anne, did a rapid drawing of her husband and his laboratory. She was there to take down his dictation as he experimented, but occasionally she sketched as well. Today, June 5, 1777, Lavoisier was starting a new experiment.

Pillar of scientific theory

He believed that he could discover once and for all the nature of 'phlogiston' – a central concept to the fledgling science of chemistry. Phlogiston was a hypothetical gas which chemists believed was given off by combustible substances when heated.

To test the theory, Lavoisier took a systematic approach. He weighed a blob of mercury and put it in a flask with a bent neck – a retort. He put a water-filled bell jar over the end of the neck and marked the volume of air on the outside of the glass with a piece of gummed paper. Then he heated the mercury in the flask until a layer of red particles formed on top of the liquid mercury. After letting the apparatus cool, he found that the volume of air had decreased and instead of growing lighter, as should have been the case if it had lost its phlogiston, the matter in the bottom of the retort was heavier. Not only had the mercury changed its nature, so had the air.

FRIENDLY RIVAL *A British chemist, Joseph Priestley, was the first to isolate oxygen, but he did not realise its significance.*

It appeared to have become so poisonous, it killed a mouse and put out a candle. Lavoisier realised that far from giving off phlogiston when heated, mercury absorbed an unidentified substance from the air, which made the mercury heavier and changed it into a red powder.

Lavoisier took his experiment a step further. He scraped the red powder together, put it into a smaller retort and reheated it in the depleted air he had collected from the previous experiment. Eventually, he was left with a small lump of mercury lighter than the powder had been, while the volume of air had increased. He tested the air and found it was now indistinguishable from ordinary air.

Further experiments showed Lavoisier that the gas released by burning the red powder had special powers: it revived a half-suffocated mouse and made a candle burn more brightly. He believed erroneously that the new gas was in all acids, and so named it oxygen from the Greek roots *oxy* for 'sharp' and *gen* for 'making'. Having deduced that the red powder he had produced was a combination of mercury and oxygen, Lavoisier named it mercuric oxide.

'It took them only an instant to cut off that head, and another hundred years may not produce another like it.'
MATHEMATICIAN JOSEPH LAGRANGE, OF LAVOISIER

Lavoisier had proved conclusively that air is a mixture of at least two gases: oxygen and nitrogen. This later led him to lay the foundations for modern chemical nomenclature.

Before he could complete his researches, Lavoisier was guillotined in 1794 by the new government, which condemned him because he owned shares in a tax collection company. The judge at the trial said, 'The Republic has no need of scientists.'

LABORATORY STUDIES *Madame Lavoisier sketched herself taking notes behind her husband as he collects the air one of his helpers is exhaling in a bell jar. His experiments with gases led him to develop the theory of respiration.*

A measurement for the modern age

HOW THE METRE WAS ESTIMATED

AS THE NIGHTS LENGTHENED INTO THE WINTER OF 1798, surveyors Jean Delambre and Pierre Méchain were reaching the end of six and a half years' labour. They had sat through hours of interrogation and spent days locked up in jails the length of France – all in pursuit of their task: to define a unit of measurement suitable for Revolutionary France.

Putting an end to the old order

In 1790, a year after the revolution, France's National Assembly was determined to get rid of anything associated with the monarchy. The Academy of Sciences was instructed to come up with a new system of weights and measures. The Academy appointed a committee.

After some debate, the committee chose a system first suggested by a priest, Gabriel Mouton, back in 1670. Mouton had proposed that the basic unit of length should be a fraction of the circumference of the Earth and that the system be based on units of 10 for easy calculation. At the National Assembly in 1791, the metre – based on the Greek word for

'measure', *metron* – was recommended. It was to be one ten millionth of the distance from the North Pole to the Equator – a quarter of the Earth's circumference. A provisional metre was calculated from the estimated distance.

A portion of the Earth's circumference through the poles was now to be surveyed and used to calculate the whole distance from Pole to Equator. The committee chose the section of the line of longitude from Dunkirk, through Paris, to Barcelona in Spain.

Languishing in country jails

Delambre and Méchain packed their bags and surveying equipment and set off into the French countryside. Delambre and his team worked south from Dunkirk and Méchain and his team north from Barcelona. As well as the hard work of measuring and calculating distances, the surveyors came across an unexpected problem. Their equipment of white flags and beacons aroused the suspicion of local revolutionaries, for white was the colour of royalty. They were both repeatedly arrested as spies.

When at last they finished, in November 1798, the surveyors extrapolated the length of a quarter of the circumference of the globe from the section they had surveyed – about one ninth of the whole. Then they divided the sum by 10 million to calculate the new metre. It was a mere 0.35 mm different from the provisional metre calculated earlier.

DECIMAL TIMEPIECE *A watch that ran the new ten-hour day next to the old one of 24 hours never really caught on even with the most ardent republicans – nor did the decimalised week, perhaps because it meant only one day of rest out of every ten.*

LONG TRIP *The distance from Dunkirk to Barcelona was measured using geometric calculations.*

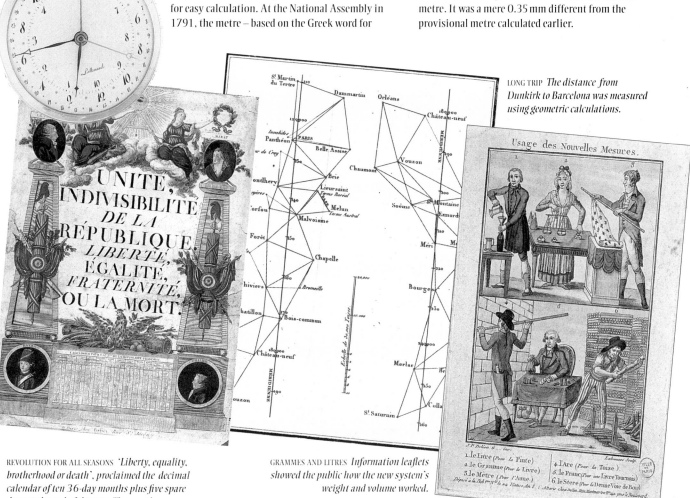

REVOLUTION FOR ALL SEASONS *'Liberty, equality, brotherhood or death', proclaimed the decimal calendar of ten 36-day months plus five spare days at the end of the year. The months were named after seasonal changes of weather, for instance foggy (Brumaire) or windy (Ventose).*

GRAMMES AND LITRES *Information leaflets showed the public how the new system's weight and volume worked.*

Flying a kite in a thunderstorm

HOW FRANKLIN DEMONSTRATED THE NATURE OF LIGHTNING

OVER THE CITY OF PHILADELPHIA, CAPITAL OF THE English colony of Pennsylvania, a thunderstorm was brewing one June day in 1752. Standing under a shelter in the middle of a field, Benjamin Franklin, local printer, journalist and scientist, checked his kite. It was no ordinary model. Franklin had designed it especially to carry out a unique experiment to find out what lightning was. It was made of a large silk handkerchief stretched over a

frame of two crossed strips of cedar. Attached to the longer crosspiece was a length of pointed wire extending a foot beyond the top of the kite. The end of the kite's line was tied to a metal key, and the key to a short length of nonconductive silk ribbon.

As the thunderstorm approached, Franklin launched his kite as high as he could to reach the storm clouds. He knew that if it was struck directly by lightning, electricity conducted down the wet line could kill him, so he took care to keep the silk ribbon and himself dry under the shelter. Conducted by water along the kite's rain-soaked line the electricity from the atmosphere would flow down to the key, and then be transferred into a Leyden jar, the only means available of storing an electrical charge.

'Some are weather wise, some are otherwise.'
BENJAMIN FRANKLIN

A Leyden jar was a water-filled glass bottle, coated outside and inside with metal foil and sealed by a cork through which a brass nail or wire for conducting electricity was pushed until it touched the water. Electricity was normally generated by rubbing a glass with cloth and touching it to the protruding head of the brass element.

Franklin did not have to wait long before the metal wire at the top of the kite conducted electricity down from the clouds and through the key. Franklin received a mild electric shock from the key and later from the outside of the jar. The shocks proved conclusively that lightning was electrical in nature – and Franklin lived to tell the tale.

News of Franklin's experiment spread rapidly to Europe where many others repeated it, some with fatal results. Franklin's theories on the nature of storms had already led him to suggest the use of lightning conductors – metal rods fixed to roofs which attract the lightning safely away from a building and allow the current to harmlessly discharge to the earth and then back up to the cloud.

VERSATILE MAN *Science was a major interest for Franklin, who played an important role in the American Revolution, helping to draft the Declaration of Independence in 1776. He was a noted writer, philosopher, inventor and diplomat.*

STORMY WEATHER *Clutching a Leyden jar, which will store the storm's electricity, Franklin shelters out of the rain. His son points to the thunderclouds in this romanticised portrayal of the famous experiment.*

THE STRANGE CASE OF THE TWITCHING FROG

AFTER CAREFULLY PEELING THE SKIN from a dead frog, Luigi Galvani, the professor of anatomy at Bologna University in the 1780s, removed its spinal cord and legs and the nerves linking them and placed them on an iron plate. He touched the tissues with an electrically charged brass hook. As expected, the muscles twitched and the legs convulsed.

Animal magic
But one day, without charging the brass hook, one of Galvani's assistants accidentally touched it on the carcass of a half-dissected frog. The muscles convulsed because the animal had received an electric shock. Tentatively, the assistant touched the frog again with the hook. The legs twitched.

Galvani concluded that a circuit between the two metals – brass and iron – and the spinal cord had conducted 'animal electricity' causing the leg muscles to twitch. He believed that electricity was generated inside the frog's brain, passed through the nerves and was stored in the muscles.

ELECTRIC FENCE *Galvani ties a dead frog impaled on a brass hook to an iron fence. He found that electricity in storms affected the two metals and made the frog's legs convulse.*

A powerful display for the great Napoleon

HOW VOLTA MADE THE FIRST ELECTRIC BATTERY

'HERE, MY GOOD DOCTOR, WE HAVE THE IMAGE OF LIFE itself,' said Napoleon Bonaparte, France's First Consul, to his personal physician, Jean Nicolas Corvisart. 'The pile represents the column of the vertebrae, the liver is the negative pole, the kidneys the positive pole.' The general and his doctor were sitting among an audience of scholars at the Institut de France in Paris.

Current from a tower

The object that inspired the great general to such a flight of fancy was a simple stack of discs, or 'pile', brought all the way from newly conquered northern Italy by the professor of physics at Pavia University, Alessandro Volta. It was the first device to generate a steady current. Previously, the only way scientists could produce electricity was through friction, such as rubbing a piece of cloth on glass, but this only produced a single shock when discharged. Volta's demonstration of his pile in 1801 was the culmination of nearly a decade of research, stimulated by his interest in the experiments of his contemporary Luigi Galvani.

Galvani had noted that a dissected frog's legs jump when touched by a brass hook and concluded that electricity was produced in the dead frog's tissues. Volta doubted this and suggested that electricity might have been generated by a chemical reaction between the frog's moisture, the brass hook and the iron plate on which the frog was laid.

A mouthful of electricity

Now he set about proving his theory. In the years that followed, Volta tried out many different pairs of metals in combination with various moist tissues. He even tested metals in his mouth, noting the unpleasant taste when a piece of tin and a silver spoon made contact with his tongue. Various combinations worked but the strongest current was generated by silver, zinc and brine. He cut out discs of the two metals and between them sandwiched a disc of pasteboard soaked in brine. The current from one such unit of three discs was very weak, so Volta stacked up to 60 of these on top of each other. He tested the strength of the current against a moistened finger, using one wire from the bottom of the pile and one from the top. The shock was stronger when he dipped the wires into different basins of water.

Volta now ceased experimenting with electricity, but other scientists put his battery to many uses. In England, William Cruikshank discovered electroplating and Sir Humphry Davy investigated electrolysis – decomposing compounds by passing an electric current through them. Batteries still rely on chemical reaction, but the combinations have been refined. Most dry batteries now use a reaction between zinc and chlorides.

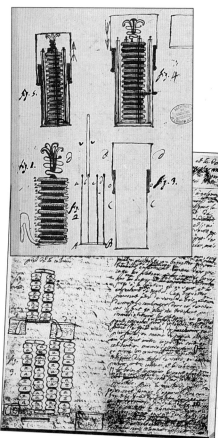

ROUGH DRAFTS *Volta's own drawings of his pile show how he assembled it and refined the same basic structure once he had found the right elements.*

SECRET REVEALED *Volta announced his findings in a detailed letter to the Royal Society in London in 1800.*

GREAT EXHIBITION *Napoleon sits poised to assist Volta with the demonstration of his new device. He was so impressed with the results that he gave Volta the title of count.*

SECRET OF THE EARTH'S AGE WRITTEN IN THE ROCK

THE EARTH WAS CREATED AT NINE IN THE MORNING ON October 23, 4004 BC, and everything upon it had come into being by sunset on October 28. This calculation, made in the 1650s by James Ussher, Bishop of Armagh and professor of Divinity at Trinity College, Dublin, was based on the sum of the ages of Old Testament patriarchs.

Vulcanists versus Neptunists

The 18th century, however, saw the rise of a new science, geology, which immediately put into question Ussher's estimate of the Earth's age. Rock strata and fossils suggested that the planet's crust was much older than Ussher's calculations. Two main schools of thought developed. 'Neptunists' believed the Earth's crust had been formed by the action of water, while 'Vulcanists' believed it had been formed through heat.

The leading Neptunist was the German academic Abraham Werner, whose lectures at the Freiberg Mining Academy attracted students from all over Europe. He argued that the Earth had been covered by great oceans and the rock strata had formed as the ocean subsided. This theory could easily be reconciled with Christian teaching about the Flood.

The Vulcanists, on the other hand, were condemned as atheists. James Hutton, an amateur scientist from Edinburgh and a leading Vulcanist, conceded that water had played some part in forming the Earth's crust through sedimentation and erosion. But he thought the internal heat of the Earth played a far more important role. The Earth's interior, he said, was made up of molten lava which erupted occasionally in volcanoes. Rocks were formed by heat and pressure. In the absence of a major catastrophe, such as a flood, the Earth had always had much the same conditions. To account for the layers of rock strata, the planet had to be millions of years old.

Decay reveals the secret

But it was not until the discovery of radioactivity that an educated guess could be made of the Earth's age. In 1905, Bertram Boltwood at Yale University found that the radioactive element uranium underwent decay and turned into a type of lead at a measurable rate. The age of the Earth could be estimated from the ratio of lead to uranium in rock strata: the more lead, the older the rock. Today's scientists, using this technique, believe the Earth is about 4000 million years old.

CLOSE RELATIVES *Darwin observed 14 species of finch – each specific to a particular island in the Galapagos archipelago. All were similar except for their peculiar beaks adapted to their diets – for example, thicker ones for crushing seeds and thinner ones for catching insects. Darwin deduced that a common ancestor had arrived from the mainland and, in isolation, without competition for food or territory, its descendants evolved differently.*

THEORY EVOLVES *Charles Lyell (left), a friend of Darwin, studied rock strata and concluded the Earth was millions of years old.*

FOSSIL FAD *Searching for the remains of extinct species (below) was a fashionable hobby in the 1820s.*

LAYERS OF AGE *Geological cross-sections (below), drawn in the 18th century, show layers of rocks deposited at different ages over many millennia. Rock strata helped scientists determine relative dates of bones and fossils.*

DRAGON ANCESTORS *The deeper the fossil, the more different it is from living creatures, as this giant lizard skull, found in chalk, shows.*

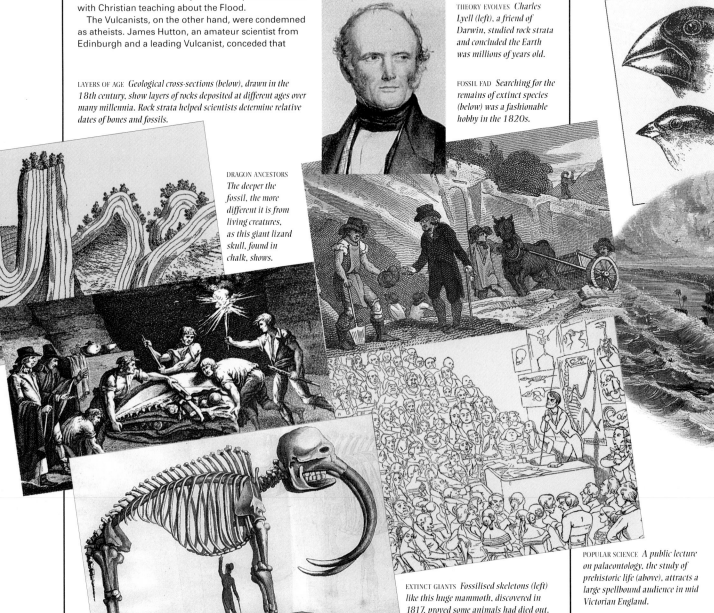

EXTINCT GIANTS *Fossilised skeletons (left) like this huge mammoth, discovered in 1817, proved some animals had died out.*

POPULAR SCIENCE *A public lecture on palaeontology, the study of prehistoric life (above), attracts a large spellbound audience in mid Victorian England.*

The key to nature's biggest mystery

HOW DARWIN FORMULATED THE THEORY OF EVOLUTION

IN THE 20 YEARS SINCE HE FIRST FORMULATED HIS theory of evolution, Charles Darwin had resisted pressures to publish his work. But now he read and reread a letter from fellow naturalist Alfred Russel Wallace which outlined a theory almost identical to his own. It was time to publish and face the consequences. A year later, on November 24, 1859, Darwin published *On the Origin of Species By Means of Natural Selection*, the most important event in the history of biology.

As a young man, Darwin cared little for education, but at Cambridge University, while training to be a priest, he became interested in natural history and geology. In 1831, at the age of 22, he was invited through his Cambridge contacts to join the Admiralty survey ship HMS *Beagle* as an unpaid zoologist and geologist. During the next five years the *Beagle* circumnavigated the globe, taking Darwin to the east and west coasts of South America, Australia, New Zealand and the Galapagos Islands in the Pacific Ocean off the coast of Ecuador.

Voyage of discovery

When Darwin set sail on the *Beagle*, he had little reason to question the accepted view that all species of animals and plants on Earth were as they had been since the Creation. But by the time he returned to England in 1836, his observations of plants and animals around the world had convinced him that they had changed as time passed, with new species appearing while others became extinct. In fact, he doubted the existence of the Creation as an event, and the reliability of the Bible as a guide to history.

Darwin knew that making his ideas public would win the enmity of anyone who accepted the Biblical account of Creation as fact. Instead of rushing into publication he decided to bide his time, and marshal the evidence to show that his theory of evolution could explain many apparently inexplicable aspects of the natural world.

His evidence came from many sources, including his own observations made during the *Beagle*'s voyage, which he recorded in extensive notebooks. In South America, he studied the bones of dinosaurs. In the Galapagos Islands he was struck by the variety of finch, tortoise and mocking bird species, which differed from one island to the next. He also used the work of fellow naturalists and geologists, the evidence of gardeners and breeders.

The survival of the fittest

But Darwin needed to find the mechanism for evolutionary change. The solution came to him after reading Thomas Malthus' *Essay on the Principle of Population* in 1838. Malthus argued that human population was kept in check by war, disease and limitations in the food supply.

Darwin extended this theory to the rest of the living world, suggesting that because even individuals within a species differed from each other, some would be better adapted to their environment and more likely to breed successfully. In short, the fittest would survive. The characteristics that favoured the parents would be passed on to the next generation, ensuring that each generation would differ slightly from its predecessor, and that different species would diverge from a common ancestor. This mechanism, which he called natural selection, underpinned the entire evolutionary process.

The *Origin of Species* caused the outcry Darwin had expected. It was denounced from pulpits all over the country. The scientific community was divided into two camps: those fiercely in favour of evolution and those adamantly opposed. Public debates across the country attracted huge audiences. By the time Darwin died, his theory was accepted as fact. He was given a state funeral and buried in Westminster Abbey.

CHANGING PLANET *Although the polyps of which coral reefs are made will not grow deeper than 120 ft (37 m) below the surface, Darwin found dead coral deeper, suggesting that the ocean floor was sinking.*

SLOW BUT SAFE *Darwin paces beside a Galapagos tortoise. He reasoned that this enormous and slow-moving creature had been able to evolve because it had no major predators.*

The flash of flintlock and a beam of light

HOW SCIENTISTS CALCULATED THE SPEEDS OF LIGHT AND SOUND

CONDITIONS WERE PERFECT ON THAT MORNING IN 1640. Without a breath of wind, the poplars surrounding the house were still; there was no hint of rain in the cloudless sky over southern France. Pierre Gassendi called to his servants to get ready for the experiment they had rehearsed so often: to ascertain, for the first time, how fast sound moved through the air.

Gassendi's most trusted servant, who acted as his scientific assistant, grabbed a saddlebag containing a pistol and a telescope, mounted a horse, and galloped off to a predetermined location. Meanwhile the scientist collected his watch, notebook and another telescope, and rode to a second prearranged spot a considerable distance from his assistant, but still within sight.

Split-second timing

Observing his assistant through his telescope, Gassendi raised a handkerchief to signal that he was ready. The assistant fired the pistol. Gassendi saw the flash of the flintlock and, moments later, heard the sharp crack of the shot. With his watch he timed the gap between the flash and the report. By dividing the time it had taken the sound to reach him by the distance between himself and the pistol, Gassendi worked out that sound travelled at about 1437 ft (438 m) per second. The principle he used was good, but the result was wildly inaccurate: the speed accepted today is about 1086 ft (331 m) per second.

In 1676, some 30 years after Gassendi's experiment, the Danish astronomer Ole Rømer noticed a curious phenomenon. When the Earth was moving towards Jupiter in its orbit, the eclipses of the planet's four moons seemed shorter than predicted. When the Earth moved away, the eclipses were longer. Rømer knew that a moon always orbited its planet in the same amount of time, and reasoned that as the Earth rushed towards Jupiter, the light reflected by the moons had less far to travel, so the 'news' of the eclipse reached an observer on Earth sooner than expected; the opposite happened as the Earth rushed away. Rømer realised he had the key to discovering the speed of light.

Making difficult comparisons

By comparing the lengths of the eclipses with the direction and distance – about 27 000 miles (43 450 km) – the Earth had travelled during them, Rømer estimated a speed of light. His equations were complicated, because he had to compensate for the rotation of the Earth and the elliptical orbiting of the moons, Jupiter and the Earth. Accurate timing was essential. The largest moon, Ganymede, disappeared from view for about 4 hours, and the difference between a 'short' eclipse and a 'long' eclipse was very small. The figure he calculated was 140 000 miles (225 000 km) per second.

Here matters rested until 1849, when the French scientist Armand Fizeau devised a simple experiment to time the speed of light. He decided to measure how long it took light to travel between his father's house at the village of Suresnes and the hill of Montmartre in Paris. The distance between the two was 5.3 miles (8.5 km). At Suresnes, Fizeau set up a machine he called his interferometer, a large wheel with 720 teeth, which could be rotated at different speeds. At Montmartre he set up a mirror opposite the wheel.

Fizeau spun the wheel and shone a beam of light through its teeth at the mirror. He increased the speed of the wheel until it seemed as if the mirror was no longer reflecting the light. When this happened, the time taken for the light to make the return journey to Suresnes was equal to the time taken for the wheel to move from a 'gap' to a tooth. Knowing that the wheel was rotating 12.68 times per second, Fizeau calculated the speed of light as 186 439 miles (300 030 km) per second. It was astonishingly close to the value accepted today: 186 291 miles, or 299 793 km per second.

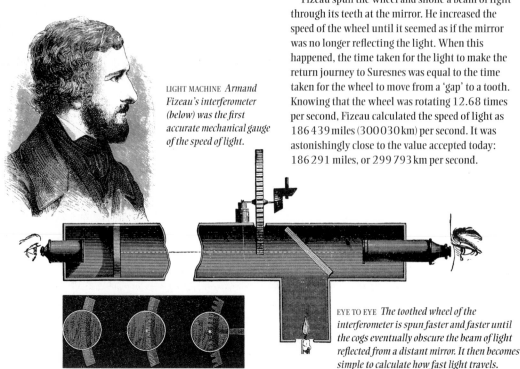

LIGHT MACHINE *Armand Fizeau's interferometer (below) was the first accurate mechanical gauge of the speed of light.*

EYE TO EYE *The toothed wheel of the interferometer is spun faster and faster until the cogs eventually obscure the beam of light reflected from a distant mirror. It then becomes simple to calculate how fast light travels.*

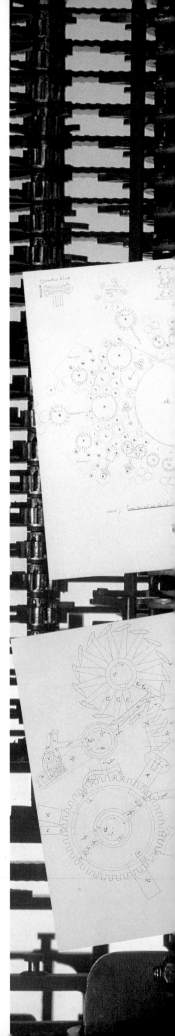

An engine fuelled by numbers

HOW CHARLES BABBAGE DEVISED A MECHANICAL CALCULATOR

AUGUSTA ADA BYRON, THE ONLY LEGITIMATE DAUGHTER of the poet Lord Byron, stood in amazement before the gleaming machine. Brass cogs whirred, steel pistons pumped, wheels turned and clicked into place to display a number, 49, the square of seven. In the drawing room next door the rest of the guests gathered around a dancing automaton. Charles Babbage's Saturday evening soirée in Dorset Street, London, was going with its usual swing.

A machine to cut out human error

The object Ada Byron was admiring was no mere conversation piece: it was a section of a much larger machine that Babbage believed would change the world. He had produced the first model of this Difference Engine, designed to compile mathematical tables, in 1822. Mathematical tables of logarithms – essential for navigating at sea – were calculated by hand and notoriously inaccurate.

Babbage announced his plan to build the engine to the Royal Society and gained enough interest to secure government funding. He hired an excellent engineer, Joseph Clement, who ran a workshop in south London. Babbage and Clement spent the next ten years painfully experimenting and piecing together the Difference Engine, which was planned to be 8 ft (2.4 m) high, 7 ft (2.1 m) long and 3 ft (1 m) deep. It was the most complex piece of precision engineering attempted during the 19th century, and would weigh several tons and contain some 25 000 parts. Babbage had to design new tools and new techniques to manufacture the parts.

Calculations were to work across eight vertical columns of toothed wheels each engraved with a figure from zero to nine. Each column was to contain 16 wheels to be read from top to bottom. The columns were to be fixed to shafts that were raised and lowered by the action of pairs of cams attached to a crank. Cranking the handle once would produce a single calculation, leaving the machine ready to produce the next number in the sequence on the next turn of the handle.

The Difference Engine worked by using the method of finite differences – a way of calculating tables sequentially that uses addition instead of multiplication and division. The result of each equation leads directly to the answer to the next one. The initial equation would be worked out on paper and entered into the Difference Engine by setting the figure wheels manually.

THE MACHINE THAT NEVER WAS *Babbage's second grand design, the Analytical Engine (top), went beyond the Difference Engine (bottom and background), and in some ways anticipated electronic computers. It was to have two parts. Calculations would be carried out in the 'mill' and data and results kept in the 'store'. Instructions were to be entered on punched cards. The machine was never made.*

The last column of wheels was a series of type wheels that would print the numbers either onto papier mâché or hot metal, for further printing. After each number had been printed, the paper or metal would move on automatically, ready for the next line.

In 1832, Clement assembled 2000 parts of the engine to show that some progress had been made. The section of just three columns, the one so admired by Ada Byron, worked perfectly, but it was to be the only concrete result to emerge from all Babbage's plans. Babbage and Clement quarrelled, and work on the machine ceased in 1833. The government officially cancelled the project in 1842.

PAPER PLANS *Detailed technical drawings of Babbage's first Difference Engine were completed although the machine was never finished. The project cost him some of his private fortune and the British government more than £17 000 .*

DISAPPOINTED MAN *Charles Babbage died believing his life's work had been for nothing.*

KEEN ADVOCATE *As Countess Lovelace, Ada Byron backed Babbage's work and wrote a clear account of it.*

The magic workshop of a modern wizard

HOW EDISON'S 'INVENTION FACTORY' WAS RUN

AT THE AGE OF 30, THOMAS ALVA EDISON DECIDED IT was time to escape from Newark, New Jersey, away from the problems he had faced with manufacturing his own inventions, and instead concentrate his energies on researching and developing new ideas. In the winter of 1876, with cash from the sale of the stock ticker, his latest invention, Edison had an 'invention factory' built at Menlo Park, a quiet hamlet of just six houses, 24 miles (39 km) from New York City along the railway to Philadelphia.

The barn-like two-storey building, the world's first research laboratory built for practical invention rather than academic research, was about 100 ft (30 m) long and 30 ft (10 m) wide. The design was simple. On the ground floor were a small library, a draughting room, a 'nook' for chemical experiments and an office for a secretary and bookkeeper. Upstairs, one large room served as the 'factory floor'. Over the years, more buildings were added and the site expanded.

HEART OF INDUSTRY *Upstairs on the 'factory floor', a team – Charles Batchelor and two assistants – experiments with electric light. Sparsely furnished with long workbenches and well lit by floor-to-ceiling windows, the workroom takes up the whole second storey. Lining the walls are long shelves bearing a strange collection: botanical specimens, lumps of ore, cases of semiprecious stones, paper funnels, wire, and bottles of chemicals in all shapes and sizes. At the back stands a pipe organ – donated by an admirer – ready to be played whenever Edison feels his team needs a little musical encouragement.*

IMPROVEMENT AND INNOVATION *Edison was fascinated by Alexander Graham Bell's invention, the telephone, and determined to improve it. He and his team, including Martin Force (left), experimented with different substances to improve its reliability. Eventually, Edison found that carbon in the transmitter was more effective, and produced less crackle, than the magnet that Bell had initially used.*

SKILLED CRAFTSMAN *Glass-blower James Hipple blows the bulb for an electric light. It will form part of the New Year's Eve display in 1879, when 3000 people will come to see the laboratory building and the village street lit up by electric light bulbs.*

Money no object

At the time, Menlo Park was one of the best-equipped laboratories in the world. No expense was spared in outfitting it at an estimated cost of between $40 000 and $100 000. Microscopes, air pumps, lathes, drills, planes and milling machines sat next to the most up-to-date electrical devices: spark generators, induction coils and machines for measuring light and electricity. Electricity generators were powered by an 80 horsepower Brown steam engine.

The workshop was to come up with 'a minor invention every ten days and a big thing every six months', Edison told a friend. The inventor himself roughed out ideas and sketches in the draughting room. These might be polished by Charles Batchelor, draughtsman and head of the workshop, whose contribution Edison considered so important that he gave him a share of the profits from each invention. The rest of the staff were paid by the hour. Francis Upton, a postgraduate in mathematical physics from nearby Princeton University, undertook any complicated calculations. 'Doc' Alfred Haid carried out chemical experiments on the ground floor, and chief machinist John Kruesi turned the drawings into working models. As Edison grew more successful he hired more workers. By the time he left Menlo Park in 1882 to set up the electric lighting system in New York City, he had more than 80 employees working in teams, each led by an experimenter and aided by specialists, for example a glass-blower or a machinist. In the six years Edison was at Menlo Park, the invention 'factory' produced more than 400 patents.

Boss who never stopped

Edison's timekeeping was erratic. In periods of peak activity, for example while he was inventing the light bulb, Edison could work around the clock and then sleep for 36 hours. At other times, he did not turn up at the workshop until noon. Perhaps one of his most useful talents was an ability to curl up and take 10 minute cat naps anywhere in the workshop and wake up refreshed. He and his team sometimes worked into the night, talking and experimenting. After putting the children to bed Edison's wife Mary often brought the team a midnight snack.

OLD MEETS NEW *A craftsman works on parts for a demonstration model. Edison valued preindustrial skills as well as up-to-date knowledge of electrical engineering.*

TOUGH TESTS *The brightness levels of light emitted by different types of electric light bulb filaments is tested in the photometric room by Edison's electrical engineer, Charles Clarke.*

PERFECT PRINTS *Draughtsman Samuel Mott makes a blueprint of a patent application to send to New York City.*

SOUND AND FURY *In the machine shop, leather belts transfer power to the machine tools from the pulleys on the overhead shafting. John Kruesi, Edison's chief machinist, oversees a team of skilled workers.*

LAB REPORTS *In the chemistry nook under the stairs on the ground floor, 'Doc' Alfred Haid, a German chemist, experiments with ways to create a good vacuum for the light bulb. 'Doc' was one of the few specialists hired by Edison; most workers turned their hands to many tasks.*

THOMAS EDISON'S MARVELLOUS TALKING MACHINE

The chief machinist, John Kruesi, shook his head in disbelief. This time Mr Edison wanted a machine that talked. As usual, Kruesi followed the inventor's sketch carefully. When he finished, on December 6, 1877, the rest of the team at Menlo Park, New Jersey, gathered round to have a look at the bizarre contraption.

Its main part was a 3.5 in (9 cm) metal cylinder turned by a hand crank along a 12 in (30 cm) shaft. Two tubes fitted with thin metal plates – diaphragms – that vibrated when hit with sound waves were mounted on opposite sides of the cylinder. Each had its own stylus. It looked absurdly simple.

Getting the feel of sound

The idea first occurred to Edison while he was working on improvements to Alexander Graham Bell's telephone. The diaphragm of the telephone vibrated to the sound of the human voice. Edison tested the strength of the vibrations by feeling them with a needle attached to the diaphragm. It occurred to him that the vibrations might be strong enough to make marks, and that the indentations might be played back to repeat the initial sound. To test his idea, he attached a diaphragm to a stylus and, shouting 'Halloo' into it, he pulled a strip of paraffined paper under the vibrating stylus. A scratch was made in the paraffin. Then he dragged the piece of paper back under the stylus. Faintly, the diaphragm whispered 'Halloo'.

Edison began experimenting furiously – trying different papers coated with chalk or wax and various metal foils. At first he tried discs, but the stylus worked better on a revolving cylinder. He settled on tinfoil as the most sensitive material. On November 29, 1877, he drew the first of a series of sketches.

When Kruesi presented the first prototype that December evening, a couple of the men bet Edison cigars that it would not work. Edison carefully wrapped a sheet of tinfoil around the cylinder and adjusted one stylus to touch it. Turning the handle slowly, he shouted a nursery rhyme, 'Mary had a Little Lamb', into one of the tubes. As he rotated the cylinder, the stylus moved back and forth making a groove in the tinfoil. Once he finished speaking, Edison moved the first stylus away from the cylinder and put the second one, attached to an amplifying horn, in its place. Then he turned the handle again. The men craned forward to hear. Out of the horn issued the rhyme. Everyone, even Edison himself, was astonished. John Kruesi turned pale. 'I was never so taken aback in all my life,' Edison said later. 'I was always afraid of things that worked first time.'

TREASURE HUNT *At the assay bench, a metallurgist tests different ores to find an economical way of extracting platinum. Edison has ores sent from all over the Americas for experiments with different techniques.*

SOCIAL HOUR *Bachelor lodgers at Sarah Jordan's boarding house in Menlo Park chat and smoke after supper, costing 30 cents. The price included one of the landlady's fruit pies. This was one of the first houses in the world to be lit by electricity, which ran across the ceiling through wires insulated with cotton thread.*

ROUGH DRAFT *Edison's tentative sketch was all the experienced machinist John Kruesi needed to make the first phonograph. The note at the bottom was added in 1927 as part of an advertising campaign.*

GENIUS AT WORK *Thomas Edison leans forward, straining to hear the first thin sounds coming from his pet invention.*

FACTORY TOWN *Edison's workshop dominated the hamlet of Menlo Park. The experimental electric railway took workers to a local fishing lake. After 1882, when Edison left, the workshop fell into disrepair.*

STRUCTURED THOUGHT *Rutherford's notebook contains his new theory on the structure of the atom, first proposed in 1911. He argued that all of an atom's positive electrical charge is concentrated in its centre.*

A LIFE OF DISCOVERY *Close to the end of an illustrious career, Ernest Rutherford, Nobel prize winner and nuclear scientist (right), consults with a colleague at the Cavendish laboratory in Cambridge in 1935.*

ATOM SMASHER *An 8 in (200 mm) sealed brass tube, into which Rutherford introduced gases and radioactive material, made possible a remarkable discovery: atoms could be split.*

COLLABORATOR *Rutherford's colleague Niels Bohr developed the 'Bohr atom' – a springboard for atomic physics.*

The building blocks of the Universe

HOW SCIENTISTS SPLIT THE ATOM

THE TWO MANCHESTER UNIVERSITY RESEARCHERS, Hans Geiger and Ernest Marsden, looked at one another in astonishment. The experiment that their physics professor, Ernest Rutherford, had asked them to perform was producing remarkable results. It was 1909 and Marsden and Geiger were using a type of radiation known as alpha particles – emitted by the radioactive material radium – to bombard a piece of thin gold foil. Most of the particles flew straight through the foil, but occasionally one was deflected to one side or even bounced back in the direction it had come from. The deflections showed up on a screen coated with zinc sulphide that flashed when hit by the high-energy particles.

Like his students, Rutherford was amazed by the deflection of the particles. He later said that 'it was almost as incredible as if you fired a 15 in shell at a piece of tissue paper and it came back and hit you'. From earlier research into radiation, he deduced that only a very strong electric field could cause such a large deflection. What must have happened, he concluded, was that alpha particles – which carried a positive electric charge – had been repulsed by a strong positive charge in the atoms of the foil.

At the heart of the matter

Rutherford's discovery of the nucleus was the latest event in a long-running scientific debate about the nature of matter. The Greek philosopher Democritus was the first to propose that the world was made up of tiny, indivisible, indestructible particles, which he called atoms, from the Greek *atomos*, or 'indivisible'. He believed that atoms were differently shaped and organised depending on the material they made up, but his theory was dismissed by Socrates and later Greek scientists and philosophers, and was not taken seriously again until the 17th century.

The first proof of the existence of atoms was put forward in 1803 by an English Quaker, John Dalton, who carried out experiments with gases in order to investigate the chemical elements – substances that cannot be broken down into anything simpler. He declared that each element had a distinctive type of atom, different from the atoms of any other element, and that atoms of each element had a different mass.

Another English scientist, Joseph John Thomson, made a vital addition to the understanding of atoms in 1897. He showed that the atom – far from being a solid structure as Dalton had believed – actually contained smaller particles. Thomson, like many other scientists of the period, experimented by passing an electric current through a glass vacuum tube. The electricity produced so-called 'cathode rays' within the tube, and these in turn produced a bright spot at one end of it. Having demonstrated that the

cathode rays were deflected both by electric and magnetic fields, Thomson concluded that they consisted of a stream of negatively charged particles, much lighter than atoms. He called the particles 'corpuscles', but they were later known as 'electrons'.

The explosion of the atom

Rutherford made his second celebrated breakthrough – smashing the nucleus of an atom – in Cambridge. The simple apparatus he used consisted of a short brass tube which contained a movable arm holding radioactive radium. Rutherford stretched thin metal foil across a window at one end and positioned a screen covered in zinc sulphide next to the foil. The tube was filled with nitrogen through an inlet pipe, and, using the zinc sulphide screen, he observed the action of the alpha particles on the gas.

The strip of foil across the window was thick enough to stop the nitrogen atoms and radiation escaping from the tube. But as Rutherford watched the zinc sulphide screen, he saw flashes, suggesting that some particles were escaping. Using his judgment as to what kind of particles would be able to pass through the foil, Rutherford concluded that they were hydrogen nuclei travelling at high speed.

He seemed to have achieved what alchemists had struggled to do for centuries, and transmuted matter. In 1919, he presented his findings in the celebrated *Philosophical Magazine*, concluding that hydrogen nuclei – later called protons – were present in the nucleus of every element. It was a huge advance in the long task of unlocking the secrets of the atom.

MINIATURE UNIVERSE *Rutherford and Bohr proposed an image of the atom resembling a tiny solar system, with a positively charged nucleus at the centre circled by negatively charged electrons.*

ESCAPING PARTICLES *Rutherford's experiments led him to a startling conclusion – under the impact of alpha particles from radioactive radium, nitrogen atoms were breaking up and so releasing hydrogen nuclei, later called 'protons'.*

ECCENTRIC GENIUS *John Dalton studied bubbles of methane – or marsh gas, found in ponds – and many other gases before putting forward his atomic theory.*

C IS FOR COPPER *Dalton drew up a pioneering table listing chemical elements and compounds by symbols.*

ELEMENTS

⊙	Hydrogen	1	⊕	Strontian	46
①	Azote	5	⊛	Barytes	68
●	Carbon	54	Ⓘ	Iron	50
○	Oxygen	7	Ⓩ	Zinc	56
☿	Phosphorus	9	Ⓒ	Copper	56
⊕	Sulphur	13	Ⓛ	Lead	90
⊕	Magnesia	20	Ⓢ	Silver	190
⊖	Lime	24	Ⓖ	Gold	190
⓪	Soda	28	Ⓟ	Platina	190
⑩	Potash	42	☿	Mercury	167

The World at Work

MAKING AND TRADING GOODS ARE UNIVERSAL ACTIVITIES. THE SPIRIT OF ENTERPRISE HAS UNDERPINNED THE GREAT EMPIRES OF THE PAST, AND COMMERCIAL NEEDS HAVE SHAPED AND TRANSFORMED ENTIRE CIVILISATIONS

THE FOUNDRY / EYRE CROWE

Making sticks burst into flame

HOW EARLY PEOPLE MADE FIRE

THE WATCHER SQUATTING OUTSIDE THE CAVE SAW THE stars fade away. Slowly, a grey light replaced the blackness, and the disembodied shapes of trees and rocks began to loom out of the mist. It was time for the nomads to move on to new hunting grounds. His vigil over, the night watchman shivered and moved back gratefully into the warmth of the cave.

Kneeling on the floor beside the fire, his companions were already preparing pouches of bark in which live embers would be embedded for the journey; more fire would be carried in the hollows of horns. Their ancestors had learned about fire and its properties when they observed lightning, or hot volcanic lava, igniting trees and grass.

Keeping the fires burning

At least half a million years ago, people living in the Choukoutien caves in China knew how to use fire to keep warm, to cook meat and to keep wild animals at bay; but they did not know how to create the precious substance, so they had to keep their fires constantly burning.

At last, sometime around 100 000 BC, early people succeeded in making fire, either by rubbing two sticks together, or by striking a lump of iron pyrites with a flint to produce sparks. Both methods probably arose from the effects such people had noticed while making tools.

Rubbing sticks together was a simple, but tedious, process. In one method, a rod of hardwood was placed point-down on softwood and rotated between the palms until the heat created by the rubbing made the softwood smoulder. By attaching a piece of dried grass or other tinder to the end of the rod, bigger fires could be made.

STRIKING *By striking a piece of dark flint against iron pyrites, a primitive man is able to make sparks. These fall on a pile of dry tinder grass, which then catches fire.*

TWISTING *Holding a dry wooden rod between his hands, an African man twists it in a hole in a wooden base. The sparks that result ignite the waiting heap of tinder.*

SPINNING *A fire-maker twists a rod – held in place with a rock – by moving his bow backwards and forwards. The sparks will set fire to the tinder.*

Masters of the art of toolmaking

HOW PRIMITIVE MAN MADE TOOLS

THE SCANDINAVIAN RIVERBED WAS LITTERED WITH chalky stones; some had split to reveal their glossy brown and black flint insides. This was exactly the type of stone the axe-maker had been looking for. The 3 hour hunt had been well worth it. Now he could begin to make a new hand-axe.

Like all early toolmakers of the period between 200 000 BC and 40 000 BC, he used flint whenever it was available. Although very hard, it flaked easily, with a sharp edge at the point of fracture. A brisk, vertical blow on the surface would knock out a solid

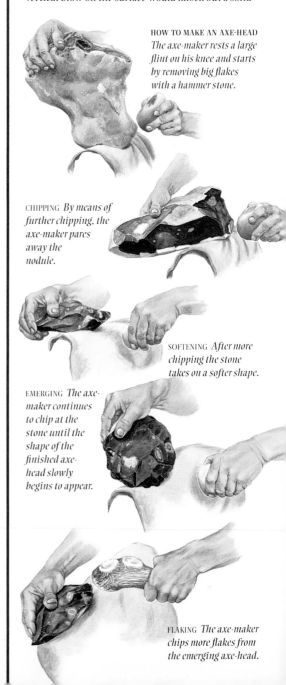

HOW TO MAKE AN AXE-HEAD *The axe-maker rests a large flint on his knee and starts by removing big flakes with a hammer stone.*

CHIPPING *By means of further chipping, the axe-maker pares away the nodule.*

SOFTENING *After more chipping the stone takes on a softer shape.*

EMERGING *The axe-maker continues to chip at the stone until the shape of the finished axe-head slowly begins to appear.*

FLAKING *The axe-maker chips more flakes from the emerging axe-head.*

cone of flint with the apex at the point of impact; an oblique blow directed near the edge of a slab would detach a chip or flake.

The axe-maker laid the stone down on the flat surface of a nearby rock. Using another stone as a hammer, and striking obliquely downwards, he sheared several large flakes off the flint. The pear-shaped core which was left was now ready to be trimmed by a more delicate flaking process.

Tools like tomahawks

Using a bone hammer, the craftsman chipped the stone all round the core until it fitted comfortably in the palm of his hand with a sharp edge projecting outwards. Now he had a tomahawk-type tool, the perfect implement for slitting open the hide of an animal and cutting up the carcass, and for grinding, pounding and scraping.

More sophisticated flaking techniques evolved as a result of trial and error. The flakes which had been detached from the core flint were used to make a wider range of tools. A hardwood baton or cylindrical bone could be used as a punch to strike at a sharp right angle on the flake, shaping it into narrow,

parallel-sided blades which were made into knives, scrapers and spearheads. Toolmakers also learned how to flake the surface of a flint in such a way that a single hard blow detached a flake which conformed to a set pattern.

One of the most important tools made in this way was the burin, or carving tool, produced by slicing both sides of a flake obliquely at one end to form a narrow chisel edge. With an engraving point at one end and a blunt, scraping edge at the other, it was ideal for making other tools. Needles were fashioned by using a burin to hone a fillet of bone or ivory into shape, which was then pierced to make the eye. The burin was also employed to shape antler, bone, ivory and flint into barbs for harpoons, spearheads and arrowheads.

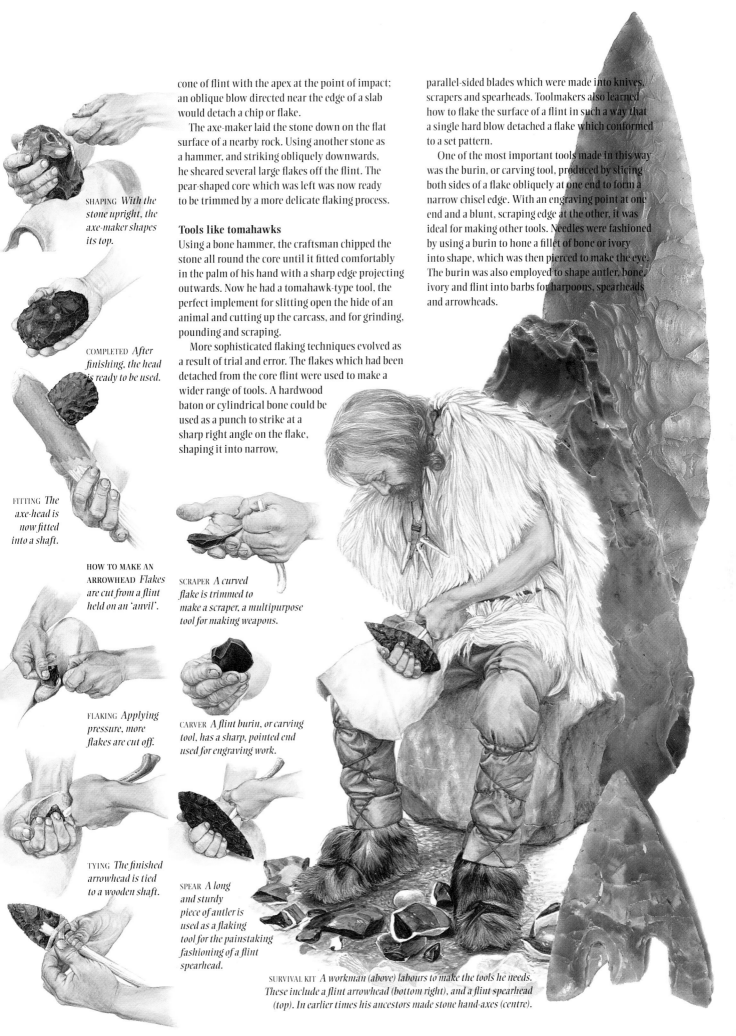

SHAPING *With the stone upright, the axe-maker shapes its top.*

COMPLETED *After finishing, the head is ready to be used.*

FITTING *The axe-head is now fitted into a shaft.*

HOW TO MAKE AN ARROWHEAD *Flakes are cut from a flint held on an 'anvil'.*

FLAKING *Applying pressure, more flakes are cut off.*

TYING *The finished arrowhead is tied to a wooden shaft.*

SCRAPER *A curved flake is trimmed to make a scraper, a multipurpose tool for making weapons.*

CARVER *A flint burin, or carving tool, has a sharp, pointed end used for engraving work.*

SPEAR *A long and sturdy piece of antler is used as a flaking tool for the painstaking fashioning of a flint spearhead.*

SURVIVAL KIT *A workman (above) labours to make the tools he needs. These include a flint arrowhead (bottom right), and a flint spearhead (top). In earlier times his ancestors made stone hand-axes (centre).*

The metal that changed the world

HOW IRON AGE SMITHS FORGED A SWORD

CHARCOAL SMELTING *One man fills a funnel-shaped furnace with iron ore, while a second hammers out impurities from the bloom – lump iron retrieved from the iron ore by smelting – on an anvil.*

WHEN IRON REPLACED BRONZE AS THE PREFERRED METAL for forging tools and weapons, humanity took a significant step forward. From its birthplace in the Middle East around 2000 BC, the knowledge of how to forge iron spread around the world. Wherever people started to use iron tools, their lives changed completely and irrevocably.

Tools for all trades

Stronger than bronze, iron's superior hardness made it possible to forge new, improved weapons which were both sharp and durable. Iron also paid other dividends. With sharper saws and better

axes, acres of forest could be cleared at speed; iron ploughs could dig up increasingly large tracts of land in preparation for planting; and with the aid of iron sickles and scythes, it was found possible to harvest more crops faster than before.

For all its versatility, iron was more difficult to work than other metals, having a high melting point of 1538°C (2800°F). Charcoal-heated Iron Age furnaces produced iron in a solid, rather than liquid, state. The iron ore was heated with charcoal for several hours until it separated itself into the usable lump iron known as a bloom, and slag – iron mixed with impurities from the iron ore.

During the smelting, much of the semi-liquid slag collected in the bottom of the furnace and was later discarded, but some remained trapped inside the bloom. Once this had been extracted from the furnace, it was hammered to expel the remaining impurities. The lump iron would be beaten down

LASTING QUALITY *The smith responsible for the sword below, manufactured between 750 and 450 BC, would doubtless be proud to find his work has survived to the present day. Although the metal has decayed, the tang – onto which the hilt was riveted – is still clearly visible at the left end of the sword.*

until it was a thin sheet, so ensuring that no slag remained at the centre. By the time the smith had finished, he was left with a block, or billet, of near-pure iron which was the raw material for a multitude of useful objects, such as spades, sheep shears and hammers, and finer items such as needles and jewellery pins. Billets were often used as a trading item or 'currency bar', passing through many pairs of hands as a form of payment for goods or services before reaching another smith altogether.

A plentiful supply of timber was needed to make charcoal for the furnaces. Two furnaces were used, the first for separating the ore into bloom and slag, and the second, the smithing hearth, for heating the bloom to remove impurities and then keeping the metal soft enough to work.

It was not until the 13th century AD – some 1700 years after ironworking first arrived in Europe – that another method of making iron was introduced. This involved melting the ore in a hotter furnace than previously used, and then pouring the liquid metal into a mould to cool. The result was cast iron.

TAKING SHAPE *A hammer is used to bevel the edges of the red-hot tipped sword.*

FROM BLOOM TO SWORD *Walls of wattle and daub provide a shelter for an Iron Age smith and his son. In one corner stands a large pile of charcoal to be used on the smithing hearth, which the smith's son is overseeing. It is his job to use his skill and experience to keep the fire at the right temperature – which varied from 800-1100°C (1500-2000°F), depending on the process being undertaken. The smith, wearing a leather apron as protection against sparks, has already been hard at work on this sword for several hours. Gripping the billet of iron – reheated earlier in the fire – in a pair of tongs, he shapes the blade with a hammer. The anvil on which he is working is made from billets hammered together into a block and balanced on a stone. A second sword lies reheating in the furnace, ready for the smith to add his finishing touches. In front of the hearth lie another pair of tongs and a fire rake – used to rake the ashes at the end of the day.*

FORGING THE TANG
The first task is to shape one end of a flat iron bar into a long, thin point. This takes several reheatings.

SHARPENING THE BEVELS
The bevelled edges of the sword are ground – to produce a sharp, cutting edge – against a flat stone, lubricated with a mixture of sand and water.

CUTTING THE POINT *A piece of waste metal is placed under the sword as a hot sett, a chisel-like tool, is used to shape the point.*

BURNISHING THE SWORD
A small piece of sandstone is rubbed up and down the sword's length to smooth both the surface and the edges.

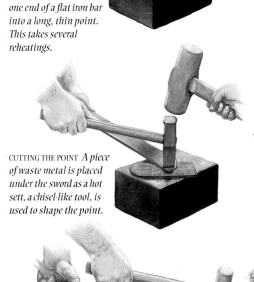

SHAPING THE POINT
Once both sides of the sword's tip have been cut, it is repeatedly reheated and hammered to give the point its correct profile.

HARDENING THE SWORD *Once the smith is satisfied with the sword, it is placed in a charcoal-filled pit dug into the ground. It is surrounded with red-hot embers and left for 8 hours before being 'quenched' in cold water. The sudden cooling hardens the iron into a strong sword.*

The right price – in volcanic rock or silver

HOW PEOPLE TRADED, FROM GIFT-GIVING TO THE FIRST MONEY

THE WARRIOR WAS PLEASED WITH THE GIFT THAT THE young man from a neighbouring tribe had brought. The shiny black lumps of obsidian would increase his power and standing in the tribe. The volcanic rock was highly prized for making blades, and presents of some of these blades would help to strengthen relationships with other members of his tribe. In return for giving them, he could expect, one day, to receive a gift or service in return.

From gifts to trade

This type of gift exchange was probably the very first method of trading goods. In time, the expectation of some type of repayment in the future developed into barter: an immediate exchange of goods, where people who wanted to trade would meet to negotiate and agree to an exchange.

Barter did have serious disadvantages, however. If a farmer with wheat to sell needed shoes, for instance, he had to find not only a cobbler, but a cobbler who wanted to buy wheat. Also, people with perishable or bulky goods to trade could not carry them long distances.

As societies became more sophisticated, increasing numbers of them agreed on a common unit of account. Primitive forms of money, such as cowrie shells, were first used during the Stone Age, but these were probably for ritual gifts rather than a form of currency. By 2000 BC, however, certain commodities were becoming the preferred means of exchange. In Mesopotamia, for instance, silver was used as a standard – though goods were still bartered. If, for

example, someone wanted to buy a bull, he might offer a variety of goods – grain, oil, cloth, honey and wood – as an exchange, but the contract would only be agreed by giving the beast and the goods a value in silver by weight. Any small differences in value would be made up in silver, and merchants carried bars of silver divided into rings that could be easily broken off for this purpose.

The first coinage was invented in Lydia – on the western edge of Asia Minor – in the 7th century BC. There, the bed of the Pactolus river yielded an alloy of gold and silver called electrum. The Lydians used lumps of electrum as money, and stamped it with marks to guarantee its value. Roughly shaped lumps with rudimentary marks soon developed into round coins, stamped with deep indentations on both sides. The denominations on these coins were very large, however; it was not until the 5th century BC, in Athens, that small change first came into use. Since purses had not yet been invented people carried their coins in their mouths. Within 100 years silver coinage was in use throughout Greece.

FIRST COIN *A lump of electrum, weighed and stamped by the Lydian mint, dates from around 630 BC.*

LION'S SEAL *King Croesus of Lydia's stamp of the 6th century BC certifies a piece of gold's value.*

TRADE WARES *Like the tools that inspired their design, Chinese coins were bronze. This 1½in (3.8 cm) high coin from c. 300 BC is based on a hoe.*

GODDESS'S BLESSING *An owl, symbol of the goddess Athena, gave this ancient Athenian coin its nickname. As the city's wealth grew, the 'Owl' gained currency throughout Greece.*

ITALIAN STYLE *In the 4th century, before coins were introduced by the Greeks, Romans and their Etruscan neighbours used lumps of copper as money.*

EASY TO CARRY *Durable and hard to fake, cowrie shells were used over a wider area, and for longer, than any other precoinage money. One cache, found in Iraq, is over 20 000 years old.*

MONEY TALKS *To boost his image as a warrior, the 5th-century-BC Persian emperor Darius I issued coins bearing his portrait armed with bow and arrow.*

PRICE LIST *An ancient Mesopotamian clay tablet from c.1800 BC is inscribed with a list of relative prices for goods from salt to gold.*

OBVERSE The temple of Juno Moneta was a mint in Rome, so her name and likeness appeared on coinage.

REVERSE The other side of the coin depicting Juno Moneta, shows the anvil, hammer and tongs that were used for striking new coins.

OFFICIAL APPROVAL A 4th-century Roman gold bar is stamped with the names of a local magistrate, Flavianus, and an assayer, Lucianus, to confirm that it is 100 per cent pure metal.

HARDWARE More than a foot long, an iron penny from Liberia could be hammered into a tool once its usefulness as money ended.

BANK NOTE Paper money, invented in 11th-century China, enabled merchants to trade without carrying heavy iron coins.

PLANT MONEY Coils of plant fibre covered with feathers are still used to pay off symbolic debts – such as dowries – on the Pacific island of Santa Cruz.

SUNLIGHT FALLING ON THE OCEAN AND solidifying in the waves was thought by early peoples to be the source of amber, one of the most prized objects of the ancient world. Men risked their lives fishing from crude dugout boats for the golden lumps – formed in fact from the fossilised resin of coniferous trees.

As early as 9000 BC, amber was used for ornaments and possibly as a talisman to ward off illness. By around 2000 BC, it had become a valuable trading commodity, and the ancient tracks known as the Amber Routes formed part of one of Europe's earliest trade networks.

Gift of the North Sea
Northern amber was especially prized. The settlements of west Jutland, where amber was commonly washed up on the coast, were the starting point for the long journey south as far as the Adriatic, where amber was in great demand for making jewellery.

Travelling south through the oak and beech forests of Denmark, amber traders also carried perishable goods, such as fish and gulls' eggs, which could be exchanged for fresh meat and furs. In northern Germany, they bartered these for metal artefacts and ingots and handed over their amber to middlemen, who in turn passed it on to traders travelling south. The great rivers of central Europe – the Elbe, the Main, and the Danube – provided the routes.

Braving alpine mountain passes
Since amber was high in value and low in bulk, it was easy and profitable to carry over the Alps. For centuries, wandering merchants used the Brenner pass route between Austria and Italy, seeking the best fords and passes, sometimes building modest wooden bridges or making a log road through marshes. The route was hazardous but well worth taking, for beyond lay the wealthy markets of Italy.

Here, in the north-west, the cultured Etruscans traded their fine pottery in return for amber carved into ornaments by the eastern alpine peoples. Farther south lay the Adriatic, with its ports visited occasionally by traders from Crete, or from the Levant – in the eastern Mediterranean – hungry for amber jewels.

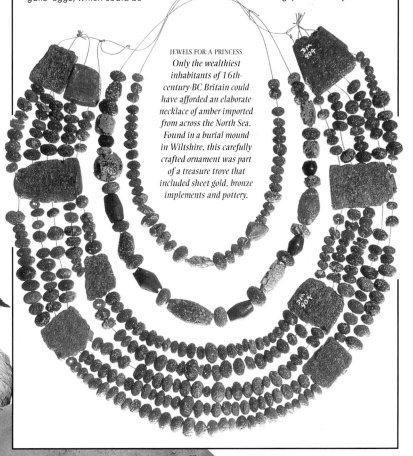

JEWELS FOR A PRINCESS Only the wealthiest inhabitants of 16th-century-BC Britain could have afforded an elaborate necklace of amber imported from across the North Sea. Found in a burial mound in Wiltshire, this carefully crafted ornament was part of a treasure trove that included sheet gold, bronze implements and pottery.

STONE MONEY Used as payment to settle tribal disputes, a stone tablet from the South Sea island of Yap is 20 in (51 cm) across.

Wheels that propelled an industry

HOW THE ANCIENT GREEKS MADE POTTERY

HEROIC LABOUR *Heracles, in a scene on an Attic pot, holds a struggling boar above a cowering king as he completes one of his 12 labours.*

TODAY THE POTTER WAS DETERMINED TO DRAW A scene of revelry. He picked up the pot and began to draw on it carefully in the style he had honed since he was a child learning the trade at his father's knee. Rapidly, the figure of a satyr took shape, a common motif on pottery in the early 6th century BC.

The potters' workshops of Athens were small family businesses, run perhaps with the help of one or two slaves. They were located in a section of the city known as the *kerameikos*, or potters' quarter, and produced ceramics of such artistic excellence that cups, jugs and vases that were once part of an Athenian's household crockery are now museum pieces, prized throughout the world.

Potters discover the wheel

The earliest potters had made pots by rolling clay into worm-shaped tubes, and building these up in rings into the shape that was required. The potter's wheel was invented in Asia Minor just before 3000 BC, but the first Greek wheel-made pottery was not produced until about 1700 BC on the island of Crete. The Greek potter's wheel was a large, heavy stone disc, set about 1 ft (30 cm) above the ground, on a turning shaft. It was still necessary to turn the wheel by hand, but its weight gave the wheel stability and momentum. To work, the potter had either to squat or to sit on a low stool. Some time after 600 BC the wheel was raised higher, so that the potter could sit on a seat. A disc was added to the bottom of the turning shaft so that the potter could turn the wheel by kicking the disc with his foot. The design of the potter's wheel has changed very little right up to the present day.

Pictures from the fire

It was in Corinth in about 700 BC that potters first discovered the method of achieving the glossy black finish that gave Greek pottery its distinctive appearance. They made a slip of clay, water and an alkali (probably leach from wood ash). When it had reached the right consistency, the artist, who had sketched his picture on the pot, painted the solution onto those areas he wanted to be black. The pots were placed in the kiln and fired.

The kiln had two chambers, one on top of the other, connected by a vertical flue; the top chamber contained the pottery to be fired and the bottom chamber held the fire. After painting, the firing of the black figure ware took place in three stages. First the pottery was fired to harden it; then the openings of the kiln were closed and firing continued, with the smoke turning the entire pot black. When the apertures were opened, readmitting air, the untreated areas of the clay returned to red, but the painted areas remained black. Potters were able to judge when the right temperature was reached at each stage.

The first pots made using this firing technique bore figures painted in black on a red clay background. By the mid 7th century, potters in Athens had taken over this technique, but from about 500 BC Athenian potters started to produce a new type of pottery decoration, known as red figure, because, in this case, the figures appeared in red against a black background.

After the pottery had been shaped on a wheel, an artist painted the background to his chosen subject using a layer of clay solution which turned black after firing. The figures were left standing out in the natural colour of the clay. Details were then added with glaze. The subject matter varied from legendary scenes to everyday events.

POT MAKING *A potter (far left) bends to add the finishing touches to his work, while his workmates carry a finished pot to the kiln. The scene, using the black figure technique that had long dominated the Athenian pottery market, was probably produced around 600 BC.*

POT PAINTING *An artist starts to decorate an elaborate Greek pot. The bowls by his side hold a slip used for painting black figures, or a black background for figures in the natural red of the clay. The pot showing this scene is an example of the red figure technique which became popular when Athenian potters began to use fine clay containing iron to achieve a rich, red colour. Red figure ware was in great demand, not only in Greece, but abroad, especially in Italy. Scenes often reflected the function of the piece, for example death and mourning on funerary vases, or drunken brawls depicted on wine jars and cups.*

EARTH WORKS *The clay for Athens' vast output of pottery was dug by miners from the many open-cast excavations just outside the boundaries of the ancient city.*

WHEEL WORK *The ancient city-state and port of Corinth in the 7th and 6th centuries BC supported small potters' workshops that often surpassed those of Athens. This clay tablet of a potter using a foot-operated wheel was found near the city boundaries.*

HOT WORK *The finished pots were fired in a multichambered kiln like the one shown in this clay tablet found near Corinth. The design allowed air to enter during critical stages of the firing process.*

Wine that travelled well

HOW THE ROMANS EXPORTED WINE THROUGHOUT THE EMPIRE

AS THE SLAVES HEAVED THE GREAT GURGLING POTS FROM the creaking wagons, Publius Veveius Papus looked on with satisfaction. The wine from his estate near Terracina in southern Italy found a good market in the European

MANPOWER *The first stage of wine-making – extracting the juice from grapes – was done by slaves. They trod the fresh fruit underfoot.*

provinces, even though the costs of transport and marketing in the 1st century AD made it expensive. This load had safely completed the first stage of its journey, by wagon to the sea, and was now to be loaded on board ship for the next stage – to customers across the seas in north-west Europe.

The vessel belonged to an experienced wine shipping agency based in Ostia, farther north along the Italian coast. Publius noted that this ship already had a small cargo of pottery at the bottom of the hold, but the bulk of the cargo would be crates of sealed ceramic containers – *amphorae* – full of wine produced by himself and other local estate owners. There was no danger of his goods getting confused with those of the other producers, because each of his amphorae bore his name stamped on the handle.

Keeping breakages to a minimum
Both for their own safety and to reduce the risk of breakage, the crew were stacking the long narrow jars – each more than 3 ft (1 m) high, and weighing 110 lb (50 kg) – extremely carefully. Once three or four bottom rows were in place, a second layer was begun, with the pointed bases of the upper amphorae tucked snugly down between the necks of those below. Branches cut from pine trees cushioned each layer.

Altogether, three layers of amphorae would be packed into the ship – a cargo of almost 6000 amphorae. To stop their valuable contents from turning to vinegar, the jars had airtight seals, either ceramic stoppers plastered into place or wooden stoppers held in with pitch.

The cost of freight
The voyage would be slow; even with a favourable wind, the heavily laden boat would do no more than four knots, and it would have to put into port several times for water and food. At the end of their long voyage the amphorae would be unloaded onto the wharf of a port in Gaul – perhaps Marseilles – and moved into one of the warehouses which stretched back from the waterfront.

There they would be stored underground to keep the wine cool until the *vinarius* (wine merchant) who was handling their marketing locally was ready to arrange their onward freight. He would already have negotiated prices for transporting the amphorae to their final destinations, using, wherever possible, the flat-bottomed barges which could carry goods into the interior by river so much more cheaply than wagons could carry them by road. Eventually the amphorae and their contents would arrive in taverns

---

TRADING STANDARDS

MANY ROMANS POSSESSED A portable steelyard, or scale, which they carried with them when trading. The steelyard, called a *momentana* or *statera*, depending on its size, was a steel beam suspended near one end from a hook. The goods to be weighed were placed in a pan or hung from a hook and a weight was then moved along the beam until it balanced. A scale could weigh up to 20 lb (9 kg), in Roman pounds of about 11½ oz (320 g) each.

SCALE *To use a steelyard, the weight of the goods was read against a scale marked on the beam.*

Flexible measures
Dry goods, sold by volume, were weighed out in a bronze, pottery or wooden *modius,* or bucket, holding about 15 pints (8.5 litres). Liquids such as wine and oil were sold by the *amphora* in quantities of about 45 pints (26 litres). However, each area had its own weights and measures, so while the foot was 11.5 in (29 cm) in Rome, for example, it measured 13 in (33 cm) in parts of Gaul.

WEIGHTS *Inlaid silver and relief work decorate bronze Roman weights.*

DUPED AGAIN *As unscrupulous traders often used doctored weights, merchants carried their own to check the scales. The* aedile, *a civic official, could fine dishonest traders.*

STACKABLE CONTAINERS *Wine was stored and transported in* amphorae, *narrow earthenware jars whose unusual shape allowed them to be stacked securely.*

and shops around Europe, where local people, especially the wealthy who had prospered under Roman rule, had acquired a taste for Italian wine.

If all went well, that is. But the sea voyage was not only slow but could also be hazardous. Publius could not have predicted that this particular boat and its cargo would founder in a storm about 10 miles (17 km) south of Toulon, in France. The wreck was discovered again by underwater archaeologists in 1972, almost 2000 years later.

KEEPING TRADE AFLOAT *A Roman wine ship carries barrels of wine, clearly visible stacked up behind the oarsmen.*

ONWARD JOURNEY *Once the wine arrived at the nearest port to its destination, it became the responsibility of a local wine merchant. He arranged for it to complete its journey by barge (above) or wagon (left).*

GOOD COMPANY *Tavern life was as important in Roman times as it is now. Often drinkers would dilute rough, strong wines with water.*

CHEERFUL AFTERLIFE *This marble headstone was made for a man who liked his drink – too much, perhaps. It dates from the 1st century.*

295

'Come and buy'

HOW MEDIEVAL TRADERS MADE THE MOST OF THEIR MARKETS

PIERCING WHISTLES AND CHEERY SHOUTS BROKE THE morning calm as the tradesmen set up their stalls and porters, sweepers and carters scurried up and down the narrow streets, throwing taunts at the fishwives with baskets of fresh fish on their heads. Wherever they were, markets in medieval Europe were run on similar lines. Most were open markets, with stalls that lined the streets; others were covered, and often sited on a river bank so that goods for sale could be delivered by boat. There were different selling areas for different goods – ranging from linen, woollens and ironmongery to poultry and vegetables – with tanners and fishmongers often sited at the edge of the market to prevent the strong smells that wafted from their stalls from upsetting other traders. Several specialised markets also evolved, such as the yarn market at Amiens in northern France.

Markets were usually held once or twice a week, and were governed by strict rules set by the town council and the guilds – associations formed by merchants in the same line of business. Stall holders rented their stalls and needed a licence to sell their goods. In Paris, for example, corn merchants could only trade without permission once they were 10 leagues (30 miles/48 km) outside the city walls.

Dusty feet on the road to profit

Most produce reached smaller market towns by cart or by foot. This could mean a hard journey, especially in the heat and dust of summer; so much so that the English court which enforced market regulations was known as the Court of Piepowders, from the French *pied-poudreux*, or 'dusty feet'. Getting goods to larger city markets, such as those in Paris, required greater organisation. Oysters and fish from Dieppe, cheese from Meaux and pulses from Caudebec in Normandy all travelled up the Seine to docks across the city.

By the 1400s, the importance of markets declined as middlemen came into their own in the food trade, acting as go-betweens in the transit of goods from countryside to town and from producer to retailer.

CLOSE EYE *Most guilds had inspectors or 'searchers' who kept a check on the quality and price of merchandise, and supervised weights and measures.*

THE ART OF SELLING *A copy of the manuscript of the Hamburg Charter, 1497, illustrates the many undercurrents at play in any market. Cattle are on sale in the annual Viti Markt ; the vendor and witnesses (left) wait anxiously for the handshake that will seal the sale, while the purchaser (right) still hesitates. In the background, the market court hears out a disgruntled trader, as the market flag – red, with a nettle leaf – flutters overhead.*

WARES FOR SALE *Market traders take part in the age-old ritual of shopping. In large towns and cities, the weekly markets were lively arenas of social activity as well as commerce.*

Medieval merchants who ruled the waves

HOW THE HANSEATIC LEAGUE MONOPOLISED TRADE

THE FIRST-YEAR APPRENTICE WAS OVERAWED BY THE burgher from Stralsund, Germany; his fellows had told him that the merchant owned a silver armchair, that his rooms were hung with rich cloths. The boy had another six years of apprenticeship before he could begin to rise through the ranks of the Hanseatic League. But perhaps one day he, too, would be a merchant and a fully fledged member of the mighty cartel that controlled trade in the Baltic.

The power of the Hanseatic League – of which Lübeck was the central city – was soon to become legendary, but the League started out modestly. It was founded in the 13th century by the prosperous merchant guilds of Lübeck and Hamburg to protect both their tradesmen and cargoes on mutual trading routes, and their economic interests overseas.

Bullying tactics bring wealth

Merchants began to travel in convoy to protect each other from bandits and pirates, and the League eventually spread far enough to dominate northern trade routes from London and Bruges to Novgorod in Russia. Raw materials from the East (including timber, firs, grain and fish) were exchanged for goods, such as wine and cloth, from the West.

As its wealth grew, the League became more ruthless in its quest for trading monopolies. Non-League merchants were threatened with death and confiscation of their goods if they traded within League 'territory', and blockades were used to exert pressure on monarchs reluctant to lose local control. Such a blockade was enforced in Norway in 1284 when League ships closed the port of Bergen to stop goods entering the country. Norway was so dependent on imported wheat that the king, to avoid a famine, was forced to grant the League absolute freedom of trade, with exemption from taxes.

SHOW OF STRENGTH *Town seals symbolise the power of the medieval port of Bergen, which prospered under the Hanseatic League's control.*

TRADE CONTROLS *In Hamburg, one of the major hubs of the League, customs officers keep an eye on incoming goods. Meanwhile, traders were prospering abroad. Hans Holbein's portrait of Georg Gisze (left), who joined the London station in 1522, shows a typically determined merchant.*

FOREIGN POWER *The riverside factory and warehouse known as the Steelyard (left) was the League's London base until Elizabeth I withdrew its special trade privileges in 1598.*

Buying themselves a part of the business

HOW MEN OF VISION INVESTED IN THE FIRST STOCKS AND SHARES

SEBASTIAN CABOT WAS CONVINCED THAT THERE WAS A north-east passage for ships along the northern coasts of Europe and Asia. The Italian navigator had gone to London to find a financier for further exploration, but even there, in 1553, it was proving impossible. But his faith in the potentially valuable trade route was unexpectedly supported by others: when a public subscription to raise money for an expedition was launched, 240 people – including 144 merchants, 20 knights and seven peers of the realm – bought 'shares' of £25 each in the venture.

This, the world's first issue of shares, raised a total of £6000, used to provision three ships for the 18-month voyage. Only one of the three reached Russia – the *Edward Bonaventure* – landing near Archangel on the White Sea; from here, Captain Richard Chancellor travelled overland to see Tsar Ivan the Terrible in Moscow, carrying with him a letter from Edward VI which outlined the mutual benefits of trade. The tsar's written consent to future contact accompanied Chancellor when he left for London.

As a result, The Muscovy, or Russia, Company was set up in 1555, with Sebastian Cabot at its head. Subscribers to this first joint-stock company shared in any profits as a return on their investment, and since the company had a monopoly on the new Anglo-Russian trade in cloth, timber, pitch and hemp, they were not complaining.

VISIONARY *Sebastian Cabot dreamt of opening up trade with Russia. He died a wealthy man in 1557.*

DIPLOMAT *In 1630, Fabian Smith of The Russia Company was appointed ambassador to Russia.*

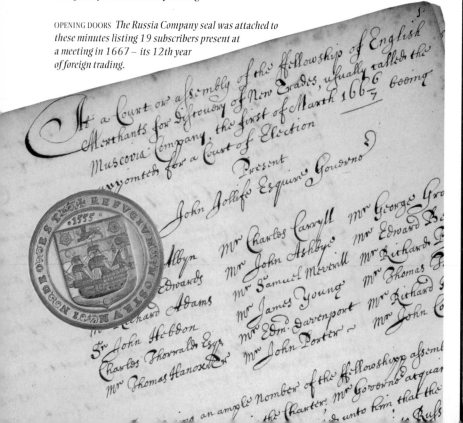

OPENING DOORS *The Russia Company seal was attached to these minutes listing 19 subscribers present at a meeting in 1667 – its 12th year of foreign trading.*

Stretching out a long arm to the East

HOW THE EAST INDIA COMPANY BROUGHT SPICES FROM AFAR

FLAGS FLAPPED IN THE BREEZE AS THE FIRST FLEET OF England's East India Company set sail for the East. The five ships carried 400 men, food and water to last for 20 months and merchandise ranging from mirrors to drinking glasses, which were to be sold or exchanged for spices. Through the voyage, the Company aimed to break the Portuguese and Dutch monopoly of the lucrative spice trade, and it was a great moment – a new fleet in search of new lands for the queen – at the dawn of a new century.

The company had received its royal charter from Elizabeth I on December 31, 1600, giving her blessing to the expedition and allowing the East

HER MAJESTY'S SERVICE *On February 13, 1601, five ships of the East India Company left London under the command of General James Lancaster, with the blessing of Elizabeth I. The wily commander won personal approval and trading rights from foreign leaders using letters from the queen.*

India Company to send six ships, six pinnaces and 500 men to the East – notably the islands of Sumatra, Java and the 'spice islands', the Moluccas – every year thereafter. The board of directors, made up of wealthy London merchants and nobles, raised the sum of £70 000 by public subscription to fund the first voyage, led by General James Lancaster.

The traders' first aim on arriving in the East was to establish good relations with the local rulers. Their protection would enable the English to trade alongside the long-established Portuguese, and the newly arrived Dutch merchants on the islands. So when the Company's fleet reached Sumatra in June 1602, for example, John Middleton, one of the captains, went ashore to present the 90-year-old king Alauddin Shah with a letter from the queen.

Competing with guile and greed

The Company's ships ventured from island to island in search of new business, leaving several men at each port of call to set up trading posts, or 'factories', and gather supplies for the crew. The English merchants faced adversity on all sides. The Portuguese attacked their ships and disguised their servants as spice sellers to spy on them, while the Chinese, who largely controlled the pepper trade, adulterated the sacks of spice sold to unsuspecting East India buyers.

When the fleet returned in September 1603 laden with spices, it faced another crisis. James I had succeeded its beloved queen, and worse still, the king had accumulated vast supplies of pepper of his own. To keep his profits high, he banned the sale of East India pepper, creating a financial crisis for the Company, although the ban was lifted later that year.

Lancaster's voyage eventually realised a profit of more than £64 000, and the Company directors persuaded its subscribers to reinvest. In 1604 a new fleet set out for both Sumatra and Amboina, a major clove island, and by 1612, East Indiamen had made eight trips to the Indian Ocean. That first expedition of 1601 had opened the door to the riches of the East.

TEA FOR TWO? *Brutal intimidation by rival Dutch traders and falling profits led the English East India Company largely to withdraw from the East Indies spice trade in 1628. It concentrated its efforts on the promising subcontinent of India, setting up factories protected by forts, such as Fort William at Calcutta (below). But the 1744 declaration of war between France and England brought an era of peaceful trading to an end. Hostilities broke out in India – the French lost their last outpost there in 1761 – and the political map of the subcontinent changed. But the company had established itself farther east, and introduced tea from China to England in the 1650s. Just over a century later, the United Company of Merchants of England Trading to the East Indies, as they were now known, had annual tea revenues of more than £6 million.*

FROM STRENGTH TO STRENGTH *The East India Company capitalised on early financial successes, and by the mid 18th century it owned 100 or so East Indiamen, as its handsome ships were known. A dozen ships or more were built annually in shipyards such as this along the Thames.*

Flower power in a golden era

HOW PEOPLE MADE – AND LOST – VAST FORTUNES THROUGH 'TULIPMANIA'

THE SILKY, VIVIDLY STRIPED TULIP – THE RAREST, MOST valuable bloom of all – seemed to shine like fire in the pale light of the conservatory. Without hesitating, the man from Haarlem paid 1500 florins for it, but in the street, he crushed the bulb underfoot. He already had a striped hybrid of his own, and was prepared to pay anything to stamp out the competition.

Flowers that bred florins
Tulips had arrived in Holland from Turkey a century earlier and now, in the 1630s, the country was in the grip of 'tulipmania' – fortunes were made and lost annually speculating on the tulip crop. At first, the trade in tulips lasted only from the end of June, when the bulbs were taken out of the ground, until September, when they were planted. After a while, trade took place all year round, with promises of delivery in the summer. The bulbs themselves rarely changed hands. Vouchers were issued and traded at spiralling prices. Fraudsters bought and sold on credit that did not exist.

By February 1637 many speculators had failed to fulfil their obligations and the number of buyers fell, causing prices to plunge. Dealers tried to raise confidence by holding auctions, but this fooled no one and thousands of speculators were ruined overnight as bulbs were rendered valueless.

PRIZE BLOOM *Variegated flowers, known as 'fine' tulips, were the most sought after. The red-and-white flared 'Semper Augustus' was particularly prized: in 1636, 30 000 florins – which would have purchased five houses – was offered for just three bulbs. As well as the flower itself, the length and strength of the stalk added value to hybrids.*

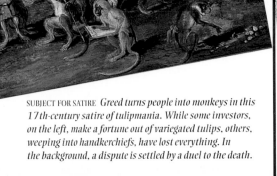

SUBJECT FOR SATIRE *Greed turns people into monkeys in this 17th-century satire of tulipmania. While some investors, on the left, make a fortune out of variegated tulips, others, weeping into handkerchiefs, have lost everything. In the background, a dispute is settled by a duel to the death.*

300

Striking a deal over a refreshing drink

HOW CITY MERCHANTS CONDUCTED BUSINESS IN THE COFFEE-HOUSES OF LONDON

THE BUZZ OF CONVERSATION SUBSIDED INTO A HUSH AND the drinkers in the upstairs room of Edward Lloyd's coffee-house settled back in their tall wooden benches. Merchants and shipowners at the tables set down their cups among auction lists and cargo inventories. A pair of underwriters – agents who negotiated insurance to cover goods in transit – looked up. All listened intently as a boy waiter climbed to a pulpit set high on the wall and started to announce the latest news from the docks.

Fortunes made and lost
A share in the cargo on its way back from India in a merchant ship carried the promise of fabulous wealth, providing its precious load of spices, silks, ivory or jewels arrived safely. That depended on the weather, the seaworthiness of the ship, the skill of its crew – and luck. English, French and Dutch ships rivalled each other for the trade, and pirates were also a threat. Merchants anxious to protect themselves against ruin took out insurance. In return for a payment, or premium, underwriters at Lloyd's agreed to reimburse heavy losses. The underwriters' business depended on paying out less in insurance

WHEELING AND DEALING *Money and shares change hands and promises are made at the Royal Exchange, the official trading centre for foreign and British merchants in 18th-century London.*

BAD NEWS *The grim expression of these coffee-house patrons suggest th news being read to them is not good*

liabilities than the sum collected in premiums. Bad news from the pulpit could mean an immediate rise in premium rates.

Edward Lloyd opened his first coffee shop in Tower Street, close to London's bustling docks, in the mid 1680s. In 1692 he moved to 16 Lombard Street, in the heart of the banking district, and when he died in about 1712 his sons-in-law took over the shop. Lloyd's was the favourite haunt of merchants, bankers, seafarers and underwriters, who met to conduct business in an informal atmosphere. It became known as the place to go for those who wanted to raise money, arrange insurance for a voyage, or auction a ship's contents.

By 1734 waiters no longer climbed to the pulpit at Lloyd's to announce the latest shipping news. *Lloyd's List*, a newsletter published every Tuesday and Friday, was available on subscription to anyone paying three shillings (15p) a quarter. It contained a vast amount of information, including details of ships' movements and the price of gold – and even the strength and direction of the wind at Deal, on the Kent coast, over the previous three days, which was a useful indication of how long ships would take to sail from the Channel coast to the port of London.

From a coffee-house, Lloyd's became an institution that has remained the centre for insurance – particularly of shipping – for more than 200 years. Officials there are still known as 'waiters'.

GOING, GONE *At Lloyd's, a taper was lit as bidding for each lot began. When it guttered, bidding stopped.*

A home of one's own

HOW ORDINARY PEOPLE CLUBBED TOGETHER TO BUY THEIR HOMES

IN 1775, A GROUP OF INDUSTRIAL WORKERS IN THE English Midlands pooled money to build their own homes and formed the earliest known 'terminating' society. It was known as Ketley's Building Society and met at the Golden Cross Inn in Birmingham, of which Richard Ketley was the landlord. The world's first recorded system of mortgage – from the Old French for 'dead pledge' – had been created.

Cooperation that paid dividends
Each member paid a fixed weekly sum of money into the society, and when enough cash had been collected it was used to buy land and build houses. As each house was completed, it was allotted to a member of the society by ballot. Once every member and his family was housed, the society was 'terminated', or wound up. Initially, most societies had about 20 members. Nonpayment of the subscription for more than three months meant expulsion from the society, and if someone resigned, his share would be sold to another member.

Before long Ketley's terminating society gave rise to similar organisations, which called themselves 'permanent' building societies as they had no intention of closing down. The first such society, Metropolitan Equitable, opened in 1845. By the end of the 19th century more than 2000 societies had sprung up throughout the country to help people to save money and to take out a mortgage on a property. By 1910, the assets of Britain's building society industry totalled some £76 million.

WORKING MEN'S CLUB *Handwritten in perfect copperplate, a contract to buy land in the village of Dilworth lists the names and occupations of the Longridge Building Society's members. Most are yeomen farmers.*

SIGNED AND SEALED *Each witness to a conveyancing contract, drawn up by the building society in Longridge, Lancashire, has his own seal by his signature. The society was started by a local curate in 1793.*

Souls for sale

HOW AFRICAN SLAVES WERE TRADED AND TRANSPORTED

THE MUD-BRICK HOUSES WERE STRUNG OUT ALONG THE river bank. Iron pots sat in their cradles over the fire, and smoke spiralled slowly upwards in the midday heat. A line of freshly caught fish hung between two sticks, and a few scrawny hens were pecking at a pile of yams. Everything looked normal, but the little village was strangely silent.

Just offshore, a party of Africans in a canoe rested on their paddles, staring intently at the deserted houses. After a brief argument, they set off downstream, disappearing round a bend in the river. They realised the villagers must have been warned that a slave collecting party was on the way, and had fled into the forest. It was not worth pursuing them.

The scourge of kidnappers
All along the west African coast now occupied by Ghana and the Ivory Coast, and for 600 miles inland, gangs of slave-traders and freelance kidnappers preyed on small peasant communities. Slavery was an important element in the African economy, with slaves being sold and exchanged across the continent. Under their African masters, they worked mostly as domestic servants but also in the gold mines and on the land. By the mid 18th century, slaves had also become a major commodity in Africa's export trade.

SLAVERY OR DEATH
Wielding shoddy muskets – made in Europe solely for the African market – raiders terrorise a west African village, rounding up potential slaves.

Europeans had been trading in Africa since Portuguese ships arrived there in the 15th century, looking for gold. But the slave trade began to assume an overwhelming importance as the nation-states of Europe sought to develop their colonial territories, where there was an abundance of land and resources waiting to be exploited. On the sugar plantations of the West Indies, in the tobacco fields and rice plantations of North America, and in the gold mines of Brazil there was an acute shortage of labour. Fit, strong slaves could be worth as much as £300 a head.

The taking of slaves soon became a well-drilled operation. Traders would surround a village with sentinels so that no one could escape. They would attack at dawn, threatening the villagers with guns brought from Europe. The captives were bound and yoked together, then forced to march to the coast. To make sure they were too exhausted to attempt escape, the younger men were made to carry heavy weights such as rocks on their heads.

Those who were too far from the coast to walk there were taken to river settlements – although this might entail a march of over 400 miles (640 km). There they were sold to agents who kept them in pens until the rainy season, which lasted from July to November; only then, when the rivers were swollen, could ships pick up their human cargoes.

Guns and bric-a-brac
The Africans rounded up slaves for export because they wanted European trade goods and most importantly firearms. Muskets, metal basins, iron bars and finished cloth from the European factories were eagerly sought – as well as a host of small items like mirrors, combs and glass beads. Rum from the Americas, French brandy, Dutch gin and Portuguese wine all made their way into the African interior.

Bargaining with the native traders was no easy task for the foreigners, however. There was no common currency in Africa, so the agents and ships' captains first had to agree the value of the trade goods brought by the Europeans, so that they could establish how many slaves the goods could buy. Each cargo of trade goods was carefully selected to suit the most recent information about local tastes. The coastal agents were hard bargainers, ready to pull out of a deal if a certain item, or the right colour of cloth, was not included in the shipment.

It could take months to put together a cargo of slaves. Some ships cruised up and down the coast, sending a boat ashore at likely points; others dealt through agents. The captain had to weigh the advantage of a quick exchange – which would

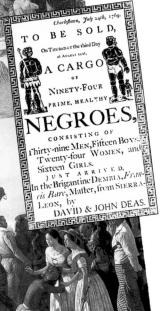

KEEPING ORDER *A white plantation overseer in Guyana in 1825 puffs nonchalantly on a pipe, while rebellious slaves are beaten by their fellows.*

HIGHEST BIDDER *Being sold away from family and friends was the common fate of slaves in the southern states of the USA.*

Charlestown, July 24th, 1769.

TO BE SOLD,

On Thursday the third Day
of August next,

A CARGO
OF
NINETY-FOUR
PRIME, HEALTHY

NEGROES,

CONSISTING OF
Thirty-nine MEN, Fifteen Boys,
Twenty-four WOMEN, and
Sixteen GIRLS.
JUST ARRIVED,
In the Brigantine DEMBIA, Francis Bare, Matter, from SIERRA-
LEON, by
DAVID & JOHN DEAS.

FREE AT LAST *Black men and women celebrate the end of slavery in the French West Indies in 1848. In France itself, slavery had been illegal for nearly 50 years.*

BELOW DECKS *So many captives were crammed into the slave ship's hold that most were forced to lie down. Berths were a maximum of 18 in (46 cm) wide.*

reduce the risk of diseases to slaves and crew along the fever-ridden coast – against a long bartering session even though this might produce lower prices, and fitter slaves. The tribal origins of the slaves were also a factor, with those from some tribes considered less troublesome than others.

By the time the slave ships set sail, it would have taken an exceptionally brave or fear-crazed captive to think of rebellion. Systematic humiliation and brutal treatment taught them the futility of resistance. Life aboard a slave ship was a daily catalogue of inhumanity punctuated by savage reprisals against troublemakers. The slaves were branded with hot irons, and forced into the fetid holds below deck, where they were kept lying down, crammed together in less than half the space allotted to ordinary seamen – or even convicts. The men were chained and shackled. The women were often raped by the crew; a slave pregnant with a mixed-race child sold for more money in the colonies. Small groups of slaves were brought on deck for exercise, and were whipped to keep them moving on their cramped, numb limbs. Many captains regularly threw sick or rebellious slaves overboard.

The most valuable cargo

However, the slave-ship captains did have a stake in the welfare of their cargoes. They had to answer to the merchants at home who had financed the expedition, and a dead or seriously weakened slave represented a serious financial loss. By the end of the 18th century French and British traders were vaccinating their slaves against smallpox. They fed them on a staple diet of rice and yams. At the first sign of sickness, the slaves were seen by the ship's doctor. Even though one in three slaves died within three years of capture, their life expectancy at sea was greater than that of English seamen of the time.

Those who survived the notorious Middle Passage – the journey across the Atlantic – were sold at quayside auctions in the Caribbean, for example, or in Portuguese Brazil. Most captains, anxious to be rid of their cargoes, sold them to local slave-traders in exchange for produce from the plantations. At last, the ships set sail for their home ports of Liverpool, Bristol, Nantes, Lisbon and Amsterdam. There, they unloaded their precious cargoes: the sugar, rice and coffee, the tobacco, gold, spices and indigo which had been produced by slave labour, and purchased with money from the sale of slaves.

SHIP SECURITY *Shackles used to chain slaves on the voyage were mass-produced in Europe.*

SCARRED FOR LIFE *Branding irons left deep scars in slaves' flesh.*

THE UNDERGROUND ROAD TO LIBERTY

AT TWILIGHT, THE HUNCHED FIGURE WAS BARELY DISCERNIBLE on the ridge of the hill. For weeks the young black woman had been on the run from the tobacco plantation in North Carolina, where she had lived and worked for all her 23 years. She strained her eyes in the fading light and could just make out the broad Ohio river – in 1856 the border between the slave states of the South and the free states of the North. In the early hours of the morning she would be met on the river bank by Harriet Tubman – one of the most prominent operators of the 'Underground Railroad'.

Helping hands across the North

The Underground Railroad was a network of safe houses and hiding places, known as 'stations', which helped slaves from the South escape to the North or to Canada. Slaves or 'passengers', hid in attics, haystacks or anywhere that was safe at the stations during the day, then travelled at night on foot, by wagon or by boat to the next station. Passengers took every precaution necessary to conceal their identity, often going in disguise.

By law in the 1850s, slave owners were allowed to pursue and recapture runaways anywhere in the United States. Police and courts had to assist. Posses of professional slave catchers roamed the country.

The Railway was run by abolitionists, both white and black, and was particularly active from 1830 until the outbreak of the Civil War in 1861. Operators risked imprisonment, fines for theft – as slaves were legally considered property – and worst of all, if they were runaways themselves, recapture. Although the Railway delivered probably no more than a few thousand slaves a year, the tales of oppression and pain told by escapees were powerful propaganda for the abolitionist cause. Harriet Beecher Stowe's novel *Uncle Tom's Cabin* was based on a runaway's true story.

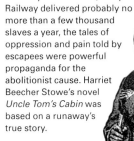

FREEDOM FIGHTER *Harriet Tubman escaped from slavery herself in 1849, but frequently risked the trip south to rescue others. She helped to liberate more than 300 slaves.*

Drudgery and danger deep underground

HOW HARD-WORKING MINERS DUG OUT COAL BY HAND

THE MINER SQUEEZED HIS SHOULDERS INTO THE NARROW seam then, lying on his side, prised the coal from the base of the seam with a stout pick. In the gloomy passage behind him, a woman loaded the coal into a corf, a large, wooden, tub-shaped sledge. She dragged the full corf away using a harness which ran between her legs from a leather belt around her waist. The tunnel to the shaft was steep and narrow and water poured in from the roof, chilling the unfortunate woman to the bone.

Families at the coalface
In the late 18th century, when huge quantities of coal were required to fuel the furnaces of Europe's Industrial Revolution, the only way to obtain coal was to dig it out by hand. Whole families worked together: men toiled at the face, while women and children moved the coal they had cut. Children were essential to the process – in some mines, the passages leading to the coalface were as low as 18 in (46 cm), forcing even a five-year-old child to bend double.

The working day began between five and six o'clock in the morning. The workers descended the shaft to the coalface in the corf itself – hooked onto a rope and lowered by turning a windlass or wheel and cog gin – or by clinging to the winding rope. Many accidents occurred: corves collided, workers lost their grip and fell, or coal dropped down the shaft onto them. At the coalface the collier hacked out the coal with a pick, then shovelled it onto a large wooden sieve known as a riddle. A worker called a

hurrier – usually a woman or an older child – shook the riddle to separate out the small coals, leaving only large pieces, which were then thrown into the corf. When the corf was full, the hurrier dragged it to the shaft at the bottom of the pit to be hauled up to the surface, then returned to the face with the empty coal sledge.

One of the biggest problems of mining for coal was firedamp – methane gas. This gas, highly explosive when mixed with air, forced miners to work in virtual darkness. To keep gas levels under control, a fireman, wrapped in wet sacking to protect him from getting burned, set light to pockets of the gas with a candle on the end of a long pole. One way to improve the airflow was to light a furnace at the bottom of one shaft to draw clean air down the other.

Drama in the dark
To prevent pockets of gas forming, and to ensure a degree of ventilation throughout the mine, engineers devised a system which relied on child labour. Children aged five or six years – known as trappers – opened airtight doors to let the workers and coal tubs pass through, and made sure they were shut at other

PUSH AND PULL *Pushed by a boy and pulled by a girl, a cartful of coal is propelled from the coalface to the bottom of the mine shaft for winding to the surface.*

HANDS AND KNEES *A young girl worker has to crawl on all fours as she drags a heavy corf of coal along a low, narrow passage more than 200 yd (183 m) long.*

OPEN AND SHUT
A trapper boy (above) opens an airtight door to let the coal through. These doors had to be opened and shut to ensure a steady flow of air.

NAKED LABOUR *Stripped to the skin, a young hewer – his neck and head supported by a wooden rest – laboriously wields his pick in the heat and damp of the cramped coalface.*

COAL CLIMBERS *Women shouldering loads of up to 170 lb (77 kg) clamber hundreds of feet up a series of near-vertical ladders to bring the hard-won coal to the surface.*

WINDING DOWN
Holding onto each other for safety (left), two children are lowered into a mine.

times so that the flow of air continued on the right path. A door left open in the wrong place could increase the risk of an explosion. It was a solitary ten-hour working day spent crouched in the darkness. A seven-year-old boy trapper said: 'I stand and open and shut the door; I'm generally in the dark, and sit me down against the door . . . I never see daylight now, except on Sundays.'

All those underground worked equally long hours, stopping only for a few minutes at midday to eat a piece of bread or oatcake and cheese. In Britain, conditions did not improve until the mid 19th century when an investigation into working

conditions led to laws in 1842 which banned women and girls, and boys under ten from working underground. Danger remained for those going down the pit, but shared adversity encouraged a great sense of pride in mining communities.

DEATH DOWN BELOW *Miners are blown helplessly into the air – some to their deaths – as a massive coal dust explosion tears through a coal mine in Anderlues, Belgium, in 1892. Tubs are overturned, a pony is knocked over and pit props are dislodged as huge chunks of coal fly fiercely through the mine roadway. The blast threatens to bring the roof crashing down on those below.*

INTO THE DEPTHS *Strapped into a harness, a fully grown workhorse begins the dangerous descent into a mine for a day's labour.*

CLEANING *Raw cotton arrives at the mill in bales and is fed into the willowing machine, where revolving teeth loosen the cotton fibres and remove stalks, seeds and dirt.*

COILING *Special machines coil the cotton fibres into loose ropes ready for the next stage – spinning.*

SPINNING *Turning a wheel by hand, a male spinner drives about 100 spindles, while a female 'piecer' twists together broken threads.*

WEAVING *Women pull threads through the loom to produce a woven pattern in the finished cloth – a process called reeding.*

RUL
TO BE OBSE
By the Hands Empl
THIS M

RULE 1. All the Overlookers shall be on the premises first and last.

2. Any Person coming too late shall be fined as follows:—for 5 minutes 2d,

3. For any Bobbins found on the floor 1d for each Bobbin.

4. For single Drawing, Slubbing, or Roving 2d for each single end.

5. For Waste on the floor 2d.

6. For any Oil wasted or spilled on the floor 2d each offence, besides paying

7. For any broken Bobbins, they shall be paid for according to their value, an
guilty party, the same shall be paid for by the whole using such Bobbins.

8. Any person neglecting to Oil at the proper times shall be fined 2d.

9. Any person leaving their Work and found Talking with any of the other

10. For every Oath or insolent language, 3d for the first offence, and if repe

11. The Machinery shall be swept and cleaned down every meal time.

12. All persons in our employ shall serve Four Weeks' Notice before leaving
shall and will turn any person off without notice being given.

13. If two persons are known to be in one Necessary together they shall be
the Women's Necessary he shall be instantly dismissed.

14. Any person wilfully or negligently breaking the Machinery, damaging th
shall pay for the same to its full value.

15. Any person hanging anything on the Gas Pendants will be fined 2d.

16. The Masters would recommend that all their workpeople Wash themsel
selves at least twice every week, Monday Morning and Thursday mornin
for each offence.

17. The Grinders, Drawers, Slubbers and Rovers shall sweep at least eight
7½, 9½, 11 and 12; and in the Afternoon at 1½, 2½, 3½, 4½ and 5½ o'clock; a
side is turned that is the time to sweep, and only quarter of an hour wi
sweep as follows, in the Morning at 7½, 10 and 12; in the Afternoon a
at the time will be fined 2d for each offence.

18. Any persons found Smoking on the premises will be instantly dismiss

19. Any person found away from their usual place of work, except for nec
own Alley will be fined 2d for each offence.

ringing dirty Bobbins will be fined 1d for each Bobbin.

wilfully damaging this Notice will be dismissed.

kers are strictly enjoined to attend to these Rules, and
rving them.

OOT MILL, NEAR HASLINGDEN,
SEPTEMBER, 1851.

J. Read, Printer, and Bookbinder,

The power behind the cotton boom

HOW LANCASHIRE MILLS SUPPLIED CHEAP CLOTH TO THE WORLD

ON THE TOP FLOOR OF A MANCHESTER COTTON MILL, tightly packed ranks of carding machines combed bales of fuzzy cotton fibre into continuous rolls, filling the air with dust. In a neighbouring building, thumping and rattling power looms spewed out yard after yard of cloth. The deafening, ear-splitting racket made by the steam-driven looms forced the operators – mainly women and children – to communicate by sign language and lip-reading.

This was only one of the hundreds of bustling cotton mills that had sprung up across Lancashire in the early 19th century. The cotton industry had changed the face of northern Britain as mill towns all over the country sprang up with phenomenal speed. The population of Manchester, the region's cotton capital, grew from 108 000 in 1821 to 316 000 in just 30 years.

The proud boast was often made that Britain's cotton mills could 'meet the demands of the domestic market before breakfast and spend the rest of the day supplying the world'. This transformation resulted from a combination of creative engineering, innovative production methods and the canny shrewdness of a new breed of enterprising merchants and investors.

A textile revolution

Until the mid 18th century, weaving and spinning had been small-scale industries, mainly carried out by people working at home. But in 1733 an inventive weaver called John Kay patented a 'flying shuttle' that shot automatically back and forth across the loom, enabling a single weaver to do the work of two. The demand for yarn rocketed and manufacturers sought ways to boost production. A Lancashire weaver, James Hargreaves, responded by improving the design of the spinning machine so that by 1771 one spinner could turn about 40 spindles at once using a hand wheel. Richard Arkwright, a wigmaker from Lancashire, took the idea a crucial stage further by driving hundreds of spindles from a water wheel in the world's first cotton factory at Cromford in

Derbyshire. By using water power and employing children as young as six from the workhouses, spinning factories cut the cost of yarn drastically.

Steam efficiency

Then in 1782 James Watt, an instrument maker, adapted the steam engine to drive factory machines. In one shift a steam-powered spinning machine could make enough yarn to circle the Earth two and a half times. Weaving was boosted in its turn when the steam-driven power loom was patented by a rector, Edmund Cartwright, in 1786. This was the final link in the chain of events that made Lancashire supplier of cotton to the world. The mills moved down from the hills, where streams had driven their machinery, to the towns where there was a cheap labour force.

New technology and factory production methods had cut the cost of Lancashire cotton fifteenfold in a generation. No one else could compete.

SHOP FLOOR *Steam power meant that one worker could oversee up to six power looms. The women and children used for this job were far cheaper than skilled weavers.*

COLOUR FAST *A steam-driven machine could print about 500 pieces of cotton a day, compared to the six a day possible by hand.*

CODE OF CONDUCT *Mill rules were posted over the door and fines were docked from the offender's pay. While the mill owners prospered, the workforce found it hard to earn enough to make ends meet.*

UNION STREET MILL *Purpose-built factories mushroomed all over Manchester. Rail and canal transport links were excellent and barges from Liverpool delivered cargoes of cotton, shipped from the USA, to the door.*

'Thar she blows!'

HOW NEW ENGLAND WHALERS CHASED AND KILLED THEIR PREY

PERCHED AT THE MASTHEAD HIGH ALOFT IN THE rigging, a lookout craned forward, scanning the horizon. In the distance he spotted a faint, single spout of spray. It was a sperm whale. This was the reward the 19th-century whalers from Nantucket in Massachusetts had been seeking since reaching the whaling grounds of the South Atlantic.

'Thar she blows' yelled the lookout, 'thar she bl-o-o-ows!' Within minutes an officer, four oarsmen and a harpooner had scrambled into each of the three wooden whaling boats fixed to the sides of the ship. The men on deck lowered these 30 ft (9 m) boats into the water, and the whalers rowed as quietly as they could towards their prey. When the whale was within striking distance the harpooner, balanced on the prow like a javelin thrower, took aim and hurled his harpoon, known to whalers as an iron, the line uncoiling in its wake like a whip.

A 'sleigh-ride' that ended in death
As the harpoon struck home, the stricken whale plunged into the ocean's depths. If luck was on the sailors' side, it would reappear in the near distance and a thrilling chase, dubbed the 'Nantucket sleigh-ride' for its exhilaration and speed – up to 17 knots (31 km/h) – would begin. Often this chase, which might last an entire day, would lead the seamen miles away from the ship. Sometimes it became too dangerous, and the whalers had to sever the line to avoid being drowned.

When the exhausted whale finally gave up, an officer used a razor-sharp lance to strike a fatal blow. The stricken whale thrashed about in the water before slowly rolling onto its back. Then the seamen passed a rope around its tail and began the long, hard task of towing the 60 ft (18 m), 50 ton carcass back to the base ship. When they reached it, they lashed the whale alongside as quickly as possible – by now the scent of blood would have attracted shoals of sharks. It was not possible to winch the whale on board, for its weight might well unbalance the ship and cause it to capsize.

Stripping the blubber
Every sailor now turned butcher as 'cutting in', the lengthy task of carving up the dead whale, began. First a cutting stage of planks was secured around the whale. Officers stood on this and used sharp long-handled cutting spades to sever the head. The body

WHALING WEAPON *Single-barbed harpoons, which cut deeper than double-barbed ones, came into use in the 1840s.*

was secured alongside the ship and blubber was then stripped from it in a series of spiral strips. Each strip was then hoisted on board on a hook fitted to blocks and tackle suspended from the mainmast. When all the blubber had been cut from the whale, the carcass was cut adrift, because it was of no further value. The head was cut in sections, each of which was hoisted on board in turn. The blubber was cut into smaller pieces which were sliced with mincing knives into thin strips and piled into two large heated cauldrons.

'The red tide now poured from all sides of the monster like brooks down a hill. The slanting sun playing upon this crimson pond in the sea sent back its reflection into every face, so that they all glowed to each other like red men.'

HERMAN MELVILLE, *MOBY-DICK*

Ankle-deep in blood and grease, the seamen stirred the blubber until it melted down into oil; they then syphoned it into 35 gallon (159 litre) wooden barrels. Next, they manoeuvred the barrels down into the hold and poured the contents into enormous casks, each holding six barrels of oil. An efficient crew could strip a whale down in about three and a half hours. Then followed a 6 hour shift of mincing and boiling, resulting in about 20 barrels of oil. A medium-sized whale could yield between 60 and 80 barrels of oil; a large whale up to 100 barrels.

In addition to the oil, which was used as lamp fuel, the whale yielded other spoils which were put to commercial uses throughout the world. Spermaceti, a white substance scooped from the head, was made into candles and ointments. But the greatest prize of all – sometimes found in the intestines – was a waxy lump of fragrant ambergris. Perfume manufacturers were so desperate for this substance that a single barrel of it could fetch $40 000 – a huge amount in the 1840s, when merchants grew rich as a result of the industry and courage and resourcefulness of the men who sailed in the Yankee whaleboats.

RICH RETIREMENT *Captain Caleb Kempton (right, with his whaling ship) made his fortune during four highly profitable trips to the South Atlantic. He retired at the age of only 34 to run his own farm in New Bedford, Massachusetts.*

MARINE ART *In 1830 an American sailor engraved this picture of the whaleship* Friends *on a whale's tooth almost 8 in (20 cm) long.*

MAN AND BEAST *Sailors sometimes became obsessed with the whales they chased day after day. The elusive beasts were extremely valuable, but also carried the threat of a sudden and violent death – one of them could smash a whaleboat with a single flick of its massive tail. This page from the log book of the Yankee whaling ship* William Baker *is adorned with detailed portraits of the whalers' giant adversaries.*

DEADLY DELIVERY *Lance poised, a ship's officer (left) steels himself to strike a fatal blow just behind the flipper of a harpooned right whale. Meanwhile, gulls wheel about in search of marine life clinging to the stricken animal.*

VICTIM'S REVENGE *The hunters become the hunted when an angry sperm whale (below) rises up to smash a whaling boat, hurling its crew into the sea.*

Prospecting for a golden future

HOW AMERICANS SCRAMBLED FOR GOLD IN CALIFORNIA

LOADED FAR BEYOND HER SAFE CAPACITY, THE STEAMSHIP *California* wallowed low in the water as she sailed through the Golden Gate into San Francisco Bay. It had been an arduous five-month voyage from New York to the gold fields of California. Many passengers had died on the journey; some had lost their possessions as the crew ruthlessly jettisoned baggage during the stormy passage through the Straits of Magellan at the tip of South America. But the fear, the stifling cabins, and the shortage of food and water were forgotten now. The travellers had reached the promised land, and every one of them dreamt of only one thing – gold.

NO TIME TO LOSE *A New York operator offers swift passage by clipper ship to the US west coast. Fast as a flying fish, the company's service leaves a rival's steamboat trailing in its wake.*

DREAMING OF RICHES *Would-be prospectors crowd a New York post office, seeking the latest news from the gold fields.*

GOLDEN PROMISE *This ¼ oz (7 g) lump was the nugget that sparked the 1848 gold rush.*

DIGGER'S DELIGHT *Forty-niner Victor Seamon painted this scene of a prospector's joy when he discovers gold at last.*

In January 1848, James Marshall, helping to build a sawmill at Sutter's Fort in the foothills of the Sierra Nevada, had seen something in the tailrace of the mill which he said 'made my heart thump for I felt certain it was gold'. Four months later, Sam Brannan, the storekeeper at Sutterville, a few miles below Sutter's Fort, ran through the streets of San Francisco, about 100 miles (160 km) away, waving a bottle of gold dust, and shouting 'Gold, gold from the American river!' The rush was on.

Harsh life in the gold fields

By December 1849, at least 60 000 people had arrived in California by overland routes or by sea, and the crews of 500 abandoned ships had joined the mad scramble for gold. Few of these 'forty-niners', as they came to be known, understood how much hard work was involved. Conditions were terrible. Pickaxes, shovels, pans and provisions were exorbitantly priced: potatoes, available in New York for half a cent a pound, were sold for a dollar a pound. The standard fare in the gold fields was limited to salt pork and 'hard tack', a kind of biscuit, washed down with water contaminated by the miners' workings. Men lived in overcrowded, dirty shanty towns.

Toiling up the foothills of the Sierra Nevada with their packs, the newcomers passed hundreds of miners. Much of the land had already been claimed, but eventually the new arrivals found unworked areas. If they found gold, miners put up stakes to mark the boundary – giving rise to the expression 'to stake a claim'. In some areas, claims were limited to 10 sq ft (1 m²) – others to 50 sq ft (5 m²).

'The whole country resounds to the sordid cry of gold, gold!, GOLD! while the field is left half planted and everything neglected.'
THE CALIFORNIAN NEWSPAPER, MAY 29, 1848

After days of digging under the hot sun, the fortunate ones found traces of gold. Now they faced the backbreaking tedium of separating the gold from the earth. Many used a shallow, flat-bottomed pan with sloping sides. They mixed earth with water in the pan, then gently rotated it until the water swirled over the sides, taking small stones and gravel with it. They repeated the process until tiny amounts of gold flakes or dust were left in the bottom of the pan. Other miners washed earth in wooden cradles and 'long toms', troughs fitted with a coarse sieve at one end. Particles of gold were trapped against the container's ridged floor when they sank, while gravel and soil washed away.

STARTING OUT
Their heads covered against the Californian sun, a group of prospectors sets off to work with their tools and donkey.

RINSING OFF
Prospectors rinsed earth from gold in their battered metal pans and buckets.

Meagre rewards for many

After months in the diggings, the prospectors returned to San Francisco, where gambling saloons, brothels and drinking parlours did their utmost to relieve the miners of their hard-won gold dust. A few men had made enough to exchange their gold for a credit note, which they sent back home; others, disillusioned, headed back east.

Those who were determined to persevere faced a heavy outlay on fresh tools, boots and provisions. Storekeepers weighed the gold on scales in their shops, and exchanged it for goods or coins. At the start of the rush, 1 oz (28 g) of gold would buy 14-15 dollars worth of whisky, tools and stores; by 1849, it would only buy 8-10 dollars worth.

The vast army of hopefuls arriving in California swelled the state's population from about 15 000 in 1848 to around 225 000 in 1852, at the peak of the craze. Prospectors found gold worth $250 000 in 1848, rising to $80 million in 1852, but the yields were spread thinly. The men who really prospered were those tending to the adventurers – storekeepers and farmers. During the 1850s, professional miners moved in, using hydraulic equipment. The lone prospector was a familiar figure for some years, but the 'forty-niners' had had their day.

ROCKING THE CRADLE *Miners stop work as a woman delivers lunch. They have been using a cradle called a 'long tom'. They wash gravel from the trough through a grid into the tray (left); any gold is caught against slats in the tray.*

DIGGING FOR GOLD DOWN UNDER

AMONG THE THOUSANDS OF FOREIGNERS WHO SAILED TO San Francisco during the Californian gold rush of 1849, there were plenty of adventurous Australians. One was Edward Hargraves, who, after noting the geological similarities between the Californian gold fields and New South Wales, returned to Australia determined to find gold.

In February, 1851, he set off for the Macquarie river about 135 miles (220km) from Sydney. He and a local named John Lister found a little gold, and Hargraves returned to Sydney to whip up enthusiasm about a gold field, leaving Lister and two others to go on searching. In May, rich deposits were found and within a week some 2000 miners had flocked to the area.

Hargraves' discovery – for which he was rewarded with appointment as Commissioner of Crown Lands – sparked a series of gold rushes over the next 50 years, notably in Victoria (south of New South Wales) and in Western Australia. Some finds were astonishing: one man at Louisa Creek, New South Wales, found 30 oz (850 g) of gold in a single panful of earth.

The scarcity of water in the near-desert fields of Western Australia made it impossible to separate gold from soil using water in a pan, as men did in California. Instead they used the wind. Holding a dish of earth up in the air, they gradually emptied it into another dish on the ground; as they did so, the lighter soil was carried away, leaving the gold dust behind.

OFFICIAL PROOF *A miner's right was a licence to prospect for gold and to seek a claim which then had to be registered. It was valid for a year.*

Nº 23 COLONY OF VICTORIA. £1
Miner's Right.
ISSUED

HIGH LIFE
A miner who struck lucky blows his money on a fancy wedding.

Greyhounds of the ocean

HOW THE GREAT CLIPPERS RACED TEA TO EAGER BRITONS

WITH THEIR SHARP-RAKED BOWS CUTTING THROUGH the water, and their lofty sails straining in the wind, the streamlined clipper ships were the epitome of grace, beauty and speed at sea. They were built to bring home tea from China in the fastest possible time – and earned their name because they clipped days off established schedules. Their captains were so dedicated to speed that anyone who fell overboard was doomed. There was no time to stop.

The clippers reached speeds of up to 20 knots (37 km/h) during their annual races across the world. And in the 1866 event – known as The Great Tea Race – they beat all records. It started in the port of Foochow in south-east China, where flat-bottomed sampans bearing loads of tea regularly pulled alongside the clippers, and barefooted stevedores competed to see who could load the cargoes the fastest. The tea was packed into vast wooden chests which were piled tier upon tier in the clippers' holds.

The fruits of victory

On arrival in London, the tea would be auctioned to brokers in Mincing Lane at up to an estimated £7 a ton. However, it was the ship that landed the first chest of tea on the wharf at London Docks that gained the winner's purse of an extra 10 shillings (50p) per ton – plus a generous bonus to the victorious crew.

At dawn on Tuesday, May 29, there was a carnival atmosphere on shore as the loaded ships left Foochow. The clippers, guided by local river pilots and towed by paddle-tugs, sailed 34 miles (55 km) down the swift-running Min river to the East China Sea. Then the race for London – some 16 000 miles (25 750 km) away – started in earnest.

Four ships – *Taeping*, *Ariel*, *Serica* and *Fiery Cross* – dashed into an early lead. They remained within sight of each other until early evening, when mist and rain closed in.

Crossing the Equator

As the race progressed, these greyhounds of the ocean, manned by 30-strong, all-British crews, became strung out. On June 15, *Ariel* and *Fiery Cross* rounded the Cape of Good Hope within 2 hours of each other, with *Taeping* some 12 hours behind and *Serica* nowhere in sight. On August 4, *Ariel*, *Fiery Cross* and *Taeping* crossed the Equator together, and on August 29, as the ships entered the English Channel, *Ariel* again led, with *Taeping* close behind.

There was high excitement ashore when the clippers were sighted, and observers on the cliffs tracked the ships' positions and hurried with the news to the nearest post house. From there bulletins were rushed on horseback to London, where tension mounted as the ships' owners realised how close the outcome of the race was likely to be.

As *Ariel* and *Taeping* pressed on, *Serica* and *Fiery Cross* made desperate attempts to catch up. But *Ariel* steadily pulled ahead. Luckily, Captain M'Kinnon of the *Taeping* was the first to spot a pilot tug – which he hired to tow him into Gravesend, the gateway to the Port of London. The *Taeping* reached London Docks at 10 am on September 6 – 12 hours ahead of *Ariel*, with *Serica* and *Fiery Cross* well behind. The *Taeping*'s crew flung a chest of tea onto the quay and proclaimed themselves the outright winners. Ninety-nine days after leaving Foochow the Great Tea Race of 1866 was over.

MAY THE BEST SHIP WIN *Racing bow to bow, canvas at full stretch, the* Taeping *(foreground) and* Ariel *surge up the English Channel as the Great Tea Race of 1866 nears its climax.*

GREAT RACE
of the
TEA SHIPS,
WITH THE FIRST
NEW SEASON'S TEAS.

PRICE OF TEAS REDUCED.

THE 'Taeping,' 'Ariel,' 'Fiery Cross,' and 'Serica' have arrived, with others in close pursuit, with something like FORTY-FIVE MILLION POUNDS OF NEW TEAS, consumption for the United Kingdom. This enormous weight coming suddenly into the London Docks, Shippers are compelled to admit it MUCH LOWER PRICES, in order to make sale.

We are thus enabled to make a Reduction of FOURPENCE in the pound.

4/0 down to . . 3/8
3/8 . . . 3/4
3/4 . . . 3/0
And so on downwards.

We may add the above Shops have brought a few lots of most unsound fine quality.

Reduction takes place on Friday the 21st inst.

116, OXFORD STREET;
87, STAMFORD ROAD; and
17, STAMFORD STREET.

BURGON & CO.,
TEA MERCHANTS.

COSTLY CARGOES *Prices tumbled when the four clippers docked with some 20 000 tons of tea, causing a temporary glut in the market.*

Bowsprit

Jackstay

Yard

BOW

Halliard

Foremast

Fore royal (furled)

Fore topgallant

Clewline

Seamen on
deck hauling
lines

Captain's boat

PORT

STARBOARD

Main royal

Buntline

Lifeboats

Main skysail (furled)

Mainmast

Mate with speaking
trumpet

Mizzen topgallant

Master

Officer of the
watch

STERN

Wheelman

Binnacle,
or compass box

Mizzenmast

Mizzen royal (furled)

WARNING NOTE *The
Cutty Sark's brass bell
was rung to announce
the time, to warn of
fog, or to mark when
the anchor was being
weighed.*

FURLING THE SAILS – IN A HURRY
*When bringing in the topsails,
the first task is to lower the yards –
the spars horizontal to the masts –
from which the sails are hung. This
collapses the heavy canvas sails. The
team on deck loosen the 'halliards' –
ropes that hold the yards up. They
can now start furling the sails. First,
they haul in the 'clewlines' fastened
to the lower corners of the sails and
running behind it to the mast. This
causes the sail to fold back on itself.
They then bring a forward fold to the
sail by tugging the 'leechlines'
attached to the outside edge. Finally,
they pull the folded sail up towards
the yard by hauling in the
'buntlines' connected to the bottom
middle of the sail.*

*Meanwhile, the men lying on the
yard hold onto the jackstay, an iron
rod connected to the top of the yard,
and haul the canvas up as far as they
can. Next, the middle of the sail is
hauled over the bundled canvas.
Finally, the sail and yard are tied
together with gaskets – strips of
woven material connected to the
jackstay. In this reconstruction, the
men have nearly finished bringing in
the fore topgallant sail, at the front
of the ship, and are still working on
the main royal (amidships). On the
mizzenmast (stern), three men try to
get a hold on the mizzen topgallant
as the clewlines, leechlines and
buntlines are hauled in.*

LOST GLORY *As a tea clipper, the* Cutty
Sark's *role was taken over by the speedier
steamships. In 1877 – eight years after
the Suez Canal opened as a short cut for
steamers – she gave up the China run,
and later carried wool from Australia.*

The world in a Crystal Palace

HOW THE GREAT EXHIBITION OF 1851 WAS ORGANISED

WHEN SIR HENRY COLE, A SENIOR CIVIL SERVANT, proposed mounting an international exhibition of works of industry to Queen Victoria's husband, Albert, the prince's imagination was fired. The exhibition, he said, would provide a 'living picture of the point of development at which mankind has arrived, and a new starting point from which all nations will be able to direct their future exertions'.

Race against time

Together, the two men set in motion plans for the Great Exhibition of the Works of Industry of All Nations. News of the event was announced at a special banquet held in London's Mansion House on October 17, 1849. A Royal Commission was formed in January the following year with Prince Albert as chairman. With the exhibition due to open on May 1, 1851, the commission had just over a year to complete its enormous task.

Government committees immediately set about writing to civic dignitaries, ambassadors and heads of state throughout the world inviting them to participate in the venture. Fifty countries and colonies agreed to send some 109 000 exhibits, including ivory and pearls from the East Indies, a stuffed squirrel from the United States of America, 'preserved fresh meats' from South Australia, a fountain gushing eau de Cologne from France, and a lump of gold weighing 3 cwt (152 kg) from Chile.

MEN OF ACTION
Prince Albert sits surrounded by other members of the Royal Commission that organised the Great Exhibition.

MAN OF IDEAS
Sir Henry 'King' Cole was a champion of industrial progress.

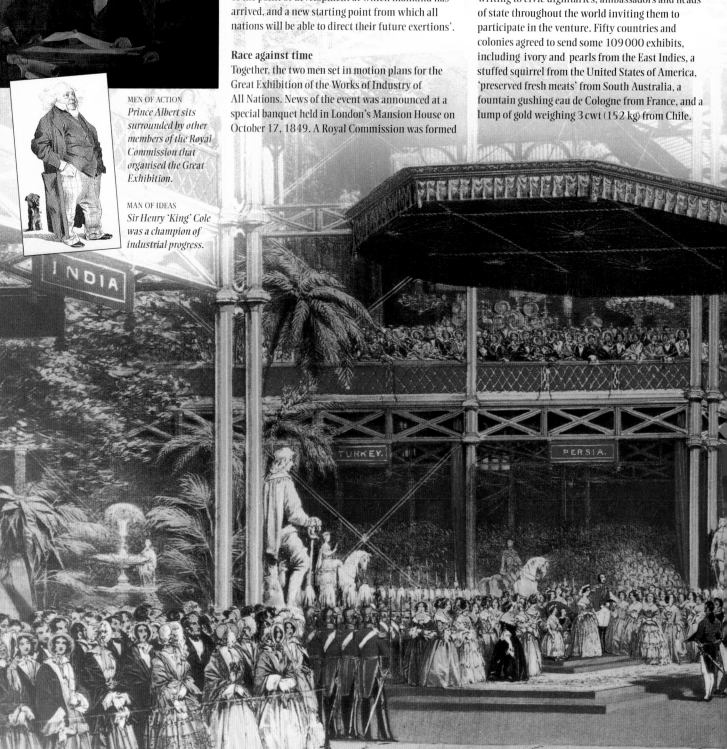

THE ARCHITECT *Joseph Paxton, designer of the Crystal Palace, started his career as a gardener's boy.*

GIANT GREENHOUSE *Workmen erect the iron skeleton of the Crystal Palace. Iron sections were cast and numbered at the factory and brightly painted in blue, red and yellow on site before being slotted into place.*

A monument to enterprise

British exhibits were mustered by the Royal Society of Arts, under the guidance of Prince Albert. Members of the society visited scores of company chairmen and industrialists, who pledged thousands of items. These included a ladies' bathing machine, elastic chest expanders, gas cookers, an artificial silver nose, a patent submarine helmet, models of flying machines, a 'defensive' umbrella-cum-stiletto, an express locomotive and a 24-ton block of coal.

The money to pay for the exhibition, which finally cost almost £336 000, was raised by voluntary subscriptions. Appeals were made to businessmen, and fund-raising banquets, circulars, pamphlets and newspaper publicity all encouraged people to contribute. The Queen gave £1000, Prince Albert £500, and thousands of workmen and ordinary citizens donated their hard-earned pennies.

A people's palace rises

The site for the exhibition, Hyde Park in central London, was chosen by Prince Albert, and the exhibition hall itself – the glittering Crystal Palace, so called because it was built with nearly 300 000 panes of glass – was designed by the renowned architect and landscape gardener, Joseph Paxton. The palace, using standardised parts, and covering some 22 acres (9 ha), was erected in only 22 weeks, just in time to receive the overseas exhibits which began arriving in London's docks on February 12, 1851 – less than three months before the exhibition was due to open.

Crated goods were taken by horse-drawn wagons to Hyde Park, where they were unloaded by men of the Royal Sappers and Miners. British and Empire exhibits were already in place in the western half of the palace, and those from the rest of the world were displayed in the east. Heavy items such as steam engines and hydraulic presses were on the ground floor, while lightweight exhibits such as silks and tapestries were displayed in the gallery.

Panic occurred when it was found that China had not sent nearly enough exhibits to fill its 300 sq ft (28 m^2) of display area. Private collections of Oriental treasures were 'plundered' by members of the Royal Commission and some 50 cases of goods were hastily assembled. Russia's 113 crates of exhibits, including tiger skins and armour, were held up because its northern ports were blocked by winter ice, and arrived after the deadline of March 1 – too late to be unpacked and displayed on the first day.

Visitors from across the kingdom

Weeks before it opened, people came to gaze at the fabulous Crystal Palace and to watch the exhibits being carried in. The railways ran cut-price excursion trains to London, and started a special cross-Channel ferry service which brought thousands of French visitors. Members of social groups and savings clubs came by specially chartered stagecoaches. Farm labourers arrived in their everyday working clothes. And 85-year-old Mary Callinack walked 300 miles (480 km) from Penzance in Cornwall, with a basket of provisions on her head, to see the spectacle.

To accommodate the influx of visitors, a register of approved boarding houses and cheap hotels was issued – including a 'Mechanics' Home', which could hold up to 1000 people at 1 s 3 d (about 6 p) a night. A special force of 6000 policemen was assigned to control the crowds in Hyde Park. Hundreds of guides

GRAND CEREMONY *On a dais near the crystal fountain, Queen Victoria and Prince Albert open the Great Exhibition. Enthusiastic season ticket holders line the galleries and nave.*

were appointed to meet trains, to take visitors to lodgings and to conduct them in batches to the exhibition to avoid overcrowding. By the time it closed on October 15, 1851, over 6 million people had visited the Great Exhibition and a profit of £168 000 had been made.

SOUVENIR OF PROGRESS *Attractive memorabilia produced for the exhibition by Britain's thriving Staffordshire potteries sold well to tourists.*

EXHIBITIONS OF WAR AND PEACE

LONDON'S GREAT EXHIBITION ATTRACTED celebrities from around the world, but it was outdone in this respect by the Universal Exhibition in Paris in 1867, which was attended by a total of 18 monarchs.

Bright star in a dark horizon
These included Wilhelm I of Prussia, who expressed pride in the display of huge Krupps cannons, reinforcing fears of coming hostilities between the two nations. The official report on the exhibition said it was 'like a meteor, bright but transient, in a horizon destined to become dark and stormy'. Three years later, Krupps cannons were to play their part in defeating the French Empire in the Franco-Prussian War of 1870-1.

German cannons were again a talking point at the Centennial Exhibition held in Philadelphia in May 1876 to mark the 100th anniversary of America's Declaration of Independence. The exhibition was opened by President Ulysses S. Grant to a 100-gun salute, and Richard Wagner wrote an 'Inaugural March' specially for the event.

CAPITAL OF FASHION *At the Paris exhibition of 1867 France's inimitable style in architecture and clothing won praise. Exotic displays from the Middle and Far East influenced European fashions for years.*

PAVILIONS OF PROGRESS *The American exhibition in Philadelphia celebrated social as well as industrial advances with special pavilions for women's art and industry and native American crafts. A display commemorated the end of slavery in the Southern states just over ten years earlier. The buildings themselves were hailed as 'cathedrals of industry'.*

A race for glittering prizes

HOW DIAMONDS WERE DUG OUT OF THE KIMBERLEY 'BIG HOLE'

A CABLE SNAPPED, THROWING A BUCKET OF RUBBLE down the vast hole. The wire lashed a miner's arm, leaving a gash inches long. At the bottom of the mine shaft, men rushed to get out of the way, but one was not fast enough. The bucket landed squarely on his head, sending him sprawling in the mud.

Driven by the promise of a fortune
With no safety regulations, there were hundreds of such accidents at the Kimberley diamond mine in South Africa, or the 'Big Hole' as it came to be called. At the peak of operations in 1873, 30000 men worked from sunrise to sunset in search of the diamonds that might make a man a millionaire.

Diamonds had been discovered at Colesberg Kopje in the north of South Africa's Cape Province two years before. By 1873, this small hill was to become the largest man-made hole in the world as prospectors flooded into the area to stake their

FIRST ARRIVALS *One of the earliest groups of diggers arriving in late 1870 pitches camp near the 'dry diggings' of the Kimberley hole. Diamonds had first been found in the wet banks of the nearby Orange and Vaal rivers.*

DIAMOND RUSH *The opening of the Kimberley mine in 1871 drew hundreds of hopefuls from around the world, all eager to find enough diamonds to return home rich.*

BLACK HOLE *Life is cheap at the world's biggest man-made hole, Kimberley mine, where the next pebble may turn out to be a perfect diamond. While some labourers dig with pickaxes, shovels or by hand in the shaft, others winch the rubble to the surface using crude windlasses or drag it up wooden ramps in carts and barrows. Some diggers have cut steps and tunnels into the earth to link their claims with the surface. Once the rubble is at the surface, labourers pour it through giant sieves to catch the stones, then sort through them for diamonds. Despite risking their lives in the mine, the black workers are exploited by their white employers, who for the most part avoid the perils of the mine shaft itself. They prefer to shout orders from the relative safety of the lip of the hole.*

claims. A boom town of more than 50 000 people sprang up on the site, named Kimberley after the British colonial secretary of the day.

The Orange Free State authorities divided the mine into 31 ft (9.5 m) square claims set out in a grid system, with 14 tracks running north to south across the site. Each plot was bought by a 'digger', who usually financed his venture by selling part of his claim to . Although only 470 claims had originally been allotted, there were soon 1600 owner-diggers at work. As the mine became increasingly crowded so the sorting tables and sieves were moved outside.

Smugglers make a killing

Diggers employed labourers to do the spadework. Black labourers received between 10 and 30 shillings (50p and £1.50) a week – considerably more than their white counterparts in Europe. White men earned a daily wage of around £2. As many as half the diamonds found at Kimberley mine may have been smuggled off the site. There were many ways of concealing the gems – labourers swallowed them or imbedded them in decayed teeth, in self-inflicted leg wounds or under toenails. Some workers fed diamonds to dogs, then killed the animals outside the mine to retrieve the stones.

Claims varied greatly in profitability – one might yield vast quantities of diamonds while the claim next door would be worthless. As digging proceeded, some miners abandoned their claims, or sold them to

others. Many of these unworked areas became unstable and started to collapse into the neighbouring ones. A flurry of lawsuits followed as owners of the abandoned claims sued for loss of property, while those whose workings had been swamped demanded compensation.

In March 1875 torrential rains caused extensive flooding and the mine collapsed into a vast hole, 650 ft (200 m) deep, leaving most claims unworkable. By 1880, the cost and technical difficulty of working the mine had become too much for individual diggers, so small companies were formed. These companies expanded until 1889, when the De Beers Company bought them all up and became sole owner of the Kimberley mine. The mine was finally abandoned in 1914, when the outbreak of the First World War led to a slump in the international diamond market.

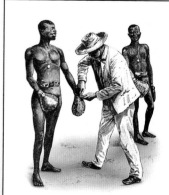

THEFT PROOF *A claim owner locks specially designed paddles on to the hands of his native labourers, making thieving impossible: all they can do is dig.*

Shopping on a shoestring

HOW FRANK WOOLWORTH REVOLUTIONISED SHOPS AND STORES

INNOVATOR *Frank Woolworth's new approach to pricing and store layout changed shopping forever.*

IT WAS ALREADY DARK ON A COLD SATURDAY AFTERNOON in February 1879 and in the small town of Utica, New York, most of the townspeople were at home or out enjoying themselves. Behind the paper-covered windows of a small store, Frank Winfield Woolworth sat nervously, waiting for customers. He had carefully set out his stock, and thousands of leaflets advertising the store had been delivered that morning. His great gamble was under way.

Just after five o'clock, a woman knocked on the door: she wanted one of the coal shovels, advertised at 5 cents each. She handed over her money and left with the shovel. She was Woolworth's first customer.

Everything at the same price

Woolworth's retailing vision was summed up in the sign above the Utica premises: 'Great 5 Cents Store'. But apart from offering a store devoted to selling every article at 5 cents, he also wanted to display all the goods – ranging from ironmongery to plants, perfume to underwear, curtains to broomsticks – on open shelves. This was a radical approach to shopping. Normally, the merchandise in a dry goods store was hidden on shelves behind the clerks, or in boxes under the counters. Customers had to ask for what they required, then steel themselves to find out the cost.

Woolworth had already helped to introduce a 5 cent counter at the small retailers in Watertown, New York, where he worked as chief clerk, and he had seen the idea catch on with rival retailers. Now he felt bold enough to leave his job and buy stock for his own store on credit.

Giving customers what they want

Shoppers loved Woolworth's sales approach. Customers were encouraged to browse at their leisure and salesgirls were friendly to them. They could look at the goods, handle them, then hand over 5 cents for each item they chose. If they changed their minds about what they had bought, they could take the goods back to the store and get a refund.

At the end of the first year, Woolworth had turned a debt of $315.41 into a net profit of $1516.60. Mass sales enabled him to buy more cheaply and provide a wider variety of items; a varied stock resulted in quick turnovers. By adding a 10 cent line, Woolworth was able to increase the variety of merchandise fourfold. In 1912, he went into partnership with his successful imitators and founded the F W Woolworth Company. Five and 10 cent stores sprung up across the country, and by 1909 the first overseas store had opened in Liverpool, England, where he anglicised the name to the 'Threepenny and Sixpenny Store'. Around 60 000 customers arrived in the first two days, amid scenes described in the press as 'bordering on the hysterical'.

SYMBOL OF SUCCESS *Designed by celebrated US architect Cass Gilbert, Woolworth's towering headquarters in New York is 792 ft (241 m) high. It was the world's tallest building when it was completed in 1913.*

EXPORT *Woolworth's were as popular with shoppers in England (bottom) as in the United States (below).*

LOCAL LANDMARK *Woolworth's shops were established in 12 towns in the north-eastern US by 1888. Because all goods cost either 5 or 10 cents, American shoppers dubbed the stores 'five and dimes'.*

The world in your hands

HOW SEARS, ROEBUCK BECAME THE FIRST MAIL-ORDER FIRM

SALESMAN *Sears knew how to appeal to his armchair readers.*

HANDYMAN *Roebuck's first trade was watch repairing.*

HEAVY LOAD *Country mailmen prepare to deliver the latest Sears, Roebuck catalogues to homes in Minnesota in 1914.*

OFFICE MACHINE *Row upon row of clerical workers process customers' orders at Sears' vast Chicago headquarters.*

DRAWING THE OIL LAMP CLOSER, AND SETTING OUT PEN and ink, the farmer's wife peered anxiously at the catalogue her neighbour had lent her. 'Don't be afraid that you will make a mistake,' it soothed. 'We receive hundreds of orders every day from young and old who never before sent away for goods.'

Taking heart, she filled in the order form. The excitement returned – the thrill she had felt when reading about china dinner sets and small household items she would never find at the local general store. But, even though it was 1906, it still seemed impossible that any single company would be able to cope with her long list of needs, let alone supply the farm tools that her husband wanted. It was even more unlikely that everything would arrive in their remote rural community in the USA's Midwest.

Beating the clock, meeting the orders

Her worries were unfounded: that year a Schedule System had been inaugurated at the vast Sears, Roebuck headquarters at Hoffman Estates, near Chicago. The company, established in 1893 by Richard Sears and Alvah Roebuck, could dispatch goods with unrivalled speed and efficiency.

An early morning shift dealt with the preliminary clerical work before the main body of workers arrived. The letters were opened by an automatic mail opener. Each order was allotted a specific dispatch time within the next 48 hours, and the time that the goods were due in the shipping room was stamped on the order. If the order was for more than one item, a separate ticket was assigned to each purchase with the dispatch time stamped on each ticket. After checking that the catalogue number quoted by the customer tallied with the goods ordered, and that the correct payment had been included, labels for dispatch were made out and attached to the order forms.

The goods were stored in huge depots, on shelves numbered to correspond with the catalogue numbers. Clerks selected the items ordered from the shelves. The rattle and rumble of mechanical devices filled the air as goods disappeared down gravity chutes or sped out of sight on elevators.

Down in the shipping room, single orders were dispatched within the specified time. Mechanical carriers took the packed goods to loading platforms, where precancelled stamps were added and the parcels loaded onto the adjacent railhead.

Three days later, the farmer's wife picked up a pile of packages at the railway station. She also bought a newspaper and, seeing an advertisement for the latest Sears catalogue, decided to send for one of her own.

VALUE FOR MONEY *Sears' catalogues promised good quality at low prices. In 1897 fashionable boots were on offer (below), while in 1902 (below left) a wide array of fob watches were for sale. An early phonograph (bottom) was the best buy in 1908.*

Absolutely American

HOW THE COCA-COLA EMPIRE WAS FOUNDED

THE CROWDED ROOM WAS BUZZING WITH ANTICIPATION. Robert 'The Boss' Woodruff, President of Coca-Cola, had called a meeting of his entire sales force. The Boss strode in, a cigar clenched between his teeth. Minutes later his staff were filing out in stunned silence. They had all been fired – but the next day, Woodruff reinstated each and every member of the sales force as 'servicemen', whose job was not just to sell Coca-Cola syrup, but also to install machinery at soda fountains, train retailers in dispensing techniques and give advice at the bottling plants.

A secret formula is born

Coca-Cola began life in a backyard in Atlanta, Georgia, in 1885 when Dr John Styth Pemberton, a pharmacist, concocted his own version of a tonic called Vin Mariani – a mixture of red wine and coca leaves. The following year, the doctor altered his formula and mixed the coca leaves with extract of cola nut, sugar and certain flavourings which he kept secret. One Saturday in early May, 1886, he went down to his local soda fountain with a jug of the syrupy concoction and mixed it with plain water, then sold the drink for 5 cents a glass. Early the next year carbonated water was used by mistake – and was found to make the beverage even more refreshing.

During the first year of sales, an average of about 13 glasses of the drink a day were sold. Before his death in 1888, Pemberton gradually sold off his business to various partners, including Asa Candler, a dynamic Atlantan businessman. By 1891, Candler had acquired the sole right to manufacture the syrup.

FAME IN GLASS *Coca-Cola's curvaceous bottle, as distinct as the trademark itself, boosted sales of the drink worldwide.*

He sold it through wholesalers to soda fountain operators, who mixed it with carbonated water. A year later he formed The Coca-Cola Company in Georgia and in 1893, registered the trademark – using the script originally penned by Pemberton's bookkeeper, Frank Robinson, in 1886.

In 1895, Candler claimed that 'Coca-Cola is now sold and drunk in every state and territory in the United States'. Advertisements and incentives, such as clocks, urns and fans, sustained his claim.

Big bucks in every bottle

But Candler missed a trick. In 1899 he sold the exclusive rights to bottle Coca-Cola to two Tennessee attorneys, Benjamin Thomas and Joseph Whitehead, for just one dollar. By 1920 there were more than 1000 bottling plants across the United States.

After Robert Woodruff took over the company in 1923, the bottle was sold in petrol stations, cinemas, parks and shops. By 1929, bottling operations in Mexico, China and all over Europe were in profit: an empire had been born.

MAN WITH A MISSION *Robert Woodruff insisted that every glass of Coca-Cola, wherever it was sold, tasted the same.*

COCA-COLA.

DELICIOUS! REFRESHING! EXHILARATING! INVIGORATING!

The New and Popular Soda Fountain Drink, containing the properties of the wonderful Coca plant and the famous Cola nuts. For sale by Willis Venable and Newman & Rawson.

FIRST SPLASH *The Atlanta Journal pops the cork on a brand new summer cooler in May 1886.*

KID'S STUFF *Three little boys in sailor suits advertise the delights of Coke in 1894 – when it still contained coca.*

TRENDSETTER *Elegant images of Coca-Cola drinkers gave the drink social appeal, and began a new age in advertising.*

WHEELS OF FORTUNE *Players place their bets and wait with baited breath for their luck to change. After 1931, when gambling in Las Vegas was legalised, around 10 000 people lived in the desert town – mainly construction workers from the nearby Hoover Dam, hungry for a slice of the action after the demanding hours of manual work. After the Flamingo hotel and casino rose out of the dust in 1946, people from all walks of life crossed the state line from California into Nevada; they were game for any gambling gimmick where money was the prize. As the 1940s and 50s progressed, 'The Strip' – a street of neon-lit hotel-casinos, bars and nightclubs – unfurled in all its gaudy splendour. Today, six-lane highways lead from all directions to a resort where year-round gambling can make fortunes or ruin players.*

DESERT FLOWER *The bright neon lights of the Flamingo hotel-casino first lit up Las Vegas, Nevada, in 1946. Inside this marble-floored pleasure dome, air conditioning cooled anxious brows in the casino – the heart of the enterprise – while outside, live flamingos walked the lush lawns and ornamental ponds.*

A gangster builds a gambling haven

HOW 'BUGSY' SIEGEL CREATED THE LEGEND OF LAS VEGAS

THE STREETS OF LAS VEGAS WERE HOT AND DUSTY, but inside the Flamingo hotel it was deliciously cool. A suave, smartly suited man with film-star good looks surveyed the scene with pride. His name was Benjamin 'Bugsy' Siegel and he liked what he saw: efficient staff meeting guests in the marble foyer, mink-wrapped women greeting one another – their expensive perfumes mingling and wafting through the plush, carpeted corridors.

Siegel, a notorious gangster, first visited Las Vegas in 1942 when the place was little more than a rail stop. By 1945, there were just two casinos there, but Siegel had met William Wilkinson, founder of the *Hollywood Reporter* and owner of a Los Angeles nightspot, who had begun plans to build a hotel-casino with 'class'. When Wilkinson ran out of money, the opportunistic Siegel moved in – after persuading gangsters to invest $1 million in the project through his vision of creating the greatest gambling mecca in the world.

GLAMOROUS THUG *Siegel's insatiable greed for money and the high life helped to build a glittering resort in the empty desert of Nevada.*

Here in Nevada, in the desert town of Las Vegas, gambling was legal. Not that legalities had ever meant anything to the poor boy from Brooklyn turned debonair mobster. From the mid 1920s, Siegel had been a fully paid-up member of the New York crime syndicate – 'the Mob' – run by Charles 'Lucky' Luciano and Meyer Lansky. By the mid 1940s the gangster was revelling in the glamour of Los Angeles, and spent much of his time with Hollywood celebrities, including Cary Grant, George Raft and Clark Gable.

The mobster plays his wildest card
In 1946 Siegel used his investors' money to purchase around 30 acres (12 hectares) of desert land, and his gangster tactics to get round any wartime restrictions that stood in his way of obtaining building materials. In December of that year, the 'Flamingo' hotel-casino opened before it was ready. Builders were still at work on the casino, so tables were set up in the lobby. Siegel flew in planeloads of guests to marvel at his dazzling showpiece. But within just a few weeks, professional gamblers had swallowed up $300 000 of Mob money and the Flamingo had to close.

When the Mob learned that Siegel had overspent the building budget by $5 million, the gangster's chips were down. On June 20, 1947, Siegel was shot dead, but the Flamingo hotel survived. Las Vegas was on the way to becoming the USA's gambling capital.

War and Weapons

*THE SWORD AND THE
SHIELD, THE LANCE AND
THE LONGBOW, THE
TRENCH AND THE TANK:
THE TECHNOLOGY OF
WAR IS A DEADLY GAME
OF LEAPFROG, WHERE
EVERY METHOD OF
ATTACK BRINGS ABOUT
A NEW DEFENCE*

Sword from the flames

HOW BRONZE AGE SMITHS MADE THE FIRST LONG SWORDS

THE SMITH CROUCHED OVER A CHARCOAL FIRE, IN WHICH lay a clay crucible containing molten copper and tin. When he was satisfied that the mixture was sufficiently liquid, he used a hook to tilt the crucible towards him. Carefully he poured the yellow mixture into a thin clay mould. This was the first stage in making a bronze sword fit for a prince.

Until people learnt to cast in metal, they made weapons and tools from wood, bone and stone. Then, about 8000 BC, the inhabitants of present-day Iran and Turkey began using copper. At first they hammered lumps of copper into shape; later, they melted them in a crucible and poured the metal into a stone or clay mould. The new swords were, like their flint predecessors, daggers designed for thrusting.

Longer, tougher weapons

These people found pure copper to be a relatively soft metal, although hammering hardened it a little. Around 3000 BC, the smiths in what is now Iran experimented by adding tin to copper, making a new material: bronze. Bronze swords, axe heads and spear tips proved much tougher than copper ones.

As a result, the sword's shape slowly changed. Smiths discovered that a bronze sword could be made as long as 3 ft (90 cm) – much longer than the 27-31 in (70-80 cm) of its copper predecessors. This was because the new, tougher alloy was less likely to bend or snap. It could also be sharpened much more effectively. Migrant smiths disseminated knowledge of the new material and bronze weapons spread into Europe, where they were in common use *c.*2000 BC.

Over the centuries, the art of forging ever-finer blades became a trade secret. Skilled metalworkers began to decorate bronze swords – and also splendid scabbards – with engraving and ornamentation.

Iron, discovered in Asia in around 5000 BC, was first used regularly *c.*1500 BC by the inhabitants of modern Turkey. More abundant than copper and tin, and as hard as bronze, it began to take over as the main material for tools and weapons. But it was not widely used in western Europe until about 700 BC.

SURE HOLD *Longer bronze swords were cast with an integral handle. This one, from the River Tyne in northern England, was made about 600 BC.*

PRECIOUS HILT *This sword, cast c.900 BC and found in Italy, has an ivory grip.*

ORNATE SHEATH *Skilled Celtic metalworkers were widely renowned. This gold and bronze scabbard of around 150 BC was found in County Antrim, Ireland.*

DECORATED WEAPON *This iron sword and scabbard from Yorkshire, c.250 BC, is inlaid with glass and studs.*

Outwitted by a horse

HOW THE GREEKS BROKE THE POWER OF TROY

AS DAWN BROKE OVER THE MEDITERRANEAN, LOOK-OUTS gazed down from the ramparts of Troy at an astonishing sight. The Greek armies that had been besieging the city for ten years appeared, suddenly and unaccountably, to have retreated. Their fleet was visible far out to sea, and amid the smoking ruins of the abandoned Greek encampment stood an enormous wooden horse. According to the legendary account of Troy's fall in around 1250 BC, the Trojans opened the city gates and dragged the horse inside. That night, as the city's defenders were sleeping heavily after celebrating their new freedom, Greek warriors hidden within the horse stealthily emerged and seized control of Troy.

Romance and commerce

The Greek poet Homer, in the 8th century BC, told this tale of the siege and fall of Troy in his epic poems the *Iliad* and the *Odyssey*. According to Homer, the Trojan War was fought over the beautiful Helen, wife of King Menelaus of Sparta. She ran away with Paris, one of the sons of King Priam of Troy, and the elopement sparked ten years of bitter fighting as the rulers of the various kingdoms of ancient Greece rallied to help Menelaus to recover his wife.

In fact, the Trojan War was a trade war. The Greeks wanted to capture Troy because of its geographical location. It lay on the south coast of the Mediterranean entrance to the Hellespont (today called the Dardanelles), the straits that control access to the Black Sea. This position allowed Troy to dominate the rich trade between Asia Minor and the Mediterranean, and the greedy Greek states seized the opportunity of launching an attack on the city.

DIVINE GIFT *Some Trojans believe the giant horse to be a gift from the gods and want to take it into the city.*

HIDDEN ENEMY *Unaware of its dangerous cargo, eager Trojan warriors pull the horse through the city gates.*

A horse with many meanings

Scholars have been unable to verify Homer's claim that the Greeks gained entry to Troy using a wooden horse. Some believe that the horse was a battering ram or a siege tower, but there is no evidence that siege towers were in use at this time. Ancient Egyptian sagas described a different trick: Greek soldiers were concealed in sacks which the Trojans thought were gifts – they carried them into the city and were later overwhelmed. Another explanation is that Homer was referring to the Greek fleet, since he sometimes termed ships 'horses of the sea'. What historians and archaeologists have established is that Troy's siege lasted for one year rather than the ten described by Homer, and that the city was finally destroyed by fire.

SURPRISE ATTACK *Carrying spears and shields, Greek warriors clamber out of the vast, wheeled horse in this relief from a Greek terracotta vase of the 6th century BC. According to Homer, among the soldiers inside the horse were Odysseus – the wandering hero of the* Odyssey *– and King Menelaus, Helen's abandoned husband. The advance party was able to overpower the Trojan sentries and open the city gates to admit their comrades, who had sailed back to Troy under cover of night. The Greeks burned and ransacked the city.*

THE MYSTERY OF JERICHO'S WALLS

THE PRIESTS BLEW A SUSTAINED NOTE ON THE TRUMPETS, the Israelite army shouted, and the city walls simply fell down. According to the Bible, this is how the Israelites under Joshua took the fortified city of Jericho in about 1230 BC – after a complex ritual in which the army had marched around the outside of the walls once a day for six days. But archaeologists have been unable to find any evidence to support this account.

Joshua was a renowned military commander who secured land beside the River Jordan for the Israelites after their exodus from Egypt in about the 13th century BC. This meant overcoming the Canaanites, who had built several fortresses, including Jericho, on the west bank of the Jordan and north of the Dead Sea. Historians have had to guess what happened in 1230 BC. It seems most likely that the Biblical story is symbolic: perhaps the Canaanites had already evacuated the city, or else they gave in without a fight, terrified by the trumpets and the shouting.

JERICHO FALLS *The Israelite priests (left) bear witness as the city's defences crumble and heavily armed soldiers prepare to attack.*

Face to face with spears and shields

HOW GREEK ARMIES JOINED BATTLE

THE OPPOSING ARMIES OF ATHENS AND SPARTA WERE lined up in the narrow plain. The densely packed formations of men began to move towards one another, advancing at a measured pace until they were close enough to jab at each other with their heavy, 8 ft (2.4 m) iron-headed spears. Some men in the front ranks fell, killed or wounded, but others immediately came from behind to pull them out of the way and take their places. Then both sides began pushing with their shields and the battle became little more than a vast scrum. After a few minutes, one side began to give ground. Soon it had been reduced to a disorganised rabble fleeing to the rear.

Between 650 and 300 BC Greek warfare was dominated by the hoplite, a heavily armoured infantryman who fought alongside his fellows in a formation known as a phalanx. The soldiers of the Greek city states – apart from Sparta – were part-timers mobilised in times of emergency who had to provide their own arms and equipment. The phalanx

CLOSED RANKS *Each man carries a heavy round shield on his left arm, protecting his own left side and his neighbour's right. Soldiers in the front lines jab with spears at their opponents' faces and necks.*

RUNNING BATTLE
When one phalanx is forced to give way, the battle turns into a series of individual skirmishes. Soldiers push with their heavy bronze-skinned wood and leather shields to try to force their opponents to the ground, where they can be finished off with spears. The shields, decorated with the heads of lions or other animals, protect the hoplites' bodies from the neck to the knees.

In the 6th century BC, when this battle scene was painted on a Greek vase, bronze body armour was replaced by the skirt-like linen tunics worn by these soldiers. They were split over the upper legs to allow easy movement, and sometimes reinforced with small metal plates.

was devised to overcome their lack of training – and, by massing the men together, to foster fighting spirit. It consisted of eight or more ranks, usually with about 120 hoplites in each, with the best soldiers positioned by convention at the right-hand end.

Outflanking the enemy
The phalanx was a powerful weapon in any battle, and was particularly effective against frontal cavalry charges and in the face of bowmen and missile throwers. But it did have drawbacks. Phalanxes of 16 ranks, each of 256 men, were not unknown, and such a dense body was difficult to manoeuvre in conflict. The Spartans, the most dedicated and highly trained warriors in ancient Greece, divided the phalanx into smaller, independent groups of 40 men.

Some commentators argued that the phalanx wasted manpower, because only the leading ranks engaged the enemy: as the Greek soldier and historian Xenophon (c.435-354 BC) said, 'When a phalanx is too deep for the men to reach the enemy with their weapons, what harm do you think they do to the enemy or good to their friends?' Another problem was that the right flank – unprotected by shields – was vulnerable to attack.

Eventually the more manoeuvrable smaller groups favoured by the Spartans took over from the large phalanx. The Romans took it a stage further by forsaking the long spear for the short sword – a much more effective weapon in close combat.

A Spartan upbringing

HOW GREECE'S FINEST WARRIORS WERE TRAINED

THE SMALL BOY ARRIVED HOME FROM SCHOOL IN A mud-stained tunic, his nose streaming blood and his face stained with tears. When his father asked him what was wrong, the child said that one of the bigger boys had attacked him for no reason. 'Bend over', his father said unsympathetically as he reached for a stick, 'you must learn that Spartans never complain.'

The boys of Sparta, a city-state in the southern Peloponnese peninsula in ancient Greece, had a tough upbringing and education. It turned them into ruthless soldiers, who were accustomed to hardship. These men were the source of the city's great power.

Sparta was the one Greek state with a professional army. It began its rise in the 8th century BC, and reached the height of its power after defeating its greatest rival, Athens, in the 5th century BC. The state's demanding military training eventually gave the word 'spartan' to the English language.

No mercy for the weak

Children of both sexes were little more than chattels of the military machine. Weak male babies were left overnight in the mountains to see whether they could survive the harsh conditions. Girls were raised in the belief that their role in life was to support their brothers and, in time, to bear sturdy warriors.

The Spartan boy's military education started at the age of seven when he began to attend school, to be taught games and physical fitness. At 12, he was removed from his family and sent to a military academy. Here life was made as harsh as possible. Underclothes were never worn, even in the depths of winter, and rushes provided the bed covering; rivers were the only means of washing. The boys were told to forage for themselves and to steal food to supplement their meagre rations. Learning to read and write took second place to military training; arithmetic was confined merely to the ability to count. Above all, students learned blind obedience to orders. The usual punishment for boys who failed to live up to the high standards was flogging.

A warrior nation tastes defeat

Sparta became a great power because of its military culture, but constant warring eventually sapped its strength. Too many of its citizens were killed in battle, and the state had to conscript foreigners – far less committed as warriors – to fight on its behalf. Other states also learnt from Spartan tactics, in particular their use of smaller, mobile phalanxes. In 371 BC at the Battle of Leuctra, to the west of Thebes in mainland Greece, the Theban general Epaminondas formed his best troops in a phalanx just 80 men across and on the left of the line rather than the right, as was customary. In this way he took advantage of the weakness of the phalanx's right flank, which was unprotected by shields. The Theban infantry and cavalry broke the Spartan line – and ended Sparta's reputation as a great military power.

KEEP FIT *Nude exercise – encouraged by the authorities – helped to keep boys and girls fit and supple. This fanciful view of Spartan pursuits was painted in the 19th century by Edgar Degas.*

BATTLE WEAR
A full-face bronze helmet gave the hoplite a fierce appearance. He could breathe easily through a long slit in the chin guard beneath the nose and over the mouth. On the battlefield he usually wore a splendid plume of horsehair on top of the helmet.

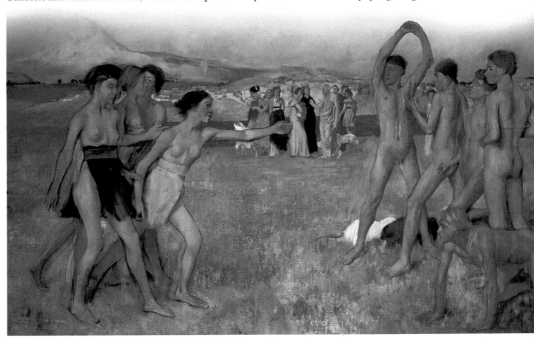

Elephants over the Alps

HOW HANNIBAL AND HIS ARMY CROSSED EUROPE'S TOUGHEST TERRAIN

TWIN HONOUR *To mark the success of Hannibal's expedition, the Carthaginians struck special coins with the general's portrait on one face, and a picture of one of his noble war elephants on the other.*

BUFFETED BY HIGH WINDS AND STRICKEN WITH COLD, the 37 war elephants and their riders toiled up the steep and slippery alpine path. Behind them came 50 000 foot soldiers and 9000 cavalrymen – with 9000 horses and pack animals – belonging to the army of the Carthaginian general, Hannibal. Deserted by their native guides, the marchers were on the look-out for the hostile mountain men, who habitually attacked any travellers who attempted the hazardous passage over the Alps.

Suddenly, as the Carthaginians entered a narrow pass overhung by sheer cliffs, a mass of long-haired Gauls emerged from a hidden position on top of the heights. They hurled rocks and boulders down on the troops and animals – yet they took care to miss the

vanguard of elephants, which the barbarians had never seen before and apparently held in superstitious awe. Within seconds the path was strewn with dead and dying infantrymen and cavalrymen. The Gauls – who had earlier greeted the Carthaginians with olive branches and promises of safe conduct – then launched their second offensive. A troop of men armed with spears, swords and bows and arrows sneaked up the path and attacked the exposed rear of the column. Thousands of men died in fierce hand-to-hand fighting. Finally, however, Hannibal's seasoned infantrymen used their swords

ATTACK FROM ABOVE *The air is filled with the screams of wounded men, horses and pack mules as hostile Gauls ambush Hannibal and his army in a narrow alpine pass.*

PANIC IN MIDSTREAM *Trumpeting with fear, Hannibal's bull elephants panic as they are ferried across the River Rhône on huge timber rafts. They break free of the ropes tethering them to the rafts, and throw their bareback riders into the chilly river. Some elephants fall overboard, but manage to wade under water to the far bank, breathing through their trunks.*

and spears to beat off the attackers, who scuttled back to their settlement in a nearby valley. The Carthaginians and their elephants were free to press on to the summit of the pass (probably Mont Genèvre), 8000 ft (2400 m) above sea level. They rested there for two days – while the stragglers, the wounded and the horses that had lost their riders gradually caught up with them.

Catching Rome by surprise

It was late October 218 BC, and 29-year-old Hannibal – one of the greatest military geniuses of the ancient world, and Rome's most formidable opponent – was marching on Italy at the start of the Second Punic War. He wanted to spoil the major attack which Rome was preparing by mounting the first strike: by crossing the Alps instead of arriving by sea, he planned to arrive from an unexpected direction and to catch the Romans unawares.

Hannibal placed great reliance on his force of specially trained war elephants, whose menacing appearance struck terror into any enemy, and whose unfamiliar smell upset their horses. Standing about 8 ft (2.4 m) high at the shoulder, the elephants – from the foothills of the Atlas Mountains in North Africa – used their specially sharpened tusks to attack opponents. They charged at the cavalry, gouging and bowling over horses and trampling men to the earth.

From its base at Cartagena, on the southern coast of Spain, the army had crossed the Pyrenees and made for the River Rhône. There the elephants had to be transported across the cold, fast-flowing water. To achieve this, Hannibal's engineers made four jetties from the trunks of trees, lashed together with rope.

'His best protection was his elephants: wherever they were placed on the line the enemy never dared to approach, being terrified of the unwonted appearance of the animals'
POLYBIUS, GREEK HISTORIAN

The jetties stretched about 50 ft (15 m) into the river, and two huge timber rafts were tied to their fronts. The structures were covered with soil to the height of the river bank, so that they resembled the broad, earthen paths which the elephants were used to.

Across the river by barge

With two cow elephants and their riders leading the way, the bull elephants obediently followed them onto the rafts, which were then cut loose from the jetties. Next, guide lines attached to barges pulled the rafts to the far bank. Halfway across, finding they were surrounded by water, the elephants panicked and tried to stampede. Terror-stricken, some of them toppled into the river, drowning their riders.

The elephants' greatest ordeal, however, began on the snowswept summit of the alpine pass which they reached some two weeks later. The pack animals

carrying their supply of palm leaves, hay and fruit had been killed in the attack by the mountain Gauls, and there was nothing for the elephants to graze on. Accustomed to eating up to 300 lb (150 kg) of vegetation a day, the elephants started to lose weight

SLIDING TO THEIR DOOM *Blinded by the driving snow, many of Hannibal's men slide off the narrow, slippery paths and skid helplessly to their deaths down the icy slopes. Meanwhile, others lose their footing and crawl along on their hands and knees, their fingers frozen to the bone. As fresh snow falls on the hard-packed snow, the heavily laden pack mules slither to the ground and are unable to struggle out of the ever-deepening drifts.*

MENDING ON THE MOVE *For three days, Hannibal's engineers busily repair a 400 yd (365 m) stretch of alpine path that has been swept away by a landslide. The men lash wooden supports together with rope, and use rubble, loose stones and their own supplies of timber to build up the road so that the epic march on Rome can continue.*

and were in danger of dying from cold and malnutrition. To make the beasts as comfortable as possible, their handlers wrapped them in woollen tent cloths, similar to the woollen leg wrappings and overgarments worn by the soldiers. Lashed by snow and driving wind, some of the troops had deserted. The rest were depressed and rebellious, and Hannibal tried to raise their spirits. Pointing to the lush green valley of the River Po, spread out far below them, he proclaimed: 'We have climbed the ramparts of Italy. What lies still for us to accomplish is not difficult!' However, his optimism was misplaced.

Blocked by a landslide

As the army began the descent fresh snow fell on steep, narrow paths which were covered with slippery, hard-packed snow from the previous winter. Only the vanguard of elephants had so far managed to avoid trouble. Before long, however, the elephants could go no farther. A landslide had swept away some 1200 ft (365 m) of the path ahead, leaving a mass of rubble. Hannibal told his engineers to dig away the snow, build up the road with loose stones, and to cover it with any timber they could find. Working nonstop it took three days to lay a path for the elephants and the army to walk along.

Sour wine to the rescue

Leaving the snow line behind them, the Carthaginians were soon faced with another major obstacle: a vast rock brought down by the landslide was jammed between sheer walls. The engineers set to work cutting down bushes and trees, which they placed on top of the rock. They ignited the pyre and, as the stone became hot, soused it with vinegary wine. The wine exploded, cracking the stone – which was then demolished by sappers wielding iron picks. The army shortly came to the verdant Po Valley, where the men rested and the emaciated animals ate their fill. But the alpine crossing had taken its toll. Hannibal had lost about 18 000 infantry, 2000 cavalry and 3000 horses, donkeys and mules. Only the 37 elephants had survived intact.

The following month – December – the great beasts helped Hannibal to rout a Roman army on the banks of the Ticino River, below the Alps. However, the strain proved too much even for the elephants, and they succumbed to exhaustion, malnourishment and disease. At last only one was left alive. Hannibal rode this as he crossed the Apennines with Rome before him.

TUSKED ATTACKERS *Urged on by their daredevil riders, Hannibal's war elephants charge ferociously at an army of foot soldiers. The great beasts rear up, ready to rip their opponents apart – or to crush them beneath their mighty hoofs. The sound of the elephants' trumpeting mingles with the agonised cries of those who are savaged to death – their weapons proving useless against the massive attackers. At the same time, the elephants' riders use their lances to harry the soldiers as they scatter and flee for their lives.*

Cogs in the mighty war machine

HOW A ROMAN LEGIONARY WAS ARMED

THE SOLDIER PULLED UP, EXHAUSTED AND COVERED with dust, after a gruelling day-long march in which his unit had covered 30 miles (50 km) of north African desert. His feet ached inside his heavy leather sandals. He was not only fully armed, but also weighed down by his rations, his personal effects and a bundle of wooden stakes. Using these, he and his fellow legionaries at once set about building the palisade which would serve to protect their camp for the night. It was 146 BC. The legion was marching on Carthage, the great Phoenician city in what is now Tunisia, and Rome's bitter enemy in the third and final Punic War.

Short-range weapons
The Roman heavy infantryman at the time of the Carthaginian wars had two basic weapons: the sword and the javelin. His sword, the *gladius*, was only 20-22 in (50-55 cm) in length. It had a sharp point, but blunt edges, and was designed for thrusting. The soldier wore it in a sheath slung from a baldric or strap over his shoulder. There were two types of javelin, or *pilum*, one heavier than the other – and the soldier would usually carry one of each. He would throw the javelins as he closed with his enemy, and then draw his sword to continue the fight at close quarters.

The legionary's shield, or *scutum*, was 2 ft 6 in (75 cm) wide and 4 ft (120 cm) long. Designed to protect the whole of his body, it was made of two thicknesses of wood glued together and covered with canvas and calf skin. The legionary was taught to fight with his left leg and shield thrust forward, using the shield to try to push his opponent over. He also often wore a chain-mail shirt, which weighed around 30 lb (15 kg). This heavy load caused many soldiers to drown as they tried to escape after being trapped by the Carthaginian general Hannibal at Lake Trasimene in 217 BC, during the long battle with Carthage.

The legionary had short hair and was clean shaven: he carried a sickle-shaped bronze razor in his kit. Shaving was a practice adopted from the Greeks, who had found that wearing a beard in battle could be dangerous. An enemy could grab it and hold the soldier captive at close quarters, where he could be stabbed repeatedly.

HEAVY LOAD *The legionary had to be fit to move around the battlefield wearing a chain-mail tunic and bronze or iron helmet, and carrying a long shield. He wore black and purple plumes on his helmet to make himself appear taller.*

Legions on the warpath

HOW JULIUS CAESAR'S ARMY CAMPAIGNED IN GAUL

UNDER THE COMMAND OF THE PROCONSUL JULIUS Caesar, the Roman army marched at speed through the French countryside in the early spring of 58 BC. Its objective was to halt the advance of the Helvetii, a Swiss tribe that had been pouring into Roman territory in southern France. Caesar's men were fully armed and ready to respond to a surprise attack.

At the head of the army column were lightly armed auxiliaries, recruited from the Italian provinces, who acted as scouts – with particular orders to watch out for the threat of ambushes. The vanguard, made up of detachments of cavalry and heavy infantry, followed. Next came the soldiers detailed to lay out the camp that would be built when the army halted for the night. 'Pioneers', whose responsibility was to find the best way over or around natural obstacles on the route, were the next group.

After the commanders, the camp followers

The creak of carts heralded the approach of the proconsul's baggage train. A clatter of hooves, and Caesar himself, his commanders and key staff officers rode by, resplendent in red cloaks, with their cavalry escort. Next were the legions, the main body of the army. At the head of each rode its commander, followed by the legion's standards and trumpeters, with the legionaries themselves marching in six-man ranks. Further cohorts of auxiliaries preceded the rearguard, which was made up of cavalry. A ragtag of camp followers, including food sellers, prostitutes and even slave dealers ready to snap up prisoners of war, trailed after the army.

Caesar – the military governor of Cisalpine Gaul (northern Italy) and Transalpine Gaul (central and northern France, Belgium and much of Switzerland) – had made careful preparations before launching the campaign. His first task had been to draw together his scattered forces. The Roman army usually went into camps during the winter months, and in Gaul the troops were widely dispersed to

ensure that people did not break the peace. Caesar also enlisted the help of local tribes who were enemies of the Helvetii, and with their aid was able to put a force of around 35 000 in the field.

Portable rations

His next job was to build up his supplies in readiness for the campaign ahead. Food was the most important. The Roman soldier on campaign lived on a diet of wholemeal biscuits, bacon and cheese – all of which could be salted or otherwise preserved – washed down with sour wine. The army did occasionally forage for food while on the march, but this was not encouraged since it led troops to spread out, making them vulnerable. The army also needed to cut large amounts of timber, because it planned to build fortifications in the path of the Helvetii, especially to stop the tribe crossing the River Rhône.

'Caesar had all the horses, starting with his own, sent away, so that everyone might stand in equal danger and nobody have any chance to flee. Then he addressed the men and joined battle.'
JULIUS CAESAR, *THE GALLIC WAR*

Caesar had a lean, well-organised military machine at his disposal. He had three legions in Gaul, each with a strength of around 5000-6000 men, divided into ten 'cohorts'. Each cohort consisted of six 'centuries' of 80-100 men, divided into sections of eight to ten. The legion also had a force of 120 cavalrymen, often largely made up of non-Roman citizens, particularly the Numidians (from modern Algeria in north Africa) and Germans. They were used as scouts and despatch riders. The legion was commanded by a 'legate', assisted by six 'tribunes', but the backbone of the legion's officer corps were the 'centurions', battle-hardened army veterans each of whom commanded a century.

At the end of each day's march, the army set up camp for the night, using guy ropes and pegs to erect a square city of leather tents. The legionaries slept eight to a tent. The camp was protected on all four sides by a ditch and an earth rampart topped with a palisade made of sharp wooden stakes, and had four large gates to allow the soldiers to sally forth quickly.

MOVABLE CAMP *A Roman army camp could be dug and constructed – to a standard design – in less than three hours. The main part is taken up with the long rows of legionaries' tents (1-6), divided by the two main thoroughfares (7,8). The officers – centurions, legates and tribunes – sleep in a line of tents opposite (9). The general has his own tent (10), and the non-Roman auxiliaries occupy separate quarters (11,12). A portable altar (13) stands in one of the thoroughfares.*

Reveille – then quick march

These fortified barracks were always constructed to an identical plan, and the familiar layout had significant advantages. Because each soldier knew beforehand exactly what to do, a camp could be thrown up very quickly. Faced with a surprise attack – even in darkness – the legionaries and auxiliaries knew exactly where they were. The camp was guarded by sentries through the night. In the morning, a bugle blared the signal for the soldiers to take their tents down; a second call gave the order to load the baggage animals and prepare to march.

These tried and tested methods worked well for Caesar. In July, his experienced soldiers decisively defeated the Helvetii in central France, and forced them to return to their homes in Switzerland.

A WALL TO DEFEND ROME'S FRONTIER

THE WILD COUNTRY OF THE ENGLISH-SCOTTISH BORDER formed the northernmost boundary of Rome's power. In AD 122 Emperor Hadrian visited Britain and ordered the building of a vast wall to protect northern England from raids by the Picts and Scots. Hadrian's Wall, parts of which still stand 16 ft (5 m) tall, ran 73 miles (117 km) from Bowness on the Solway Firth in the west to Wallsend, east of Newcastle upon Tyne. Along its top was a walkway for patrols.

Forts to house 1000 men

The wall was occasionally attacked, but it was not intended as a static line of defence. Instead, it was used as a base from which legionaries – alerted by patrols – emerged at speed to attack the northern tribes in open country. There were 16 garrison forts at roughly 5 mile (8 km) intervals, each with space to house 1000 men. Every Roman mile (0.92 miles, or 1.48 km) was a smaller fort known as a 'milecastle', with barracks for up to 100 troops.

Both types of fort had gates opening to the north, from which soldiers could march to take on raiders. Two turrets – used mainly as watchtowers – stood on the wall between the milecastles. On the northern side ran a ditch 27 ft (8 m) wide and 9 ft (2.7 m) deep, and on the south was a second, flat-bottomed ditch 20 ft (6 m) wide and 10 ft (3 m) deep, as well as a military road used for moving troops, equipment and supplies. Settlements grew up near the garrison forts, with shops, brothels, taverns where the soldiers could drink and play dice, and bathhouses where they could escape from the winter cold into a hot tub.

A second wall to the north

Hadrian's death in 138 brought a new strategy. The Scottish lowlands were reconquered and the Antonine Wall, named after Emperor Antoninus Pius, was built between the Firth of Clyde and the Firth of Forth. Completed in about 142, it was only used for 30 years and after about AD 170 Hadrian's Wall became the true Roman frontier in Britain.

ARMY OF LABOURERS *Roman legionaries did more than just fight. After the conquest of the Scottish lowlands, those stationed in northern Britain spent eight years building Hadrian's Wall.*

BATTLE TO THE DEATH *A long-haired, long-sleeved barbarian takes on a Roman soldier. At this time the legions were mostly manned by Roman citizens, augmented by auxiliaries from overseas.*

Siege by 'belfry', tunnel and catapult

HOW THE CRUSADERS WORE DOWN A SARACEN STRONGHOLD

HEAVY BOMBARDMENT *The rain of stones fired by long-range siege machines like the trebuchet wears down the city walls and the resistance of the defending troops.*

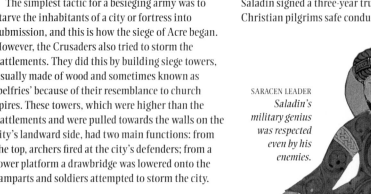

UNDER COVER *Sappers protected from missiles by movable shelters hack at the foundations of the city walls.*

A 'TREBUCHET', THEY CALLED IT. IT WORKED LIKE A giant catapult. At the left-hand end of the machine's huge beam was a sling, into which two men, struggling forward, dropped a large stone. Close to its right-hand end, the beam rested on a horizontal crosspiece. Following shouted orders, a group of soldiers ran forward and leapt onto the right-hand end of the beam, forcing it down and causing the sling to swing violently upwards. The stone flew over the city walls and crashed into the streets of Acre.

A target for the Crusaders

Lying about 100 miles (160 km) north of Jerusalem, Acre was the main harbour in the Holy Land and a vital strategic target for the soldiers of the Third Crusade – the holy war called by Pope Gregory VIII and mounted by armies from England, France and the Holy Roman Empire. Their aim was to capture Jerusalem, which had fallen to Saladin, leader of the Muslims or 'Saracens' (derived from the Arabic *sharqïyïn*, 'Easterners') in October 1187. In August 1189, a force of Christian knights dug in on the hill of Turon, a mile (1.6 km) east of Acre, which was held by a Saracen garrison. Saladin raised a relief army and for a time trapped the Christian force on the hilltop. But by mid September Crusader reinforcements had arrived and driven the Saracens off. Saladin's army remained at the Crusaders' backs, but he was not strong enough to pose a real threat.

The simplest tactic for a besieging army was to starve the inhabitants of a city or fortress into submission, and this is how the siege of Acre began. However, the Crusaders also tried to storm the battlements. They did this by building siege towers, usually made of wood and sometimes known as 'belfries' because of their resemblance to church spires. These towers, which were higher than the battlements and were pulled towards the walls on the city's landward side, had two main functions: from the top, archers fired at the city's defenders; from a lower platform a drawbridge was lowered onto the ramparts and soldiers attempted to storm the city.

Undermining the city walls

With the arrival of substantial reinforcements under the leadership of Philip II of France in April 1191 and Richard I of England in June, the siege finally reached its climax. The two kings were convinced that the best way

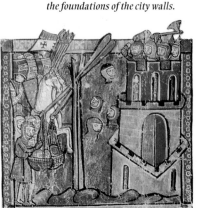

RAIN OF DEATH *The Crusaders unleash a hail of severed human heads. Rotting animal corpses were also popular missiles.*

to take the city was to breach the walls – said to be so thick that two chariots could pass each other on the top. Teams of tunnellers dug beneath the foundations, propping up the walls with timber. They then set fire to the wooden props – causing that section of the wall to collapse – and rushed into the breach. But at this point, alerted by the sound of warning drums within the city, Saladin attacked the Crusaders from behind. The western armies were forced to fight on two fronts and had to give way. The besiegers tried tunnelling again in other parts of the defences, but Acre's defenders dug down to the new excavations, foiling further attempts to undermine the walls. Meanwhile, the bombardment continued.

The garrison falls

By early July the situation inside the city was becoming desperate. After nearly two years food was running out. The Saracens' only hope was that Saladin would force the Crusaders to lift the siege. But Saladin knew that his army would not succeed. Using carrier pigeons, he ordered the garrison to break out of the city and flee along the beach while his army distracted the western forces.

The plan did not work. This time, the French and English armies simultaneously kept Saladin at bay and, on July 12, 1191, attacked the garrison. Acre surrendered. The Crusaders did not retake Jerusalem, but in September, after a skirmish near Arsuf, Saladin signed a three-year truce guaranteeing Christian pilgrims safe conduct to the holy city.

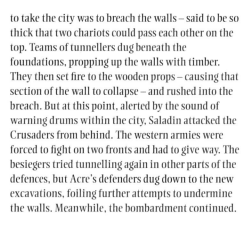

SARACEN LEADER *Saladin's military genius was respected even by his enemies.*

WINGED MESSENGERS *Throughout the siege, Saladin used pigeons to send messages to the defenders of the city.*

A FLYING WALL OF FIRE

IT WAS A SECRET WEAPON AND THE FORERUNNER OF THE modern flamethrower. It burned furiously and could only be doused by sand and, supposedly, urine. 'Greek fire' – which was liberally used by the Acre garrison to spread terror among the attacking Crusaders – was a viscous liquid which burst into flames when it came into contact with water.

The substance was first used by the forces of the Byzantine emperor, Constantine IV. The formula was probably based on naphtha and quicklime. The Byzantines packed it into brass-bound wooden tubes and pumped water at high pressure both to project it and to set it alight. An Arab chronicler of the siege of Acre, Ibn al-Athir, describes how pots containing naphtha were thrown at the siege towers. They had no effect on their own, but were followed by flaming pots which ignited the naphtha and set the towers ablaze. Greek fire was widely used by the Saracens during the Crusades. It was eventually superseded by guns, which were easier to use.

SEA OF FLAMES *Byzantine forces set fire to the Arab fleet during the siege of Constantinople in AD 673.*

SCALING THE WALLS *In 1099, during the First Crusade, Christian knights besieged and captured Jerusalem. In this 14th-century manuscript, the siege is portrayed as an attack on the Heavenly City – with the crucifixion of Christ depicted in the upper-storey windows. At the right, Crusaders pound the walls, while above them soldiers climb onto the battlements. Siege towers were a key piece of an attacker's equipment, but were vulnerable to fire. Defenders could set the towers alight using naphtha.*

Rites of passage

HOW A KNIGHT WON HIS SPURS

THE YOUNG NOVICE KNIGHT FELT VERY TIRED AS HE WAS escorted into the presence of King Edward III. The previous evening he had bathed himself to wash away his sins. He had spent the whole night keeping silent vigil in chapel, praying to remain pure. This morning, after years of training, he was finally to be created a knight. It was January 1348, and the ceremony was taking place during the Great Tournament at Eltham Palace, south-east of London.

Joining the brotherhood

The ritual was elaborate. First two knights fixed gilt spurs to the novice's heels – a mark of his status as a mounted warrior, which set him apart from a mere foot soldier. Then the king stepped forward and fastened the young man's sword and scabbard around his waist before embracing him. 'Be a good knight', he said. The group moved to the chapel, where the novice placed his new sword on the altar for blessing by the Bishop of London. Afterwards, the bishop handed the sword back, first tapping the young man on the shoulder with it – the origin of the current practice of a monarch 'dubbing' new knights.

The ceremony was not always so grand. Sometimes it was carried out amid the mud, blood and chaos of the battlefield. Even so, it gave membership of an exclusive international brotherhood.

Page, squire, knight

No man was born a knight: a soldier could be knighted to reward him for acts of bravery, or the son of a high ranking family could be brought up to be a knight, but he would have to earn his position. The aspiring knight began his preparation at the age of about seven when he became a page to a noble or knight. He learnt how to ride while acting as a personal servant to his master; he was also taught to read and write, and encouraged to take an interest in popular court pastimes such as chess, singing and poetry or playing the lute. When a little older, he began instruction in using the sword and lance.

A wealthy knight might have several pages. Each aspired to become a squire – a word derived from the Latin *scutiger* (shield bearer) – by about 16 or 17. He would have the honour of accompanying his master on campaign and of keeping his weapons sharp and clean. Only when the squire had proved his bravery and skill at arms in battle would he be considered for a knighthood.

PURIFYING *The evening before the ceremony that will make him a knight, the candidate takes a ritual bath, according to custom. It cleanses him in both body and spirit in preparation for his new life.*

GIRDING *A belt carrying the sword in its scabbard is tied around the new knight's waist. This was the central part of the ceremony.*

DUBBING *The earliest account of a royal dubbing dates from 1128: Geoffrey of Anjou was knighted in the presence of King Henry I.*

SWORD PLAY *Young pages practise their swordsmanship. Even at the age of seven, knights' pages spent hours studying the arts of war.*

Fighting qualities

A knight was expected to display love of combat, strength, endurance and bravery in battle. He should also be humble, show respect for the Church and be unquestioningly loyal – to his king and to the cause for which he was fighting. His behaviour was governed by a set of ideals expressed in the 'code of chivalry', inspired by the legendary court of King Arthur and the real deeds of the emperor Charlemagne (*c.*742-81). A knight might also dedicate himself to a lady at court, through the conventions of a passionate 'courtly love', as expressed by the French troubadour poets.

OUT TO IMPRESS *The pageantry of a medieval tournament made a magnificent spectacle. In early tournaments mock battles took place between two teams of knights. In the 13th and 14th centuries jousting became popular: two knights charged at each other with levelled lances, each trying to unseat the other. This view of knights parading prior to action is from a 15th-century manuscript.*

KNIGHTLY SYMBOL *This ceremonial shield was used in tournament parades. The knight pledges himself to his lady under the motto* Vous ou la Mort *– 'You or Death'. The grim figure of Death is seen creeping up close behind the knight.*

Men of steel

HOW PLATE ARMOUR PROTECTED KNIGHTS IN BATTLE

THE YOUNG SQUIRE THREW HIMSELF FORWARD AND turned a double somersault on the grassy ground. One of his friends copied him, and several others cheered and clapped. The boys were knights in training, getting the feel of their new suits of armour.

At this time, in the 13th century, chain mail – a handmade skin of individually riveted iron links – was the only armour. It was laborious to make, and expensive: sometimes there were 100 000 links in one garment. Foot soldiers who could afford it wore a mail shirt that reached to the hips; knights wore a longer version, the 'hauberk', which came to the knees. But increasingly they found they needed greater protection than mail could provide: arrows and lances could rip the links apart, and blows through the metal shirt caused bad bruising.

EARLY DESIGN *This Italian armour of around 1400 is partly covered with fabric – a characteristic of early plate. The cloth was cut by specialist tailors.*

CLASSIC STYLING *Four Milanese craftsmen collaborated to make this suit of about 1450, with its distinctive elbow and foot defences.*

GERMAN PLATE *Fluted armour, as worn by the Duke of Württemberg about 1530, was a speciality of the Nuremberg and Augsburg armourers.*

NOBLE IN THE SADDLE *Warriors on the battlefield wore thick breastplate armour, ridged to deflect a lance; the lower leg was protected by a metal shin guard called a 'greave', and the upper leg by plates, which included a specially shaped steel cap for the knee. Metal guards called 'vambraces' protected the upper and lower arms, while steel caps covered the elbows. To cushion his head against blows, the knight wore a barrel-shaped helmet over a close-fitting linen skull cap with ear flaps, tied with strings under the chin. A knight's horse might be armoured even more ornately than the knight himself. The magnificently gilded ceremonial horse's armour pictured here, made in the 16th century, was a gift to the Holy Roman Emperor Charles V from his grandfather, Maximilian I. Charles used it in tournament jousts.*

JUNIOR MODEL *This ceremonial plate, worn by the Holy Roman Emperor Charles V as a boy, is designed to imitate the fabrics of court dress.*

FIGHTING GEAR *Archduke Ferdinand II, Governor of Bohemia, wore this foot-combat armour in tournaments. It is decorated with Austrian eagle motifs.*

PRECIOUS DRESS *The Royal Armourers of Greenwich, in London, made this elaborate suit for the Earl of Cumberland in about 1585.*

CRAFTSMAN AT WORK *Armour plates were hammered out by hand by the armourer, then repeatedly heated in a forge and allowed to cool. The breastplate and the front of the helmet were made especially thick to shield these vulnerable areas. Different blocks or 'stakes' were used as moulds to shape the various components of a suit.*

From strength to strength

The solution was found in suits of armour that incorporated metal plates. In the first half of the 14th century, knights began to wear the 'coat of plates', a leather jerkin inside which were riveted small rectangular iron or steel pieces. They also sported protective shoulder discs and leather gloves covered or lined with metal plates. Armour became increasingly decorative. By the mid 15th century, Italy – in particular Milan – had become the world's finest producer of plate armour, and the armourers themselves were lauded as some of the most skilled craftsmen of the age.

DEADLY DARTS *The broad arrowheads (above) made a large cut and were used to disable war horses; the needle-shaped tips (left) could pierce plate armour.*

A deadly shower of arrows

HOW THE LONGBOW OUTSHOT THE MORE POWERFUL CROSSBOW

FIVE THOUSAND ENGLISH LONGBOWMEN MARCHED forward. About 300 yd (275 m) away from the French they halted. Each man took the stake that he had been carrying and thrust it at an angle into the ground. Protected by this rough palisade, each bowman drew the 24 arrows from his sheaf and stuck them point first into the ground in front of him. Orders were shouted and each archer picked up an arrow, strung his bow and pulled back the string, grunting with the effort. The bowmen loosed their strings, and clouds of arrows whistled upwards. They fell onto the serried ranks of the French – the first shots in Henry V of England's celebrated triumph at Agincourt, northern France, in October 1415.

Bows taller than a man

The battle was a great victory for the English archers. They were equipped with yew longbows about 6 ft (1.8 m) long and tipped with horn at both ends. These shot an arrow of around 3 ft (1 m), and could kill at 600 ft (180 m). The bows are thought to have originated in Wales in the 12th century. The first ones were made of elm, but this soon changed to yew, a softer wood which bends more easily. Such huge bows were very difficult to draw. The archer pushed the wooden shaft of the bow away with one hand, while with his other he pulled the string and arrow back using all the muscles of his shoulders. When the feathered rear end of the missile was level with his ear – or sometimes his cheek, depending on the arrow's length – he loosed the arrow.

Aiming was a matter of experience – a veteran knew exactly how high to shoot to make the arrow fall on a targeted area. All the archers loosed at once. If they ran out of arrows they were able to get more from carts wheeled along the lines by boys.

Strings drawn by machine

Many French bowmen at Agincourt used the crossbow, a much older weapon. These were used in Asia for hunting as early as the 6th century BC, but were not taken seriously as military weapons until the 12th century. A bow – initially made of wood, then of bone and finally, in the 14th century, of steel – was attached across the end of a wooden shaft, or 'tiller'. The first crossbows were drawn by hand, but as the bows became stronger and more difficult to bend, mechanical ratchets were developed to draw the bowstring. The projectile – a stone, an arrow or a smaller bolt – rested in a groove on the top of the tiller and was shot when the archer released the string by pulling a trigger on the bottom of the tiller.

The first significant test of the longbow against the crossbow came at the Battle of Crécy between the English and French in 1346. The English, with their longbowmen, triumphed. Although the crossbow was generally more accurate than the longbow, it was much more difficult to operate. The ability of the longbowmen to maintain an almost constant rain of arrows completely demoralised their enemies.

CITY UNDER ATTACK *A besieging army's artillery pound a city's walls. The wheelless cannons (right) are hidden behind wooden screens, which are moved or tilted when the guns are ready to fire. More modern wheeled guns fire between the screens or, at the left, between cylindrical wicker barriers filled with rocks. The first gun carriages had been invented by the 15th century. Handles, or 'trunnions', were cast on each side of the barrel: the gun could be placed in a wheeled carriage and aimed more accurately. The highly respected 'master gunner' commands the artillery: skilled gunners aim and fire the weapons; the dirty work of moving and loading the guns is left to unskilled 'matroses'. Albrecht Dürer's woodcut of 1520 is accurate in detail, but is not to scale. In reality, the tents of the commanders would be pitched much farther away from the din of the battle.*

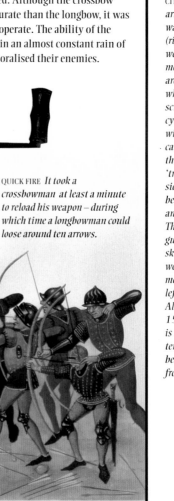

LETHAL WEAPON *The very high tension of its mechanically drawn string gave the crossbow more penetrative power than the manually drawn longbow.*

QUICK FIRE *It took a crossbowman at least a minute to reload his weapon – during which time a longbowman could loose around ten arrows.*

'Black powder' and stone balls

HOW EARLY CANNON WERE LOADED AND FIRED

IN 1415, WHEN HENRY V OF ENGLAND INVADED FRANCE, he took ten pieces of artillery with him. His chief engineer ordered all of them to be aimed at one of the gates of Harfleur, a port on the Normandy coast. Each gun was placed on an earth bank. The ammunition – limestone balls for the larger guns, and round lead missiles for the smaller ones – was piled beside the guns. The master-gunners opened a wooden barrel in which they kept the 'black powder' (gunpowder), made from saltpetre (potassium nitrate), sulphur and charcoal. Ammunition and powder were pushed into the gun, and a match held to a touch-hole on top of the barrel. The bombardment of Harfleur began.

From fireworks to firearms

The Chinese had invented gunpowder before the 10th century, and used it to propel fireworks, but European soldiers did not use the explosive mixture in firearms until the 14th century. Guns were first fired in action during the siege of Metz in 1324.

The first guns were little more than bronze pots that shot arrows or darts. They often blew up, and were just as dangerous to the user as to the intended target. More practical guns were devised in the 14th century. They were made by wrapping heated wrought-iron rods around a solid metal core, hammering them together, then clamping them with red-hot hoops that shrunk as they cooled, binding the barrel tightly. The core was then knocked out, leaving a gun barrel that was open at both ends. The projectile was inserted into the barrel from the back end and a removable cylindrical powder chamber was secured behind it with wooden wedges.

These early guns were fired using a red-hot iron bar or wire, heated in a fire on the battlefield and inserted through the touch-hole to detonate the powder. After firing, the wedges were knocked away and the powder-chamber refilled. This could take ten minutes. Because the powder-chamber was only wedged in place it had no proper seal with the walls of the barrel, so each time the gun was fired there was a backblast of hot gases.

The invention of the large 'bombard' at the start of the 15th century was a great advance. Its powder-chamber had twine wrapped round it for a tighter fit with the barrel, which greatly reduced the backblast. Gunmakers also learnt the art of casting the barrel in one piece, using bronze. These guns were loaded from the muzzle rather than from the back. Gunpowder was rammed home first, followed by the cannonball; they were lit by holding a flame to the touch-hole. The new guns were more effective than their predecessors, and by the early 16th century were playing a significant role on the battlefield.

BACKBLAST *When the gunner fires, a wooden stake at the back end of the gun absorbs the strong recoil.*

HAND GUN *A knight fires from the saddle in this fanciful view. In reality, he would dismount first.*

War on the hoof

HOW THE MONGOLS CONQUERED A VAST EMPIRE

FLAMES ROARED THROUGH THE PALACES AT THE HEART of the city. The defenders of Chung-tu, the capital of the Chin empire in northern China, knew that they were beaten. They were terrified, because the invading Mongol armies had a reputation for ruthlessness and wild plundering that travelled before them. The Mongols took Chung-tu (modern Beijing) in 1215, after a four-year campaign. The palaces burned for more than a month.

Discipline and organisation

Genghis Khan, the Mongol leader, built on this success. In the next decade, while his generals subjugated China and raided Persia and Russia, he defeated the powerful Muslim Khwārazm empire (modern Uzbekistan). He died in 1227 aged about 65, but his successors created through conquest an empire which at its height – in the early 14th century – embraced China, the whole of central Asia and much of eastern Europe.

The key to Genghis Khan's success was a sophisticated military machine. The army was highly organised and disciplined. The basic unit was a 10-man troop known as an *arban*. Ten troops formed a 100-man squadron, or *diaghoun*, and ten squadrons formed a regiment, or *minggan*, of 1000 men. Each regiment was based on a clan, or part of a tribe. Ten regiments formed a *touman*, of 10 000 men. Rules of conduct were extremely strict – death was the penalty for all but the most minor offences. Personal loyalties were strong, partly because fellow soldiers were members of the same clan. In Genghis Khan's army, all the generals were his own sons or placemen and were unquestioningly loyal.

A diet of milk and meat

Mobility and speed of attack were the essence of Mongol strategy. Mounted on small, sturdy ponies, the army could cover up to 50 miles (80 km) a day. It would advance on a broad front in several columns, accompanied by pack camels, ox-drawn wagons and thousands of spare ponies. Each soldier had as many as seven of these hardy animals, which did not need to be fed since they lived by grazing; fresh animals ensured that the army could keep moving fast.

The soldiers could survive for long periods on a basic diet of curdled mare's milk – which they took from their horses – and dried meat. They were even known, when desperate, to drink blood drained from a pony's neck. When the enemy was located, the army temporarily abandoned its supplies and the columns came together to form a large spearhead. When battle was joined, the troops moved at speed, following signals from black and white flags – or torches when fighting in bad light.

A network of spies

Each Mongol soldier carried one or more bows that he fired from the saddle. These had a maximum range of about 1000 ft (300 m), but were used mostly for rapid firing from short range. In addition, he used a sabre, a lasso and a lance. The Mongols also had a highly efficient intelligence network. Spies operated under the guise of traders in merchant caravans travelling throughout the world. News was returned rapidly to the Mongol ruler, or Khan, using a courier system with posthouses on the roads.

But the greatest Mongol weapon was terror. Genghis Khan was known to order the slaughter of prisoners who refused to submit to him. As a result, the battle was often half won before it had begun.

IN HOT PURSUIT *With swords raised ready for the kill, helmeted Mongol warriors (above) spur their horses on as they try to catch a dishevelled, desperately fleeing enemy. Most of the Mongol horsemen wore iron or toughened leather armour, but some preferred chain mail shirts.*

IN COLD BLOOD *No one is spared as the Mongols close in and capture a strongly defended Chinese town in 1205 (left). Armed with swords, axes, spears and metal war clubs, the Mongols swarm the parapets, killing the men and forcing themselves upon the defenceless women.*

TOWER OF SKULLS *The gruesome Mongol practice of making towers studded with the severed heads of their enemies was carried on by their Mughal descendants. In the 14th century, Mughals demolished the Hindu Kush fortress of Mikrit and used the debris and skulls to erect a grim warning to any future foes.*

LAYING SIEGE *Heavily armed soldiers of Kublai Khan, grandson of Genghis Khan, cross a pontoon bridge and lay siege to the fortress of O-Chou during their ruthless conquest of China in the second half of the 13th century.*

MISSILE ATTACK *Mongol attackers use their latest 'secret weapon' – a huge catapult – in an all-out assault on a walled city in Russia. The catapult, and others like it, was specially made by Chinese and Persian siege engineers whom the Mongols had captured and put to work. Such war machines played a large part in establishing the might of the Mongol empire.*

HAMMERED FOR STRENGTH AND TEMPERED FOR SHARPNESS

THE JAPANESE BELIEVED THAT THEIR BEST SWORDSMITHS were inspired by the gods. The finest blades seemed to defy time, and a samurai often used swords that were hundreds of years old.

A medieval warrior in battle wore two weapons, the *tachi*, a sword over 2 ft (60 cm) long, and the *tantō*, a dagger 8-12 in (20-30 cm) long. The tachi was used both for chopping and for parrying blows, and needed to be strong enough to pierce armour. Its long, curved blade made it a formidable weapon and the samurai needed both hands to wield it. The blade had to combine a very sharp edge with a resilient centre. To achieve this, the smiths used a tough steel core inside a casing of the metal that had been folded and hammered up to 15 times to make it harder still.

For the final tempering, the smith covered the blade with a malleable mixture of clay, charcoal, ashes and sand, and then scraped the cutting edge bare. He heated it, until the edge had turned a precise shade of red – which he judged by eye. Then he plunged the blade into water. The edge cooled quickly, which made it brittle and meant that it could be sharpened to a very high degree on a grindstone. But the clay mix retained heat, so the rest of the blade cooled more slowly – making it stronger and more resilient.

Macabre test

Once sharpened, the blade was passed on to other craftsmen for polishing and for the fitting of the hilt, guard and grip. Swords used in battle were carried in plain bamboo scabbards, but ceremonial swords had lacquered and decorated sheaths.

Finally, the sword was tested for sharpness. The blade was sometimes tried on the bodies of criminals, usually after the wrongdoers had been executed but occasionally when they were still alive. The results were recorded on the *tang*, or stem of the blade, after the handle had been specially detached for the purpose.

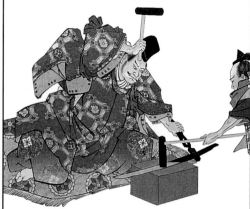

AN EMPEROR'S REVENGE *Exiled on Oki Island, the Emperor Go-Toba (1183-98) said that the making of swords was work for princes. He spent years forging a 'perfect' sword to kill the man who had deposed him – the regent Hojo Yoshitoki.*

BATTLE WEAR *Dressing for war, a samurai pulls up his thigh guards, or* haidate

Next he dons his body armour

Then he fastens on his pair of swords

He ties on a head towel; this helps to support the helmet

Then he ties on his iron faceguard, or mempo

Finally, he puts on his kabuto, *or battle helmet*

The way of the warrior

HOW JAPAN'S SAMURAI ELITE LEARNED TO FIGHT AND KILL

THE CORRECT WAY TO BOW TO SUPERIORS MIGHT seem an unlikely subject of instruction for youths wanting to become samurai warriors. But the lesson had a hidden purpose. Each time a pupil bowed, his teacher struck him sharply on some part of his body with a wooden sword. Eventually the pupils started to realise what the lesson was about. As they bowed, they began to duck and weave in order to avoid the teacher's blows. They were learning one of the most valuable of samurai skills – *zanshin*, or alertness.

Japan's samurai class emerged in the 10th century AD from a bitter power struggle between Japan's three major clans. The warring tribes realised that the clan with the most loyal and skilled warriors would prevail. As the war progressed, local military governors, or *daimyo*, became increasingly powerful at the expense of the emperor. The soldiers who served the daimyo, the samurai, evolved into a separate and distinct military elite. Membership of this group passed from father to son.

The first samurai fought mainly with bow and arrow, but by the 15th century the sword had become their principal weapon. Schools of martial arts, or *ryu*, were set up, each run by a master swordsman. The samurai learned to develop absolute obedience and to achieve a state of *mushin* – or 'no mind' – when fighting. The warrior was told that if he emptied himself of fear and pride he would wield his weapon instinctively and with greater flexibility and power. The samurai code of conduct became known as *bushido*, or 'the way of the warrior'.

ON THE ATTACK *Wielding a long and a short* bokutō, *or wooden sword, a renowned 16th-century swordsman named Musashi attacks his tutor, Kasahara – who deftly wards off the blows with the wooden lid of a rice pot.*

Fearless to the end

The samurai had to overcome fear of death. Rather than suffer shame through capture, he was expected to commit ritual suicide, or *seppuku* (the Japanese for 'self-disembowelment'), also known as *hara-kiri* ('belly-cutting'). He thrust his sword into his own abdomen, dragged it across his stomach and twisted it – a very painful method designed to demonstrate his courage. Some even stabbed themselves again just below the chest and then cut their throats.

In the 17th century, the warrior class took up more peaceful pursuits. Poetry and calligraphy became as important as swordsmanship, which was seen as more a mental than a physical exercise. The samurai rank was abolished in 1871, but the spirit of bushido lived on in the Japanese Armed Forces, especially in the belief that to surrender was dishonourable – and in the kamikaze suicide attacks against the ships of the Western allies during the Second World War.

HORNED AGGRESSOR *A samurai horseman wearing a helmet with a traditional* waki-date, *or crest of horns, leads a headlong attack against defenders of the fortified port of Osaka in southern Japan. The city – the centre of Japan's rice trade – was captured in 1614-15 by the shogun Tokugawa Ieyasu, who made Edo (modern Tokyo) his capital.*

BLADE BEAUTIFUL *This elegantly curved* tachi, *or cavalry sword, was made in the 13th century by a master craftsman, Kunimune – who, while tempering the steel, drew the fine wavy line, or* hamon, *on the cutting edge.*

MUSCLE MEN *Renowned for their feats of martial strength, two of Japan's fierce warrior heroes of the 12th century – Asahina Yoshihide (left) and Hatakeyama Shigetada (right) – display their fighting spirit with staff and with sword.*

345

Cat warriors feed victims to the sun god

HOW AZTECS WAGED WAR AND DISPOSED OF THEIR PRISONERS

THE SOLDIER'S HEADGEAR WAS MADE FROM THE CARCASS of a wild cat. Like a hood, the muzzle lay over his forehead and the skin hung down his back. The top of the soldier's head was crowned with a multicoloured plume of feathers, and he wore a heavily padded cotton jerkin over a short-sleeved tunic. In one hand he brandished a *chimalli*, a small round wood or reed shield, and in the other a *macquauitl*, a wooden club with sharp cutting edges made from obsidian, a volcanic rock. The soldier's weapons and colourful garb marked him as a member of an Aztec warrior elite, the 'Jaguar Knights'.

Thirst for blood

The Aztecs lived in what is now central and southern Mexico. They were a warrior people who, beginning in the 14th century, conquered an empire extending from the Gulf of Mexico to the Pacific Ocean in the space of 200 years. All male babies were dedicated to war by the midwife who delivered them; their umbilical cord was buried with a set of arrows and a shield. Some boys were later educated to be priests or state officials, but most were sent to a *telpochcalli* ('house of young men'), a school run by warriors. Here they mixed with experienced fighters and learned the arts of war. Only those who proved themselves early on by killing enemies or capturing prisoners could become professional warriors.

The Aztecs went to war to conquer land and expand their empire. But battles also had a more sinister purpose – prisoners of war could be used in sacrificial rituals. An almost constant supply of victims was needed because Huitzlipochtili, the Aztec sun god, had to be placated by daily human sacrifices. The need to take live prisoners, therefore, dominated Aztec tactics on the battlefield.

Music before battle

Before battle the opposing sides would line up in dense masses to the accompaniment of a cacophony of sound – made by conch shells, trumpets, flutes, drums and gongs – intended to maintain the warriors' morale and to strike fear into the opposition. Battle usually began with a hail of arrows and javelins, after which the Jaguar Knights and other prize warriors advanced, fighting hand to hand with swords and spears. Behind them came more soldiers, with ropes to tie up prisoners.

When one side was defeated, peace negotiations were opened at once. Neither the Aztecs nor their opponents believed in long, drawn-out campaigns, and the losing side did not quibble over paying tribute. Since no monetary system existed, this took the form of merchandise, jewels or forced labour. The vanquished did not expect the return of their prisoners, who were now rounded up for sacrifice.

CAT ATTACK *In fearsome array, a Jaguar Knight wields a heavy club encrusted with obsidian blades. He ranks just below the emperor in status.*

SALUTE TO THE SUN *Aztec sacrifices could take three forms. Most commonly, the victim would be spreadeagled on an altar and his heart cut out and held up by the priest. But he might be forced to fight several warriors at once, or be tied to a rack and pierced with arrows until he bled to death.*

BEAUTIFUL – BUT DEADLY *A short knife with an ornate warrior-shaped handle of turquoise and shell mosaic was used by the Aztecs in human sacrificial ceremonies.*

Overthrowing an empire

HOW PIZARRO TRICKED ATAHUALPA AND DEFEATED THE VAST INCA ARMY

THREE SHIPS LAY ANCHORED IN THE BAY OF GUAYAQUIL on the Pacific coast of Ecuador. Small boats, heavily laden with soldiers and horses, ploughed towards the shore. Some of the men were armed with pikes, and others with arquebuses – guns that were fired by holding a match to a hole on the top of the barrel. The soldiers also wheeled ashore two small cannons called 'falconets'. Unseen eyes had been watching the men as they landed, and messengers were already on their way to warn the local Inca chieftain that the 'bearded ones' had returned.

In search of Inca gold

The invaders were Spanish conquistadores (conquerors), chasing wealth and fame in Central and South America. Europeans had realised that a fortune could be made from these mineral-rich lands when they heard reports of the voyages of Christopher Columbus and Amerigo Vespucci in the late 15th century. Now Francisco Pizarro, a Spanish adventurer, was leading an expedition in the name of Charles V of Spain to conquer the Incas, whose empire stretched from present-day Ecuador to Peru.

Pizarro had been inspired by earlier Spanish military exploits in South America – in particular Hernán Cortés' defeat of the great Aztec empire in Mexico in 1518-21. In 1527 Pizarro had landed at the main Inca port of Guayaquil and opened negotiations with the locals. He returned to Europe to seek the king's backing for an expedition against the Incas and in January 1531, with official approval, set sail once more from Panama to Peru. His total military strength, besides his two falconets, amounted to 180 men and 27 horses. After landing he awaited the arrival of a fellow adventurer, Diego de Almagro, with 100 more men. Then he set off inland for the Inca city of Cajamarca, 600 miles (970 km) to the south-east in the Andes Mountains.

Spanish ambush

Cortés had brought the Mexican Aztecs to their knees by seizing their leader Montezuma, and Pizarro planned to use the same trick. He sent a message to the Inca emperor Atahualpa suggesting a meeting in Cajamarca. The conquistadores arrived there on November 15, 1532, and arranged by messenger to meet the Inca leader in the main square. But when Atahualpa arrived with his retinue of around 4000 soldiers he found the square deserted: Pizarro's men were hiding in the surrounding buildings. The Spanish leader sent Father Vicente de Valverde to present the Inca emperor with a Bible and ask him to embrace Christianity. When Atahualpa refused, Pizarro ordered an attack. The Incas – armed only with clubs and slings – were overwhelmed.

The Inca ruler pledged to fill his cell with gold higher than a man could reach in turn for his release. Pizarro agreed, but when Atahualpa proved able to fulfil his promise, the Spaniard kept him in prison anyway. Eventually Atahualpa was strangled on Pizarro's order, on August 29, 1533.

The Spanish went on to occupy the country's capital, Cuzco. The Incas – although far superior in numbers – found their mostly wooden weapons were no match for the Spanish guns and fled. Two and a half centuries of Inca power and prosperity were over.

STYLISH WEAPON *Pizarro's dress sword had a velvet grip.*

FACE TO FACE *Atahualpa addresses Francisco Pizarro's half-brother, Hernando, from a litter. His cloak, softer than silk to the touch, was made from the skins of hundreds of bats.*

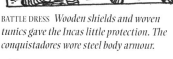

BATTLE DRESS *Wooden shields and woven tunics gave the Incas little protection. The conquistadores wore steel body armour.*

AN ERA ENDS *This 17th-century drawing shows Atahualpa's head being severed: in fact, the Inca chief was strangled and given a Christian burial.*

CONQUISTADOR *Francisco Pizarro was 50 years old when he began his voyage.*

347

The army that did not believe it could lose

HOW FREDERICK THE GREAT TRANSFORMED PRUSSIA'S ARMY INTO AN INVINCIBLE FORCE

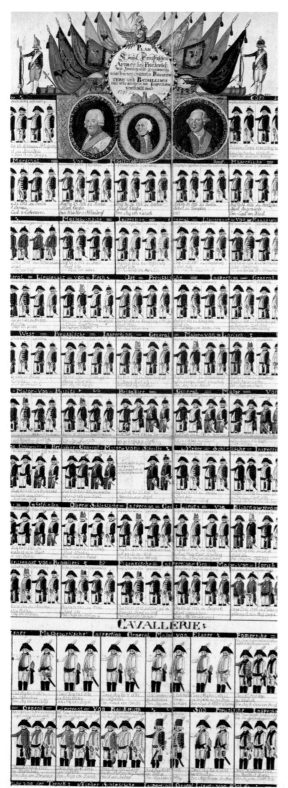

'IF ANY REGIMENT OF CAVALRY SHALL FAIL TO CRASH straight into the enemy when ordered, I shall have it dismounted immediately after the battle and turned into a garrison regiment. If any infantry battalion so much as begins to waver, it will lose its colours and its swords, and I shall have the braid cut from its uniform.' Frederick II of Prussia was addressing his generals on the eve of his greatest victory – at Leuthen in Poland on December 5, 1757, when 43 000 Prussians outmanoeuvred and overwhelmed an Austrian army nearly twice its size.

Defending Prussia's independence

Frederick succeeded to the throne of Prussia on the death of his father Frederick William in 1740. His father had regarded him as a flute-playing fop, but the new king was as much a resolute military leader as he was a man of culture. His kingdom was one of several small states fighting for dominance in what is now Germany. He inherited a strong and disciplined army, but also a state with no natural defensive positions near its borders. Frederick realised that if Prussia was to survive attacks from France, Russia and Austria, its army would have to be the finest in Europe.

Frederick took control of all military matters. To ensure the unquestioning loyalty of the officer corps, which was largely made up of nobles and gentry, he awarded it a status above all other social groups. He decreed that no officer could be tried by a civil court, on the ground that the court members, being civilians, did not have 'honour'. He expected the highest standards of behaviour from his officers.

'Battles must be decisive. Have at the enemy! Give him a good salvo in the nose at twenty paces, and again at ten. Then bayonet him in the ribs.'
FREDERICK THE GREAT

As for the rank-and-file soldiers, Frederick strengthened the harsh code of discipline that had already been imposed by his father. Even for the most trivial offence, a soldier could be brutally flogged with up to 1500 lashes over a period, or made to 'run the gauntlet' between two ranks of fellow soldiers beating him with sticks. Recruits learned to advance unflinchingly in the face of the enemy's fire. The loose-fitting waistcoat that was part of the army's blue uniform was redesigned and cut much more tightly to force every soldier to hold his body upright. The uniform was also made more elaborate, with the addition of white belts and polished buttons giving each infantryman an imposing appearance.

Frederick knew that his men would often be outnumbered by larger armies, so he made sure they were able to move fast to attack weak points in the enemy's line. The soldiers were drilled hard and

UNIFIED STRENGTH *Frederick expanded the Prussian army in the course of his military reforms. This pictorial roll call, dated 1796 – ten years after his death – includes generals (top), grenadiers, artillery, cavalry, engineers and cadets.*

STRATEGY TO WIN *In this view of the Battle of Rossbach, Saxony (1757), the French and Austrian armies are marked by red flags and the Prussian army by blue. The Prussians' training – superior to that of any other European army – helped them to win the day despite being outnumbered by almost two to one.*

practised manoeuvres and turns while marching at up to 120 steps a minute. As a result, Prussian soldiers could march, load muskets and change formation in battle more quickly than their enemies.

Frederick's insistence on battlefield mobility and speed of response made him an exceptional commander. He also placed unusual emphasis on the use of cavalry. Most commanders of the day used cavalry mainly for reconnaissance. Frederick, however, used his cavalry to charge his enemies from a flank at a critical moment in an attempt to throw them off balance. The artillery of the day was difficult to move. The Prussian army used a new type of horse artillery, light 6 lb (2.7 kg) cannons that could be galloped around the battlefield.

Feeding an army

Frederick also transformed arrangements for the supply of the army in the field. Instead of relying on supply depots or living off the land as they marched, each soldier carried three days' food and supplies in his knapsack, while a supply column for each regiment carried food and ammunition for a further eight days. Another column for the whole army carried a further month's supplies.

The army won resounding victories against the French, Austrians and Russians during the Seven Years' War (1756-63), despite being heavily outnumbered. Other nations were so impressed that they adopted many of Frederick's army reforms.

ROYAL STANDARD BEARER *During the Battle of Zorndorf in August 1758, Frederick leads his men forward, regimental flag aloft and sword in hand. This fanciful view captures the essence of the king's ability to inspire a tired army to victory.*

Breaking the enemy line

HOW ADMIRAL NELSON TRIUMPHED AT TRAFALGAR

THE BRITISH AND FRANCO-SPANISH FLEETS MET OFF CAPE Trafalgar, near Cádiz in southern Spain, on October 21, 1805. Two days earlier, the 33 ships of the Franco-Spanish fleet commanded by Admiral Pierre Villeneuve had sailed south from Cádiz heading for the Mediterranean. British frigates, or light scouting vessels, had sighted them and signalled to the remainder of the British fleet, under Admiral Horatio Nelson, waiting beyond the horizon. The frigates pursued Villeneuve's fleet until, on the morning of October 21, 1805, he realised he could not escape and ordered his ships to turn and sail in a straight line back to Cadiz. The 27 ships of the British fleet formed two parallel columns and sailed towards the enemy line from the west, aiming to break it in two places. The Battle of Trafalgar had begun.

During the 18th century sea battles were fought by parallel lines of ships – as a result, the vessels were called 'ships-of-the-line'. They were categorised by the number of guns they carried – 'first-raters' had 100 or more, while 'fourth-raters' mounted 50-64. The larger ships had three gundecks and the smaller

had two. Frigates were lighter and smaller single-decked ships, with just 24-38 guns. The admiral sailed in the flagship, one of the ships of the line which signalled tactics and orders to the others.

Blasting the rigging, cracking the hull

The naval strategy of the time differed between the French and British fleets. Each had a distinct philosophy as to how ships should be attacked: the French fired at a ship's masts and rigging, while the British aimed lower, at the ship's hull. Once the ships were locked together, the victor sent across a boarding party to force the other to surrender by 'striking' (lowering) her colours.

It was a Scottish amateur naval strategist, John Clerk, who first suggested new tactics. He argued that if squadrons of ships turned through 90 degrees and sailed directly towards the enemy line they would

The Victory's *jib boom, shot away*

The wooden Cupid on the Victory's *bow lost an arm in the fighting*

Foremast

In the face of fierce musket fire, only a few men remain on the Victory's *upper deck*

MOMENT OF TRUTH *About an hour into the battle, HMS Victory – which cannot steer properly because its wheel and tiller ropes have been smashed by French shot – has collided with the French 74-gunner Redoubtable (left). As Nelson lies dying in the cockpit, the Victory's guns briefly cease firing because Captain Hardy thinks the Redoubtable has struck its colours. Jean-Jacques Lucas, the French captain, orders his men to board the Victory, but they are repelled. The Royal Navy will win the day.*

The Victory's *massive bower anchor*

A chain of men pass cartridge cases up to the waiting powder monkeys

Light room

A sailor takes a barrel of powder out of the magazine

Cartridges are made in the filling room

Marines stand at the head of every ladder, charged with ensuring that only powder monkeys and the wounded go below

Many of the dead are thrown overboard through the gun ports

A powder monkey, a boy whose job is to deliver powder to the gun crews, takes a cartridge

initially present a much smaller target. As they broke through the line they could bring all their guns to bear on the ships to both port and starboard. But the enemy ships could only use the guns mounted on the side from which they were being attacked. This manoeuvre became known as 'crossing the T', since it involved moving at right angles across a line.

It was the tactic triumphantly employed by Nelson at Trafalgar. He himself led one of the attacking divisions in his flagship, HMS *Victory*. By nightfall, 17 Franco-Spanish ships had been captured and another had been blown up; the British lost none. It was a decisive victory, but it was marred by the death of its architect, Nelson, who was shot by a French marksman as he stood on his quarterdeck.

DOUBLE BREAKTHROUGH *Guns blazing, the English fleet breaks the Franco-Spanish line in two places during the Battle of Trafalgar. Sailing with the wind, Nelson's* Victory *(foreground) leads the main division, while the* Royal Sovereign *(background), under Admiral Cuthbert Collingwood, heads the attack from higher up. Once the line was broken, the battle became a series of individual skirmishes.*

One of the 24-pounders has broken free from its tackle, crushing the crew that fired it

The Redoubtable's main yardarm has been cut to make a bridge for boarding

Mainmast

Entry port

Mizzenmast

Mizzen channel

Davit to hold the ship's boats

Sheet anchor

Sailors used rolled up hammocks to protect them from enemy fire

Main channel

24 lb cannon

12 lb cannon

Operating table

The ship's wheel was shot away at the very start of the battle, so the tiller is manned by 40 sailors

Amputated limbs are thrown into half barrels

The wounded are tended in the cockpit

Pumps

32 lb cannon

A team man the pumps

Shot locker

Bosun's stores

Main hold

Nelson, shot, dies in Hardy's arms

Spare shot

Beaten by distance and the rigours of winter

HOW NAPOLEON'S GRAND ARMY WAS DESTROYED IN ITS PUNISHING RUSSIAN CAMPAIGN

'AS A RESULT OF ALL MY MOVEMENTS 400 000 MEN WILL be concentrated in one spot; we can therefore expect to find nothing in the country itself and we shall have to carry everything with us.' So wrote Napoleon Bonaparte, emperor of the French, before he invaded Russia. On June 24, 1812, after months of careful planning, the main body of his *Grande Armée* ('Grand Army') crossed the river Niemen – which marked the border between the Grand Duchy of Warsaw (as the French puppet state of Poland was called) and Russia. Napoleon's relations with the Russian tsar Alexander I were once good, but they had deteriorated sharply after Alexander broke his promise to keep British goods out of Europe.

ROUGH GOING, EASY GOING *While most of Napoleon's troops lived on horseflesh – and drank melted snow and animal blood – during the disastrous retreat from Moscow, the emperor himself – roasting a rabbit on his sword – went short of very little. He travelled in style in a warm and comfortable fur-lined sleigh. He had plentiful supplies of his favourite foods – such as mutton, beef, lentils, rice, beans and wholesome white bread – as well as a selection of the finest French wines. The Russians' scorched-earth policy, plus the freezing conditions, meant there was little or no food to be obtained from the land over which the French had previously advanced. Thousands of troops lost toes and fingers from frostbite, while others collapsed in their tracks and froze to death. Napoleon, however, proceeded with all speed to Paris to show its citizens that – contrary to rumour – he had not been killed. His health, he declared, had never been better.*

Preparing for war

In preparing for invasion, the emperor's first task was to build up a large force to match the 420000-strong Russian army. Eventually Napoleon's troops numbered more than 600000 men, although only around 450000 actually crossed into Russia; the rest were used to secure the French base in Poland.

Napoleon knew that this vast army would perish unless it was well supplied. In previous campaigns, his soldiers had largely lived off the land. But finding adequate supplies in the desolate steppes of Russia would be no easy matter. To solve the problem, Napoleon organised depots in Poland and East Prussia where ammunition, weapons, food and clothing was stockpiled for his troops. The army's transport organisation was doubled in size to 26 battalions for the invasion. In all, some 10500 wagons were used. Most of these were four-horse vehicles capable of carrying 33000 lb (1500 kg).

Each soldier in Napoleon's army carried four days' food in his knapsack. A further 20 days' supplies – rice, biscuits, vegetables and brandy – followed in the wagons, together with fodder for the herds of cattle that accompanied the army to supply it with meat. The total food allowance would therefore keep the troops for just 24 days – a very short time, given the fact that the distance from the river Niemen to Moscow is over 500 miles (800 km) as the crow flies, and that the average rate of march was about 15 miles (24 km) a day.

Scorched earth and burning city

When the Grand Army invaded, the Russians retreated, burning the land as they went. The French had to endure extreme heat alternating with heavy rains. Hundreds of soldiers fell ill with exhaustion. Worse still, the rains turned the roads in western Russia – many little more than dirt tracks – into quagmires, and the heavier wagons could not be used. In mid August Napoleon reached Smolensk, 280 miles (450 km) west of Moscow. After fierce fighting he drove the Russians out of the city and set up a forward supply depot. On September 7, he engaged the Russians in a pitched battle, at Borodino, 70 miles (112 km) west of Moscow. It was a brutal and bloody conflict for both sides – some 50000 Frenchmen were killed or wounded, including 47 generals, while at least 44000 of the 120000 Russians taking part were killed, wounded or captured. The Russians withdrew, and Napoleon entered Moscow a week later. Almost all the population had fled eastwards; also a huge fire started by the Russians was eventually to consume most of the city. For a month Napoleon stayed in the Kremlin, the tsars' palace – which had escaped the flames. He expected Alexander, who was in St Petersburg, the Russian capital, to beg for peace. He did not, and by October 18, worried by reports of growing unrest in France, Napoleon decided on retreat.

Freezing and starving

When Napoleon reached Smolensk, on November 9, winter was setting in, and he found almost all the food at the supply depot there had already been eaten by the large force he had left behind to garrison it. Bitter cold was as bad an enemy as hunger. Many men collapsed on the march, freezing to death where they fell. Constantly harried by the Russians, the French marched towards Borisov, beyond the Berezina river.

Between November 25 and 29, they crossed the river on ramshackle pontoon bridges, under heavy fire from Russian artillery. One of the bridges collapsed, and some 50000 men died – drowned, crushed in the panic or slaughtered by the Russians. The survivors straggled on until they reached the river Niemen once more; but by then a mere 30000 walking skeletons were all that remained of the once proud and glorious Grand Army.

FIERY ADVANCE *Seated on a white horse, Napoleon urges his battle-weary troops on towards Moscow. Behind them, the old quarter of Smolensk – which the advancing French had occupied and then set alight – illuminates the night sky. Meanwhile, camp followers bid tearful goodbyes to their menfolk; but some of the army's horses are too ill or too exhausted to travel any farther.*

SHADOWY RETREAT *Morose and bitter, Napoleon quits his temporary quarters in the Kremlin as fire threatens to engulf it. Soldiers of the Imperial Guard cluster around the emperor, muskets at the ready. They keep a close look-out for any Russian snipers or incendiaries who may be lurking in the shadows.*

Squaring up for battle

HOW THE DUKE OF WELLINGTON'S INFANTRY FOUGHT AT WATERLOO

MOMENT OF VICTORY *Members of Wellington's staff triumphantly indicate the disorganised French retreat. Napoleon's bruising day-long onslaught has been repulsed.*

THE FRENCH CAVALRYMEN SPURRED THEIR HORSES' flanks, forcing the frightened animals forward into the fire of the British muskets. Ahead of them through the smoke the riders could see solid blocks of red – each of which was a battalion of British infantry formed up in a square.

The front rank of each battalion was kneeling, its bayonet-topped muskets held with butts resting on the ground and at an angle of 45 degrees to deter the French from riding their horses into the square. The ranks behind were standing and firing in turn at the cavalry. Within the square – which usually measured around 150 ft (45 m) on each side – were the battalion commander and adjutant on their horses, and the fifes and drums that accompanied the soldiers into every battle. The formation also gave refuge to the wounded and to horse artillerymen, who had retreated from the French cavalry charge.

The return of Napoleon

The British redcoats were facing up to Napoleon Bonaparte's army near the village of Waterloo in Belgium on June 18, 1815. Earlier in the summer, Napoleon – who had been defeated by the British, Russian, Austrian and Prussian allies in April 1814 – had defiantly returned to France from exile on the

Mediterranean island of Elba. The French rallied to him and he marched against two allied armies – a combined force of British, Dutch and Germans commanded by the Duke of Wellington, and the Prussian army under Field Marshal Gebhard Blücher. Napoleon defeated the Prussians near the village of Ligny on June 16, and two days later turned on the Duke of Wellington's army at Waterloo.

Facing the French onslaught

Napoleon followed an initial artillery bombardment and a massed infantry attack with a cavalry offensive in mid afternoon. Wellington ordered the infantry battalions, who were formed in a line two ranks deep, to regroup in squares. Each 550-man battalion followed a well-practised drill. First the men re-formed in another line four ranks deep, and then, while three companies – around 150 men – stood firm to form the front of the square, the remaining seven companies fell back to form the other three sides, standing four ranks deep. The squares formed a staggered defensive line, so that the attacking cavalry found themselves under fire from the flanks as well as the front. The French horsemen suffered heavy casualties and were forced to withdraw.

Prussians to the rescue

Fearing another cavalry attack, Wellington ordered the squares to stand firm. The French cannon cut swathes through the allied troops, but at this crucial point Blücher's army, which had regrouped after its defeat at Ligny, came to the rescue. Napoleon was forced to throw in his final reserve, the Imperial Guard – the army's finest infantry. However, volleys of musketry depleted their leading ranks and they were hit by the British in the flank. The French fled.

The allied victory was a close thing, and hung on the resilience of Wellington's infantry squares – and Blücher's arrival. Wellington commented: 'In all my life I have not experienced such anxiety, for I must confess that I have never been so close to defeat.'

IRON DUKE *The Duke of Wellington – whose prominent nose inspired his nickname of 'Old Nosey' – went on to become Britain's Prime Minister.*

IMMOVABLE OBJECT *A well-drilled British infantry square stands firm against the attacking French cavalry. The horsemen are an easy target for the trusty 'Brown Bess' muskets of the British redcoats. Each soldier fires at the rate of just three shots a minute, but the three ranks fire in turn – creating a terrifying hail of balls. If any enemy cavalry do get close, they can be quickly repulsed with thrusts of the bayonet by members of the kneeling front rank. Within the square, guarded by an escort, flutter the flags of the two battalion colours, King's and Regimental – the standards that represent the battalion's honour and which the soldiers would defend to the death.*

DEFEATED EMPEROR *Napoleon's hopes of winning back power were shattered at Waterloo. He spent the rest of his life in exile on St Helena.*

Clash of the iron ships

HOW THE US CIVIL WAR BLOCKADE BROUGHT ARMOURED SHIPS INTO ACTION

WHOOPING AND YELLING ENCOURAGEMENT, SPECTATORS crowded along the banks of the Elizabeth River and onto the decks of moored ships to witness the strange encounter in the open waters of Hampton Roads, Virginia. Two heavily armoured naval vessels, or 'ironclads', manoeuvred slowly and clumsily in the still waters. The *Monitor* was fighting for the northern states of the Union in the US Civil War, and the *Virginia* for the breakaway southern states of the Confederacy. The date was March 9, 1862; on the previous day the *Virginia* had triumphed over three ships of the Union's wooden fleet, which was blockading the Confederate states' seaboard.

A stranglehold on trade

President Abraham Lincoln had ordered the naval blockade in April 1861, just eight days into the war. His intention was to squeeze the Confederate war effort dry by stopping both the lucrative export to Europe of cotton – the staple on which the southern states' prosperity had been built – and the import of weapons and war material.

The blockade was not completely effective. Confederate 'blockade-runners', fast vessels commanded by experienced skippers, carried cargoes of cotton to Nassau, Bermuda and Havana. Here they met merchants from Europe and exchanged the cotton for the supplies that the southern states so desperately needed. Privateers and later commerce raiders did their best to harry northern maritime trade, forcing the Union navy to divert some of its best ships to hunt these vessels down.

Nevertheless, by the spring of 1862 the South's economic isolation was beginning to bite. The existing Confederate fleet was not strong enough to force the Union navy to lift the blockade, but the Confederates' new weapon – the ironclad *Virginia* – brought them new hope. It was built from a wooden steam frigate, the *Merrimac*, which Union sailors had scuttled in the navy yard at Norfolk, Virginia, at the beginning of the Civil War; the Confederates salvaged the ship, tore away the upper hull and replaced it with iron armour. After the ironclad's victory over the Union ships *Cumberland*, *Congress* and *Minnesota* on March 8, Confederates believed that they had a weapon capable of smashing the blockade.

However, the Union navy had an answer to hand in the *Monitor*. This custom-built vessel was shaped like a barge, a completely revolutionary design for a warship, with a low armoured hull topped with a gun turret which contained two 11 in (28 cm) guns. The armour on its sides was 5 in (13 cm) thick.

OUTGUNNED *Union sailors look on in horror from their lifeboat as the ungainly Confederate ironclad* Virginia *wreaks havoc. The 24-gun Union sloop* Cumberland *sank after being rammed, and its sister ship, the 50-gun frigate* Congress, *went up in flames with the loss of 120 crew. The* Virginia's *victory caused panic among Union supporters.*

HEAVYWEIGHT CONTEST *The* Monitor *and the* Virginia *slug it out at such close quarters that at times they are actually touching. The armour of both vessels proves strong enough to withstand the barrage of missiles for more than 4 hours.*

Slugging it out

The two ships fought for hours. Eventually the *Virginia* hit the *Monitor*'s pilot house, and the *Monitor* retired into shallow water; the *Virginia*, however, had to withdraw with a leaking bow. The Union blockade was intact, but the *Virginia*'s success on the previous day transformed naval history. Navies around the world began to invest in heavily armoured metal warships. The days of large wooden fleets were numbered.

TEAM WORK *Covered with dirt and surrounded by smoke, a Union gun crew fires one of the two guns in the* Monitor's *steam-driven revolving turret.*

Disaster for the redcoats on an African hillside

HOW ZULU SPEARS PROVED MIGHTIER THAN BRITISH GUNS

WARRIOR CHIEF *Cetshwayo went to war to drive back the British invaders. Exiled from his homeland, he visited England and met Queen Victoria, who commissioned this portrait of the Zulu leader.*

TWENTY THOUSAND ZULU WARRIORS SAT QUIETLY ON the valley floor, about 5 miles (8 km) from the rock outcrop of Isandlwana, where soldiers of the British army were encamped. The bare-chested Zulus were ready to fight to the death to defend their homeland. It was the morning of January 22, 1879, and the British had invaded Zululand 11 days earlier.

Although they were so close to their enemy, the Zulus had no intention of attacking, because the phase of the moon did not augur well. But when a British mounted patrol rode up to the lip of the valley, saw the Zulu encampment and fired a few shots, the African warriors knew they had to respond immediately. They surged forward, carrying spears and shields, at a fast trot. Coming on the camp they felt victory was certain, for the British had not built defensive positions and were vastly outnumbered.

Ultimatum to the king

The invasion of Zululand was unprovoked. In 1877 the British had annexed Transvaal – a republic established by the Boers, descendants of the Dutch settlers in southern Africa – which lay to the north-west. They inherited a long-running border dispute with the Zulus, and Sir Henry Bartle Frere, British High Commissioner for southern Africa, determined to resolve it by overrunning the Zulu homeland. The British issued Cetshwayo, the Zulus' warrior-king, with an ultimatum they knew he could not accept – disbandment of his army, or war.

Cetshwayo did not want to fight, but he had no choice. He mobilised 30 000 highly trained warriors. The British army invaded in three widely dispersed columns. The central column, of around 1750 British and 2850 native troops, set up camp below the spur of Isandlwana on January 20, and in spite of advice from Boers well versed in the Zulu way of fighting, made no effort to fortify the position. Early on January 22, Lord Chelmsford, the British commander, led a large contingent away to reinforce another British camp – unaware that part of the Zulu army was already in place close by. At the time of the attack on Isandlwana, only about 1700 troops were defending the position.

No escape from the assegai

Some Zulu warriors carried the traditional *assegai*, a short stabbing spear with an 18 in (45 cm) blade and 30 in (75 cm) shaft; others had muskets, rifles, wooden clubs or axes. All wore a stretched cowhide shield on one arm. They were supremely fit, capable of covering 50 miles (80 km) at a fast trot in a single day, and advanced at speed on several fronts, almost entirely surrounding the British.

The redcoats drew up in defensive lines in front of their camp, but were overwhelmed on the right flank, where they quickly ran out of ammunition. They began to retreat. Some tried to form defensive squares, but they were eventually beaten back. The Zulus followed them into the camp, fighting hand to hand. Only a few of the British soldiers escaped.

The victory was overwhelming, but it was the Zulus' only large-scale triumph of the war. They suffered heavy losses at Khambula and Gingindlovu, overcome by British rifles and cannon – and, dispirited and disheartened, they were decisively defeated at Ulundi six months later, on July 4.

DEFIANT TO THE END *As the attacking Zulus overrun the British position, the redcoats make a desperate last stand. This fanciful view of the battle, painted by the English artist Charles Fripp, places the fighting some way from Isandlwana, but in reality these hand-to-hand struggles took place in the camp at the foot of the rock.*

An army without drills or roll calls

HOW THE WORLD'S FIRST COMMANDO FORCE OUTSHOT AND HUMILIATED THE BRITISH ARMY

THE BOER GUERRILLAS CROUCHED AMONG THE ROCKS, their rifles at the ready, watching a cloud of dust approaching over the plain below. Behind them, hidden from view by an outcrop, were mounted men dressed in broad-rimmed hats and drab civilian coats. Through the dust the guerrillas could make out a British supply column: a convoy of ox-drawn carts with a small mounted escort. The Boers behind the rocks opened fire and their mounted comrades galloped round the outcrop and headed for the carts. The outriding British soldiers fell, shot dead or wounded, and their colleagues, seeing that they were at the mercy of the sharpshooters, surrendered. The ambush was simple but effective. Another British supply column would fail to reach its destination.

Independent farmers

The Boers were descended from South Africa's original 17th-century Dutch settlers, who made their living by farming – 'boer' is Dutch for farmer. They had an uneasy relationship with the British, who had annexed the Boers' land early in the 19th century, forcing the settlers northwards. British attempts to force the independent Boer Orange Free State and Transvaal to join the British Empire sparked conflict. The Boers won a short war in 1881, but in October 1899 fighting broke out once more.

The Boers did not appear well organised for war. Their only regular troops were the few hundred soldiers of the Transvaal State Artillery, and the bulk of their forces consisted of volunteers, formed in *kommandos* – mobile groups of armed horsemen. But these units proved highly effective against the slow-moving British army. Commandos could vary in size from 300 to more than 3000 men, and were raised locally. Discipline, as conventional armies knew it, was virtually unknown among these troops: there was no drill, no saluting and no roll calls.

There were pitched battles early in the war, but increasingly the Boers fought only when they could choose the ground, and harried the British from a distance. The invading army brought in massive reinforcements, but it was not until May 1902 that the Boers surrendered – overcome only by the most ruthless tactics. Vast fences of barbed wire were erected, and mounted British columns moved across the land, burning Boer farms and driving the Boers into a smaller and smaller area. Huge concentration camps were built to contain Boer men, women and children. Conditions in them were appalling.

The dogged Boers lost the war, but won the lasting respect of the British Army. When in June 1940 the British prime minister, Winston Churchill, ordered the creation of bands of troops to raid German-occupied Europe, they were dubbed commandos.

REINFORCEMENTS *British troops leave Southampton for South Africa. The Boer commandos were outnumbered seven to one by the end of 1900.*

AMBUSH *Boers derail and overcome a military supply train. Such shock raids wreaked havoc with British communication lines.*

SHARPSHOOTERS *Boer marksmen regularly outgunned the British.*

COMMANDO *The Boer soldiers were superb riders. They carried German Mauser rifles, and ammunition bandoliers across their chests.*

COMMANDER *Boer general Christiaan de Wet was an inspirational leader. He kept his guerrillas constantly on the move.*

Digging in for battle

HOW TROOPS LIVED, FOUGHT AND DIED IN THE TRENCHES OF THE FIRST WORLD WAR

STANDING ON A STEP CUT INTO THE FRONT WALL OF THE trench, a British sentry watched as a flare lit up the night sky. It was followed by others in quick succession – and then by bursts of machine-gun fire, the crack of rifles and the thump of artillery. The sentry could only guess what was going on. Either it was a raid on a German trench or some unfortunate patrol had been spotted on reconnaissance in no-man's-land, the muddy and desolate area between the front lines of the opposing armies.

DELOUSING *A soldier patiently uses a candle flame to force body lice from the seams of his uniform.*

STORMING THE DEFENCES *While their comrades fight in the front-line trench below, British troops throw a portable bridge over the top and rush across it to 'take out' the second line of trenches beyond.*

SURRENDER *A cornered German is forced to crawl out of his last refuge – a muddy hole.*

RESCUING THE WOUNDED *A stretcher-bearer helps carry a wounded soldier away to a first-aid post in the support trenches.*

LIFE SAVER *Early in the war gas masks were primitive hoods. New designs arrived in 1915.*

PIGEON POST *A Canadian private sends a message winging back to the support trenches.*

SNIPING *Using a home-made periscope fixed onto his rifle, a Scottish Highlander fires without exposing himself to enemy bullets.*

AT EASE *A batman pours tea for a group of officers in a dugout. Ragtime tunes or sentimental songs floating from the gramophone remind them of life far from the boom of the front-line guns.*

HAND WEAPONS *Soldiers used spiked clubs in close combat.*

Bogged down in stalemate

When the First World War broke out in August 1914, the combatants planned a series of pitched battles bringing a swift victory. They expected the fighting to be over by Christmas. But none of the generals fully appreciated that technological advances during the past 50 years had made defence stronger than attack. New rapid-fire weapons such as the machine gun, the magazine rifle – fitted with a 'clip' carrying several bullets – and quick-firing artillery had been developed. They meant that concealed soldiers defending a fortified position could mow down attackers in a hail of bullets and shells.

During the autumn of 1914 the war was fought on two fronts – in France and Belgium in the west, and in Poland and western Russia in the east. By the end of the year the Germans had decisively driven back Russian advances on the eastern front, but in the west, where neither the French and British nor the Germans had been able to break through, there was stalemate. Far from going home, at Christmas 1914 the soldiers on the western front found themselves facing one another in two vast lines of trenches stretching around 450 miles (725 km) from the Swiss border to the Channel coast.

Trenches were not new to the First World War. Soldiers had dug themselves into field fortifications as long ago as the Peninsular War, fought in 1808-14 by Britain, Portugal and Spain against France. In 1914 on the western front, the trenches grew out of makeshift refuges hurriedly dug in the early weeks of the war. The Germans and the British-French allies

SUDDEN ATTACK *After weeks of eyeing the enemy warily across no-man's-land, the British infantry launch a devastating onslaught on the German trenches. The attackers cut their way through the defensive barbed wire lining no-man's-land – but one man has been snared.*

DEATH'S CORRIDOR *Men scrabble for their lives in the narrow trench.*

OUTDOOR LIFE *In the front line, the lower ranks live and sleep in holes cut into the muddy trench walls. A sniper and an officer with binoculars watch for enemy movement.*

LOOKING OUT *Protected from snipers' bullets by the sandbags piled on the forward edge of the allied trench, an Australian soldier scans the German lines and no-man's-land through a periscope. He can snatch up his rifle quickly if he spots trouble.*

MINING *Sappers tunnel under no man's land, laying explosives close to the enemy trenches. Their efforts could create havoc among the enemy, but they were sometimes blown up by their own mines.*

tried to outflank each other in a desperate 'race to the sea' across France and Belgium, but when they failed both sides determined to defend the ground they held. The earthworks were gradually expanded.

The front lines were usually around 750 ft (230 m) apart, although they were sometimes much closer, and were linked by 'communication trenches' to safe areas in the rear. Long tangles of barbed wire, sometimes 150 ft (45 m) thick, defined the borders of no-man's-land. The trenches were dug in a zigzag pattern so that defenders would be less vulnerable to being shot straight along the trench by an intruder. They were wide enough to allow two men to pass, and around 4 ft (1.2 m) deep, with a thick parapet of earth or turf and sandbags built up on the front edge. This protected the men from sniper fire when they stood up. To return fire they stood on the 'firestep' cut into the front wall of the trench.

The world of the trenches

Trench layout varied from place to place and army to army. In British areas, the communication trenches usually ran back from the front line to support and reserve trenches containing stores, field kitchens, latrines, mortar positions and underground dugouts. Through the communication trenches, reserve troops brought forward food and water, ammunition and materials to repair the parapets. Telephone cables ran through them back to battalion headquarters and to the artillery gunners far behind the front line.

The British and Germans defended the whole of their sections of the front, but the French divided their sections into active areas – heavily armed positions – and passive areas, defended by wire and guns firing along the line. German trenches had deep, shell-proof dugouts and heavily fortified machine-gun bunkers, and were generally constructed much better than the Allies', because they were defending a fixed line; in contrast, the Allies saw their trenches as temporary fortifications from which they would emerge to push the Germans back.

The length of service in the front line varied greatly, but most troops averaged seven to ten days, alternating front-line service with spells as reserves. Periodically they would be withdrawn well behind the lines for rest and retraining.

Rats, cold, lice, dirt and gas

Officers had a degree of comfort in their dugouts, but life for the lower ranks was much tougher. When the weather was bad all soldiers suffered terribly from the cold, wet and mud. Standing for long periods in water gave them 'trench foot', which turned the feet green and swollen. Rats were a constant menace. The men lived in fear of sudden death by a sniper's bullet or a shell. Poison gas – first used on the western front by the Germans at the Second Battle of Ypres in April 1915 – could blind and kill. Gas alerts, when the men scrabbled for their protective masks, were common – and often proved false alarms.

'I shall not easily forget those long winter nights in the front line: 16 hours of blackness broken by gun flashes and punctuated by the scream of a shell or the sudden heart-stopping rattle of a machine gun.'
FREDERICK NOAKES, BRITISH PRIVATE

Lice infested the soldiers' clothes and spread the dreaded and debilitating trench fever. It was almost impossible to keep clean. Often the only chance to wash in the front line was using a 2 gallon (9 litre) container of water carried up from the reserve trenches – to be shared among as many as 40 men. The British soldier's rations were canned stew or 'bully beef' (corned beef), bacon, bread, biscuits, jam and tea. The Germans were given sausages and black bread, while the French survived on stew and a daily wine ration of around a pint, or half a litre. The British had a raw service rum to take the edge off their suffering, but it was doled out in tiny amounts. Rum was always issued before an attack.

Large-scale attacks were rare and in most sectors there were long periods of quiet. At dawn and dusk – the most likely times for enemy activity – the parapet was manned. The sentries stood in groups of five about every 600 ft (180 m). The nights were used for repairs and reconnaissance. Small patrols of two or three men would spy on the opposing front line. Occasionally larger groups were sent on 'trench raids' to take a prisoner for interrogation.

WATCH TOWER *A tree stump built up with camouflaged corrugated iron serves as a look-out. By telephone, the observer directs artillery fire onto enemy positions.*

FRONT-LINE AMBULANCE *A bullet-proof metal stretcher cover protects wounded soldiers from snipers' bullets.*

HELPING HAND *Temporarily blinded by a German mustard gas attack, two South African privates (below) rely on a friendly guide.*

IRRESISTIBLE FORCE *The British developed the 'tank' – an armoured vehicle with tracks, capable of ploughing across uneven no-man's-land – in 1915-16. It was first used in combat on September 15, 1916, during the Battle of the Somme, and first proved its worth on November 20, 1917, when 474 tanks broke the German line at Cambrai.*

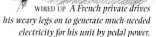

WIRED UP *A French private drives his weary legs on to generate much-needed electricity for his unit by pedal power.*

Over the top – after the shells

In theory, a full-scale infantry attack was preceded by a prolonged bombardment of the enemy trenches. This was supposed to punch holes in the defensive wire and put selected positions out of action. After assembling in no-man's-land under a smokescreen, the soldiers would advance towards the enemy trenches at walking pace while shells from their own artillery gave them cover. The reality was sometimes very different. At the Somme on July 1, 1916, waves of attacking infantry were mown down because a week-long British artillery attack had failed to destroy the well-fortified German defences.

Trench warfare dominated the war in the west from 1915-18. Efforts to break the deadlock cost both sides horrific casualties at Verdun and on the Somme. Politicians had claimed that it would be 'the war to end all wars'. The soldiers who survived prayed that their leaders were right, but vowed that war must never be fought in this way again.

REST AND RECUPERATION *Far back from the front, a British army ambulance delivers wounded soldiers to a peaceful military hospital set up in a French château.*

SPY IN THE SKY *Observation balloons float serenely high over the front line, but are the target for enemy aircraft. One observer parachutes to safety from a damaged balloon.*

EMERGENCY CARE
Grim-faced orderlies help load a badly injured British infantryman onto a stretcher. Wounded men were patched up quickly in the support trenches before being taken for further treatment to a field hospital situated well behind the lines.

HORSE DOCTOR *A member of the Blue Cross calms an injured horse. All the armies used horses to haul artillery and supply wagons.*

LOVE FROM HOME *Eagerly awaited parcels and letters bring extra food, knitted socks and news of loved ones.*

WIRING UP *A British team construct a barrier of tangled barbed wire across the edge of no-man's-land.*

FRONT-LINE FOOD
Rations parties bring food forward from the support trenches each day. They carry stew and porridge in screw-top containers on their backs.

PORK & BEANS

OXFORD'S PUDDINGS

CHOICE BLENDED TEA

Warriors of the open skies

HOW ARMED AIRCRAFT TOOK WAR TO THE AIR

THE GERMAN PILOT HEARD THE WARNING RATTLE OF gunshots, indicating that he was being attacked from behind. Flying about 13 000 ft (4000 m) over the front lines, he pulled his sleek Fokker E III monoplane sharply up and looped the loop in the clear sky above the British fighter, a Vickers FB 5 'Gunbus'. Then he swooped down and attacked the FB 5 from the rear before climbing steeply in front of the British biplane, turning sharply in the sky above it and coming screaming down to attack it once more. The British pilot and gunner slumped forward, and their stricken aircraft fell away into a steep dive. The German pilot flew on, in search of another victim. He was Max Immelmann, dubbed 'the Eagle of Lille' after the city in which he was based, and his unstoppable attacking manoeuvre – the 'Immelmann turn' – was the talk of German airmen in late 1915.

Arming the scouts

When Europe went to war in 1914, the aeroplane was just 11 years old. At first, unarmed planes were used for reconnaissance over the front line in France. But generals soon realised that it was essential to prevent enemy scouts from flying over their lines, and began to arm the aircraft. These new 'fighting scouts' carried rifles, revolvers and even metal darts. Rapid-fire machine guns would be far more effective, but there were problems in mounting them on the aircraft. In a conventional front-engined plane, a gun attached to the nose would simply shoot the propeller off. Guns were mounted on the side of the aircraft fuselage, but here they were difficult to fire with any accuracy. Rear-engined 'pusher' aircraft could carry a gun on the nose, but were slower.

FIRING FROM THE TAIL *With gunners blazing from the rear cockpits, British de Havilland DH 4 bombers attempt to shake off a chasing pack of German fighters. Heavily laden bombers were too big and slow to turn and fight successfully, and on long-range missions they could not spare the time or fuel to do so, but flying close together enabled them to concentrate their fire on the attackers. Bombers were often protected by a fighter escort.*

Airborne terror

By October 1915, German fighters were winning so many aerial battles that shocked Allied pilots were talking of the 'Fokker scourge'. The new front-engined Fokker E III was fitted with a nose-mounted machine gun that fired through the propeller. A device synchronised the firing of the machine gun with the movement of the propeller blades so that the gun did not fire when the blades were in front of it.

Skill at aerobatics was vital for the fighter pilot. Each air force had its aces – those who shot down five or more enemy aircraft. In France and Germany these men were treated as heroes, but the British authorities stressed the achievements of the squadron and did not publicise those of individual pilots. The most successful ace of the war was the German Manfred von Richthofen who had 80 air victories, and the top Allied ace was the Frenchman René Fonck, with 75.

The pilots' exploits were often romanticised, but the reality was that they lived every day under threat of death. The strain took a terrible toll and some men had to be retired from combat. Once an aircraft was disabled in battle there were no second chances: neither side issued parachutes to aircraft crews until the Germans began to do so in 1918.

ATTACK FROM BEHIND *German Albatros C-type fighting scouts swoop onto a pack of British Martinsyde fighters high over the winding river Somme in northern France. In 1914-15 most fighters flew their missions alone, and fought one-to-one 'dogfights'. From 1916 groups of six or more began to fly in formation, giving pilots better all-round protection. When enemy formations encountered each other, a large-scale aerial battle followed. Surprise was the key element in attack: the best tactics were to come in from behind, above or below, or – best of all – out of the blazing sun. Survival depended on the ability to outwit – or outfly – the enemy. The pilots depended for their lives on their aircraft, but although improvements made these faster and more manoeuvrable during the war, they remained fragile constructions of wood and canvas that could break up in midair if handled too violently.*

A 'BARON' – AND A CIRCUS THAT FLEW

MANFRED VON RICHTHOFEN KEPT A COLLECTION OF SILVER cups, each inscribed with details of one of his victims. In June 1917, Germany's leading air ace was given command of four squadrons in a new elite grouping, Jagdeschwader Nr 1 ('1st Fighter Wing'), made up of experienced pilots and deployed to critical sectors of the Western Front to ensure local air superiority. Von Richthofen usually chose to fly an all-red Albatros or Fokker fighter and this quirk led the Allies to nickname him the 'Red Baron' – while his squadrons, because of the bright colours they painted their aircraft and the fact they constantly moved up and down the front line, were dubbed the 'Flying Circus'.

Stealthy attack

These men, flying sleek Albatros D-type fighters fitted with twin machine guns, dominated the skies wherever they were stationed. Von Richthofen passed on to them the tactics he had earlier learned from his mentor, Oswald Boelcke. The Flying Circus pilots learned to attack stealthily like hunters and to go for older and slower machines, so that they could make the most of the superior power of their D-types.

Von Richthofen did not survive the war. On April 21, 1918, he was flying at low altitude over the Somme in pursuit of a British fighter when he was attacked from behind by Canadian pilot Captain Roy Brown in a Sopwith Camel and from below by Australian guns. The German ace was shot in the chest and his Fokker crashed. Brown was credited with the 'kill', but it was never proved who shot down the Red Baron.

SCOURGE OF THE BRITISH *Von Richthofen won the ungrudging respect of Allied pilots.*

THE EAGLE FALLS *Australian gunners carry the Red Baron's corpse away from the wreck of his famous Fokker triplane.*

THE EAGLE RISES *A German magazine pays tribute to the country's top air ace.*

The ghostly bombers

HOW GERMAN ZEPPELINS RAIDED ENGLAND IN THE FIRST WORLD WAR

THE MASSIVE PALE BLUE OBJECT SHAPED LIKE A GIANT cigar was cruising through the cold, dark night over the North Sea. This ghostly apparition was German naval Zeppelin L31. With its sister airships L32 and L33, it was heading for London on a bombing mission on the night of September 23, 1916.

The Zeppelin consisted of a lightweight metal framework of crisscrossing aluminium girders 650 ft (198 m) long, 91 ft (28 m) high and 78 ft 6 in (24 m) in diameter, covered with a cotton skin. Laced to the girders within the hull were 19 hydrogen-filled bags, which made the airship lighter than air. It was flying at just over 60 mph (95 km/h) and at an altitude of 12 000 ft (3650 m). On board were 20 crew – including mechanics, gunners, two pilots and a navigator – commanded by Kapitänleutnant Heinrich Mathy, an experienced airship captain.

The Zeppelins were following a detailed flight plan and each had a precise target to hit. But they would have to contend with British searchlights and attacks by ground artillery and fighters – and might be forced to drop their bombs haphazardly and beat a retreat.

Giants of the air
Zeppelins took their name from their creator, Count Ferdinand von Zeppelin, a former general in the German army. He did not invent the airship, but pioneered a new construction. Previously, airships had been 'non-rigid' – little more than a gasbag made of rubberised cotton or linen – or 'semirigid', with a wooden or metal keel attached to the bottom of the envelope. Zeppelin developed a fully rigid airship, the first of which, LZ1, made its maiden flight on July 2, 1900. It had a metal or wooden frame surrounding 17 hydrogen bags, holding a total of 400 000 cubic feet (11 300 m³) of gas. The German government was impressed with the LZ1 and by the start of the First World War the country's army and navy had a fleet of 13 airships, including ten Zeppelins.

The Germans launched the first bombing missions on England in January 1915. Initially these were against east coast towns: the Kaiser, unwilling to bomb the royal family in Buckingham Palace, forbade attacks on London. But in the spring he relented – and the Zeppelins hit the British capital. To combat the airships, the British strengthened London's anti-aircraft defences and brought fighter aircraft back from the western front. The Germans tried to raise the airships above the range of guns and fighters, introducing the L30 series – of which L31, L32 and L33 were examples – in 1916. These had a ceiling of 13 000 ft (almost 4000 m), but still found it difficult to evade the British defences.

Caught in the searchlights' beam
Only one of the three German airships reached home in the raid on September 23. After bombing Essex, L32 was shot down by Royal Flying Corps pilot Lieutenant Frederick Sowrey, with the loss of its entire crew. L33 bombed dockland oil depots in the East End of London, but was caught in the searchlights of the artillery and disabled by shells; it crash-landed in Essex, and the crew were rounded up by a local policeman. Mathy's L31 was the lucky one. It bombed housing in south London, then confused and silenced the ground artillery by firing flares that mimicked those used by British fighters. It escaped, partly hidden by a ground mist, back to Germany.

A ventilation shaft allows air into the heart of the ship. The aluminium framework is strengthened with steel bracing wires and covered with a cotton skin. A walkway runs along the inside bottom of the frame – crewmen use it to move from one gondola to another. A long rope ladder allows access to two machine gun posts at the top of the Zeppelin. There is plenty of room inside the ship, and off-duty crewmen relax in hammocks slung from the frame. At altitudes of 13 000 ft (4000 m), the cold is punishing; the crewmen wear two sets of underwear, fur-lined jackets and hats, strong leather boots and gauntlets

A life raft offers an escape from crashes at sea

Bombs hang head down, ready for despatch through hand-operated bomb doors

The wing gondolas each house a single machine gunner as well as one of the massive 1440 hp Maybach engines. There are two other identical engines, one at the back of the control gondola and one in the rear gondola

There are two sets of bomb doors, one here and one on the other side of the twin wing gondolas

Crew in the rear gondola take orders from the captain through a speaking tube running inside the ship

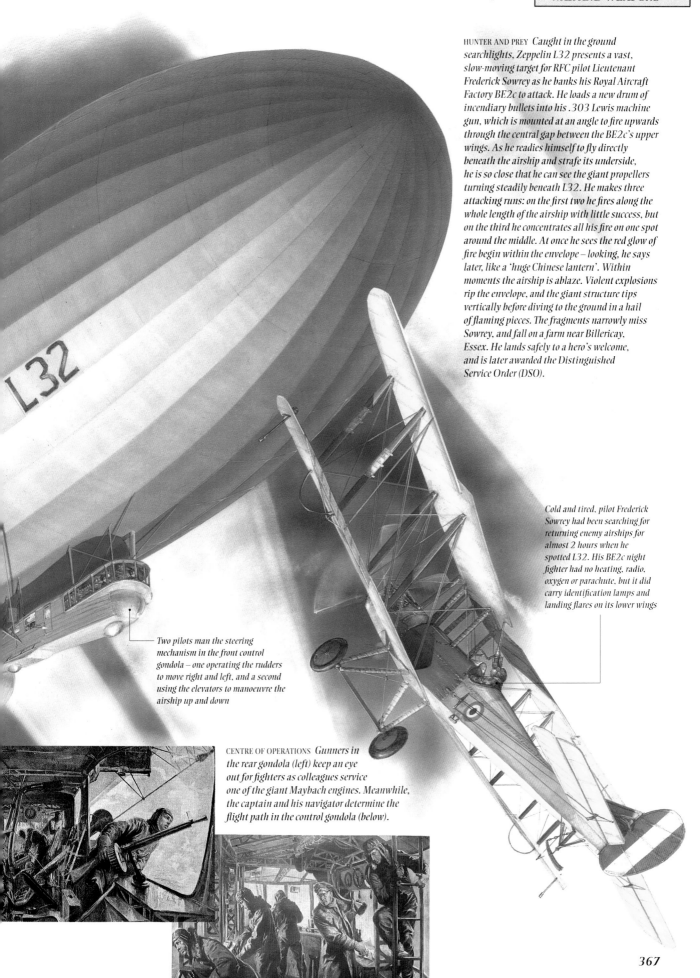

HUNTER AND PREY *Caught in the ground searchlights, Zeppelin L32 presents a vast, slow-moving target for RFC pilot Lieutenant Frederick Sowrey as he banks his Royal Aircraft Factory BE2c to attack. He loads a new drum of incendiary bullets into his .303 Lewis machine gun, which is mounted at an angle to fire upwards through the central gap between the BE2c's upper wings. As he readies himself to fly directly beneath the airship and strafe its underside, he is so close that he can see the giant propellers turning steadily beneath L32. He makes three attacking runs: on the first two he fires along the whole length of the airship with little success, but on the third he concentrates all his fire on one spot around the middle. At once he sees the red glow of fire begin within the envelope – looking, he says later, like a 'huge Chinese lantern'. Within moments the airship is ablaze. Violent explosions rip the envelope, and the giant structure tips vertically before diving to the ground in a hail of flaming pieces. The fragments narrowly miss Sowrey, and fall on a farm near Billericay, Essex. He lands safely to a hero's welcome, and is later awarded the Distinguished Service Order (DSO).*

Cold and tired, pilot Frederick Sowrey had been searching for returning enemy airships for almost 2 hours when he spotted L32. His BE2c night fighter had no heating, radio, oxygen or parachute, but it did carry identification lamps and landing flares on its lower wings

Two pilots man the steering mechanism in the front control gondola – one operating the rudders to move right and left, and a second using the elevators to manoeuvre the airship up and down

CENTRE OF OPERATIONS *Gunners in the rear gondola (left) keep an eye out for fighters as colleagues service one of the giant Maybach engines. Meanwhile, the captain and his navigator determine the flight path in the control gondola (below).*

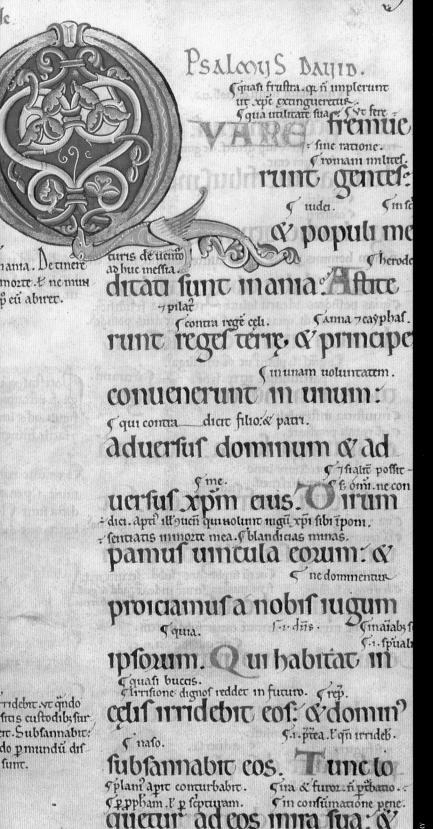

Getting the Message

THE URGE TO SHARE
OUR IDEAS HAS TAKEN
MANY FORMS OVER THE
CENTURIES. BY WRITING
WORDS AND THOUGHTS
DOWN, WE CAN SEND
THEM ACROSS SPACE OR
PRESERVE THEM FOR
POSTERITY, WHILE
RADIO AND TELEVISION
HAVE USHERED IN THE
ERA OF INSTANT MASS
COMMUNICATION

Letters drawn on clay and carved in stone

HOW ANCIENT CIVILISATIONS DEVISED WRITING SYSTEMS

RECORDS MADE OF KNOTTED STRING

HIGH IN THE ANDES MOUNTAINS, AN OLD MAN FIDDLED WITH a bunch of coloured, knotted strings and racked his brain to interpret the meaning of the knots. The strings shook as he relayed the figures to the chief statistician. If the old man made even the slightest error of interpretation, his punishment could be death.

Message of the belt
The Inca peoples who inhabited what is now Peru from around AD 1000 used these knotted cords, which they called quipus, to record important data from each district in the Inca empire, then added these figures together to achieve an up-to-date census. The excellent memories of the *quipucamayocs* – 'keepers of the quipu' – helped them to interpret the knotted cords. The knots at the end of a string represented single units, above them tens, above these hundreds, and so on. Each differently-coloured cord represented a commodity, such as gold, maize, cattle or spears.

In North America too, native peoples used available materials to convey information. Wampum belts, so called after the polished clam shells from which they were made, were exchanged between tribes to mark agreements, to record events or to send messages.

The belts, made from shell beads strung together in rows, could be as long as 6 ft (1.8 m). Designs varied from tribe to tribe, but white belts generally expressed good tidings, such as peace and friendship, whereas black belts (made from the purple shell) meant bad news. A black wampum marked with a hatchet in red paint signified war.

AIDS TO MEMORY *Knotted strings and belts made of shells carried messages across the Americas.*

CLAY WAS MOULDED INTO THE RECORDS OF THE ancient world, as well as its bricks, tiles and pots. From before 3000 BC, the Sumerians of ancient Mesopotamia kept records by drawing pictures on soft clay tablets with a pointed stylus, then leaving the tablets to harden in the sun. During the next 200-300 years, the pictures evolved into symbols composed of triangular or wedge-shaped imprints in the clay, made using the straight-cut end of a reed stylus. This script is known as cuneiform, from *cuneus*, Latin for 'wedge'.

Building words out of syllables
Cuneiform symbols could represent abstract ideas, as well as objects. This sometimes caused confusion. For example, a circle not only represented the sun, but also the concept of heat and time associated with it. Instead of creating even more symbols, scribes began to join two or more symbols together to form more complex words. It is as if, in English, to make the word 'treason', we joined the words 'tree' and 'sun'. In this way, short words came to represent the syllables of a longer word with a different meaning, and came to have a phonetic value. Initially the system included around 1200 symbols.

The Chinese developed a different system of writing. The earliest examples, dating back to about 1700 BC, are simple pictures of everyday objects and activities. Each character represented a whole word, rather than just a sound; more than 9000 characters were in use by AD 220.

Making paper out of pulp
The art of paper making had been known in the Far East since AD 105 when a Chinese court official, Tsai Lun, soaked a mixture of rags, hemp and bark in water, then drained the pulp over a flat sieve. The fibres formed a sheet of wet paper which dried to form a writing surface. By the early 2nd century the Chinese were making paper from bamboo. They soaked the shoots, then boiled, pounded and washed them to form a pulp. They squeezed out the water in a press, and hung the sheets on walls to dry. From the 6th century the Japanese were practising a similar method using bark. In western Europe, meanwhile, scribes relied on parchment and vellum – the specially treated skins of sheep and goat. It was not until AD 751 that some Chinese soldiers revealed the secret of paper making to their Arab captors, who set them to work making paper in Samarkand, whence it spread throughout the Arab world and into Europe.

ANCIENT CRAFT *An 18th-century illustration shows Japanese craftsmen making paper by dipping shallow wooden-framed moulds into vats of crushed tree pulp, then shaking them so that the fibres mesh together.*

1 Stripping the bark

EARLY SIGNS
A Mesopotamian cuneiform tablet from c.2500 BC records the hire of donkeys. The sign for a donkey (right, top) eventually developed into a more abstract symbol (right, below).

ANCIENT NOTE *Chinese characters are still discernible on the world's oldest piece of paper, made during the Han dynasty (206 BC-AD 221).*

2 Soaking the bark in water

3 Beating the fibres to a pulp

4 Dipping mould in pulp vat

5 Hanging the paper to dry

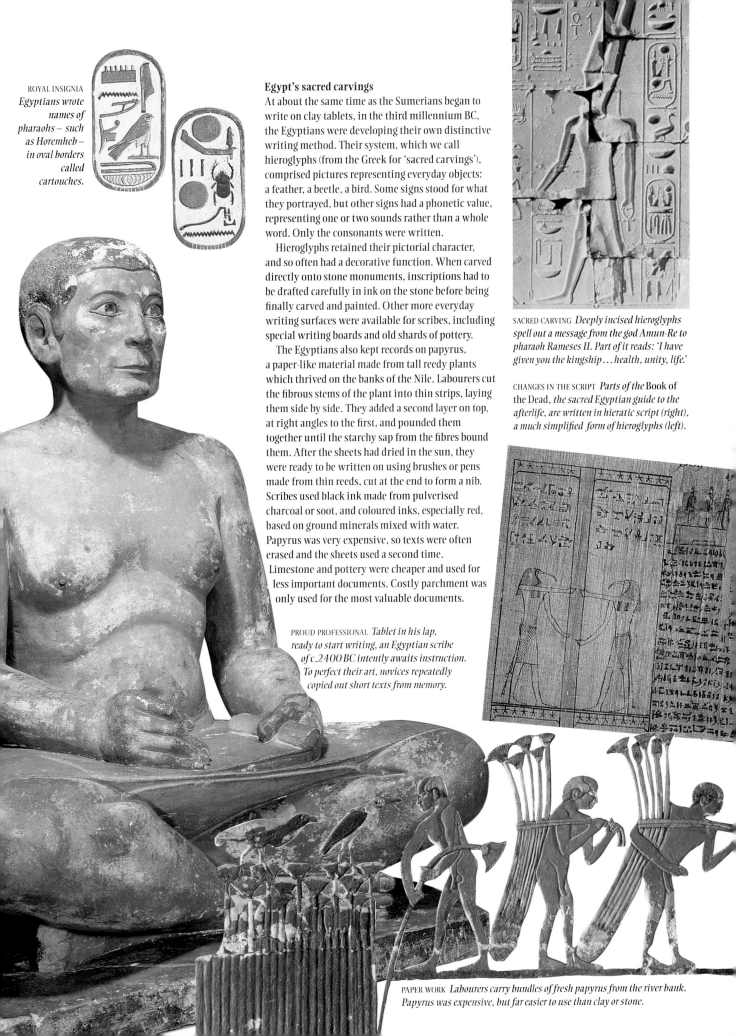

Egypt's sacred carvings

At about the same time as the Sumerians began to write on clay tablets, in the third millennium BC, the Egyptians were developing their own distinctive writing method. Their system, which we call hieroglyphs (from the Greek for 'sacred carvings'), comprised pictures representing everyday objects: a feather, a beetle, a bird. Some signs stood for what they portrayed, but other signs had a phonetic value, representing one or two sounds rather than a whole word. Only the consonants were written.

Hieroglyphs retained their pictorial character, and so often had a decorative function. When carved directly onto stone monuments, inscriptions had to be drafted carefully in ink on the stone before being finally carved and painted. Other more everyday writing surfaces were available for scribes, including special writing boards and old shards of pottery.

The Egyptians also kept records on papyrus, a paper-like material made from tall reedy plants which thrived on the banks of the Nile. Labourers cut the fibrous stems of the plant into thin strips, laying them side by side. They added a second layer on top, at right angles to the first, and pounded them together until the starchy sap from the fibres bound them. After the sheets had dried in the sun, they were ready to be written on using brushes or pens made from thin reeds, cut at the end to form a nib. Scribes used black ink made from pulverised charcoal or soot, and coloured inks, especially red, based on ground minerals mixed with water. Papyrus was very expensive, so texts were often erased and the sheets used a second time. Limestone and pottery were cheaper and used for less important documents. Costly parchment was only used for the most valuable documents.

SACRED CARVING *Deeply incised hieroglyphs spell out a message from the god Amun-Re to pharaoh Rameses II. Part of it reads: 'I have given you the kingship … health, unity, life.'*

CHANGES IN THE SCRIPT *Parts of the* Book of the Dead, *the sacred Egyptian guide to the afterlife, are written in hieratic script (right), a much simplified form of hieroglyphs (left).*

PROUD PROFESSIONAL *Tablet in his lap, ready to start writing, an Egyptian scribe of c.2400 BC intently awaits instruction. To perfect their art, novices repeatedly copied out short texts from memory.*

PAPER WORK *Labourers carry bundles of fresh papyrus from the river bank. Papyrus was expensive, but far easier to use than clay or stone.*

All the books in the world

HOW THE GREAT LIBRARY AT ALEXANDRIA WAS CREATED

'HOW MANY BOOKS NOW, DEMETRIUS?'
It was a question Egypt's great king Ptolemy I often asked and it was one Demetrius knew, whatever answer he gave, would be followed by another:
'And is that all the books in the world?'
'No, my lord.'
'But we must have them all, all!'
A former personal bodyguard and general of Alexander the Great, Ptolemy had seized power in Egypt after Alexander's death in 323 BC. Although he was one of the world's great military leaders, and ran a mighty kingdom, he also sought a different kind of power: to conquer the world of knowledge.

He set out to build a library containing everything that had ever been written. The city of Alexandria, founded by Alexander the Great in 332 BC, was the major political and commercial centre of Alexander's vast empire, which stretched from the Indian Ocean to the Mediterranean Sea. Ptolemy I was determined that it should also be the world's cultural capital.

Books taken from visiting ships

Ptolemy made a Greek scholar, Demetrius Phalereus, responsible for creating a great place of learning centred on the royal palace. Together the two men started to build a storehouse of knowledge. While Ptolemy wrote to emperors, kings and princes asking them to send their nations' books, Demetrius dispatched agents to buy books abroad. All ships sailing into the port of Alexandria were searched for books, and any found were copied by a team of scribes. The copies were returned, and the originals stayed on the shelves.

The great enterprise continued for many years. During the reign of Ptolemy II (284-246 BC) the poet Callimachus catalogued the library and recorded 90 000 original manuscripts and 400 000 copies; a further 42 800 rolls were stored in a smaller library elsewhere in the city. Ptolemy III (246-221 BC) borrowed the original manuscripts of the plays of Aeschylus, Sophocles and Euripides from Athens and returned mere copies – gladly forfeiting the money he had left as a deposit. But the library's vast collection was lost to later generations.

STORE OF KNOWLEDGE *A scholar reads aloud from one of the library's manuscripts. Such scrolls were often cumbersome and special low tables were built to support them.*

CENTRE OF LEARNING
A lecturer speaks to a group of students, while a slave holding a manuscript waits until he is needed.

FOREIGN TONGUE *A linguist dictates his translation of a Hebrew manuscript to a scribe. All the books collected for the library were laboriously translated into Greek.*

SCHOLAR KING *Ptolemy I was a writer as well as a lover of books, a skilled diplomat and a renowned soldier. At the end of his reign he wrote a history of Alexander the Great's campaigns. After his death he was worshipped as a god by his subjects.*

THE BURNING OF THE GREAT COLLECTION

HOW THE ANCIENT WORLD'S GREATEST library came to be destroyed is an event shrouded in mystery. One theory is that the library was burnt in *c.*47 BC, during the bitter struggle between Ptolemy XII, his sister Cleopatra and her ally, Julius Caesar. The Roman army set fire to Ptolemy's fleet, and the blaze spread from the harbour to the waterfront buildings and on to the library, defeating all efforts to extinguish it.

In AD 391 at least part of the library's annexe was ruined when the Roman emperor Theodosius I ordered the destruction of pagan temples and other buildings. Another story is that the main library was demolished when Arabs conquered Alexandria in AD 646. The victors, fervent Muslims, thought that the only true knowledge was contained in the Koran, and that all other books should be burned. Legend has it that the priceless manuscripts served as fuel to heat the public baths for months.

LOST CAUSE *Forced by raging flames to abandon the library's precious store of manuscripts, scholars flee for their lives into the courtyard.*

THE WORLD'S WISDOM *In one of the library's storerooms, slaves replace scrolls on the shelves while readers browse. The books are handwritten on rolls of papyrus or vellum and classified by subject. Each book – whether one, two or more rolls in length – has its own pigeonhole in the library stacks. A tag hangs from the end of each roll detailing its contents. At its largest, the library's collection numbered around 700 000 books, collected over three centuries.*

Sacred texts and aching backs

HOW MEDIEVAL MONKS COPIED MANUSCRIPTS

'A MAN WHO KNOWS NOT HOW TO WRITE MAY THINK this is no great feat', wrote Prior Petrus as he sat hunched over his angled desk in a chilly Spanish monastery in about 1100. 'But only try to do it yourself and you will learn how arduous is the writer's task. It dims your eyes, makes your back ache, and knits your chest and belly together – it is a terrible ordeal for the whole body.'

Until about 1200, many of the books made in Europe were for religious use, most of them Bibles and psalters (books of psalms). Most monasteries had a library, some containing hundreds of volumes. Each of these books had been written out by hand and most were copies, taken line by line from an existing book, often on loan from another monastery.

Copying was sacred work. The monks' aim was to preserve and transmit the holy texts: hence books were produced to the highest standards possible. It took a scribe years of training and painstaking practice to achieve the confident and elegant touch that sets the finest medieval manuscripts apart.

Parchment, pen and ink

One of the most expensive components of a book was the material from which the pages were made. Both parchment, thin sheets of sheepskin or goatskin and the finer vellum, made from calfskin, needed extensive preparation. First they were thoroughly washed in cold running water, then soaked for up to ten days in wooden or stone vats of lime solution. Any hair was scraped off the skin before it was washed again. It was then stretched on a wooden frame and the scraping repeated, and once dry, the skin was rubbed with chalk and pumice stone until smooth. For a work of 340 leaves, such as the 8th-century masterpiece of Celtic art, the *Book of Kells*, around 200 calfskins were needed.

The scribe marked up the first page of parchment with a master grid that indicated where the lines of text and margins would fall. He might copy the grid by making small score marks with a knife or awl through the first page onto a pile of pages beneath.

Next he had to prepare his pen. Although pens made from metal were available, the most common writing implement was the quill, made from a flight feather of a large bird, such as a goose. To prepare a quill, the scribe soaked it in water and then heated it to make it hard and flexible. He then cut the nib to

BUYING *A monk inspects the colour and texture of a sheet of parchment before buying it from a craftsman.*

TRIMMING *Back at his desk, the scribe picks up his knife and cuts the parchment to exactly the right size.*

> '*It crooks your back, dims your sight, twists your stomach and your sides. Three fingers write, but the whole body labours*'
> ANONYMOUS 8TH-CENTURY SCRIBE

the correct shape for the thickness of the lettering he wished to use. Ink was kept in a hollow animal horn. Black ink was made from lampblack, oak galls or the bark of trees mixed with gum. Red ink, also called 'red lead' or 'red vermilion', was made from toasted lead or mercuric sulphide respectively, and was mostly used to write some initial letters or the first lines or titles of some texts.

The book to be copied, called the exemplar, was laid open on a lectern arranged close to the scribe's desk, and copying could begin. One of the great skills of the scribe was the ability to make the text fit the line, and to make blocks of text fit the page, without obviously bunching or squeezing the letters. The best handwritten texts have a measured regularity, giving a sense of orderliness and poise.

RULING *Before he can start copying, the monk carefully lays out the page, marking the margins and lines with a stylus.*

TOOL KIT *A scribe's tools for writing might include a metal pen with a grooved tip to hold the ink, and a metal-tipped bone stylus, such as the two below, used to score the parchment.*

PREPARING PARCHMENT *A 15th-century stationer cuts parchment into oblong sheets, while his colleague erases the writing on another sheet so that it can be reused. Rolls and stacks of parchment sit on the shelves behind.*

Tricks of the trade

For most of the Middle Ages, scribes generally used the 'Gothic' script, which was designed to be compact so as to save parchment. For the same reason they used abbreviations – such as the ampersand (&) for 'and' – and joined up letters. The untrained eye often finds medieval manuscripts very hard to decipher.

If a scribe made a mistake, he would scratch away the offending letter or passage with his knife – one of the advantages of a resilient surface like parchment. A second scribe or reader checked the text against the exemplar, and further corrections were then made. Working in the library or, if fortunate, a special scriptorium (writing office) with large windows of clear glass, a skilled scribe could produce up to four sides of a page a day, depending on how complex the work was. A Bible could take a team of monks many months to produce.

Once the text was finished, the scribe passed the pages to the artists for illumination. When this was finished, the completed pages were folded and arranged in the correct sequence for binding between boards covered with leather.

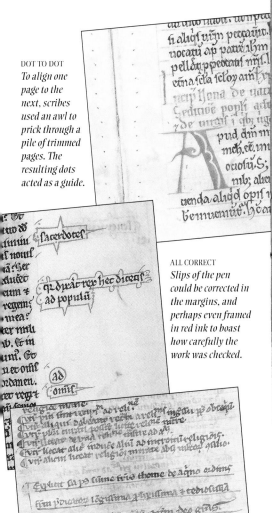

DOT TO DOT *To align one page to the next, scribes used an awl to prick through a pile of trimmed pages. The resulting dots acted as a guide.*

ALL CORRECT *Slips of the pen could be corrected in the margins, and perhaps even framed in red ink to boast how carefully the work was checked.*

POSTSCRIPT *Copyists often finished on a personal note. One exhausted monk, completing a work by Thomas Aquinas, wrote that it was 'the longest, wordiest and most boring to write: thank God, thank God and thank God again'.*

SPREADING THE WORD *A 10th-century ivory book cover depicts Pope Gregory the Great (540-604) writing at his desk while three monks – sharing a single inkhorn – make copies of his manuscript. These copies will in turn be copied, and so on, until Gregory's words are eventually disseminated throughout the Christian world.*

Golden letters that lit up the page

HOW MEDIEVAL ARTISTS ILLUMINATED MANUSCRIPTS

FLORAL MASTERPIECE *The illuminator could exercise his art to the full in a Book of Hours – a selection of texts and prayers compiled for wealthy patrons. Flowers and insects light up the 15th-century Hasting Hours.*

SAINTS SIT ENTHRONED, EMBLAZONED IN VERMILION, gold leaf and azure blue, and ringed by a giant capital letter. Entwined garlands of briar roses and meadow flowers climb the margins. Armoured knights pursue outlandish dragons and serpents across the foot of the page. The word 'illumination' aptly describes the art of medieval book illustration: the elegant, handwritten texts are complemented by brilliant images that light up the pages with a dazzling mixture of delicacy, colour and imagination.

Applying gold and precious stones

Illumination generally took place only after a scribe had finished writing the text. It was usually carried out by one or more specialist artists – although some scribes were also illuminators. Some were monks, others professional artists, and a few were women.

Using the spaces left blank by the scribe, the illuminator would draw a delicate outline of the design, employing a compass and ruler for geometric shapes. Next, the artist would apply gold leaf by painting gum or gesso onto the areas to be covered, then rubbing a delicate sheet of the thinnest beaten gold onto the page with a smooth stone.

After this, the artist applied paint with brushes, usually in two or more stages. First he put on a layer of background colour, then the final colour at full strength. The colours were made up of natural plant and mineral pigments, such as iris sap (green) and madder root (deep red). Some were imported from afar: ultramarine blue, for example, was made from semiprecious lapis lazuli, which came all the way from Afghanistan. To make the pigments stick to the page, and to make the colours fast, they were mixed with water and egg white.

Illumination was already a highly sophisticated art by the 8th century AD; later in the Middle Ages wealthy private clients created a demand for ever more lavish picture books, such as illustrated Lives of the Saints, Apocalypses (visions of the end of the world), Bestiaries (encyclopaedias of animals, real or imagined), and the collections of private devotions known as 'Books of Hours'. The illuminators' art became ever more ambitious, encompassing scenes of domestic life and biblical narrative.

DANCE OF LIFE *Elegantly dressed dancers pass before a blindfolded Lady Fortune in a symbolic illustration from a 15th-century French manuscript.*

WORK IN PROGRESS *An unfinished page shows how the lettering was completed before the artist added colour to the preliminary outline of the illumination – in this case, a motif of leaves, berries and a bird.*

TEAM WORK *Master illuminator Hildebertus chases away a mouse as his apprentice practises drawing in this 12th-century picture.*

FINE ART *Gold leaf was used to give spectacular highlights to precious, beautifully made books.*

WORK COMPLETED *In the late 8th century, at least three scribes and four artists laboured on the island of Iona to create the sumptuous Celtic gospels known as the* Book of Kells. *One intricate page consists almost entirely of the Greek monogram of Christ's name, the* Chi-Rho.

ḣ ɢ ᴇ ᴜ ᴘ ᴀ ᴍ ᴏ

Casting the type that would spread the word

HOW GUTENBERG CREATED EUROPE'S FIRST PRINTING PRESS

VERSATILE BLOCKS
*Metal punches –
individual letters
carved in relief
and reverse –
made it possible
to cast type with
ease and speed.*

IT WAS ONE OF THE GREAT TRIUMPHS OF THE WESTERN world. In 1455, Johann Gutenberg of Mainz, Germany, produced the first printed book in Europe using movable type – rows of reusable metal letters. Up until the mid 15th century, all books had been handwritten – a highly labour-intensive and time-consuming process. By making it possible to produce multiple copies of books, Gutenberg's printing presses enabled new ideas in politics and religion to spread rapidly across Europe. His great achievement went beyond technological ingenuity and opened the floodgates of literature to a Renaissance world eager for knowledge.

Gutenberg's first major work was no ordinary book, but a Latin Bible – a vast project that had taken 15 years of toil, most of which was spent trying to raise the money. Furthermore, he was determined to keep his revolutionary invention a secret until his work was complete.

Striking the master-set
Gutenberg's first challenge was to produce a complete master-set of letters and characters – some 270 in all, including capitals, lower case (small) letters, punctuation marks and various special characters and abbreviations. The letters and characters were carved in relief and in reverse on the tips of steel rods, which were later hardened by heating and then plunging into cold water.

This master-set was used as a set of punches. Each character was hammered into a copper block called a 'strike'. This was fitted into the base of a larger mould and used as the eventual printing block. Each letter was given the appropriate spacing: 'i', for example, needs less space than 'w'. A metal alloy, probably composed of lead, tin and antimony, was poured into the mould to produce a cast block of metal with the raised impression of the letter at its tip. Gutenberg could now cast individual letters and characters comparatively quickly.

The next stage was to sort the letters into compartments in a tray. The printer then picked out the letters and assembled the words line by line, ensuring that each line was the same length. This process could take a whole day to complete. When the text was ready, it was fitted into a frame to produce a stable, portable 'forme' which was then laid on the base of the press.

CHINESE PRINTING BOOMS BEFORE THE BIRTH OF CHRIST

IN AD 868, NEARLY 600 YEARS BEFORE Gutenberg, Chinese printers produced the *Diamond Sutra* – a collection of Buddhist scriptures and illustrations – and the earliest known printed item. It was part of an established printing tradition, the origins of which can be traced back to the small jade, ivory or bamboo seals, or 'chops', that were used to authenticate documents much as a signature is used today. The raised characters embossed on these seals were smeared with ink and then printed onto the documents.

Letters carved in wood
This same principle was later adapted for woodcuts. Chinese characters and pictures were delicately carved in reverse out of a block of wood, then inked. Paper – another Chinese invention dating back to c.AD 100 – was rubbed over the woodblock to transfer the image. By the 9th century AD, sheets of text and illustrations were frequently prepared on single blocks of wood and used to produce books (in the form of scrolls), calendars, religious images, greetings cards and even newsletters.

Letters baked to order
However, Chinese printing was a slow process. Because each written word in Chinese consists of an individual character or symbol, rather than a series of letters corresponding to sound, printers needed thousands of letters. An early form of movable type, invented in c.AD 1040 by Bi Sheng, solved the problem. Baked clay was used to create a bank of the most common characters, and additional characters were fashioned and baked when they were needed. The characters were assembled in a wooden frame for printing. It was, nonetheless, a cumbersome process, and the crude print failed to reach the standards of traditional calligraphy.

ORIGINAL BLOCK *The
ancestor of movable
type, seals such as
this porcelain
example, were first
used 13 centuries
before Christ.*

BUDDHIST SCROLL *Each page of the* Diamond Sutra *was printed from a single woodblock, then glued to form a scroll. Later Chinese printers selected type from a special revolving table.*

MASTER PRINTER
*Johann Gutenberg
is portrayed
holding a piece of
metal type in his
left hand. He may
have heard that the
Chinese could
print multiple
copies, but he did
not know how. His
method was
entirely his own
invention.*

Printing the pages

The next job was to ink the forme, using a blend of clear linseed oil varnish – which helped the ink to stick to the metal type – and lampblack for the black colour. A pair of 'inkballs' – leather pads, filled with wadding, attached to wooden handles – were lightly coated in the ink, then dabbed onto the type. The printer laid the paper over the forme, then turning a handle on the press, moved it into position. He then pulled a bar on the press, which operated the screw and applied even pressure over the forme. He then released the bar, removed the forme, peeled back the paper and stacked it next to the press. Gutenberg printed around 16 copies an hour in this way, and some 20 men kept six presses working full time for more than a year to produce the Bible.

While printing was under way, Gutenberg's main creditor, Johann Fust, called in his debts. Awarded in settlement all the materials and presses, Fust reaped the rewards of a brilliant invention, but Gutenberg's technological triumph was his own.

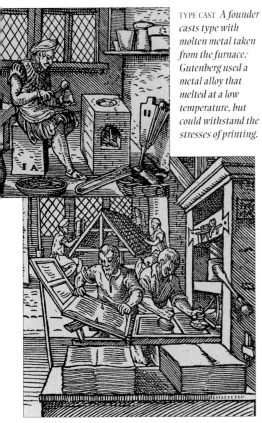

TYPE CAST *A founder casts type with molten metal taken from the furnace. Gutenberg used a metal alloy that melted at a low temperature, but could withstand the stresses of printing.*

TYPE SET *While compositors assembled type into pages, two printers operated each press. Gutenberg's workshop also employed proofreaders to check the finished result.*

TYPE FACE *A page from the Book of Leviticus displays the Gothic lettering developed by Gutenberg to imitate the handwritten script that printing was poised to usurp. Artists added the elaborate illuminations later, before the pages were bound. Gutenberg printed around 450 copies of the Bible, but only 48 are known to have survived.*

A code for spies and plotters

HOW RENAISSANCE THINKERS DEVELOPED A RANGE OF COMPLEX CIPHERS

CODE MASTERS *Leon Battista Alberti (left) substituted one set of letters for another. Giovanni Battista Porta (below) used cryptic symbols instead of conventional script.*

MEANINGFUL SYMBOLS *Users of Porta's system divide their message into pairs of letters and trace the place where they intersect on the key chart to produce a symbol. The recipient unscrambles the code using an identical key – but so can the enemy, should he obtain a copy of the chart.*

BY THE LATE 15TH CENTURY, THE GREAT CITIES OF Europe were thriving cosmopolitan centres. The major nations had ambassadors at each other's courts; the big trading companies and banks had representatives in foreign commercial centres. Voyages of exploration were opening lucrative new trade routes. The nations of Europe were engaged in tense games of economic rivalry and squabbles for territory. Spies, plots and intrigue abounded.

Merchants and politicians communicated through a network of private messengers and public couriers. These services, however, were easily intercepted. All important, confidential documents were sent in code. By the middle of the 16th century major European courts had secretaries employed specifically to create, break and send messages in code.

Caesar writes gobbledegook

The most basic codes, in which one set of letters is substituted for another, are known as ciphers. In the simplest of these, the normal alphabet is just replaced by another alphabet. In the first century AD, Julius Caesar used this code to send secret messages: by substituting all the letters of his text with letters three places down the alphabet. But these ciphers are comparatively easy to break. Because the substitution of letters is always the same, the code breaker can identify familiar patterns. In English 'e' is the most common letter, so whatever letter appears most frequently in the message is likely to be 'e'. Similarly, certain letters are used regularly in combination, such as 'ch' or 'st'.

By the 15th century, more complex tables of signs and symbols were being used to represent the letters of the alphabet. Yet even these codes could be deciphered. Some of the brightest minds of the time were engaged in perfecting an impenetrable code. The first of these was the Florentine artist, musician, architect and athlete, Leon Battista Alberti. He believed the answer lay in a 'polyalphabetic' cipher – in which the letters of the alphabet were represented by a series of other alphabets that changed as the message progressed.

A code they could not crack

Alberti developed a disc made up of two dials fixed in the centre. Twenty letters of the alphabet (excluding H, J, K, U, W, and Y) and the numbers 1-4 were written around the edge of the outer plate. A jumbled Latin alphabet was written around the edge of the inner plate. To create a message, the sender took a letter from the outer plate, then read off the cipher equivalent from the inner plate. After a given number of letters, the inner plate was turned an agreed number of positions so that the alphabet was represented by a different set of ciphers. The recipient, of course, had to be equipped with the same cipher disc and know when to turn the plate. Another polyalphabetic system involved writing out

MOVABLE DISC *Alberti invented the cipher disc, a series of dials which could be reset during encryption and encoding for greater security. This sophisticated version incorporates numerals, letters and symbols.*

CODE BOOK *French cryptographer Blaise de Vigenère revealed his methods in his 1586 treatise* On Codes. *He based his 'undecipherable cipher' on a single keyword.*

the letters of the alphabet repeatedly in a table. At its simplest, the first line was the normal alphabet (a-z); the second line began the alphabet with 'b' and ended with 'a', the third ran from 'c' to 'b' and so on. The sender and recipient agreed on a secret 'keyword' – usually a term or phrase. To encipher a message, the sender wrote down the plain text (the message in normal writing), then above it wrote the keyword, repeated as necessary. For example, if the keyword was DIFFICILE, and the plain text was 'leave immediately', he wrote:

Keyword: D I F F I C I L E D I F F I C I
Plain text: l e a v e i m m e d i a t e l y
Cipher text: n m f a m k u x i g q f y mn g

The cipher text was created by first seeing which keyword letter appears above the plain text letter. If, for example, the keyword letter is 'c' then the alphabet from which the cipher letter is taken begins with 'c', so the plain text letter 'l' is enciphered as 'n'.

The message contains the key

To stop keywords from falling into the wrong hands, French diplomat Blaise de Vigenère developed the 'autokey'. The message itself contained the keyword, signalled by a 'priming key' – a letter or letters recognised as such by the recipient. In this way the keyword changed with every message.

Polyalphabetic ciphers were considered uncrackable. However, if either the sender or the recipient made an error, the text became nonsense. It was not until the age of machine-generated codes that they came into their own. The Enigma system, used by the Germans during the Second World War, was based on such a system, as are many of the computer-generated codes used today – which, because of the speed at which they can be altered, are virtually impossible to break.

BUSINESS SECRETS
Conquistador Hernán Cortés encoded details of New World riches to keep them secret from pirates and commercial rivals.

POLITICAL SECRETS *Mary, Queen of Scots wrote coded messages to Catholics conspiring against Elizabeth I. In 1586, Elizabeth's agents intercepted and decoded her letters, which were used as evidence in her trial for treason.*

DOMESTIC SECRETS
In his diaries, Samuel Pepys combined shorthand with French, Latin, Greek, Spanish and symbols of his own invention to disguise accounts of the saucy shenanigans in his private life and the political intriguing at his work.

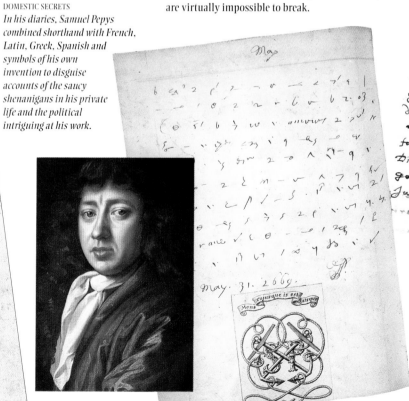

CODED CONSPIRACY
This secret letter from Louis XIV of France was decoded by Britain's top cryptanalyst, John Wallis, in 1693.

The world in so many words

HOW THE FIRST NEWSPAPERS WERE PUBLISHED

ONE CHILLY NOVEMBER EVENING IN 1588, A VILLAGE tavern in the heart of rural Somerset was busier than usual. Many people were speculating about the outcome of the battle between the English fleet and Philip II's Armada. One gentleman in the crowd was a wealthy lawyer who read aloud a newsletter he had received by courier from London that morning. When he confirmed the good news that the English had in fact defeated Spain, the villagers burst into cheers – their fears of an invasion were over.

Gossip in the cathedral

Only the wealthy few could afford to enjoy the benefit of such a news service in the 16th and 17th centuries. For an annual fee of around £5, subscribers employed intelligencers – private news writers – to keep them informed of the latest city gossip while they were at their country retreats. The intelligencers gathered their information at the meeting places in the city: the nave of St Paul's Cathedral was one of the main haunts, hence the later establishment of printing offices in nearby Fleet Street. Once they had pieced together the most important stories of the day, they returned to their offices, where they dictated their reports to a group of clerks who wrote out the news by hand. Depending on the circumstances, a subscriber could receive as many as three newsletters from the same intelligencer in a week. Others had to rely on sporadic newsletters, often religious in nature, which focused on battles, or natural disasters, such as fires and floods, and were illustrated with vivid woodcuts.

In England the publication of news was regarded as an interference with the affairs of the state and was not allowed without royal permission. For foreign news people had to turn to the so-called newsbooks, or Corantos, that were exported from the Netherlands to England, having been translated into English first. The first of these arrived in 1620, but soon English publishers began to produce their own versions. Only a year later, Thomas Archer, a London stationer, printed translations of foreign news in England's first regular publication, *Corante or Weekly Newes from Italy, Germany, Hungarie, Spaine and France.*

Other newsbooks followed, but in 1632 they were all banned by the Court of Star Chamber – a royal court which protected the security of the state – after a news item had offended both the Spanish and Austrian ambassadors. The ban lasted until 1638 when, for the first time, parliament allowed the publication of domestic political news.

During the Civil War (1642-9) an average of ten newsbooks appeared each week. News of the dramatic clashes between the Cavaliers and Roundheads was spread throughout the country by partisan publishers: for example, John Birkenhead founded the *Mercurius Aulicus* to champion the Royalist cause, while the Parliamentary *Diurnalls* gave the Roundhead viewpoint.

Daily newspapers appeared from the beginning of the 18th century. The first of these, *The Daily Courant*, published between 1702 and 1735, cost a penny. The news was printed in two columns on one side of a sheet. At the same time, some of Britain's finest periodicals first appeared. In 1704 Daniel Defoe, the author of *Robinson Crusoe*, founded the weekly *Review*. Defoe and his contemporaries, essayists Richard Steele and Joseph Addison who wrote for both *The Tatler* and *The Spectator*, set new standards in political and satirical writing.

Taxed and banned, but never silenced

Despite the increasing numbers of journals, the government continued to restrict access to news, particularly of home politics. In 1712, a stamp tax added a penny per sheet to the price of newspapers, making them too dear for most people. However, London's coffee houses – the popular meeting places for businessmen – displayed papers which the customers could read free.

In 1738, the House of Commons passed a law forbidding the publication of parliamentary reports; it remained in force until 1771 when the radical publisher John Wilkes engineered a showdown between the press and parliament and successfully contested the law. In the intervening years, publishers used fictitious names and anagrams to thinly disguise news of parliament, while journalists, who were not allowed to take notes in the public gallery, wrote the news from memory. The astonishing powers of recall of the editor of the *Morning Chronicle* earned him the nickname of William 'Memory' Woodfall.

SOCIAL ARBITER *Richard Steele made* The Tatler *the arbiter of the arts and social etiquette.*

ART FORM *Joseph Addison focused on literature, rather than news, in* The Spectator.

OLD NEWS *The first known English newsletter chronicles the Scottish defeat at Flodden in 1513.*

BAD NEWS *Typically, this early 17th-century newsletter combines a strong picture with sensational reporting.*

REGULAR NEWS *The weekly 'Corantos' of the 1620s, forerunners of Britain's newspapers, were 24-page pamphlets devoted to foreign news and sold in markets and fairs for 1 groat – 4 old pence.*

Numb. 8

Mercurius Civicus.
LONDONS
INTELLIGENCER:
OR,
Truth impartially related from thence
to the whole Kingdome, to
prevent mif-information.

From *Thursday, July* 13. to *Thursday July* 20, 1643.

STYLE SETTER *The civil war weekly Mercurius Civicus was often illustrated, and was one of the first newspapers to summarise its contents at the top of the front page.*

READ ALL ABOUT IT *A 19th-century broadsheet describes a brutal murder in gory detail – complete with a grisly illustration of the crime.*

DREADFUL
MURDER
Committed by Nicholas Scrudaugh, on the BODY of his
Wife and four Children.

The Daily
½d. ILLUSTRATED ½d.
Mirror
A PAPER FOR MEN AND WOMEN ONE HALFPENNY.
THURSDAY, JANUARY 28, 1904
DEATH IN THE CIGAR A LAST SMOKE

GOOD NEWS *In a German newsbook of 1648 – later coloured in – a delighted courier announces the end of the Thirty Years War.*

The Sun.
God Save
Victoria

FALSE START *At first the Daily Mirror, founded in London in 1903 as a ladies' newspaper, was a flop. It found success by broadening its appeal, using pictures and dropping its cover price from 1d to ½d.*

DISMAL NEWS *Constrained by the 1765 stamp tax, an American paper grimly announces its temporary closure.*

SPECIAL ISSUE *The Sun marks Queen Victoria's coronation in 1838 with a 'gold' medal printed in ink made of bronze dust mixed with varnish.*

The TIMES are
Dreadful,
Dismal,
Doleful,
Dolorous, and
DOLLAR-LESS.

Thursday, October 31, 1765.

THE
PENNSYLVANIA JOURNAL
AND
WEEKLY ADVERTISER
NUMB. 1195.

EXPIRING: In Hopes of a Resurrection to LIFE again.

THE
Universal
DAILY
Register
Printed Logographically
Numb. 1.]
SATURDAY, JANUARY 1, 1785. [Price Two-pence Halfpenny.

To the Public.

WORTH THE WAIT *A 19th-century gentleman waits for his turn with The Times. Founded in 1785 as The Daily Universal Register, the newspaper earned its nickname 'The Thunderer' between 1817 and 1841, during the editorship of Thomas Barnes (left).*

Strips of pictures full of meaning

HOW EUROPEAN SCHOLARS UNRAVELLED THE MYSTERIES OF EGYPTIAN HIEROGLYPHS

PATIENT CODE BREAKER *Jean-François Champollion compared the Rosetta Stone's hieroglyphic symbols with its other scripts as part of his patient detective work. It took him from 1808 to 1822 to crack the code.*

PICKAXES GLINTING AND BODIES PERSPIRING IN THE unforgiving August sun, a group of French soldiers hacked at the rocky ground near Rashid, or Rosetta, a port on the western delta of the Nile to the east of Alexandria. They were part of Napoleon Bonaparte's expeditionary force which had invaded Egypt in 1799, and were hard at work on a piece of military engineering – perhaps a trench, a road or a fortress.

As he dug, one soldier found a dark granite stone, 3 ft 9 in (1.1 m) high by 2 ft 4½ in (72 cm) wide, decorated with strange writing and pictures. He called Lieutenant Pierre Bouchard, commander of the detachment, who immediately recognised that some of the carved inscriptions were in the angular script of ancient Greece. At the top of the stone were several lines of hieroglyphs, the enigmatic pictorial writing found on carvings and papyri throughout Egypt. When the scientists on the expedition saw the stone, they guessed that it could be the key to unlocking the hitherto indecipherable hieroglyphic script.

Struggling with the code

Deciding that the stone held the key to the meaning of hieroglyphs and cracking the code were two different things, however. It was to take 23 years and the work of several different scholars before the Rosetta Stone finally gave up its secrets. The Greek inscription revealed that the stone itself dated from 196 BC, and was inscribed by priests who had gathered at Memphis. The stone was inscribed with a decree composed in honour of the 12-year-old pharaoh Ptolemy V. The priests, it is thought, wrote in Greek, which was then translated into Egyptian demotic (popular) script and from that into hieroglyphs. This discovery, the first stage in the decoding process, was made by David Akerblad, a Swedish diplomat, in 1802. But although Akerblad went on to identify three words of the demotic script, he could make no further progress.

In 1814, Sir Thomas Young, a British linguist, doctor and physicist, took up the challenge when he, too, obtained a copy of the inscriptions. Young progressed to the stage where he could isolate and interpret one single hieroglyph – the one representing the name 'Ptolemy' enclosed in a cartouche, an oval decorative frame. But although Young put forward the idea that some hieroglyphs were what he termed 'alphabetic', he clung to the existing belief that the majority were, in fact, visual symbols and used alphabetically only to write non-Egyptian names.

TELLTALE SLAB *The Rosetta Stone praises King Ptolemy V of Egypt in three different languages. The Greek (bottom) and demotic (centre) scripts gave Jean-François Champollion the clues to decode the hieroglyphic pictograms (top). There are only 66 different pictograms in the hieroglyphic text, corresponding to 500 words of Greek. Champollion concluded that hieroglyphs represented individual sounds, not whole words.*

The great breakthrough

It took the genius of Jean-François Champollion, a young French scholar, to make the breakthrough. He first saw a copy of the Rosetta Stone's mysterious inscriptions in Paris in 1808 and, after studying them for several years, established that it was the cartouche that in all probability held the key to the mystery – a discovery which Young, for his part, was also making independently in England. But it was Champollion alone who had the intuition to take that one speculative step further forward, arguing that all the hieroglyphic signs represented individual sounds, rather than simply being word-pictures.

To test his revolutionary theory, Champollion started work on deciphering the meaning of each sign in the cartouche. Here, he was aided by a visual clue the priestly scribes had left behind centuries before. They had decorated the cartouche with a drawing of a lion. Champollion realised that the direction in which the lion was looking indicated which way the particular inscription should be read – from right to left, not left to right. Working from this, and using a Greek text that mentioned Ptolemy as a basis, he managed to interpret three of the hieroglyphs making up the word Ptolemy correctly, and also to match them to their Greek equivalents.

What Champollion now needed was further data against which to check his deductions. This he found in 1822, when he first saw a copy that Young had made of another cartouche – this time containing the word Cleopatra – on an obelisk at Philae. Excitedly, he discovered that his interpretation of the three Rosetta Stone hieroglyphs corresponding to the sounds 'p', 'o' and 'l' was confirmed: these sounds appear in both Ptolemy and Cleopatra. Armed with these three syllables, he deduced the meanings of the rest from their positions in both words.

Knowledge of the sound syllables enabled Champollion to decipher 80 more names from different cartouches. He then tried to complete a grammar of hieroglyphs, but died of exhaustion aged only 42, two years before its publication.

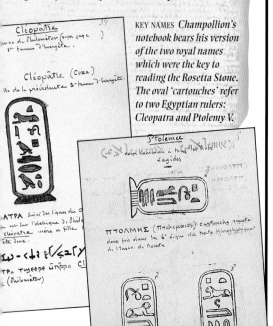

KEY NAMES *Champollion's notebook bears his version of the two royal names which were the key to reading the Rosetta Stone. The oval 'cartouches' refer to two Egyptian rulers: Cleopatra and Ptolemy V.*

Cauldron of secrets

HOW A BRITISH ARCHITECT DECIPHERED THE 'LINEAR' SCRIPT OF ANCIENT CRETE

SIR ARTHUR EVANS WAS MYSTIFIED BY THE CLAY TABLETS his team kept digging up at Knossos, Crete, in 1899. The eminent British archaeologist could not make sense of what he called the 'linear' writing on the tablets, used by the Minoan civilisation of ancient Crete, from the 16th to the 12th centuries BC. One of the scripts, known as Linear A, still remains undeciphered, and it was to be over 50 years before the other – Linear B – gave up its secrets.

After Evans's finds, thousands of tablets with linear inscriptions were unearthed at other sites in Crete and elsewhere in Greece. They were all lists of articles used in daily life. The script comprised about 90 symbols representing the syllables of the spoken tongue, rather than individual letters. By comparing the symbols with a similar script found in Cyprus, scholars were able to match some of the symbols with sounds – but were unable to extract the meaning of the words.

The answer in the ears

Michael Ventris, a gifted British architect, had been fascinated by these ancient Mediterranean languages since the age of 14, when Sir Arthur had given a lecture about them at his school. At first he thought they were related to the lost language of the ancient Etruscans, but in 1952 his ideas were turned upside down by a systematic analysis of a tablet found at Pylos, in the Peloponnese, in Greece, 12 years previously. Ventris found recurring groups of symbols which referred to place names in Crete, and a series of pictures of three-legged vessels, or tripods.

Each picture was labelled with a word. Ventris deciphered the Minoan for the vessel as *ti-ri-po-de*, which is closely related to the Greek word *tripodes*, meaning 'tripods'. Pictures of vases bore separate labels, according to the number of handles on each vase. Again, the labels were recognisably linked to Greek words, meaning 'no-eared', 'three-eared' and 'four-eared'. Armed with this vital link, Ventris and his successors embarked on the complete decipherment of Linear B.

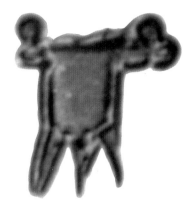

TRIPOD *This picture of a three-legged cauldron, labelled* ti-ri-po-de – *similar to the Greek* tripodes, *meaning 'tripods' – was the link between Linear B and Ancient Greek.*

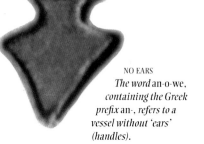

NO EARS *The word* an-o-we, *containing the Greek prefix* an-, *refers to a vessel without 'ears' (handles).*

THREE EARS *The tablet calls this three-handled vessel* ti-ri-o-we-e; *the Greek for 'three' is* tris.

FOUR EARS Ke-to-ro-we *echoes the Greek prefix* tetra-, *which means 'four'.*

WRITTEN IN CLAY *Michael Ventris had a photographic memory and a brilliant ear for languages. A tablet found in the Peloponnese provided the vital clues to deciphering Linear B: descriptions of different vessels, complete with pictures.*

Electrifying a chain of monks

HOW THE TELEGRAPH WAS DEVELOPED

IN 1746, THE FIRST EXPERIMENT TO SEND SIGNALS USING electricity was carried out by the French physicist Jean Nollet. He sent an electric current from a Leyden jar – a device for storing an electric charge – along a chain of 700 Carthusian monks who were connected to each other by the iron rods each held in his hand. Nollet found that the monk at the end of the chain reacted to the current almost as soon as the first. However, he now faced the same problems as other early experimenters: there was no steady source of electric current, nor any wire capable of carrying it over a worthwhile distance. In the absence of either, scientists concentrated on developing visual telegraphy.

Message in a bubble

Alessandro Volta's invention of the electric battery in 1800 was a great step forward. At last, telegraphers had an adequate, sustained power supply. One early idea, developed in Germany by Samuel Thomas von Sömmering in 1812, used electrolysis to send signals. Sömmering's receiving set was a bath of acidulated water containing a series of metal rods, each of which represented one letter of the alphabet. Every rod was attached to a wire. To send a message, the operator connected the appropriate wire to the terminal of a battery and completed a circuit. The chemical reaction released hydrogen, which appeared as a stream of bubbles from the rod. The vast amount of wiring involved made this system impractical.

In 1816, an English inventor, Francis Ronalds, developed a single-wire system. Two users were seated at each end of the wire: each had a clockwork device fitted to a circular brass dial on which the letters of the alphabet were revealed, and a small ball suspended on thread at the end of the wire which was

set in motion by the electric current. Both clocks were carefully synchronised. When the sender saw the letter that he wished to transmit, he interrupted the current, and at the receiver's end the ball stopped moving for a split second. The receiver would note the letter that appeared on the clock at that moment. Although it was slow, the system worked provided that the clocks were precisely synchronised. But the British authorities were not interested; they were happy enough with visual telegraphs, such as the semaphore, and the idea went no further.

Pointing the way forward

A Russian nobleman, Baron Pavel Lwowitch Schilling, who had worked with Sömmering, produced another system in 1830, harnessing the ability of an electric current to deflect a compass needle. This idea was taken up by a British inventor and entrepreneur called William Cooke, who in 1836 went into partnership with the scientist Charles Wheatstone. Together they produced the first successful electric telegraph system, which was swiftly adopted by the railway systems for signalling and for sending messages for customers.

The Wheatstone telegraph involved five compass needles and an ingenious lattice grid of 20 letters (C, J, Q, U, X and Z were omitted). Just five wires were needed to send signals. This was indicated by the movement of two needles at the other end, which pointed at the letter in question. The receiver would read off each letter in turn and an assistant would write down the message.

SHOCK TACTICS
In the absence of wires, Jean Nollet daringly sent a pulse of electricity through a human circuit.

INDICATOR BOARD *The moving needles of a Wheatstone telegraph pointed to one letter at a time. Easy to operate, these telegraphs were the first in use.*

FAST MESSAGE *Rail companies invested in new technology.*

TALK OF THE TOWN *By 1859, telegraphs were part of daily life. At the busy offices of the Electric and International Telegraph Company in London, people could send 20 words for a penny per mile for the first 50 miles.*

Morse dashes ahead

At about the same time another system was being developed in the USA by the artist, inventor and entrepreneur Samuel Morse. His system used only one wire, and depended on a code that represented each letter and number in a sequence of dots and dashes. These were created by short and long pulses of electricity – a dash being three times as long as a dot. At the receiver's end the code was originally written out by an electrically activated stylus on a moving strip of paper, and then read off. Skilled operators soon found that by listening to the clicks made by the stylus as it descended they could interpret the message and write it down immediately.

Morse's method had numerous advantages over Cooke and Wheatstone's system. It used just one wire, so was much cheaper to install; it needed just one well-trained operator at each end; and it was fast. Here lay the future of telecommunications – until the invention of the telephone in the 1870s.

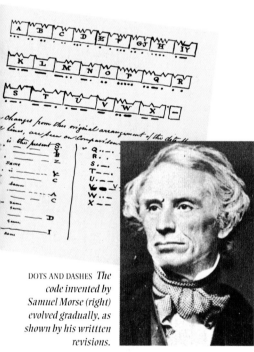

DOTS AND DASHES *The code invented by Samuel Morse (right) evolved gradually, as shown by his writtten revisions.*

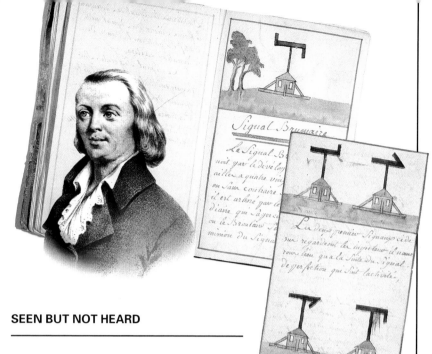

FRENCH ARMS
Claude Chappe's system used mechanical arms to convey 9999 separate words. He kept a careful record of all his signals in a notebook, complete with colour diagrams.

SEEN BUT NOT HEARD

ON TOP OF A SMALL STONE TOWER IN PARIS, A PAIR OF mechanical arms jerked rapidly into a succession of strange angular positions. The arms were attached to the ends of a movable crossbar fixed to a tall mast. Each separate signal, controlled from inside the tower, represented a letter, number or code word forming part of a message. A few miles away observers in another tower viewed the message through a telescope, then passed it on to the next tower in the chain. It took just 2 minutes to send the message across 16 stations – a distance of 150 miles (240km) – from Paris to the last station just outside Lille.

Britain rushes to catch up

This startling demonstration of rapid, long-distance communication was carried out in 1794 by a French engineer and clergyman named Claude Chappe. News of his 'visual telegraph' soon spread to England. The British government, alarmed by the military implications of this development at a time of war with France, hastened to find its own system. The first task was to connect the Admiralty in London to the ports of southern England.

By the end of 1796 the British had perfected the shutter telegraph, invented by George Murray and installed in stations between London and Portsmouth and Deal in Kent. On the roof of each of these stations was a frame in which six hexagonal boards, or shutters, were mounted in a three-by-two pattern and pivoted on a horizontal axis. A system of pulleys and cables enabled them to be opened or shut to create various configurations representing letters and numbers, and a few whole words such as 'and' and 'sail'. The stations were set on a series of hilltops, rooftops and other prominent places, some as far as 10 miles (16km) apart.

Bad weather seals semaphore's fate

At the end of the Napoleonic Wars in 1815, the Admiralty adopted another system, the semaphore (from the Greek *sema*, meaning a 'signal', and *phoros*, meaning 'bearing'). It was similar to Chappe's system, but used a mast with two arms.

The great drawback of visual telegraphy was that it depended on good weather; in England, conditions were only right for about 200 days a year. By the 1850s, Morse code was being transmitted through the electric telegraph at a rate of 25 words per minute, fog or no fog. The days of visual telegraphy were over.

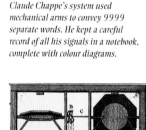

BRITISH SHUTTERS
The Admiralty's six-board system was capable of 64 permutations. The stations were manned by four men: two to observe, one to write messages and one to operate the shutters. George Murray (above) was paid £2000 in 1796 for his invention.

RUGGED RIDERS *Gritty young men such as John Hancock (left) and William Fisher (right) endured tremendous hardships to provide a vital link between the USA's east and west coasts. The average age of Pony Express couriers was 20, and most weighed less than 126 lb (57 kg).*

NIGHT RIDER *A postal courier loses no time in changing horses at a country inn as he takes the mail from London to the provinces. An ostler brings out a fresh mount, and a stable boy offers the rider a topcoat against the chill night air.*

WHEN BRITAIN'S POSTMEN ALWAYS DEMANDED CASH ON DELIVERY

WHEN HENRY VIII'S COURT TRAVELLED FROM PLACE TO PLACE in the early 16th century, its mail was delivered to it by mounted messengers who rode from one staging inn to the next. These Royal Posts, as the inns were called, were supervised by a Master of the Posts, who appointed postmasters – usually the innkeepers themselves – to make sure the mail moved quickly and safely onwards.

Paying for the privilege
To begin with, the mail service was not intended to make a profit, and not available to the public. But in 1635, Charles I decided to turn it into a money-making business. He issued a proclamation opening the postal service to public use, as long as the people were prepared to pay for the privilege. For example, it cost twopence (the price of a light meal) to send a one-page letter for 80 miles (129 km) or less, while letters to Scotland cost eightpence and those to Ireland ninepence. The money was paid not by the senders of the letters, but by those who received them.

In 1660, Charles II appointed the first Postmaster General to oversee the growing number of post roads that radiated from London. But Londoners themselves did not have an internal postal service. Then, in 1680, an enterprising merchant named William Dockwra started a local 'penny post'. Letters were prepaid and stamped to indicate place of posting, and deliveries were made hourly. There were several sorting offices and hundreds of receiving offices throughout the capital. Dockwra's scheme was so successful that two years later it was taken over by the government.

The Penny Black is born
By the 1790s local penny posts operated in most of Britain's main cities and market towns. However, the charges for letters, calculated according to distance and the number of sheets used, were still high. The Post Office was called upon to reform the entire system. In 1840, Rowland Hill introduced a nationwide penny post, making prepayment of the postage possible: the sender stuck an adhesive stamp – the Penny Black, bearing a picture of Queen Victoria's head – onto the envelope.

PIGEONHOLING *Clerks in 1809 diligently sort mail in one of the large general post offices that opened up throughout England.*

COURTSHIP BY POST *A young lady gives a letter carrier the money for a missive to her sweetheart in the early 1800s.*

POSTAL PROTEST *MPs received a petition in 1839 calling for a penny postal service. People could not afford to pay sixpence or more for a personal letter, and saw the pricing system, based on distance, as unfair.*

FAST, BUT COSTLY *Posters proclaiming the speed of the Pony Express appeared throughout the USA in the early 1860s. Initially, the service was very expensive at $5 per ½ oz (15 g), and customers resorted to using tiny handwriting on wafer-thin tissue paper.*

'The mail must get through!'

HOW THE RIDERS OF THE PONY EXPRESS LINKED THE EAST AND WEST COASTS OF THE USA

YOUNG RIDER *William Cody was aged 14 when he joined the Pony Express, and this photograph was taken.*

LYING LOW ON THE BACK OF HIS SPEEDING PONY, 15-year-old William Cody – later to become famous as the Wild West showman 'Buffalo Bill' – tried desperately to outrun a war party of 15 American Indians. According to an account he wrote later, bullets flew low over his head as the mounted attackers gradually got closer. Their chief drew his large war-bow and fired an arrow which grazed the youngster's buckskin shirt. Cody swung round in the saddle, drew his Colt .45 and fired back. The leader's face twisted in agony; he fell from his horse and lay motionless on the ground.

It was April 1861 and Cody was one of the Pony Express riders carrying news of the outbreak of the American Civil War from the east coast to the west. This life-or-death pursuit was typical of the risks run by these riders who, at various times, charged through outlaw ambushes, swam flooded rivers, led their ponies through blinding blizzards in mountain ranges, and suffered agonies in sun-baked deserts.

Risking death daily

Cody had been one of scores of youngsters who had answered the ominously worded advertisement issued by the newly formed Pony Express company: 'WANTED Young skinny wiry fellows not over eighteen. Must be expert riders willing to risk death daily. Orphans preferred. Wages $25 a week.' The successful applicants took an oath not to drink, not to swear and not to fight among themselves. Their motto was: 'The mail must get through!'

The first east-west mail delivery in the United States left the Pony Express stables in St Joseph, Missouri, bound for Sacramento, the capital of California, almost 2000 miles (3200 km) away, on April 3, 1860. Until then, it had taken the eastbound mail about a month to travel from coast to coast. It came by train from the eastern seaboard to Tipton, Missouri, south by stagecoach over the rough territories of Arkansas, Texas, New Mexico, Arizona and California, and finally north along the west coast to Sacramento. The Pony Express aimed to reduce the journey time by half, and to establish an equally speedy service from west to east.

Riding round the clock

On changing ponies at relay stations every 10 to 15 miles (16-24 km), the rider simply slung the mailbag on the saddle of a fresh horse, remounted and galloped off again. The changeover took less than two minutes. After being passed from rider to rider every 75 to 100 miles (120-160 km), the mail reached Sacramento on the afternoon of April 13, after a journey of ten days. From then on, riding all day and all night, the 200 Pony Express riders – who, when in the saddle, did not stop to eat or drink – averaged some 200 miles (322 km) a day.

But despite its speed, the Pony Express was outclassed when, in October 1861, it was possible to send and receive messages in minutes by the coast-to-coast electric telegraph. Overnight, the Pony Express riders seemed slow and old-fashioned compared with the 'talking wires', as the Indians called the telegraph. After only 19 eventful months the Express's 500 ponies were no longer needed, and their youthful riders found themselves out of work.

TRAVELLING LIGHT *At the beginning of each journey, a Pony Express rider was issued with a lightweight leather mailbag called a* mochila *– Spanish for 'knapsack' – in the corners of which were weatherproof leather pockets fitted with small padlocks. The rider sat on the* mochila *to keep it in place.*

Eyewitness to carnage

HOW A JOURNALIST AND A PHOTOGRAPHER BROUGHT NEWS DIRECT FROM THE FRONT LINE IN THE CRIMEAN WAR

'THE SILENCE IS OPPRESSIVE; BETWEEN THE CANNON bursts one can hear the champing of bits and the clink of sabres in the valley below. The Russians on their left drew breath for a moment, and then in one grand line dashed at the Highlanders. The ground flies beneath their horses' feet; gathering speed at every stride, they dash towards that thin red streak topped with a line of steel.'

Uncovering incompetence and scandal

William Howard Russell, a young Irish barrister and journalist, scribbled his description of the Battle of Balaklava in 1854 as he watched the event from a ridge above the battlefield, close to the British commanders. The battle took place on October 25, and an account of it appeared in *The Times* 20 days later. Russell's startling reports of the conflict, based on his observations of combat, were written with a quill pen in accounts books which he bought locally

and sent regularly to the editor of *The Times*, John Thadeus Delane. They usually travelled by steamer to Marseilles, then by train to Calais and on to London – a journey of some 10-14 days.

There was a ready and eager readership awaiting news of the Crimean War of 1854-6, fought by Britain, France and Turkey against Russia. Whereas reports of Trafalgar and Waterloo, earlier in the century, had been dispatched by the commanders themselves, in the Crimea the British press had, for the first time, been allowed to send journalists to report from the war zone itself. *The Times* was then at the height of its power, with a circulation at 40 000, five times greater than any other English paper. It had a worldwide network of correspondents, and the popular belief was that Delane heard the news before even Lord Aberdeen, the prime minister.

What Russell found in the Crimea was a story more complex and alarming than anyone could have foreseen. During nearly 40 years of peace since Waterloo, the British army had barely evolved: at the start of the Crimean War it was badly equipped and incompetently led. Before long, British forces were embroiled in a lengthy stalemate, which resulted in thousands of soldiers suffering from malnutrition and illness. Indeed, only one in eight casualties of the war died as the result of combat.

The embarrassed establishment

Russell admired the bravery of the soldiery, such as that two-deep 'thin red streak' of red-coated Highlanders who repelled the charging Russian cavalry (the phrase, for some reason, entered popular imagination as 'the thin red line'). But his view of the overall conduct of the war was damning. When the Light Brigade was ordered to charge during the Battle of Balaklava, he wrote despondently: 'A more fearful spectacle was never witnessed than by those who, without the power to aid, beheld their heroic countrymen rushing to the arms of death.' Russell was anxious to retain the cooperation of the military authorities, so the reports he prepared for *The Times* always fought shy of being openly critical. His private letters to its editor, however, were far more candid. Delane used this information as the basis for a devastating series of articles in the *The Times* condemning the government's handling of the war. Public disquiet engendered by the newspaper was largely responsible for the fall of the government in February 1855 and the coming to power of Lord Palmerston.

NEWS HOUND *William Russell poses for Roger Fenton. The two men, working separately, sent home the first ever independent view of the realities of war. Senior commanders treated Russell with suspicion, and offered him no protection.*

RELIVING THE ACTION *A Victorian father enthusiastically reads out Russell's account of the Battle of Balaklava to his family, whose reactions vary from excitement to sorrow.*

TENT CITY *Roger Fenton's view of an army camp gives a deceptively serene impression of life in a war zone. Technical problems made it impossible to photograph battle.*

FRESHLY ARRIVED *Local workers stand behind a mass of cannonballs, newly unloaded from a supply ship at Balaklava wharf.*

DEADLY AFTERMATH *Cannonballs litter a wasteland in* The Valley of the Shadow of Death – *one of the few scenes by Fenton that comes close to summing up the horrors of the war.*

Russell returned to London that December, having reported the events of the war for nearly two years. The reports which had made him well known had also embarrassed the establishment. Queen Victoria spoke of 'the infamous attacks against the army which have disgraced our newspapers', and the government was anxious to find a means to counter the depressing and negative descriptions of the war.

The government turned to the infant art of photography. Financed by the print dealer Thomas Agnew and under the patronage of Prince Albert and the Secretary of State for War, Roger Fenton was sent to the Crimea. He arrived in March 1855, with two assistants, a mobile darkroom and 36 cases of equipment. He was immediately confronted by the same chaotic and harrowing conditions that Russell had seen, but Fenton's task differed from Russell's. Travelling on a government mission, Fenton was instructed to produce positive images of the war – with the full cooperation of the military authorities.

War photography – a new art

Fenton used a new technique called the wet collodion process, involving exposure times that were much shorter than those of previous methods, taking just a few seconds rather than minutes. Nevertheless, the technique was cumbersome and messy, and the process of coating a glass negative with collodion – a sticky solution of guncotton (cellulose nitrate) – and exposing it before it dried was especially difficult to carry out in the intense Crimean heat. Action images

of battle were impossible. The Russians also mistook Fenton's van for a munitions wagon and took pot-shots at it. After witnessing the failure of the combined British-French assault on Sebastopol in June 1855, Fenton was struck down by the cholera epidemic that killed thousands of troops. Before fever had a chance to take hold, however, he was on board a ship heading for home.

'My van grew so hot towards noon as to burn the hand when touched. As soon as the door was closed to commence the preparation of a plate, perspiration started from every pore.'
ROGER FENTON

In just four months, Fenton had successfully exposed some 360 plates – mainly portraits, groups of soldiers and landscapes. Despite popular acclaim, and widespread publication as engravings, many of the prints went unsold. But they did not pass unnoticed. As *The Literary Gazette* reported, Fenton's photographs 'illustrate in a variety of scenes the stirring and deadly business of the period . . . It is obvious that photographs command a belief in the exactness of their details which no production of the pencil can do.' Fenton had given the public new insights, and the art of war photography was born.

IN THE FIELD *Roger Fenton's mission was to show people at home a heroic and romantic image of war. An assistant hands Fenton a fresh photographic plate from the mobile darkroom, converted from a British wine-merchant's van. Inside were water cisterns, chemicals, workbenches and stores – all the facilities a front line photographer needed.*

PHOTOGRAPHIC VAN

Linking two continents with a single cable

HOW THE FIRST TRANSATLANTIC TELEGRAPH LINK WAS LAID

WITH ITS ENGINES STOPPED, THE MIGHTY *GREAT EASTERN* drifted slowly and silently downwind throughout the night, dragging a five-pronged hook along the floor of the Atlantic Ocean, 2½ miles (4 km) below. The ship had been engaged in laying a telegraph cable under the Atlantic to link Britain and the USA, but the cable had first broken and then slipped overboard while being mended. Now the *Great Eastern* was trying to retrieve the cable and haul it to the surface.

Early the following morning, the hook caught on something on the seabed and a winch on deck began hauling it up. But the rope broke and the unknown 'catch' fell back into the sea. Several more attempts were made but all were in vain. Eight days later, on August 11, 1865, the expedition was abandoned. The cable was lost.

Hopes crushed as cable fails

Laying a cable across nearly 2000 miles (3200 km) of the Atlantic in depths of up to 3 miles (5 km) was proving to be a stern challenge. The first attempt, in 1857, failed when the cable broke after more than 300 miles (485 km) had been laid. The following year, HMS *Agamemnon* and the USN *Niagara* set out from Plymouth, England, each bearing half the cable.

In mid Atlantic, they spliced cables and turned their bows east and west. After two unsuccessful attempts, on August 5, 1858, the first transatlantic telegraph link was established. But two months, and only 732 messages later, the link failed.

The *Great Eastern*'s 1865 voyage was a determined attempt to succeed where others had failed. The steamer, designed by the British engineer Isambard Kingdom Brunel, was the largest the world had ever seen, weighing 32 000 tons and 700 ft (213 m) long.

Mission accomplished

Almost as soon as the 1865 expedition was abandoned, a new challenge was being planned. On the inauspicious date of Friday, July 13, 1866, the *Great Eastern*, equipped with a new cable and improved hauling machinery, set sail once again. The voyage was comparatively trouble free and the cable was laid out over the stern of the ship, with complete success. The new cable earned £1000 on its first day of operation, but the *Great Eastern*'s mission was not complete until the cable which had been lost the previous year had been recovered. This operation was also successful, enabling a second transatlantic cable to be established.

WASHING LINE *A cartoonist of the time depicted Father Neptune hanging out his laundry on the Atlantic telegraph cable.*

FAILED ATTEMPT *In 1858, before the massive steamship* Great Eastern *was adapted to lay cables, a joint Anglo-American venture laid the first telegraph line across the Atlantic. From the middle of the ocean, the US naval ship* Niagara *laid a copper cable protected by tar-soaked hemp and iron wires west to Trinity Bay, Newfoundland. HMS* Agamemnon (*pictured here*) *took it east to Valentia Bay, in County Kerry, Ireland. Public voices as diverse as priests and poets hailed the cable as an immense step forward in human history. But on the very day its makers were honoured for creating a 'loving girdle round the earth' by a procession and banquet in New York, messages through the cable became extremely hard to read.*

REELING IT IN *The* Great Eastern *towers over a frigate delivering the cable at Sheerness. The ship was very easy to manoeuvre, thanks to her 58 ft (17.5 m) diameter paddle wheels.*

MUSIC AND MUSCLE *Encouraged by a merry shanty on the violin, the crew of the* Great Eastern *haul in the anchor off Maplin Sands, Essex.*

REELING IT OUT *Crewmen uncoil some of the 2490 miles (4000 km) of steel-armoured copper cable from one of the* Great Eastern's *three vast tanks. On land, the cable was stored in salt water to check for any adverse reaction to its watery surroundings.*

ELUSIVE GOAL *The Great Eastern's elaborate paying-out machinery (left) was a vast improvement on that carried by the Agamemnon. A series of wheels, brakes and weights controlled the speed and tension of the 7000 ton cable to avoid placing unnecessary strain on it. Despite these precautions, the cable snapped – and after two days spent fruitlessly dragging a grappling hook on the sea floor, the crew lowered a buoy (above) to mark the site. The cable was later found, but broke again while being reeled in. A few days later, the mission was abandoned.*

Electricity harnessed to carry spoken words

HOW ALEXANDER GRAHAM BELL INVENTED THE TELEPHONE

'MR WATSON, COME HERE, I WANT YOU!' IT WAS March 10, 1876, and the speaker was a young Scottish-born Canadian inventor called Alexander Graham Bell. At the other end of the line, in another room, was his assistant, an electrical machinist called Thomas Watson, who heard the first intelligible words transmitted by telephone.

Since 1872, Bell, professor of vocal physiology at Boston University, had been trying to develop a multiple-message telegraph, using a single wire and matching sets of vibrating metal reeds, each set tuned to an identical pitch. When Bell sent an electrical charge through the wire, the transmitter reed set off an identical vibration in its harmonic twin – the receiver reed. Bell wanted several such matched sets, each set at a different pitch, to be able to transmit several messages over the wire simultaneously, but his experiment did not work.

During a test in June 1875, Watson noticed that a 'receiver reed' failed to sound. When he plucked the reed to loosen it Bell, in another room, heard his corresponding 'sender reed' emit an identical sound, even though there was no electrical connection. Watson's plucked reed had created and sent a tiny flow of electric current through the wire into Bell's reed. By early 1876, Bell and Watson had developed a transmitter with a sound-sensitive diaphragm which picked up the varying timbre of the human voice and sent a flow of current along the wire. When it reached an identical diaphragm at the other end, the process was reversed: the currrent made the diaphragm vibrate, reproducing the original sound.

HANDS FULL *A wealthy subscriber, who could afford to buy two identical telephones, uses the pair to avoid shifting one from ear to mouth.*

SOUND PROFESSION *The mechanics of the spoken word fascinated Bell, a speech therapist whose mother and wife were both deaf.*

SKY LINES *The enormous demand for telephones created aerial clutter in busy American cities by the 1890s.*

MOVING LINES *A lady makes a telephone call from a train – a rare privilege in 1910 – during a stop at a Utah station.*

SOUND CHECK *Bell regales a Boston audience, listening to Watson calling long-distance from Salem – 18 miles (29 km) away.*

EMERGENCY LINES *Policemen and doctors were among the earliest telephone users. The first coin-operated telephone went into use in Hartford, Connecticut, in 1889.*

Wired up for the telecommunication age

HOW THE FIRST TELEPHONE NETWORKS WERE SET UP

THE WORLD'S FIRST TELEPHONE DIRECTORY – LISTED on a single sheet of paper – opened in New Haven, Connecticut, in early 1878. It had just 50 subscribers. Britain's first exchange, in London, was even smaller, with just eight subscribers when it opened in August the following year.

The earliest exchanges were staffed by operators who connected one caller to another. Initially, teenage boys were hired to do the job, but some callers found them to be slapdash and cheeky. Women – judged to be more reliable and to have a more reassuring telephone manner – took over, and soon came to dominate the profession.

Making connections

Typically, a subscriber in the 1880s made a call by turning the handle on the telephone set to ring the operator at the exchange. The operator inserted a plug to ask the subscriber: 'Number, please?' She could then connect the caller's line with the line of the number required – unless it was engaged. The usual way for the operator to check when a call had finished was to reconnect and listen in.

At first, telephone networks were restricted to a local district, because no wire was efficient enough to carry the weak currents of speech farther than about 30 miles (50 km). Long distance messages went by telegraph. However, after the American Bell Company acquired Western Electric in 1881, and improvements were made in the amplification and wiring of the system, telephone equipment became standardised. The first commercial long-distance line, connecting Boston to Providence, Rhode Island, was made possible by the use of 290 miles (467 km) of hard-drawn copper wire in 1884. The first transcontinental line, linking the eastern seaboard of the United States to the West Coast, opened in 1915.

Buttons and dials take over

However, not everyone was impressed with the network. Almon Brown Strowger, an undertaker in Kansas City, Missouri, became so infuriated with the operators on his network – whom he suspected of diverting business to his rivals – that he set about inventing an automatic exchange, which he patented in 1891. Subscribers could call up a number using two push-buttons, sending a signal to the exchange, where the right connection was made automatically. Later models used a dial – the origin of the dial-based telephone, and ancestor of the push-button machine.

Strowger's automatic exchange was established in La Porte, Indiana, in 1892, but automation was slow to catch on. It was not until 1919 that automatic switching and dials became a feature of Bell telephones, and it took several decades to complete the change. Yet the use of the telephone spread steadily: in 1880 there were 61 000 telephones in the United States, and by 1900, 1.4 million.

HOLD THE LINE *Women had to be fit, agile and at least 5 ft 3 in (1.6 m) tall to reach the highest switchboard plug. They also had to be single and free of speech impediments.*

PARTY LINE *Guests at a New Year's celebration see in 1882 by listening to messages on the telephone.*

First with the news

HOW REUTER CORRESPONDENTS SCOOPED THE WORLD

THE TELEGRAM ANNOUNCING THE ASSASSINATION OF President Abraham Lincoln reached New York too late for the Liverpool-bound mail steamer *Nova Scotian*. New York Reuter agent James McLean arrived at the dock just in time to see the ship set sail. Without losing a moment, he hired a tugboat and set off in pursuit. He carried his hastily-written report of Lincoln's death in a watertight pouch and as the tugboat drew level with the *Nova Scotian*, he threw his dispatch on board.

In it, McLean told how the President and his party had attended a performance of *Our American Cousin* at Ford's Theatre in Washington on the previous evening – April 14, 1865. At around half-past nine a swashbuckling actor named John Wilkes Booth walked into the President's unguarded private box, drew a short-barrelled brass pistol, and shot Lincoln once through the back of the head. Booth then jumped onto the stage, brandished a large dagger, and cried: '*Sic semper tyrannis*! [Thus always with tyrants!] The South is avenged!'

Booth, who had sworn to take vengeance on Lincoln for the recent defeat of the Confederate forces in the American Civil War, escaped through a backstage door and

UNAWARE *With murder in mind, John Wilkes Booth creeps up behind Lincoln as he watches a play.*

SETTING UP THE NEWS *Reuter telegraphers in a cable hut in Ireland have the grim task of relaying the news of Lincoln's violent death.*

READ ALL ABOUT IT! *Shocked Victorian gentlemen, avid for the latest information, scramble to buy copies of* The Times *from a newsvendor in central London, and to read the complete, dramatic account of President Lincoln's assassination.*

rode off into the night. Lincoln died a few hours afterwards without regaining consciousness. The assassin made for Virginia, where 12 days later he was cornered in a barn by Union troops. During the ensuing siege, the barn was burned to the ground and the 27-year-old Booth was shot dead.

Dispatched by private telegraph

Meanwhile, the *Nova Scotian* ploughed on towards the telegraph station at Greencastle, near Londonderry, in the north of Ireland. Many Reuter dispatches were put on mail ships bound for Southampton, in the south of England, which sailed by way of southern Ireland – where they passed the tiny harbour of Crookhaven. There Reuter had set up a private telegraph line to the cable office at Cork, 60 miles (96 km) away, to convey reports of the American Civil War. As the mail ships neared Crookhaven, the dispatches were placed in canisters and lowered into the sea. They were picked up by a Reuter steam tender, and taken to Crookhaven.

From there, the news travelled down Reuter's private line to Cork – and then, by ordinary telegraph line, on to London. They reached Reuter headquarters in the City long before the mail ships approached Cork. Even without using this 'short cut', McLean's dispatch on Lincoln's death, telegraphed from Greencastle, appeared in the London newspapers on the morning of April 26, 1865; his enterprise won him the nickname of 'Tugboat' among his fellow newsmen.

Another major scoop for Reuters – an account of the relief of Mafeking on May 17, 1900 – came during the Boer War between Britain and the Boers, or Dutch colonists, in South Africa – at a time when British troops under Colonel Robert Baden-Powell had been besieged in the town of Mafeking for almost seven months.

Message in a sandwich

On learning that a British relief force was on its way, the Boers retired. Word of their withdrawal soon reached Pretoria, capital of the Boer republic of Transvaal, where it was picked up by a Reuter correspondent, William H. Mackay. To avoid censorship, Mackay hurried to the frontier with Portuguese Mozambique, where he gave an engine-driver £5 to hide the message in one of his lunch sandwiches and deliver it to the offices of the Eastern Telegraph Company in the capital of Mozambique, Lourenço Marques (now Maputo).

The message was then telegraphed to Reuters in London, two days before the official Government announcement. A copy of Mackay's telegram was rushed to the Lord Mayor, who read it out to jubilant passers-by from the steps of the Mansion House, in the heart of the City.

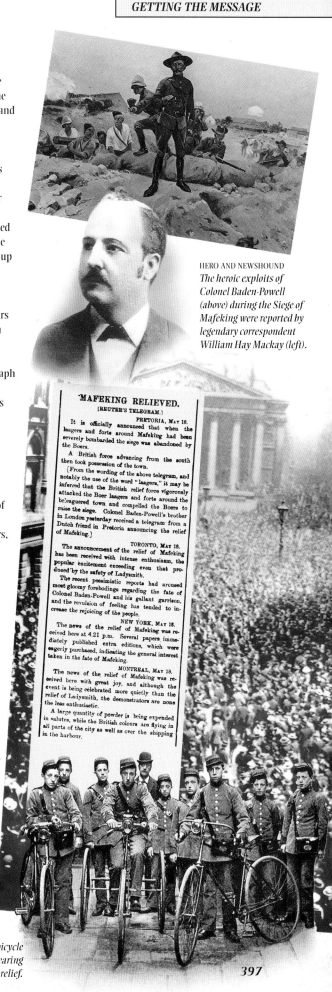

HERO AND NEWSHOUND *The heroic exploits of Colonel Baden-Powell (above) during the Siege of Mafeking were reported by legendary correspondent William Hay Mackay (left).*

MAFEKING RELIEVED.
[REUTER'S TELEGRAM.]

PRETORIA, MAY 18.
It is officially announced that when the laagers and forts around Mafeking had been severely bombarded the siege was abandoned by the Boers.

A British force advancing from the south then took possession of the town.

[From the wording of the above telegram, and notably the use of the word "laagers," it may be inferred that the British relief force vigorously attacked the Boer laagers and forts around the beleaguered town and compelled the Boers to raise the siege. Colonel Baden-Powell's brother in London yesterday received a telegram from a Dutch friend in Pretoria announcing the relief of Mafeking.]

TORONTO, MAY 18.
The announcement of the relief of Mafeking has been received with intense enthusiasm, the popular excitement exceeding even that produced by the safety of Ladysmith.

The recent pessimistic reports had aroused most gloomy forebodings regarding the fate of Colonel Baden-Powell and his gallant garrison, and the revulsion of feeling has tended to increase the rejoicing of the people.

NEW YORK, MAY 18.
The news of the relief of Mafeking was received here at 4.21 p.m. Several papers immediately published extra editions, which were eagerly purchased, indicating the general interest taken in the fate of Mafeking.

MONTREAL, MAY 18.
The news of the relief of Mafeking was received here with great joy, and although the event is being celebrated more quietly than the relief of Ladysmith, the demonstrators are none the less enthusiastic.

A large quantity of powder is being expended in salutes, while the British colours are flying in all parts of the city as well as over the shipping in the harbour.

SPREADING THE NEWS *Reuter bicycle messengers raced through London bearing the news of Mafeking's relief.*

397

The spark that set the world talking

HOW SCIENTISTS DISCOVERED RADIO WAVES

TO A LAYMAN, THE SPARK WHICH ARCED BETWEEN THE two metal balls with no apparent energy to power it would have looked like magic, but to German physicist Heinrich Hertz it was the culmination of three years' work. What Hertz was doing was trying to prove the hypothesis that electromagnetic waves could be transmitted through the atmosphere. The experiment, conducted in 1888, was based on a theory propounded 24 years earlier by James Clerk Maxwell at Cambridge University. In 1864, Maxwell proposed that when an electric current passed through a conductor, such as a piece of wire, it produced atmospheric disturbances in the form of invisible electromagnetic waves.

It was already known that if an electric current were passed through an induction coil – two coils of wire wrapped round an iron core which ended in two metal knobs – a spark jumped across the narrow gap between the knobs. To test Maxwell's theory, Hertz made a second 'spark gap' a few feet away by holding a copper loop – also with a ball at each end. When Hertz operated the coil to make a spark jump across the first set of knobs, the second set also produced a spark. The first spark had sent out electromagnetic waves which were received by the copper loop, causing the second spark.

Messages fly through the air

In an attic workshop in his parents' house in Bologna, the 20-year-old Guglielmo Marconi read about these experiments after Hertz's death in 1894 and set to work to try to repeat them. His early transmitters and receivers were similar to Hertz's coil and loop. Marconi added to the receiving loop a Branly coherer – a small glass tube containing metal filings – which could detect far weaker signals than the crude copper knobs. In 1894, he succeeded in sending his first radio signal – a simple electric pulse. Within 30 years, however, 'talking' wireless was to become a significant part of Western society's everyday life.

THE THEORIST Physicist James Clerk Maxwell argued that radio waves could be produced, but lacked the means to put his theory to the test.

THE PIONEER The experiments of Heinrich Hertz proved Clerk Maxwell's theory and paved the way for the pioneers of radio.

PRACTICAL PROOF
Marconi carries out early experiments in the garden of his family estate.

THE VISIONARY Guglielmo Marconi used Hertz's work as the basis for the first radio transmitters and receivers.

'Can you hear anything, Mr Kemp?'

HOW WIRELESS SIGNALS WERE FIRST SENT OVER THE ATLANTIC

CROUCHED OVER A PRIMITIVE WIRELESS RECEIVING SET, the Italian physicist Guglielmo Marconi listened eagerly through a pair of earphones, anxious to catch the first wireless signal sent 3000 miles (4800 km) across the Atlantic. It was December 12, 1901, and Marconi's makeshift radio station, perched on a hill above St John's, Newfoundland, was being battered by fierce winds and freezing rain. Nearby, the kite carrying the station's aerial was in danger of being blown from its moorings.

Achieving the impossible

Even so, at 12.30 pm, Marconi heard the prearranged Morse signal – three easily keyed dots standing for the letter 'S' – coming faintly through the atmospheric crackle and hiss from a powerful transmitter station at Poldhu Cove in south-west Cornwall, England. The signal was the first of a series being sent by one of Marconi's British assistants.

Not trusting his own hearing alone, Marconi passed the earphones to his chief assistant, George Kemp. 'Can you hear anything, Mr Kemp?,' he asked. Kemp listened intently and confirmed that he, too, could hear the signals. The series of three clicks and then a pause came through twice more during the next 2 hours. The Italian inventor had achieved what his fellow scientists had said was 'impossible'. The next day Marconi repeated his experiment. Despite gale-force winds and driving snow he picked up more signals – increasingly fainter and harder to hear.

Proving the impossible

The press hailed Marconi's breakthrough as an unqualified triumph. But his scientific rivals were not so impressed. Thomas Edison said that he 'didn't believe a word of it', and in London the renowned physicist Professor Oliver Lodge wrote to *The Times* newspaper describing Marconi's announcement as 'incautious'.

To confound the doubters, Marconi decided to give a demonstration in front of neutral witnesses. After returning to England, he set sail for New York on the liner *Philadelphia*. The ship was fitted with an aerial and ultrasensitive receiving equipment, including a special printer to record Morse messages on tape. At a distance of more than 2000 miles (3200 km) from the transmitter station at Poldhu Cove the Morse letter 'S' was heard and recorded in the ship's log. 'It proved,' declared Marconi, 'what I had known all along – that radio could encompass the world!' It was another 20 years, however, before speech could be transmitted.

WIRELESS TO THE RESCUE *During a storm in 1899 a steamship smashed into the East Goodwin lightship (right). The crew of the lightship sent out distress signals, resulting in the world's first sea rescue made possible by wireless.*

FROM SINGLE LETTERS TO SPEECH

MARCONI'S PIONEER WORK WITH WIRELESS SIGNALS – IN which single letters were sent between England and North America – was taken a step further in December 1906 when the world's first radio broadcast featuring speech and music was made at a wireless station in Massachusetts by Reginald Aubrey Fessenden. A Canadian-born former university professor, Fessenden used a special transmitter which varied the radio waves to correspond with the sound waves from his voice. His Yuletide programme was heard by ships' radio operators off the nearby coast.

Marconi talks to Australia

Then, in 1915, speech crossed the Atlantic by radio for the first time when the American Telephone and Telegraph Company sent a programme from a US Navy wireless station in Virginia to a French military radio station on top of the Eiffel Tower in Paris. Two years later Marconi began experiments on what became known as 'very high frequency' transmissions – above the frequency range of previous equipment. He received signals from a station in Caernarfon, Wales, on a receiver 20 miles (32 km) away.

By the early 1920s, short-wave wavelengths were being used for long-distance communication in both Britain and the United States. By the end of 1923, broadcasts from a more powerful short-wave transmitter were picked up in London, some 3500 miles (5600 km) away. And by 1924 short-wave radio had improved so much that Marconi was able to make the first speech broadcast from England to Australia.

Buying and building cat's whiskers

By then many listeners had bought or made their own cut-price crystal sets on which to hear their favourite programmes. Schoolboys as well as adult wireless buffs spent hours creating and refining their cat's whisker sets – named after the thin wire which found the right point on the surface of the crystal to detect radio waves. The enthusiasts exchanged tips and information, as well as trading the odd bits and pieces needed for the sets' construction – for which radio magazines gave detailed instructions. In order to get good reception, the sets had to be no more than some 60 miles (96 km) from the nearest transmitter. By 1927 more than 2 million cat's whisker sets – either shop-bought or homemade – were in daily use in Great Britain alone.

YOUNG EARS *A worried mother tells a friend that she cannot tear her schoolboy son away from his new-fangled wireless set.*

LIFESAVER *Mr Punch of magazine fame (above) thanks Marconi for the many lives saved at sea by his great invention.*

SAVING OUR SHIPS *In April 1912 the first SOS message was sent by the liner* Titanic *before she sank after striking an iceberg on her maiden voyage across the North Atlantic. A New York cartoonist of the day shows Marconi informing Father Neptune that a wireless SOS message will bring lifeboats to the scene of any disaster at sea.*

Moving pictures from a hatbox lid

HOW BAIRD CREATED THE FIRST WORKING TELEVISION

THE IMAGE FLICKERED, SWAM, DISAPPEARED AND CAME back. Suddenly it was recognisable as a man's head – and it was moving. John Logie Baird had at last transmitted, from one attic room in London's Soho to another, the world's first live television image. It was a picture of William Taynton, an apprehensive junior from the solicitors' office downstairs, sweating under a battery of bare bulbs.

By 1925, when Baird made that transmission, inventors around the world were racing to find a way to transmit moving images electrically. The wild-eyed, long-haired Scot, one of whose greatest creations so far had been the 'Baird Undersock' to keep feet warm and dry, had beaten them all.

The 'televisor' in action

Baird's 'televisor' divided an image into strips of light and shade by scanning it with a spinning disc based on an earlier German invention. The disc's 30 lenses, arranged in a spiral, 'saw' only a small strip of the image at a time; more revolving discs, one slotted, and another with a thin spiral aperture, transformed these strips into a single beam of light of varying intensity.

Immediately behind the array of discs lay a selenium cell, which converted the light into electric current. This was amplified by a set of valves and sent by wire to the receiver, where it made a neon bulb glow with varying intensity. Viewers peered at the flickering light through another disc, which span at the same speed as the transmitter. It reversed the earlier process, reassembling the image into a tiny picture, about 1⅓ × 2 in (3.8 × 5 cm), made up of 30 vertical stripes.

Outdated by electronics

The first version of Baird's television sat on an old tea chest. One of the discs was the lid of a hatbox with holes cut out by scissors, mounted on a knitting needle and driven by a motor taken from an electric fan. The photoelectric cell ran on torch batteries.

In 1927, two years after Baird gave his first public demonstration in London, he succeeded in sending images via telephone between London and Glasgow, a distance of some 390 miles (625 km). Despite constant improvements, the mechanical scanning device was crude, and always produced a blurred image. Baird's great problem was that he never had enough light. The photocell was not very sensitive, and anyone appearing on the televisor had to sit under an enormous bank of hot lights and wear thick make-up to add contrast to their features.

The future of television lay in a different technology. Six years after Logie Baird's pioneering experiments, the Radio Corporation of America developed a way of scanning pictures electronically. RCA's breakthrough, the 'iconoscope', was to render Logie Baird's crude mechanical method redundant.

DISCS THAT SEE *Although Baird's television apparatus – an unwieldy collection of whirling discs and electric motors – was eventually superseded by a more reliable and sharper electronic system, his tireless enthusiasm for his invention earned him the title of 'father of television'.*

The Emitron cameras need strong overhead lighting. The large metal reflectors housing the powerful tungsten bulbs give a chorus of cracks and bangs as they heat up and expand

The microphone is fixed to a telescopic extending boom. The operator can adjust its position by manipulating an array of strings and pulleys

Floor lights supplement the overhead lights, giving depth and modelling

FIRST SUCCESS *The world's first televised image was a fuzzy, 16-line rendition of a simple mask (right) in March 1925. Baird later used a puppet called 'Stooky Bill', and eventually transmitted images of people.*

War of the screens

HOW THE WORLD'S FIRST TELEVISION BROADCASTS WERE MADE

GERMANY'S RULING NAZI PARTY CLAIMED IT WAS THE opening night of the world's first regular television service. On March 22, 1935, about a hundred people viewed a 90-minute broadcasting demonstration. The blurred screen in the Berliner Funkhaus – headquarters of Reich broadcasting – showed a series of speeches hailing the miracle of television.

Many countries, including Britain, France, Germany and the USA, had long been striving to improve the technical quality of television by increasing the number of scanning lines used to create the picture on the receiver screen. Scanning with a few lines was fairly simple, but produced a picture which had very little detail.

Germany takes up the challenge
In Britain, an experimental service, which produced a fuzzy image composed of 30 vertical scanning lines, had been broadcasting since 1932. When the BBC announced early in 1935 its intention to inaugurate a 'regular high-definition television service', Germany took this as a challenge, and was determined to be first.

Early German television was an uninspiring mixture of films and propaganda. Most Germans only became aware of television as a result of the coverage of the Berlin Olympics in 1936. As many as 150 000 Berliners went to one of the city's 28 *Fernsehstuben* – public television parlours – to watch up to 8 hours of live transmissions daily. They were not told that the superior electronic cameras used to cover the events had been specially imported from the United States.

Although it was the stated aim of the Nazi Party to use television as a means of mass propaganda, very few Germans had their own sets. Manufacture of the mass-market *E-1 Volksfernseher*, the 'people's television receiver', had only just started when the outbreak of war in 1939 halted production.

Britain's two-system compromise
On November 2, 1936, the BBC unveiled the first regular television service from Alexandra Palace, in north London. To begin with, broadcasts alternated between the Baird system, which mechanically scanned images into 240 lines, and the Marconi-EMI system (similar to the 'iconoscope' invented by RCA) which used an electronic camera tube to create a sharper 405-line picture.

Those wealthy enough to have televisions (which retailed for up to £150 each) could settle down to gaze at the screen for just 2 hours a day – from 3 pm to 4 pm and then from 9 pm to 10 pm. That first afternoon, the opening ceremony was followed by Chinese jugglers and an American comedy duo called 'Buck and Bubbles'.

In February 1937 the BBC abandoned the Baird system altogether. Its technicians far preferred Marconi-EMI's electronic camera, which was movable and quiet in operation, to Baird's noisy contraption, which had to be bolted to the floor and housed in a soundproof cubicle.

LIVE ENTERTAINMENT *Shortly after 9 o'clock one evening early in 1937, 'Picture Page' goes out live from Studio 'A' in Alexandra Palace, north London. Over a thousand viewers within 25 miles of the transmitter tuned in to this popular magazine programme, presented by Joan Miller. One of the regular features was a look at the latest dance steps. Scene hands have arranged the simple wooden boxes, pillars and drapes – all painted a uniform grey – to turn one corner of the cramped studio into a 'ballroom'. The brilliantly lit scene before the cameras contrasted with the surrounding darkness, where the technicians and orchestra all prayed that the electronic cameras would finish the evening's transmission without breaking down.*

Without zoom lenses, the cameras are simply pushed towards the performers for close-ups. As camera no. 2, fitted with a lens hood to prevent flare from the studio lights, moves in, the assistant walking behind the cameraman takes special care of its fragile cable

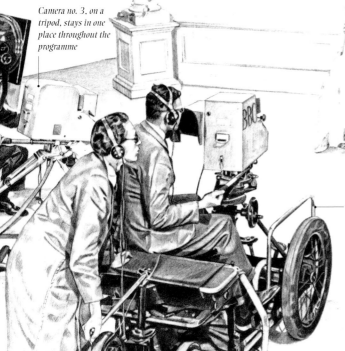

Camera no. 3, on a tripod, stays in one place throughout the programme

Camera no. 1 is mounted on a wheeled 'dolly'. The cameraman, looking into his side-mounted viewfinder, and his assistant, who pushes the dolly, receive instructions from the show's director through their headphones

Banishing the curse of Babel

HOW A POLISH DOCTOR CREATED AN INTERNATIONAL LANGUAGE

GROWING UP IN 19TH CENTURY POLAND, WHERE constant tension arose between four ethnic groups speaking Polish, Russian, German and Yiddish, Ludwik Zamenhof dreamed of a universal tongue. He believed that if people could speak a second, common language in addition to their own, international understanding would be encouraged and a new, more peaceful world would be created.

Zamenhof spoke Russian, Polish and Yiddish with his family and learned Latin, Greek, German, French, English and Hebrew at school. He had qualified as a doctor when, in 1887, he published his first book in the new language he had devised, under the pen name 'Dr Esperanto'. 'Esperanto', meaning, 'one who hopes', was also the new language's name.

Easy to say, easy to spell

Zamenhof based his grammar on English, but for his vocabulary he drew on all the major western European languages. He created an extensive system of prefixes and suffixes which, linked with root words, give shades of meaning. The largest Esperanto dictionaries contain roots from which 150 000 words can be made. English speakers are likely to recognise about 70-75 per cent of Esperanto words, Slav speakers about 50 per cent.

There are 28 letters in the Esperanto alphabet, and each is sounded in only one way. To make things easy, all words are spelt as they are pronounced and the emphasis falls on the last syllable but one. Adjectives end in 'a', nouns in 'o' and plurals of these are made by adding 'j'. There are six simple verb endings – for example, *lerni* ('to learn'); *lernas* ('learns'); *lernis* ('learnt'); *lernos* ('will learn'); *lernus* ('could learn'); and *lernu* ('learn!').

Esperanto has its own literature – there are more than 30 000 books written in or translated into Esperanto – and hundreds of thousands of people worldwide still speak the language.

LIVING LANGUAGE
Shakespeare's works (top) and cheques have been printed in Esperanto, and pictures of its creator, Dr Zamenhof (second top), have appeared on stamps and coins.

The president who taught his country the way to speak

HOW THE TURKISH LANGUAGE WAS REINVENTED

A THOUSAND DELEGATES PACKED THE ORNATE THRONE room in Istanbul's Dolmabahçe palace. Loudspeakers broadcast their debate to the crowds outside, and radio relayed it throughout the country. The scholars and politicians inside the palace were discussing a question of vital importance for every citizen – what the country's language was to be. Presiding over the first meeting of the Turkish Language Association in 1932 was the republic's first president, Mustafa Kemal – known later as Kemal Atatürk.

Over the centuries, Turkey had developed two separate languages. The official language, used by the educated, was a hybrid of Persian, Arabic and the ancient Turkish tongue. The second was the language spoken by ordinary people. Kemal reasoned that one way to unite the country was to create a truly Turkish language for all its citizens.

An important step away from the past, and towards the West, was the replacement of the old Arabic script with a Latin alphabet based on a phonetic transliteration of the Turkish. The new 'Turkey-Turkish' closed the space between the written and the spoken word, and was eventually understood by any citizen who had mastered the alphabet. One of the main benefits of the reform was replacing the former ornate, roundabout way of speaking with a direct and simpler style which got straight to the point.

MIDDLEMAN *The 'new' Turkish of Kemal Atatürk (above) bridged the gap between the old-style letter-writer (top) and the modern newsvendor (bottom).*

Calculating to win the war

HOW THE FIRST ELECTRONIC COMPUTERS WENT INTO ACTION

THE SECOND WORLD WAR WAS AT ITS HEIGHT, AND SCIENTISTS IN Britain, America and Germany were striving to build an electronic computer that would help to achieve victory. A team of British scientists, led by mathematician Alan Turing, set the pace in December 1943 with the world's first practical computer, Colossus, at Bletchley Park, 47 miles (76 km) northwest of London.

Success and failure
This huge machine was called Colossus because of its vast size. It contained a photoelectric punched-tape reader which processed 5000 characters a second. Its job was to try to decipher top secret German military messages, known to the Allies under the name of 'Fish', that were sent by various radio teleprinters. Colossus had its successes – notably against the Lorenz cipher machine nicknamed 'Tunny'. It also had its failures: one formidable adversary was a fiendishly clever cipher machine, the Siemens 'Sturgeon', or *Geheimschreiber* ('Secret Writer').

Meanwhile, the Americans were working on ENIAC – the Electronic Numerical Integrator and Computer. This, the first fully electronic computing machine, was built to calculate the trajectory of artillery shells. The 30 ton monster was completed at the University of Pennsylvania in 1945, and went straight into action.

War and peace
Meanwhile, the Germans, led by an outstanding scientist named Konrad Zuse, had made their own computers. In 1934 Zuse was at the Technical University of Berlin, training to be a civil engineer. This involved a great deal of laborious calculations, and he turned to mechanical ways of doing them. He built two prototype computers – Z1 and Z2 – which stored numbers in binary form. This system has only two numbers – 0 and 1 – rather than the ten of the decimal system.

Impressed by Zuse's ideas, the German Aeronautical Research Institute in Berlin sponsored the production of his third computer: the Z3 – used to calculate the vibration of airframes under stress. Next, Zuse built a fourth computer, known as Z4, for war service. It was used by the Aerodynamic Research Institute in Göttingen to study the interaction between airflow and the flight of aircraft.

In April 1945, as victorious Russian troops advanced on the city, Z4 was taken by truck to a house in the village of Hinterstein in the Bavarian Alps. There it was discovered in a cellar by British and United States soldiers, and was later put to peaceful use. In 1950 the computer was moved to the Federal Technical University in Zürich, Switzerland, and for the next five years it carried out calculations on aerodynamics.

Zuse went on to set up his own small computer manufacturing company. His Z22 model was one of the first in the world to use transistors. The company was later taken over by Siemens, and Zuse went back to research work. In the 1960s his pioneering work was finally recognised throughout the Western world and he received honorary appointments at the Technical University of Berlin and at Göttingen University.

BRITISH BRAINS *Alan Turing, a brilliant mathematician, was one of the brains behind the highly secret code-cracking efforts at Bletchley Park. His work on a device called the Bombe helped decode the German military cipher 'Enigma'.*

GERMAN GENIUS *At the end of the Second World War, Konrad Zuse, the German computer wizard (left), started his own small manufacturing company. He later concentrated on research work, and decades later his pioneering genius was recognised throughout the world.*

TEAMWORK *In the late 1940s two of Turing's colleagues, Tom Kilburn and G.C. Tootill, devised a computer program (left) to find for any given numeral the highest number which will fit into it an exact number of times. For this, they employed a binary system similar to Zuse's, using two numbers, 0 and 1.*

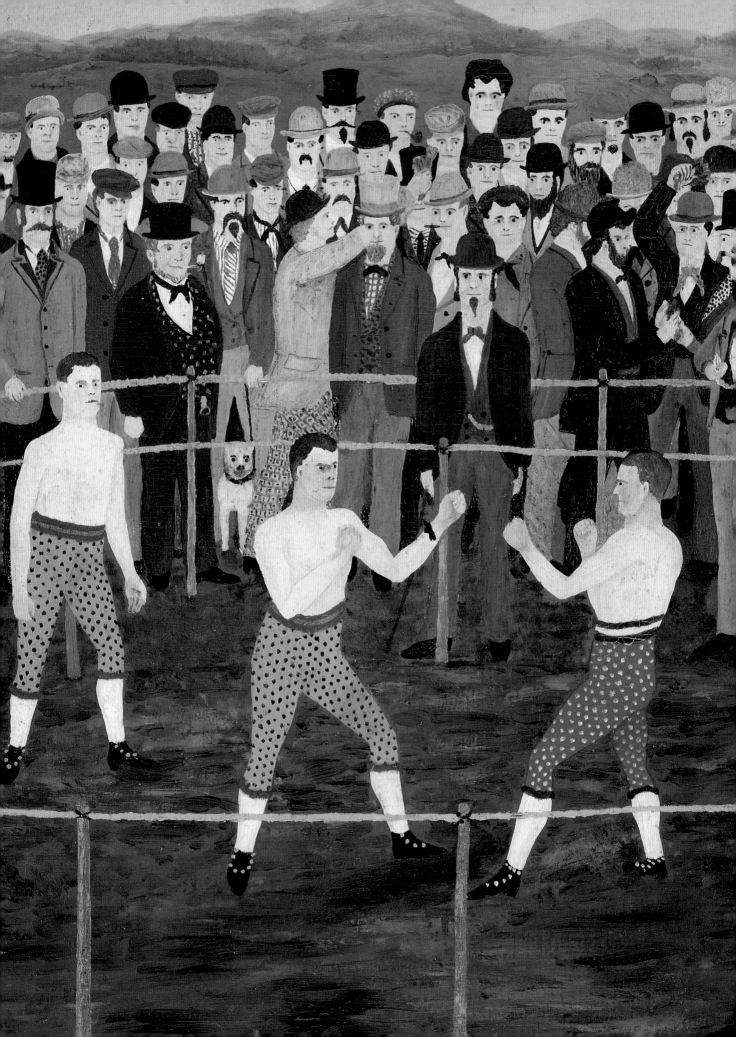

Arts and Leisure

FROM THE EARLIEST TIMES, PEOPLE HAVE GATHERED TO ENJOY SPORT AND OTHER ENTERTAINMENTS. WHAT WE DO WITH OUR SPARE TIME REFLECTS THE WORLD WE LIVE IN: TODAY'S CINEMA AND PHOTOGRAPHS ARE SUCCESSORS TO THE OPEN-AIR THEATRES AND CAVE PAINTINGS OF THE DISTANT PAST

A canvas made of rock

HOW THE FIRST ARTISTS DEPICTED THE ANIMALS IN THEIR WORLD

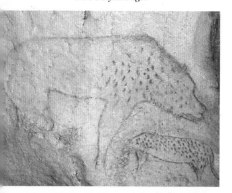

ON THE PROWL *A prehistoric creature, either a hyena or a bear, prowls the cave walls in a mural painted some 30 000 years ago.*

EARLY BIRD *This lively owl is one of the earliest known portrayals of a bird. The ancient artist drew it with his fingers, allowing the yellow and white of the underlying rocks to show through.*

GIGANTIC SHADOWS DANCED ON THE WALLS of the cave as the flames from torches and lamps of burning animal fat flickered. The artists worked on in the half-light, one crouching on a primitive wooden scaffold to reach the highest part of the wall with the smoothest surface. Another engraved a bold line on the dazzling white limestone with a sharpened flint, tracing the outline of a woolly rhinoceros, the curve of the cave wall fleshing out its belly. Mounds of coloured powder on the floor of the cave were the raw material that the artists mixed with animal fat to create red and black paints, which they stored in hollow bones.

Animal figures roam the walls

In time, large panels all over the cave were covered with dramatic depictions of Ice Age animals: deer, horses, bears, lions, bison, mammoths, ibex, a panther, an owl, and enormous wild bulls. There were also a large number of rhinoceros and large cats, which are unknown in other cave paintings. Fluid lines brought the shapes to life, an exaggerated horn here, an extra leg there, a configuration of red spots filling in the outlines. The paintings – more than 300 works – remained undisturbed for more than 30 000 years until, in December 1994, they were rediscovered by chance by amateur archaeologists in the cave of La Combe d'Arc, near Vallon-Pont-d'Arc in the Ardèche region of France.

Stone Age spray painting

The amazing cavalcade of animals at La Combe d'Arc appears to have been produced by a combination of hand and a form of spray painting which had been devised by the Stone Age artists. Having ground up charcoal and rocks containing iron carbonate and manganese oxide to create their pigments, the artists mixed them with their own spittle and then blew, or spat, onto the cave wall.

The presence in the pigments of manganese oxide, which is known to affect the body's central nervous system, may account for the unusual depiction of some of the animals in the paintings. Perhaps mixing paint in the mouth before projecting it onto the wall led to the artists experiencing a hallucinatory effect, which may account for some of the distortions in the cave images and their powerfully mystical ambience. The masterpieces certainly had some kind of religious or social importance, but exactly why they were created is still a mystery to historians, although it is clear that many hands, and maybe mouths, were involved in producing them.

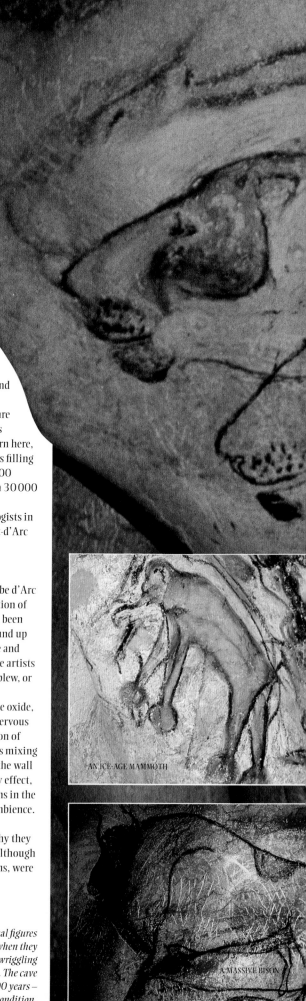

AN ICE-AGE MAMMOTH

A MASSIVE BISON

A MURAL STAMPEDE *This colourful array of animal figures was one of the first things archaeologists saw when they played their lamps on the cave walls after wriggling through a tiny gap in the rocks at La Combe d'Arc. The cave had been closed by a landslide for some 18 000 years – which had kept the paintings in pristine condition.*

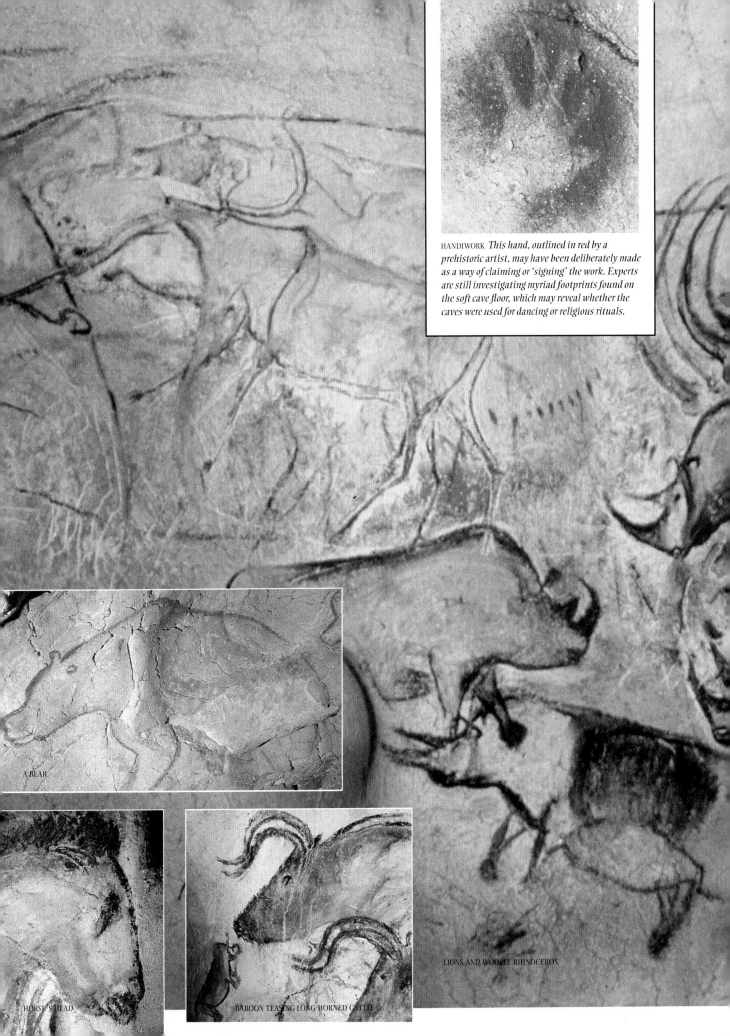

HANDIWORK *This hand, outlined in red by a prehistoric artist, may have been deliberately made as a way of claiming or 'signing' the work. Experts are still investigating myriad footprints found on the soft cave floor, which may reveal whether the caves were used for dancing or religious rituals.*

A BEAR

LIONS AND WOOLLY RHINOCEROS

HORSE'S HEAD

BABOON TEASING LONG-HORNED CATTLE

A chorus line of clouds

HOW ATHENIANS SPENT A DAY AT THE THEATRE

TAKING A DEEP BREATH TO CALM HIS NERVES, THE ACTOR sat up in bed, centre stage. 'Almighty Zeus! What a night!' he declaimed through the tiny megaphone mouthpiece of his mask. 'It's going on for ever. Will daylight never come?'

As the March dawn of 5th century BC Athens broke over the distant cypress-framed Aegean, the audience settled back in their seats, reassured that *The Clouds*, Aristophanes' latest play, was up to his usual standard. They were an expert audience – weaned on the comedies of Cratinus and Eupolis, whose plays do not survive, and the tragedies of Aeschylus, Euripides and Sophocles. They sat through 20 or more plays a year. Even so it was a surprise when the chorus came on dressed as clouds.

The huge open-air theatre on the southern slope of the Acropolis was filled with a capacity crowd of 30 000. Everyone had been encouraged to go. Prisoners were let out of jail, trade and politicking was suspended and a public holiday was declared. Those who could not afford to pay had been let in free and reimbursed a day's wages. Today would be a long day; this was only the first of five plays on day two of the Great Dionysia, the most important of the two annual theatre festivals in honour of the god Dionysus. The other, the Lenaia, which had taken place in January, was devoted entirely to comedy. The show would continue until sunset.

Admission was by lettered token covering all of the day's plays and indicating a particular seat. In the front row special seats were reserved for the ten judges and other dignitaries. Everyone brought their own cushions. Officials with staves were at hand to keep order in case the crowd got out of control.

The stage was backed by the *skene*, a stage building that provided all the scenery. It had a central door which opened, revealing the inside of a house or temple. A dais on wheels could be pushed through the door out on to the stage. It was often used in tragedies to reveal the body of a character slain off stage.

Behind the stage and *skene* were buildings used as dressing

SAVED BY A GOD *A scene perhaps from Euripides, in which Alkmene, a woman condemned to be burned alive for infidelity, is showered with rain by Zeus (top left) as she sits on the pyre, decorates a wine vase from the 4th century BC. Their child, Heracles, became a Greek legend in his own right.*

THEATRICAL EXPRESSIONS *The mask (left) is a clay copy of one worn in tragedy, while the statuette (below) is based on a comedy character. Both date from the 2nd century BC. Comic actors often wore padded costumes.*

SETTING THE SCENE *In Greek theatres, the* skene *was a stone building which not only provided changing areas and storerooms, but the permanent framework for decorative wooden sets. The remains of the Theatre of Dionysus (right) can still be seen at Athens, complete with marble seats where the spectators sat.*

rooms and stores. There could be found the crane or *mechane* that enabled some characters, usually gods, to fly on to the stage.

In Greek drama, the three main actors took all the parts, male and female. Because of the huge size of the theatre, gestures had to be stylised and exaggerated and actors wore platform shoes and huge elaborate wigs to make themselves easier to see from the back rows. They wore large masks made of stiffened linen and painted with fixed expressions to show mood and temperament. Costumes were in colours appropriate to the characters – bright ones for happy people, dark ones for tragic. In comedies some characters might wear grotesque, padded outfits.

Organising the big event

On the first day of the Dionysia festival songs were sung to the Dionysus. All the plays were performed in honour of Dionysus, the savage god whose followers, the *bacchae*, became one with him through an orgy of drink and ecstatic dancing. Each of the ten tribes that formed the Athenian state sent in two choruses, both 50 strong – one of men, one of boys. The second day was devoted to the performance of comedies. Five authors were commissioned by the *eponymous archon* – the chief magistrate of Athens who served as festival organiser – to write them. Playwrights submitted their work to the *archon*, who chose the ones he liked most. Even the most celebrated writers, such as Sophocles and Euripides, had to

go through this process. Then the *archon* assigned the play to a rich citizen – a *choregoi* – elected to pay for the production. The *choregoi* was allocated a chorus, which he was expected to train in their songs and dances, and a troupe of professional actors. The actors were usually a three-man team, with possibly a couple more to take the bit parts. By the end of the 4th century there were theatrical stars who specialised in comedy or tragedy.

Over the final three days tragedies were performed. Each day was devoted to one author's work – three tragedies and a satyr play, which was a less formal production. The chorus in the satyr play dressed as followers of Dionysus, wearing horses' tails and ears.

The chorus performed on the circular *orchestra* below the stage, moving the plot along and explaining the action with songs and dances. Usually, they had the last word. 'Lead the way out,' they sing at the end of *The Clouds*, 'No more dancing today. It's all done. We've reached the end.' That year, 423 BC, Aristophanes' new play won third prize at the Great Dionysia. The best actor and the best playwright were, as usual, given wreaths of ivy.

ROLL OF HONOUR *A piece of marble, inscribed with the names of some of the Great Dionysia competition prizewinners, includes among its list of victors the name of the playwright Aeschylus, known as the father of Greek tragedy. Dated 458 BC, it records that the prize was for his trilogy* Oresteia.

IN STEP *Athletes compete in a running race. Each* Olympiad, *or period of four years between games, was named after the winner of the sprint race.*

To honour Zeus

HOW THE GREEKS ORGANISED THE OLYMPIC GAMES

THE ATHLETES MARCHED SLOWLY DOWN THE GREAT WAY to the grove of Altis at Olympia, the most sacred site of Zeus, the ruler of the gods. In two days the Olympic Games would start. So important was the event that a halt was always called to the feuding between the Greek states. Heralds were sent out around the Mediterranean, giving the official dates of the festival and announcing a truce. This year, 376 BC, the 100th games were to take place.

After the procession, the athletes and their entourages pitched their tents around the Olympic complex. Artists, souvenir sellers and vendors of food and wine shouted their wares; horse-traders made deals; politicians talked peace. As night fell, fires were built up, the smell of spit-roasts filled the air and singers chanted tales of battles won and lost.

The games took place every four years, and preparations began about a year in advance. Ten judges, the *hellanodikai*, were chosen by lot from the ruling families of the nearby city of Elis. They presided over the preparations as well as the events themselves. Tracks and accommodation were made ready. About a month before the games, athletes with trainers in tow arrived and went into a training regime under supervision of the judges, who made sure the competitors were of Greek descent, fit, properly trained and that they followed strict diets.

Sporting prowess

Over five busy days, the games combined athletics with sacrifices to the gods. The programme varied over the centuries, but long jump, boxing, chariot races, foot and horse races were usually included. Most taxing was the pentathlon, in which athletes had to prove their prowess at five disciplines: running, jumping, discus, javelin and wrestling.

Because horses were expensive to keep and to train, horse and chariot-races were events for the Greek elite. Bareback horse racing in the hippodrome was especially dangerous, because it was held after chariot races which damaged the track. The biggest crowd-puller was the *pankration* – wrestling with no holds barred. Eye-gouging and biting were forbidden, but a favourite trick was to break an opponent's fingers.

A win at the Olympics was the greatest triumph a Greek athlete could achieve. Although the prize was initially just a wreath of wild olive leaves, it was

LONG SHOT *An athlete prepares to hurl his javelin into the air. The Greeks used a leather thong wrapped around the javelin which unwound as the weapon was launched, making it spin and therefore fly straighter.*

VICTOR'S WREATH *A victorious athlete is crowned with a circlet of wild olive leaves. At Olympia, sacred olives grew wild in a grove behind the great Temple of Zeus.*

ROUGH AND TUMBLE *A wrestler acknowledges defeat by raising his index finger. Wrestlers, like all the other athletes, competed naked and covered with olive oil, but they also coated themselves with a layer of dust to enable opponents to get a grip.*

SACRED STADIUM *The main stadium at Olympia, built around 350 BC, stands outside the temple complex, but all races were arranged to finish at the end nearest the temple. The track was straight, instead of oval, so athletes ran back and forth in longer races.*

FLYING LEAP *An athlete takes part in a jumping contest holding halteres or jumping weights. The record was 53 ft (16 m), so the event must have been a triple jump – a hop, a skip and a jump. To help the athletes to concentrate during the long jump – considered to be the most difficult event – music was played.*

accompanied by adulation. An individual's win was a win for his city. Athletes returned home to triumphal processions and civic banquets. Poets were hired to write odes in their honour, statues were erected to them and an Athenian winner could look forward to dining at the city's expense for the rest of his life.

By 376 BC, the pressure to win had already begun to erode the ideal of competing to honour Zeus.

Though the games were open to all, only the wealthy could afford to make the trip and hire a trainer. Professional athletes travelled around the ancient world, going from one competition to the next. The games stopped being held – until their modern revival in 1896 – at the end of the 4th century AD, possibly because the Christian emperor, Theodosius I, objected to their setting at the heart of a pagan festival.

WEIGHT AND BALANCE *Hand-held weights used by athletes to give them extra momentum in the long jump were either cast in bronze (above right) or carved from stone (above left) with finger holds. They weighed between 2 lb and 10 lb (1-4.5 kg).*

AGILITY AND GRACE *Twisting his torso, an athlete (left) begins to throw his discus. Greeks made only a single three-quarter turn instead of spinning twice like modern athletes. A typical bronze discus (below) from the 6th century BC weighing 3 lb (1.35 kg) is dedicated to Castor, the son of the god Zeus, renowned for his discus throwing.*

FIT FOR BATTLE *Contestants wearing helmets and leg protectors and carrying shields (left) run in the hoplitodromos, the race-in-armour, which was introduced in 520 BC.*

Necklaces fit for the gods

HOW THE CELTS MADE BEAUTIFUL GOLD TORQUES

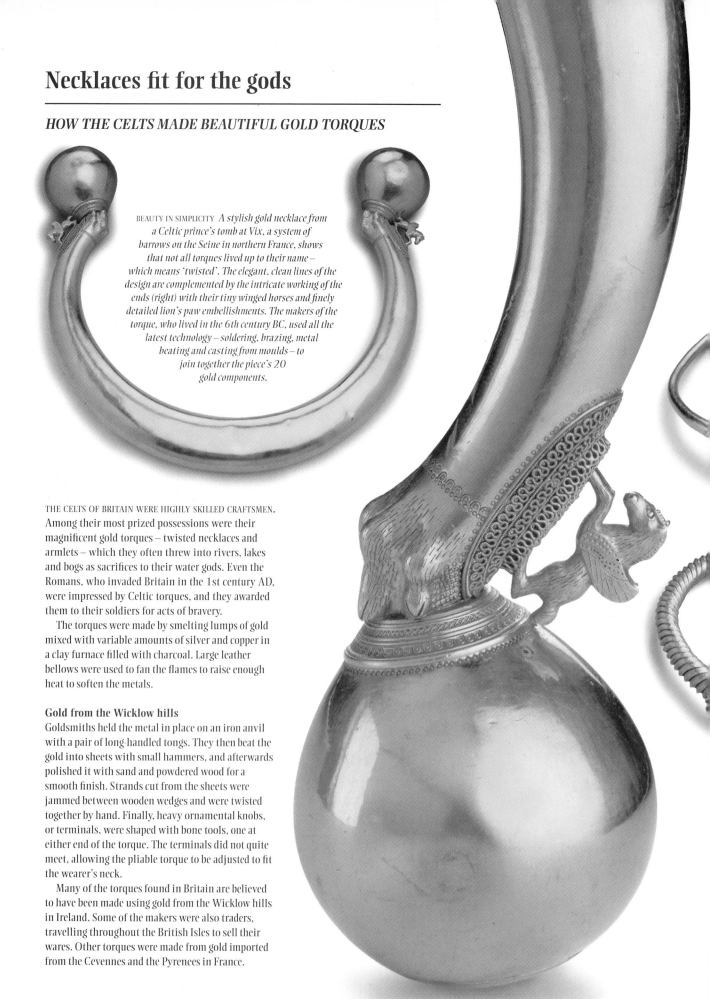

BEAUTY IN SIMPLICITY *A stylish gold necklace from a Celtic prince's tomb at Vix, a system of barrows on the Seine in northern France, shows that not all torques lived up to their name – which means 'twisted'. The elegant, clean lines of the design are complemented by the intricate working of the ends (right) with their tiny winged horses and finely detailed lion's paw embellishments. The makers of the torque, who lived in the 6th century BC, used all the latest technology – soldering, brazing, metal beating and casting from moulds – to join together the piece's 20 gold components.*

THE CELTS OF BRITAIN WERE HIGHLY SKILLED CRAFTSMEN. Among their most prized possessions were their magnificent gold torques – twisted necklaces and armlets – which they often threw into rivers, lakes and bogs as sacrifices to their water gods. Even the Romans, who invaded Britain in the 1st century AD, were impressed by Celtic torques, and they awarded them to their soldiers for acts of bravery.

The torques were made by smelting lumps of gold mixed with variable amounts of silver and copper in a clay furnace filled with charcoal. Large leather bellows were used to fan the flames to raise enough heat to soften the metals.

Gold from the Wicklow hills

Goldsmiths held the metal in place on an iron anvil with a pair of long-handled tongs. They then beat the gold into sheets with small hammers, and afterwards polished it with sand and powdered wood for a smooth finish. Strands cut from the sheets were jammed between wooden wedges and were twisted together by hand. Finally, heavy ornamental knobs, or terminals, were shaped with bone tools, one at either end of the torque. The terminals did not quite meet, allowing the pliable torque to be adjusted to fit the wearer's neck.

Many of the torques found in Britain are believed to have been made using gold from the Wicklow hills in Ireland. Some of the makers were also traders, travelling throughout the British Isles to sell their wares. Other torques were made from gold imported from the Cevennes and the Pyrenees in France.

GOLDEN RELIC *Celtic craftsmanship at its finest shines through in the metalwork of a gold torque of the late 4th century BC, buried with a noblewoman at Waldalgesheim in the German Rhineland. Other objects found in the tomb included a funeral dinner service and a complete two-wheeled war chariot.*

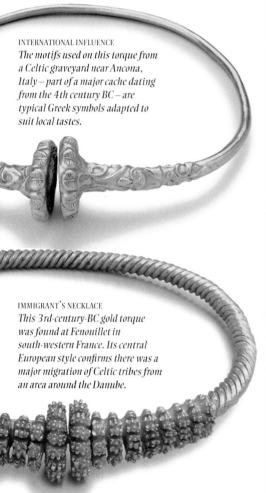

INTERNATIONAL INFLUENCE
The motifs used on this torque from a Celtic graveyard near Ancona, Italy – part of a major cache dating from the 4th century BC – are typical Greek symbols adapted to suit local tastes.

IMMIGRANT'S NECKLACE
This 3rd-century-BC gold torque was found at Fenouillet in south-western France. Its central European style confirms there was a major migration of Celtic tribes from an area around the Danube.

Torques first appeared in the 5th century BC, and are part of a distinctive Celtic culture known as La Tène, after a town in north-west Switzerland where many important objects were found. Torques varied greatly in weight and size: some were as large as 15 ft (4.5 m) long. As well as being worn by warriors, torques were frequently buried in graves, placed around the necks of women and girls. Sometimes used as armlets, and – very occasionally – as girdles, these beautiful objects were not just worn for personal ornamentation: they were also used to decorate shrines and sacred groves.

The white horse on the hill

HOW PAGAN ARTISTS DREW A GIANT PICTURE

STRAINING NOT TO SLITHER DOWN THE STEEP SLOPE, a group of Late Bronze Age Celts gathered on a hill close to what is now the Ridgeway, which runs along the top of the Oxfordshire downs in southern England. The date was somewhere between 1600 and 1400 BC, and their job was to cut a silhouette of a horse into the hillside.

First they picked out the outline, probably marking it out on the turf with cords and stakes. The artists may have worked from a small-scale drawing on wood, leather or fabric. The White Horse of Uffington, as the monument is now called, is 370 ft (110 m) long and 130 ft (40 m) tall.

Once this job was complete, filling in the outline of the white horse was a simple but laborious business. The horse was not merely carved out of the chalk, but created by cutting trenches, using bronze scrapers and wooden spades, and filling them with chalk quarried from the hill.

The image is highly stylised: an unnaturally thin body, and legs no more than a series of bold strokes across the hill. The result looks surprisingly modern, impressionistic rather than realistic.

LANDMARK *The massive White Horse of Uffington may have been cut as a totem to mark a hill-fort which was begun at the same time just above it. The community may have wanted to advertise its presence at a time of increasing competition for resources due to a rise in population and a more hostile climate. The detail of the head (above) shows the eye, which measures 4 ft (1.2 m) across.*

Building up the picture

HOW ROMAN CRAFTSMEN MADE INTRICATELY PATTERNED MOSAICS

TRUE TO LIFE *In the hands of a master, mosaics could be ambitious in both subject matter and execution. This scene of flooding on the River Nile is so highly detailed that it could almost be mistaken for a painting. A good mosaicist took care always to have a plentiful stock of naturally coloured stone so that his apprentices could readily saw and chop a fresh supply of the square* tesserae *– the pieces of stone that made up the picture.*

THE CUSTOMER COULD NOT DECIDE. HE HAD CHOSEN the border motifs and patterns representing the four seasons for the corners, but he was clearly in two minds about the central panel – should it be circular or hexagonal, and should it contain a portrait of Diana or Venus? He flipped through the pattern book again to compare designs. It was the 2nd century AD and, in search of the perfect mosaic for his dining room floor, a wealthy villa owner had come to the nearby province of Sousse to employ the master mosaicist Macarius, well known for his elaborate mosaic designs. At last he made his choice – a circular panel with a portrait of Venus. Macarius arranged for an assistant to measure the room where the mosaic was to be laid.

The mosaic takes shape

Now there remained the important question of how the mosaic would be laid. Should it be created on site, or would the customer prefer to have it prefabricated in the workshop? This time there was no indecision. The client did not want workmen tramping around his house for weeks, so he asked for the mosaic to be made in the workshop.

After the site had been measured, Macarius worked out the design himself using the pattern book. Then he gave two apprentices the patterns to follow, so that they could create the mosaic in the workshop. The design was first sketched out on sand in wooden trays about 3 ft sq (1 m²). In each square they bedded small squares of coloured stone, called *tesserae*, into the sand, carefully building up the design. When a panel was finished, glue was spread across its surface, and a sheet of linen pressed on top to help to hold the

PERFECT HARMONY *The poet and musician Orpheus charms the animals with his harp. Portraits of gods or goddesses, and images of animals or plants are among the typical subjects of Roman mosaics.*

tesserae in place. After the glue had dried, the panel was turned upside-down and a mortar of lime and sand poured onto its underside. Once the mortar had set, the panel was turned the right way up again, the glue and linen on the front were removed with hot water and the mosaic was polished. When all the panels were ready, they were carried on carts to the river where they were loaded onto a barge for delivery to the customer's villa. There, Macarius supervised their laying into a bed of mortar, fitted together like the pieces of a jigsaw puzzle.

Thrills and spills of the hippodrome

HOW A ROMAN CHARIOT RACE WAS RUN

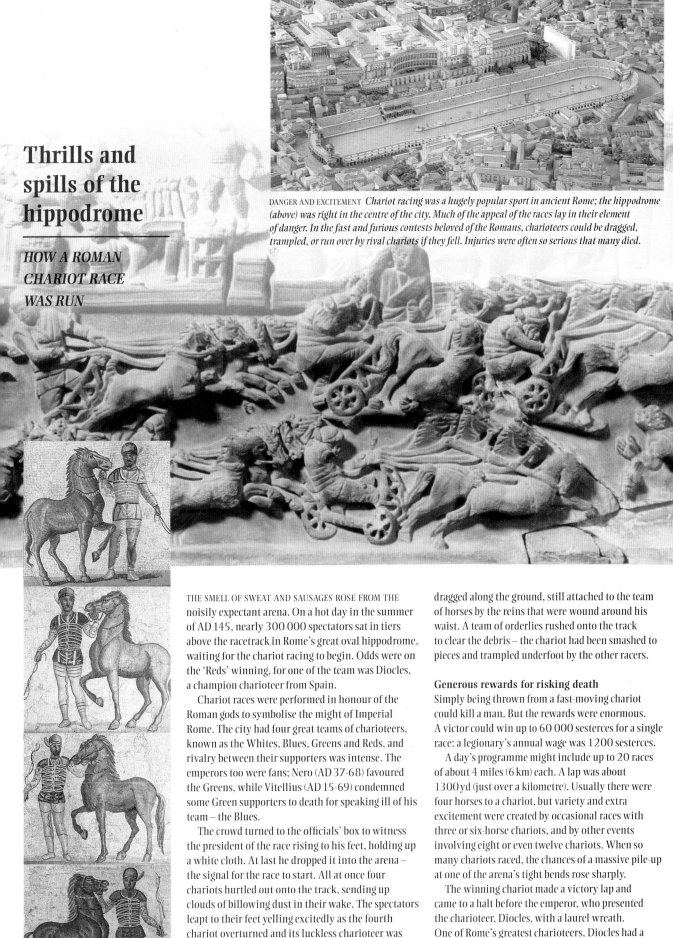

DANGER AND EXCITEMENT *Chariot racing was a hugely popular sport in ancient Rome; the hippodrome (above) was right in the centre of the city. Much of the appeal of the races lay in their element of danger. In the fast and furious contests beloved of the Romans, charioteers could be dragged, trampled, or run over by rival chariots if they fell. Injuries were often so serious that many died.*

THE SMELL OF SWEAT AND SAUSAGES ROSE FROM THE noisily expectant arena. On a hot day in the summer of AD 145, nearly 300 000 spectators sat in tiers above the racetrack in Rome's great oval hippodrome, waiting for the chariot racing to begin. Odds were on the 'Reds' winning, for one of the team was Diocles, a champion charioteer from Spain.

Chariot races were performed in honour of the Roman gods to symbolise the might of Imperial Rome. The city had four great teams of charioteers, known as the Whites, Blues, Greens and Reds, and rivalry between their supporters was intense. The emperors too were fans; Nero (AD 37-68) favoured the Greens, while Vitellius (AD 15-69) condemned some Green supporters to death for speaking ill of his team – the Blues.

The crowd turned to the officials' box to witness the president of the race rising to his feet, holding up a white cloth. At last he dropped it into the arena – the signal for the race to start. All at once four chariots hurtled out onto the track, sending up clouds of billowing dust in their wake. The spectators leapt to their feet yelling excitedly as the fourth chariot overturned and its luckless charioteer was

dragged along the ground, still attached to the team of horses by the reins that were wound around his waist. A team of orderlies rushed onto the track to clear the debris – the chariot had been smashed to pieces and trampled underfoot by the other racers.

Generous rewards for risking death

Simply being thrown from a fast-moving chariot could kill a man. But the rewards were enormous. A victor could win up to 60 000 sesterces for a single race: a legionary's annual wage was 1200 sesterces.

A day's programme might include up to 20 races of about 4 miles (6 km) each. A lap was about 1300 yd (just over a kilometre). Usually there were four horses to a chariot, but variety and extra excitement were created by occasional races with three or six-horse chariots, and by other events involving eight or even twelve chariots. When so many chariots raced, the chances of a massive pile-up at one of the arena's tight bends rose sharply.

The winning chariot made a victory lap and came to a halt before the emperor, who presented the charioteer, Diocles, with a laurel wreath. One of Rome's greatest charioteers, Diocles had a monument erected in his honour. His achievements include winning 1462 of his 4257 races.

TEAM COLOURS *The main teams of charioteers wore coats glittering with their distinctive colours so that they could be identified easily from a distance on the racetrack.*

Saintly Guido, the father of music

HOW MUSIC WAS WRITTEN DOWN IN THE MIDDLE AGES

THE SINGING MASTER AT THE BENEDICTINE MONASTERY of Pomposa in Italy, Guido of Arezzo, listened in despair as his choir sang a cacophony of conflicting notes. He sang the correct tune again and listened as they struggled to copy him. Back in his cell at the end of a hard day, Guido wondered if there was an easier way to teach his choir new songs.

Without more ado Guido, who lived from *c.*990- *c.*1033, began creating a system of notation that would simplify the job of teaching. His success in doing so earned him the name *Beatus Guido, inventor musicae* 'Saintly Guido, father of music'. He named the notes after the first syllables from six lines of a hymn to St John the Baptist:

Ut *queant laxis* • **Re***sonare fibris* • **Mi***re gestorum*

Fa*muli tuorum* • **So***lve polluti* • **La***bii reatum*

Each syllable stood for one note in the scale. *Ut* was changed to *Do* from about 1600 and by this date a seventh syllable *Si* – from **S**ancte **I**oannes, the seventh line – was added to make the modern eight-note scale and later changed to *Ti*. Guido's next step was to devise a clear way of writing the notes down.

Musical notation has two functions: it locates the pitch of each note and it shows how long each note ought to be sung or played. In the early Middle Ages,

signs that looked like accents were placed above the words of a chant. They gave the vaguest indication of pitch but they said nothing about the length of a note.

Saving music for the future

Guido's idea, which marked a revolution in the story of music, was both brilliant and simple. He placed the notes on lines (called the 'stave') or in the spaces between the lines so that the exact relationship of each note to the others could be defined. The system for indicating the length of notes, which was not required to sing medieval plainchant, was gradually perfected after Guido's death.

His fame spread. So envious were his brothers of his success that they harried him out of the monastery. He took refuge in Arezzo, from where Pope John XIX summoned him to Rome to demonstrate his new system.

Guido presented the pope with a sheet of music written according to the new method and explained its working to him. After this, his Holiness, according to a witness, 'to his great astonishment was able to sing a melody unknown to him, without the slightest mistake'.

SOCIAL CLIMBING *Troubadours who could not support themselves found a noble patron. Bernard de Ventadour (above) is said to have been the son of a blacksmith. Legend tells that he seduced his noble master's wife and took refuge at the court of Eleanor of Aquitaine – whose lover he became. About forty of his works still survive. Many were collected and written down in songbooks in the late 13th century (right) when the songs of the troubadours became fashionable all over Europe. One of his most celebrated songs ran:*

'How can I go on without baring my love to her. When I see her and the limpid look in her eyes, I can hardly stop myself from running to her. And I would run to her, but fear holds me back; for I never saw a body better formed for love, that was so languid and so slow to awaken.'

FINGER EXERCISES *A tonsured monk demonstrates Guido's method of learning the names of musical notes using the fingers of one hand.*

'My heart is singing like a nightingale'

HOW A TROUBADOUR ENTERTAINED

THE LILTING VOICE OF THE TROUBADOUR FILLED THE fire-lit hall. The ladies of the court listened intently to his plaintive song, which he sang in the *langue d'oc*, the language of southern France in the Middle Ages: 'When the nightingale sings, who delights us with her song, my heart is singing like a nightingale for my fair sweet lady.' His song conveyed the joy and pain of love as he related the story of a chivalrous knight and a pure lady.

Music of love in the wealthy south

The art of the troubadour originated in the cosmopolitan courts of southern France. Situated at the crossroads of the rich trade routes from northern Europe to the Mediterranean and Moorish Spain, wealth bought spare time. In the great halls, at Toulouse and Avignon for example, travelling players helped to while away the long winter evenings. After dinner, the fire was stoked up, the rushes on the floor were swept aside and the *jongleurs* – acrobats, musicians and singers – cartwheeled, somersaulted and sang songs about battles and kings. Jesters told lewd jokes and did conjuring tricks. When the tapers began to burn low, the courtiers gathered closer together and prepared to listen to the new art form – troubadour's love songs.

Unlike the other musicians, troubadours wrote their own lyrics – they were poets – and often of noble birth. Guilhem IX, Duke of Aquitaine and Count of Poitiers (1071-1127), is the earliest known troubadour. He wrote robust verse about lovers' trysts and the pleasures – and pain – of adultery.

'That man is dead who does not feel in his heart the sweet savour of love,' wrote Bernard de Ventadour, one of the greatest troubadours of the 12th century, expressing the essential theme of all troubadour lyrics. Through them the ideal of courtly love eventually spread throughout Europe. Courtly love was not always adulterous, but the songs expressed a heady mixture of passion and respect.

The music was strongly influenced by religious plainsongs. Sometimes, the troubadour accompanied himself on the lute or the harp, and occasionally he was supported by a group of musicians. Songs were generally passed on by word of mouth and those that survive have only a rough musical notation which gives neither the rhythm nor accompaniment. The pitch of the notes is indicated, however, and scholars have recreated the melodies by adapting the rhythm from the lyrics. In this way the great songs of the age of courtly love survived to be heard again.

ROMANTIC TRAGEDY *The troubadour Jaufre Rudel of Blaye dies in the arms of the Countess of Tripoli. Legend has it that he fell in love with her before they met because of all the good things he heard about her. After sailing across the sea to find her, he was struck down by a mortal illness but was able to die happy, for he had found his beloved.*

LOVERS' BOWER *Tales of love, and scenes such as this one of two lovers from a German manuscript, provided inspiration for wandering balladeers all over Europe.*

SONG OF LOVE *A lady sings to the accompaniment of a lute. One of the most talented poets of the era was the Countess of Die, who wrote to her lover, Raimbaut d'Orange, 'My handsome friend, gracious and charming, when will I hold you in my power?'*

BOOK OF LOVE
Romantic love songs belong in a heart-shaped book.

A mural for the monks' dining room

HOW LEONARDO PAINTED 'THE LAST SUPPER'

THE ARTIST'S FOOTSTEPS ECHOED ON THE REFECTORY floor as he walked to the scaffolding before the half-finished painting on the far wall. He mounted the scaffolding, selected a brush, made one stroke, and climbed down. Without speaking or looking at anyone, he left the way he had come.

Leonardo da Vinci – painter, architect, engineer – was known throughout Italy as a genius. His painting of *The Last Supper* in the monastery of Santa Maria della Grazie in Milan was one of a number of works he was engaged on in the city. Since he had started the painting the year before in 1495, his working methods had become well known. Sometimes he would paint nonstop from sunrise to sunset, without pausing even to take food or drink. Then days would pass without the work being touched. A brief flying visit to make one or two minor changes might follow. But he never failed to spend hours every day in front of the painting, staring at it in silence.

At table with the son of God

Christ's last meal with his disciples was a common subject for monastic dining rooms. Leonardo, however, brought his own genius to the 30 ft (9 m) painting, bending the rigid Renaissance rules of perspective. The walls in the painting do not align properly, and the angle of the table has been skewed so that it seems as if Christ and the Apostles are actually in the room. Monks sitting at the refectory table might almost have felt that they were attending the Last Supper themselves.

Innovative methods

Leonardo was an inveterate experimenter, and he tried a new technique with *The Last Supper*. Instead of painting *al fresco* – on damp plaster with water-based paints – he used a mixture of oil and tempera on a dry wall. Painting *al fresco* requires decisiveness and speed, because each section of the work has to be completed while the plaster is still damp, usually in the same day. But Leonardo liked to have the leisure to make little improvements as he went along.

When it was completed in 1497 the mural was a wonderful sight, blazing with colour. Had the painting been done as a traditional fresco, it would have survived intact. But it was already badly decayed by 1517 and was restored a number of times over the years. What can be seen today is a shadow of the original. Fortunately there are well-preserved copies, some by Leonardo's students, and some of his own sketches.

FACE OF CHRIST *The serene figure of Christ, already aware of his terrible fate, contrasts with the dismay of his followers. Leonardo made Christ's head the vanishing point of the painting, so that the viewer's eye is automatically drawn towards the face of the Saviour.*

PRELIMINARY SKETCH *Leonardo's dramatic use of gesture to convey emotion was an innovation. He described painting as 'silent poetry'.*

MOMENT OF TRUTH *'One of you will betray me,' says Christ to his disciples gathered around the table at the Last Supper. They respond with horror and disbelief. Judas (left) was traditionally shown alone on the other side of the table, but Leonardo increased the tension by including him in the group. He spent nearly a year trying to find a model with suitably wicked features, but ended up making a composite portrait.*

Beauty and grace from a lump of stone

HOW MICHELANGELO BUONARROTI CARVED 'DAVID' OUT OF A BLOCK OF MARBLE

THE SCULPTOR WORKED ON, OBLIVIOUS OF THE burning Tuscan sun, the white cloud of dust covering his body, or the marble chips that flew through the air as his chisel gouged the block before him. He worked with demonic energy, utterly absorbed as he moved from level to level of the scaffolding around the marble slab, which stood 17 ft (5 m) high.

A vast block of fine marble had been lying in the Duomo (cathedral) workshops in Florence for nearly 40 years, left untouched since Agostino di Duccio abandoned his attempts to carve it in 1464. During this attempt, the marble had been damaged so badly that it was feared the block would split in two if moved or jarred. On July 2, 1501, members of the *Opera del Duomo* – the cathedral works committee – decided to raise the block upright and ask local artists to suggest how it might be used.

The 26-year-old Michelangelo petitioned for the work, possibly submitting a model of his idea for a figure of the biblical giant-slayer, David, to win the contract. The deal was signed on August 21, and work began on September 13.

A model from the quarries

First, the marble was moved to a specially built enclosure, put on a turntable base and tilted to an inclined position. A gesso model of the sculpture – perfectly in proportion though only a few inches tall – was set nearby. Its bone structure and well-defined musculature were almost certainly based on a young marble cutter from the town of Carrara.

This miniature vision of the proposed David differed only very slightly from the life-size figure that emerged over the next three years, and it played a crucial part in keeping the sculptor on course during his difficult task. To transfer the proportions of the model accurately to the block of marble, the model was divided up – on paper – with horizontal and vertical lines, and the resulting 'squares' were then enlarged. These were worked into wax or clay models which were individually copied.

Michelangelo worked for 20 hours a day, toiling under the sun in summer and beside burning braziers in winter. Under his chisel, a figure came to life, starting from the edge of the block. The emerging sculpture had a sense of enormous strength, based on its solid vertical core from the right foot, leg and thigh up to the massive head.

STRENGTH AND BEAUTY *The boy David, harpist and armour-bearer to King Saul, stands poised with a rock in one hand and a slingshot over his shoulder, ready for single combat with the giant Philistine, Goliath, who had been terrorising the Jewish army.*

Fine detailing

The statue's powerful right hand, hanging beside a muscled thigh, held a rock, and the left hand reached for the slingshot strung over the shoulder. To finish the more delicate areas of eyes, nostrils, lips and hair, Michelangelo exchanged the chisel for an augur and needle.

The sculptor had ignored the traditional depictions of David, who was usually shown triumphant with the bloodied head of Goliath at his feet. He portrayed the young David resolute and watchful before the fight, combining the suppleness of youth with great physical power. The work was completed in 1504 to general acclaim, when 40 men pushed the sculpture in a carriage through Florence to its site outside the Palazzo della Signoria.

The varnish that vanished

HOW STRADIVARI MADE THE SWEETEST-TONED VIOLINS

IN 1704 THE 60-YEAR-OLD MASTER VIOLIN MAKER Antonio Stradivari, living in Cremona in northern Italy, decided to write down the secret behind his success. This, he said, lay in the composition of the golden-red varnish that Cremonese craftsmen applied to their stringed instruments. He duly inscribed the formula of the varnish inside the cover of his Bible. Perhaps for commercial reasons, use of this varnish was gradually abandoned, so that, when Stradivari's Bible was mislaid years later, the secret was lost.

Since then, violin makers have striven, unsuccessfully, to rediscover the secret of the Cremonese varnish. What they do know is that, unlike other violin varnishes of the time, it was not thick, oily and fast-drying – which limited the instruments' range and sound. Instead, it was thin, and non-greasy, drying slowly to form a delicate elastic skin over the woodwork, and allow the instruments to produce their mellow tone.

'It is perfection and must be played to perfection'
SIR YEHUDI MENUHIN

At the age of 22, ten years after he was first apprenticed to the celebrated violin maker Nicolo Amati, Stradivari – or Stradivarius, to give him his Latin name – began to put his own label on the instruments he had made. To begin with, he copied Amati's small, solidly constructed models, but from about 1684 he started to produce broader, longer models – known as 'Long Strads'. These had shallow, only slightly arched bodies which yielded a fuller tone. This proved essential as classical music spread from chambers to concert halls.

Today there are over 600 Strad violins in existence, along with some 60 cellos and 17 violas. They are treasured by the world's greatest string players, and change hands for more than £1 million.

FINE GRAIN
The master violin maker favoured maple for the back, neck and bridge; willow or pine for the internal structure; pine for the belly; poplar or maple for the sides; and ebony for the fingerboard.

SIGN OF QUALITY
The master's initials are on his instruments.

STRADIVARI'S MESSIAH
No two Strads are identical, and the best of them have their own characters, histories and nicknames. The 'Messiah', made in 1716, is one of the 'Long Strads'.

FINE DETAILS
Stradivari drew his own designs for ivory inlays.

EARLY WORK
This delicately inlaid violin (right) made by Stradivari for a child in 1683, shows how much he had learned from his teacher, Nicolo Amati.

LEGENDARY LABEL
The hand-printed label is a sign of a true Stradivarius.

Antonius Stradiuarius Cremonenfis Faciebat Anno 1699

'I will seize fate by the throat...'

HOW BEETHOVEN COMPOSED AFTER HE WENT DEAF

CROWNING ACHIEVEMENT *Beethoven's mighty* Choral *was the first-ever symphony to use a full orchestra, a large chorus and a group of soloists.*

JOTTINGS OF A GENIUS *Beethoven wrote thoughts and musical ideas in a long series of notebooks. As he lost his hearing, these became ever more important. To converse, friends wrote in the notebook and the composer replied verbally. Underneath the cover of his diary of 1820, an idea for the opening of the 'Credo' from his* Missa Solemnis *is sketched out next to the address of a dentist.*

WITH A FINAL FLOURISH OF HIS ARMS, THE CONDUCTOR brought to a close the first performance of Ludwig van Beethoven's majestic 9th Symphony, the *Choral*. The audience in Vienna's Kärntnertor Theatre on that evening in May 1824 gave the work a standing ovation. They stamped, clapped their hands and yelled 'Bravo!' But Beethoven, who was standing next to the conductor, with his back to the audience, was unaware of the noisy acclaim. One of the soloists plucked the sleeve of the composer's black frock coat, and turned him round to see the rousing reception he could not hear.

Slowly enveloped in silence

Beethoven was 27 years old when, in 1798, he noticed that he occasionally had difficulty hearing. Two years later he saw a doctor about the problem. In 1802, as his hearing gradually grew worse, he feared he would become totally and incurably deaf. In an anguished letter to his two brothers he wrote of committing suicide, stating: 'I could not prevail upon myself to say to men: speak louder, shout, for I am deaf...the humiliation when someone heard a flute...and I heard nothing.'

For the next few years, despite his affliction, the great musician continued his joint career as a brilliant solo pianist and a composer of unprecedented power and profundity. He wrote the magnificent 3rd and 6th symphonies, the *Eroica* (1804) and the *Pastoral* (1808) – as well as the 4th and 5th – with badly impaired hearing. By 1820, however, he was completely deaf. He stopped performing in public, but refused to give up composing. In a letter to his boyhood friend, Franz Wegeler, he wrote dramatically: 'I will seize fate by the throat and so defy it.'

Beethoven began to take long country walks around the picturesque villages outside Vienna — jotting down the musical themes and melodies that he could hear as clearly in his head as he had before he became deaf. Afterwards he painstakingly transformed the jottings into finished compositions. In his final, silent years, Beethoven – who died in March 1827, aged 56 – composed some of his greatest works, including the last five string quartets, the *Missa Solemnis*, and the 9th Symphony.

EARLY STARTER *A professional musician by the age of 11, Beethoven – seen here in a portrait from 1818 – continued to compose right up to his death.*

MAGIC WORKSHOP *In this room Beethoven composed many of his greatest masterpieces, trying them out on the Graf piano. The sketch dates from soon after his death.*

UNFINISHED SYMPHONY *The last music Beethoven ever wrote was a sketch for the 10th Symphony, which he never completed.*

'Well played, your grace!'

HOW CRICKET WAS PLAYED AT THOMAS LORD'S ORIGINAL GROUND

IN THE FIELD *Marylebone Fields, north London, was a popular cricketing venue in the mid 1700s. The game was played with curved bats, and the wickets had only two stumps.*

LORD BY NAME *Born in Yorkshire, in 1755, Thomas Lord made his money in London as a wine merchant. He had to move his cricket ground twice. The first move was forced by rising rents, and the second because the Regent's Canal was to be cut through the playing area.*

THE BOWLER MOVED IN AND SLUNG THE LEATHER-CASED ball underarm at the waiting batsman, who struck out with his curved bat and sent the ball flashing to the fence. 'Well played, your grace!' the bowler, Thomas Lord, shouted. The batsman doffed his high hat in acknowledgment and dusted his white breeches. 'Deuced civil of you to say so, Thomas,' he replied, with the courtesy expected of a duke.

Lord had begun his cricketing career as a slow bowler and general attendant at the select White Conduit Cricket Club in London. Then, in 1787, he opened his own private cricket ground, Lord's, at Dorset Fields in rural north London. The matches held there were mostly waged by blue-blooded amateurs, who wore white shirts and trousers, high collars, black ties or neckerchiefs, broad cotton braces and shiny high hats.

The origin of cricket is not certain, but it may have first been played around the 14th century by shepherds in southern England, who used their crooks, or 'cricks', as bats. The wickets were formed by the small gates, or wickets, of the sheep-pens; the bails were the movable crossbars on top of the gates. As the game grew in popularity it moved from the country to the town. By the time Thomas Lord made his name, cricket was played by all classes.

A bowling revolution

Gradually, the make-up of the game changed. New, straighter bats were made, and boundary lines were introduced. Then there was the growing use of round-arm – with the arm swinging round at shoulder height – instead of underarm bowling, a style hated by members of Lord's Marylebone Cricket Club (MCC), who considered it unsporting.

In 1814 Lord moved his ground, turf and all, to its present home in St John's Wood, in north London. By then the MCC had become the arbiter of the game. In 1835 it revised the rules and at last officially sanctioned the use of round-arm bowling.

CRICKET'S VALHALLA *Lord's ground, which moved to its present site in 1814, is still the home of the Marylebone Cricket Club. The pavilion, which burned down, was replaced by a larger one (below). The MCC organised the first Tests against Australia; cricket hero W. G. Grace (below, right) often led the side.*

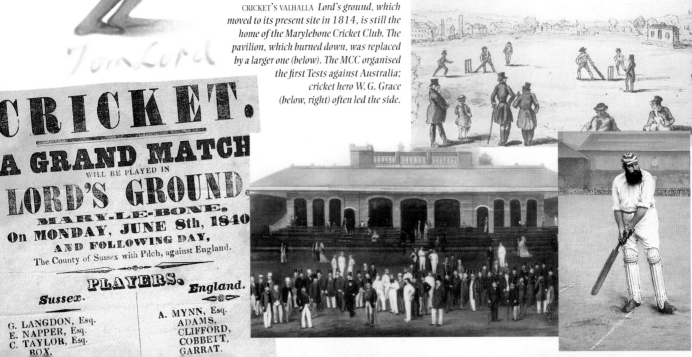

CRICKET.
A GRAND MATCH
WILL BE PLAYED IN
LORD'S GROUND,
MARY-LE-BONE,
On MONDAY, JUNE 8th, 1840,
AND FOLLOWING DAY,
The County of Sussex with Pilch, against England.

PLAYERS.

Sussex. | England.
G. LANGDON, Esq. | A. MYNN, Esq.
E. NAPPER, Esq. | ADAMS,
C. TAYLOR, Esq. | CLIFFORD,
BOX. | COBBETT,
 | GARRAT.

Fighting with 'pickled' fists

HOW BOXERS FOUGHT BEFORE THE QUEENSBERRY RULES

IN APRIL 1741 A 36-YEAR-OLD PUGILIST NAMED Jack Broughton took part in a bare-knuckled prize fight that changed the face of British boxing for ever. Broughton, an inch under 6 ft (1.8 m) tall and weighing almost 14 stones (90 kg), was matched against George Stevenson, a Yorkshire coachman. Stevenson, quicker on his feet than Broughton, threw punches faster than his opponent, but Broughton's superior weight and reach proved the crucial factor. The match finally ended when Broughton struck the Yorkshireman a hammer blow under the heart and sent him reeling to the ground. Three weeks later Stevenson died from his injuries, and Broughton vowed to do all that he could to prevent such a tragedy from ever happening again.

Brutal battles
Two years later, at his newly opened boxing arena near Tottenham Court Road, in London, Broughton drew up a set of rules aimed to bring 'science and humanity' to the sport. Until then, boxing had been a brutal mixture of bare-knuckle fighting and all-in wrestling. Broughton's Rules banned boxers from hitting opponents below the belt or when they had fallen. A round ended when a man was put down. The fallen boxer was then given half a minute in which to recover, after which he had to toe a chalk mark scratched in the centre of the ring. If he failed to 'come up to scratch', he was deemed to have lost the contest. Wrestling holds were only allowed above the waist.

FATHER FIGURE *Regarded as the father of British boxing, Jack Broughton came to the sport by accident. He discovered that he was a born boxer after getting involved in an argument that turned into a fist fight. This encouraged him to turn professional.*

FIGHTING FIT *Bare-knuckle fighters, such as Daniel Mendoza, champion of England in the late 1700s, were a tough breed, and fights could go for 30 or 40 rounds.*

Gloved fists
Broughton also introduced the first boxing gloves, which he called 'mufflers'. These, he hoped, would spare pugilists 'the inconveniency of black eyes, broken jaws and bloody noses'. However, the gloves were worn only during sparring exhibitions; for prize fights, bare fists – toughened by 'pickling' in a solution of soda shortly before a fight – were still the style. When large sums of money were at stake, supporters of individual boxers often hurled themselves into the ring if their man looked like losing. To prevent this, Broughton set up wood-and-canvas rings, or stages, standing some 6 ft (1.8 m) above the ground.

With some changes, Broughton's Rules remained in force until 1838, when they became known as the London Prize Ring Rules. Then, in 1867, the Marquess of Queensberry, a keen follower of boxing, introduced his rules, which still apply today. These prohibited wrestling in the ring, called for gloves to be worn and ordained that rounds should last for 3 minutes, with a minute's break in between.

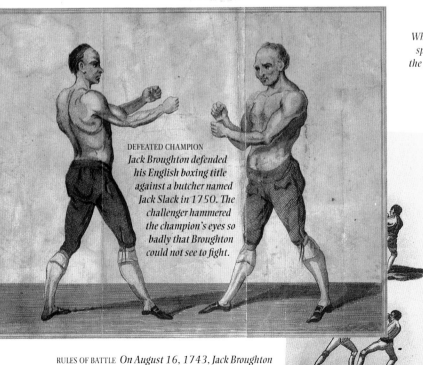

DEFEATED CHAMPION
Jack Broughton defended his English boxing title against a butcher named Jack Slack in 1750. The challenger hammered the champion's eyes so badly that Broughton could not see to fight.

RULES OF BATTLE *On August 16, 1743, Jack Broughton introduced his rules for boxing. These banned anyone except the fighters and their seconds from getting onto the raised boxing stage, and provided for two umpires – to be selected by the contestants – to settle any disputes that might arise.*

SAFETY PRECAUTION
When the gentry took to sparring for exercise in the 19th century, gloves became popular as a means of preventing damage to the face.

THE RING

RULES
TO BE OBSERVED IN ALL BATTLES ON THE STAGE

Catastrophe on canvas

HOW GERICAULT PAINTED THE VAST 'RAFT OF THE MEDUSA'

THE SCANDAL THAT INSPIRED THEODORE GERICAULT'S great *Raft of the Medusa* began on July 2, 1816, when the French frigate *Medusa* sank off the west coast of Africa while escorting a convoy taking settlers and soldiers to the colony of Senegal. The captain and senior officers took the best lifeboats – leaving 150 passengers and crew to fend for themselves on a makeshift raft some 65 ft (20 m) long and 28 ft (8.5 m) wide. Only 15 men were still alive when, after 13 days, the raft was found. Two survivors published an account of their ordeal, which included murder, starvation, and most shocking of all, cannibalism.

Tragedy inspires a masterpiece
These grim events, news of which emerged despite strenuous government attempts at a cover up, fired Géricault's imagination, even 18 months after the event. Géricault built up a dossier of documentary evidence, spoke to survivors, and engaged the *Medusa*'s carpenter to build a scale model of the raft. He even visited Paris hospitals to observe the expressions on the faces of dying patients.

In January 1819, after six months of preparatory sketching, Géricault began work on the monumental canvas, measuring 23 ft (7 m) across and just over 16 ft (5 m) high. Working in a studio hired specially for the purpose, he started by drawing an outline of the action, then boldly painted the figures directly onto the canvas from life models. He used fast-drying oils which meant that each section, once started, had to be completed within a single day.

Eight months later, the painting was completed. It captured the moment when, for the handful of survivors on the floating raft, despair turned to hope. The *Raft of the Medusa* was shown at the Louvre's Salon, but the French government accused the artist of deliberately fomenting public disquiet. In disgust, Géricault took the painting to England, where he exhibited it for two years. In January 1824, the artist died at the age of 33 in Paris. Later that year his work was bought by the French government, and it is now back where it belongs: in the Louvre.

SURE TOUCH *Géricault's self-portrait at the age of 17 shows a precocious mastery of paint. He claimed to have 'enthusiasm that overcomes and masters every obstacle'.*

MUTINY ON THE RAFT *On their second night of clinging desperately to the overcrowded raft, delirious soldiers and sailors tried to murder their officers and destroy the vessel. Some 65 people died in the mayhem that followed. In an early study of his tragic theme, Géricault used black chalk, wash and gouache to convey the intense drama of the mutiny: swords fly as corpses cascade into a merciless sea.*

CANNIBALS *Soon after the mutiny, ravenous hunger drove the survivors to eat the flesh of those killed in the fighting; a few days later, to conserve their small rations of wine, the strongest threw the weakest and the badly injured overboard. Géricault's study of cannibalism on the raft captures the full horror of the scene: some men feed on a corpse, while others abandon themselves to despair.*

ACTION SKETCH *Alexandre Corréard, author of a grisly account of his sufferings, points to the rescue ship.*

ADRIFT ON A STORMY SEA *The rough pen-and-ink composition (left) evolved over 18 months into Géricault's haunting masterpiece (below). A carefully orchestrated jumble of bodies, living and dead, emerges from the deep shadow around the mast. Muted colours and a dynamic asymmetrical composition draw the viewer in to share in the exhilaration, and exhaustion, of the moment. The sighting of the rescue ship unleashes hope and excitement in all the survivors bar one – a father holding the corpse of his son.*

Birth of the blockbuster

HOW THE FIRST BESTSELLERS WERE PRODUCED

IT WAS THE FIRST OF THE MONTH, and readers all over Britain were eagerly awaiting another instalment of the latest work by Charles Dickens. The year was 1836, and the serialisation of Dickens' *Pickwick Papers*, published in 20 monthly parts, was well into its stride. The 24-year-old author, and his publisher, Chapman & Hall, could congratulate themselves on the success of their innovative idea – to sell the novel in monthly instalments costing just one shilling each.

Novels had been published in England since the early 18th century, but the potential of books to attract a mass audience really become apparent only with the publication of Sir Walter Scott's hugely successful *Waverley* novels in 1814. By the 1830s, some 50000 Britons were regular readers of fiction; just 20 years later, thanks to serialisation, readership had quadrupled to around 200000.

The reign of the 'three-decker'

Until the advent of serialisation, most novels had a first print run of about 1000 copies and were published as a three-volume set, a format which soon became known as the 'three-decker'. Each volume cost half a guinea, while a complete set was 31s 6d (£1.58). This was far too much for many readers, but they had an alternative. Rather than buy the books outright themselves, they could borrow them from a subscription library.

Joining such a library cost up to three guineas a year, in return for which subscribers could borrow up to eight books at a time, and exchange them as often as they liked. Their overwhelming success – one, Mudie's Select Library, boasted an annual income of £40000 from subscriptions by the 1860s – showed publishers that there was a great demand for novels if they could find a cheaper way of getting them into the hands of the public. Serialisation was the answer, as it meant that publishers could count on

MULTIVOLUME TYRANNY *Books by authors such as Charles Dickens, Thomas Hardy and George Meredith were published in three-volume sets (above). This practice boosted the takings of the subscription libraries like Mudie's in London (right). Readers could exchange books in person or by post from among the 4 million volumes available at Mudie's. Subscription was a guinea a year 'and upwards' (above right).*

selling the equivalent of tens of thousands of copies of a book before printing a single bound volume. Selling first publication rights to a magazine greatly reduced the risk of committing cash to printing expensive books that might not sell. Wilkie Collins' *The Woman in White*, the first crime-detection novel, sold 100000 copies in serialised form, while most of the novels of George Eliot, Anthony Trollope and Thomas Hardy were serialised before their appearance in book form.

By the 1860s the most successful authors of the day were earning generous advances – payments made before the book had been written. After scoring major successes with *Adam Bede* and *The Mill on the Floss*, George Eliot was offered £10000 in payment for *Romola*. This was a phenomenal sum in an era when a single person could live comfortably on £300 a year. By the time Charles Dickens died in 1870, he was an extremely wealthy man, with an estate valued at £93000.

Despite the profits that flowed from the immense popularity of serialisation, publishers remained dependent on the large orders placed by the big subscription libraries. As a result, they were forced to continue to produce the old-fashioned 'three-deckers'.

SERIAL RIGHTS *Serialisation of books became common after the success of the 20-part issue of the* Pickwick Papers *(below) by Charles Dickens. A sketch from 1840 (below, right) shows booksellers' men laden with bundles of the new Dickens weekly,* Master Humphrey's Clock, *streaming from the offices of the publisher to distribute the supplement, which serialised his novels and short stories.*

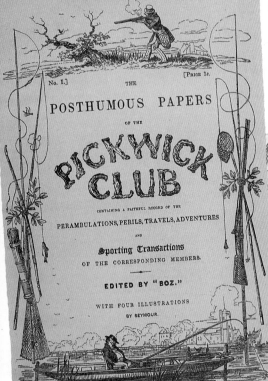

No. I.]   [PRICE 1s.

THE

POSTHUMOUS PAPERS

OF THE

PICKWICK CLUB

CONTAINING A FAITHFUL RECORD OF THE

PERAMBULATIONS, PERILS, TRAVELS, ADVENTURES

AND

Sporting Transactions

OF THE CORRESPONDING MEMBERS.

EDITED BY "BOZ."

WITH FOUR ILLUSTRATIONS
BY SEYMOUR.

MUDIE'S SELECT LIBRARY,

(LIMITED.)

30 TO 34, NEW OXFORD STREET.

BRANCH OFFICES { 241, BROMPTON ROAD, S.W.
2, KING STREET, CHEAPSIDE, E.C.

SUBSCRIPTION.

One Guinea Per Annum and upwards.

THIS BOOK IS THE PROPERTY OF MESSRS ANDER BROTHERS AND FORMS PART OF THEIR WORKPEOPLE'S LIBRARY

THE WORKPEOPLE EMPLOYED IN OUR VARIOUS FACTORIES ARE AT LIBERTY TO TAKE OUT BOOKS, ONLY ONE BOOK CAN BE WITHDRAWN AT ONCE AND MAY BE KEPT A REASONABLE TIME, BUT MUST BE RETURNED IN ANY CASE, IMMEDIATELY ON REQUEST OF THE LIBRARIAN

Taking a risk on a new writer

Sometimes a large proportion of a book's first edition would be pre-sold before publication, giving the publisher the financial security to allow it to take on new authors – many of who might not make their mark until their third or fourth novel. This investment in fresh talent was a crucial factor in this remarkable publishing era. A host of talented British novelists emerged in the 19th century, many of whom became as popular in the United States as they were in Britain. But success in America did not bring much of a financial reward until the final decade of the century. Before the United States signed the International Copyright Convention, novels were simply pirated by unscrupulous publishers. The authors received nothing from the sales, and no laws existed to help them.

COMPANY LITERATURE *Benevolent employers sometimes provided libraries for their workers to use.*

FRUITLESS QUEST *A dispirited mother and her son resume their search for lodgings after another rejection, in a scene from a serialisation of Thomas Hardy's novel* Jude the Obscure.

THE FINAL CHAPTER *A woman is totally unaware of her surroundings as she turns the final pages of a gripping new novel. By the mid 19th century serialisation of novels had made fiction into something exciting, popular and new that almost everyone wanted to enjoy.*

PENNY DREADFULS

'SHE STRETCHED FORTH HER LONG TALONS, YELLING in the fierce grip of *delirium tremens*, as she advanced to where Jack was, to fling him into the flames, shrieking: "Now, boy, you shall be my victim. Ha! Ha! Ha!".' This extract comes from a serialised story published in 1880 under a title that misleadingly conjures up a vision of Victorian sentimentality – *The Darling of the Crew*.

Ghoulish tales

Other works had titles such as *The Death Ship* or *Varney the Vampire* which were a fairer reflection of what lay between the covers. From the 1830s such stories were churned out by hack writers by the thousands to satisfy those who found the novels of writers like Charles Dickens too tame, too highbrow and too expensive. Instead, for just a penny per eight-page issue, the 'penny dreadfuls' thrived on tales of witches, ghouls, pirate queens and murderers.

STOLEN IDEAS *'Penny dreadfuls' such as* Turnpike Dick *and* The Penny Pickwick *(right) were plagiarised parodies of popular novels, sold by street traders and hawkers. Respectable publications such as* The Boy's Own Paper, *however, tried to steer young readers towards more wholesome and educational tales.*

A gift from the Old World to the New

HOW THE STATUE OF LIBERTY WAS DESIGNED, BUILT AND ERECTED

THE IDEA OF PRESENTING A GIFT FROM the French Republic to the citizens of the USA came to the sculptor Frédéric-Auguste Bartholdi while dining at the home of a French historian in the summer of 1865. Bartholdi proposed that the gift should take the form of a huge statue to be called 'Liberty Enlightening the World'; it would mark the centenary of American independence, which French soldiers and arms had helped to win.

Six years later, while visiting New York, the 37-year-old Bartholdi found the perfect site for the statue: Bedloe's Island in Upper New York Bay, visible to all shipping using the harbour. Once back in Paris Bartholdi began work on the imposing edifice that would become popularly known as the 'Statue of Liberty'. He was inspired, he said, by the Colossus of Rhodes, a vast bronze statue of the Greek

GRAND GESTURE *The man who created the Statue of Liberty, Frédéric-Auguste Bartholdi, had already made a reputation as a creator of giant sculptures, many with patriotic themes.*

sun god Helios which had dominated the entrance to the harbour at Rhodes in the 3rd century BC – and which was one of the Seven Wonders of the Ancient World.

Bartholdi's first step was to design and fashion a 4 ft (1.2 m) clay maquette, or small preliminary model, of Liberty. He then made thousands of precise measurements with a plumb line to scale the model up: first to 9 ft (2.7 m), then to 36 ft (11 m) high. Finally, he made plaster sections that formed a replica of the full-size statue – some 46 ft (14 m) higher than the Colossus of Rhodes. The money for the statue was raised by public subscription, and it eventually cost some $400 000. One of Bartholdi's first major problems was choosing the sort of face Liberty should have. He decided to model it on the stern features of his mother, a Protestant zealot who had literally driven her other son mad by refusing to let him marry the girl of his choice – a Jewess.

Poignant symbol of freedom

Madame Bartholdi was living in the province of Alsace in north-east France, which had been ceded to Germany after France lost the Franco-Prussian War of 1870-1. The German authorities prohibited Bartholdi from visiting her, so, paradoxically, he saw her as a symbol of Liberty. He gave her face a classical, Grecian mould, emphasising her somewhat sullen lips, hooked nose and dark, hooded eyes. Next, Bartholdi needed to make the statue strong enough to resist the elements, yet light enough to be

GIANT'S WORKSHOP *Craftsmen fit the slats into place for the plaster cast of the Statue of Liberty's left hand (left). The 152 ft (46 m) high plaster model of the statue was divided into sections. Once each section was cast, carpenters created accurate wooden moulds over the plaster. Then metal workers (below) hammered copper sheets into shape inside the moulds to make the statue's metal skin.*

LARGER THAN LIFE *Two finished sections of Bartholdi's enormous statue, a finger and the left ear, lie in the Paris workshop of Gaget, Gauthier and Company, where the statue's copper skin was beaten out in sections.*

FIRST FLAME *To help raise money for this modern colossus, the only completed section of the statue – the torch – was displayed at the Centennial Exhibition in Philadelphia in 1876. Visitors paid to climb up the staircase inside the arm for a view from the balcony.*

TEMPORARY MONUMENT
Before being shipped to the US, the statue was erected in the yard of the workshop where the sections were being created (right and below) to ensure that the parts fitted correctly. The internal framework – four spines from base to neck, with horizontal ribs – anticipated the structure of modern skyscrapers.

transported by sea. Again he sought inspiration from the Colossus of Rhodes, which had been hollow with an outer shell covering an inner framework. The shell of the Colossus had been made of heavy bronze; for his outsize creation Bartholdi chose 300 sheets of light, flexible copper only ⅒ in (2.5 mm) thick.

A gift delayed

Work proceeded slowly and methodically, and by July 4, 1876 – the 100th anniversary of the United States' Declaration of Independence – the giant birthday present was still not complete. By the beginning of 1881 the statue's inner framework had been created by the brilliant engineer best known for Paris' famous tower, Alexandre Gustave Eiffel. From it sections of copper skin were hung.

Finally, in June 1884 – 19 years after Bartholdi had first thought of her – Liberty rose majestically above the streets of Paris. On July 4, Independence Day, the statue was formally handed over to the American ambassador. The following summer it was dismantled, packed into 210 vast crates and transported to New York. There a fresh problem presented itself.

Liberty was to stand on a pedestal of about her own height, so that she would tower some 305 ft (93 m) above ground level. However, only half the money needed to build the pedestal had been raised. The publisher Joseph Pulitzer appealed for donations through his popular newspaper, the *New York World*, in which he printed the name of every contributor, great and small. In the end, 121 000 people gave more than $100 000.

By the time the pedestal was complete, Liberty had been in crates for 15 months. The components were unpacked and assembled on Eiffel's structure, working from the bottom up without the use of external scaffolding. The workmen simply climbed up inside the structure, leant over the top and riveted the copper plates to springy iron bars attached to Eiffel's framework. On October 28, 1886, Lady Liberty at last raised her lamp above her new home.

The man who froze time

HOW EADWEARD MUYBRIDGE TOOK HIGH-SPEED PHOTOGRAPHS OF ANIMALS IN MOTION

IN 1872, CALIFORNIAN HORSE LOVERS WERE IN THE midst of a dispute. Leland Stanford, ex-state governor and powerful head of the Central Pacific Railway, and a group of his friends maintained that at certain times when a horse was travelling at a fast trot or a gallop, all four of its hoofs left the ground at once. Others, including James Keene, the head of the San Francisco Stock Exchange, were sure this was impossible. But there seemed to be no way of settling the argument, until Leland Stanford thought of a simple experiment.

The fastest camera in California
He suggested that a local photographer, an Englishman named Eadweard Muybridge, should try to capture the movement of Stanford's racehorse Occident on film. Though dubious, Muybridge agreed to photograph Occident trotting at some 22 miles an hour (35 km/h) on the Sacramento racecourse. He borrowed yards of bed linen from people living in the area; these were draped across the track to form a background against which the horse would be silhouetted. Muybridge did not try to take correctly exposed pictures. He felt that a silhouette would be enough to decide

the issue. His first attempts to capture Occident in motion failed because the camera's manually operated shutter was too slow to produce a short enough exposure, so he devised a mechanical shutter consisting of two wooden blades which slid vertically in a grooved frame to reveal an 8 in (20 cm) slit through which light could pass. This gave an exposure time of 1/500th of a second. The result was a series of pictures of Occident which proved Stanford right: a horse doing a fast trot did leave the ground entirely.

Stanford was sufficiently impressed to commission a photographic study of locomotion that would capture each stage of the horse's movements.

PHOTOGRAPHIC PIONEER
By the time he was 42, Eadweard Muybridge had established his reputation with a series of pictures of 'types', like these natives of Alaska photographed in 1868, and with panoramic landscape photographs. But the work of his later career, the series of humans and animals in motion, ensured his place in history.

SPLIT SECOND *Muybridge's first successful attempt to photograph motion was a silhouette of the racehorse Occident, clearly showing all four feet off the ground.*

OUTDOOR STUDIO *Muybridge photographed horses in motion (above) at the stock farm of Leland Stanford (far left). He installed a battery of 12 cameras at 21 in (53 cm) intervals in a 'camera house'. They were fitted with 'double slide' shutters in which two pairs of blades were attached to springs – one pair sliding up as the other slid down, so exposures were taken at a previously unheard-of speed of less than 1/2000th of a second. The shutters were attached to threads stretched across a track of grooved rubber. A moving horse would break the threads in quick succession, so triggering the camera shutters.*

GRAND OLD MAN *Muybridge was one of the greatest figures of the new art of photography by the time he retired to his place of birth, Kingston upon Thames, where he died in 1904.*

The racehorse experiments were resumed at Stanford's ranch in the summer of 1878. Although slightly underexposed, the resulting series of pictures – of a Kentucky racehorse named Sally Gardner – clearly showed her different gaits. Muybridge painted the negatives to show only the horse's silhouettes, which revealed the horse's legs in positions thought to be impossible. The sequence of 12 pictures had been taken in about half a second, the speed at which a present-day film camera exposes its frames.

INCREDIBLE TRUTH
Journalists were sceptical of Muybridge's series of photographs of horses in motion, until the girth of the horse Sally Gardner (above) broke and was recorded on film. The public, however, generally continued to believe that the photographs must be some kind of hoax – the horse's gait looked physically impossible.

MOVING PICTURES *The zoöpraxiscope (right), devised by Muybridge, projected light through hand-painted photographs on rotating glass discs, with the selected images appearing to move as a result. The machine was innovative in two ways: first, because it projected the image, and second, because it used sequential photographs.*

Tricks of the human eye

In October the magazine *Scientific American* published six engravings from enlarged negatives of Muybridge's photographs of a horse walking and 12 engravings of it trotting. The magazine invited its readers to cut out the drawings and mount them on a zoetrope, a cylinder which produced the illusion of movement when it was spun and the pictures viewed through a slot in the side. This relied on the phenomenon called 'persistence of vision'. When the human eye sees a series of images in rapid succession – ten or more a second – they appear to show continuous movement. Without interruptions between frames, however, the eye would simply see a blur. After reading the article Muybridge felt that better results could be obtained by projecting his images onto a screen and he invented a light projector, which he dubbed a zoöpraxiscope. In 1888 Muybridge showed his horse pictures to Thomas Edison, the inventor of the phonograph. Muybridge suggested that his projector and the phonograph could be combined to show 'talking' pictures. Although the idea came to nothing, Edison ran a set of the horse pictures through his 1890s Kinetoscope – a peep-hole machine that was the forerunner of today's film projector.

This success led Muybridge to record the movements of humans – from athletes to cripples – and more animals, this time from the Philadelphia zoo. The resulting book of photographs, *Animal Locomotion*, published in 1887 with the help of the University of Philadelphia, is still the leading authority on human and animal movement.

Larger than life

HOW REAL PEOPLE INSPIRED SOME OF FICTION'S GREATEST CHARACTERS

THE SHARP-EYED SURGEON LOOKED CLOSELY AT THE NEW patient standing before him. 'Well, my man,' said the doctor in a harsh, high-pitched voice, 'you've served in the army.' The patient, who was wearing civilian clothes, nodded his head. 'Aye, sir,' he replied. 'Not long discharged?' the doctor queried. 'No, sir,' came the answer. 'A Highland regiment?' the doctor continued. 'Aye, sir,' confirmed the patient. 'A noncom [noncommissioned] officer?' enquired the doctor. Again the man replied, 'Aye, sir.' The doctor thought for a moment and said, 'Stationed at Barbados?' Dutifully, the man responded, 'Aye, sir.'

The watching medical students were awestruck as the surgeon – Dr Joseph Bell of the Medical Faculty of Edinburgh University – completed his interrogation. No one was more impressed than Bell's medical clerk and pupil, the 17-year-old Arthur Conan Doyle – who was later to base his fictitious detective, Sherlock Holmes, on the deductive powers of Dr Bell.

The first scientific detective

'You see, gentlemen,' Bell said, as the patient left the room that day in 1876, 'the man was a respectful man but did not remove his hat. They do not in the army, but he would have learned civilian ways had he been long discharged. He has an air of authority and is obviously Scottish. As to Barbados, his complaint is elephantiasis [enlarging and hardening of the skin], which is West Indian and not British.'

Doyle never forgot the surgeon's unique working methods. 'His strong point was diagnosis, not only of disease, but of occupation and character... I tried to build up a scientific detective who solved cases on his own merits and not through the folly of the criminal,' wrote the author afterwards.

A SHILLING SHOCKER *One wild winter night in 1885, Robert Louis Stevenson (above) began to scream and shout in his sleep. His wife Fanny shook him awake, and drenched in sweat Stevenson demanded: 'Why did you wake me? I was dreaming a fine bogey tale!' Unable to sleep, Stevenson feverishly started to write what he called a 'shilling shocker'. The Strange Case of Dr Jekyll and Mr Hyde was based on the bizarre double life of Deacon William Brodie (right), who worked as a cabinet-maker by daylight, and as a burglar after dark. His haul included £800 taken from an Edinburgh banking house, and a silver mace from Edinburgh University. Hyde's evil nature was portrayed by Henry Brodribb Irving (below) in a lurid dramatisation of the chilling tale in 1910.*

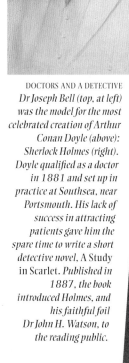

DOCTORS AND A DETECTIVE *Dr Joseph Bell (top, at left) was the model for the most celebrated creation of Arthur Conan Doyle (above): Sherlock Holmes (right). Doyle qualified as a doctor in 1881 and set up in practice at Southsea, near Portsmouth. His lack of success in attracting patients gave him the spare time to write a short detective novel,* A Study in Scarlet. *Published in 1887, the book introduced Holmes, and his faithful foil Dr John H. Watson, to the reading public.*

The double life of Deacon Brodie

In 1886 the Scottish author Robert Louis Stevenson published his macabre masterpiece, *The Strange Case of Dr Jekyll and Mr Hyde*. In the short novel the hitherto law-abiding Dr Jekyll develops a drug which transforms him at will into the demonic murderer, Mr Hyde, and an antidote which, for a time, succeeds in turning him back into his 'real' self.

Like Conan Doyle, Stevenson was born and bred in Edinburgh, where, as a boy, he heard tell of the infamous Deacon Brodie, who in the late 18th century had led an incredible twin existence in the city. By day, William Brodie, who was born in the Scottish capital in 1741, worked as a successful cabinet-maker. His title of 'Deacon' referred to his professional status: he was a Deacon of the Incorporation of Wrights, or carpenters. As well as fashioning exquisite cabinets for leading bankers and businessmen, he was a frequent and honoured guest in their homes. By night, however, the Deacon broke into his customers' houses and offices and, using skeleton keys he had made while ostensibly at work, got away with gold, silver, jewels and money.

His criminal career ended in a daring, but disastrous, raid on the General Excise Office for Scotland on the night of March 5, 1788. Brodie was arrested and put on trial. Found guilty of burglary, he was hanged at Edinburgh's Tolbooth prison later that year.

THE MAN WHO BECAME ROBINSON CRUSOE

EARLY IN OCTOBER 1704 THE privateer *Cinque Ports* put into Más a Tierra, one of three small uninhabited islands forming the Juan Fernández chain, some 400 miles (650 km) off the coast of Chile. The island was used by pirates as a repair station, and it was there that the ship's sailing master, 28-year-old Alexander Selkirk – after expressing his lack of confidence in the captain of the *Cinque Ports*, Thomas Stradling, and in the seaworthiness of the ship – chose to be left behind with his belongings.

Winter accommodation

The privateer – a privately owned vessel commissioned by a government to seize and plunder enemy ships – duly sailed without Selkirk. As winter approached he built a hut as his living quarters. The island abounded in goats, which he shot for food. When his powder ran out, he pursued the animals on foot, learning to run and climb as nimbly as the goats themselves. He made knives from some old iron hoops, and learned to fashion a crude suit of clothes out of goatskin.

Finally, on January 31, 1709 – after four years and four months of isolation – Selkirk was picked up by another privateer: the *Duke*, commanded by Captain Woodes Rogers. After further adventures at sea, Selkirk reached London in October 1711. There Captain Rogers wrote a book, *A Cruising Voyage Around the World*, about the castaway's experiences.

Cannibals and a demonic goat

Various pamphlets and articles on Selkirk also appeared. Then, in 1719, the journalist and pamphleteer Daniel Defoe wrote what is generally considered to be the first novel published in Britain: *The Life and Strange Surprising Adventures of Robinson Crusoe, of York, Mariner*, or *Robinson Crusoe* for short.

In Crusoe, Defoe created an imperishable character whose name conjures up images of idyllic desert islands, intrepid, self-sufficient heroes, timely rescues and happy endings. But it appears that Defoe and Selkirk, who was born in Largo, Fifeshire, in 1676, never met. *Robinson Crusoe* was a highly dramatised version of existing accounts of Selkirk's life – with embellishments such as fights with cannibals, and mistaking a dying goat for the Devil.

FACT AND FANTASY *Goats and cats were the main companions of Alexander Selkirk (above) during his spell on a desert island. He is said to have taught the animals to dance and joined in their antics. When Daniel Defoe (right) wrote his fictitious account of Selkirk's adventures, he invented a native servant named Man Friday (below), who worked alongside his master, building a boat while the castaway made fishing nets.*

The theatre promoter who cleaned up

HOW TONY PASTOR MADE AMERICAN VAUDEVILLE RESPECTABLE

A STRING OF CARRIAGES DREW UP OUTSIDE TONY Pastor's newly acquired vaudeville theatre on New York's 14th Street and discharged the passengers into the swelling crowd outside the entrance. Men ushered their families into the brilliantly lit interior, where the sounds of the orchestra tuning up mingled with the buzz of anticipation from the audience.

The lights dimmed, and a roll of drums signalled that the show was under way. Acts were announced on a board at the side of the stage. They came and went in quick succession: an acrobatic pantomime act, a comedy duo, speciality dancers, a humorous monologue and a glamorous singing duet. It was October 24, 1881, and 44-year-old Tony Pastor had opened the first 'respectable' vaudeville show in the United States.

The audiences liked what they saw, and flocked to Pastor's Theatre. They were decent folk dressed up for the occasion: husbands arm-in-arm with their wives and daughters, young men with their sweethearts. It was the kind of show, they said, that any child could take its parents along to see.

The term 'vaudeville' was introduced to the United States by French immigrants – it may have derived from the words *vau*, or *val de Vire* (the valley of Vire in Normandy, famous for popular songs). It was first

SETTING STANDARDS *Shocked by the crudity of variety acts, Tony Pastor (left) set out to raise their tone and to woo respectable people into the theatre.*

YOUNG MAMMA *Brassy-voiced Sophie Tucker (below), later known as 'The Last of the Red-hot Mommas', made her singing debut at the 14th Street theatre.*

LOW SKILLS, HIGH SKILLS *Comic double acts such as Bert Williams (above, left) and George Walker (above, right) were one of the staples of vaudeville, where audiences lapped up their good-natured crosstalk routines. Another crowd-pleaser were Europe's Novelty Aerial Gymnasts (left), whose daring made people gasp.*

BELLES AND BELLS *The Cycling Kaufmann Beauties (above) were a popular stage attraction. A man (right) leads his performing dogs in a musical novelty act.*

TOP OF THE BILL *Statuesque American singer Lillian Russell (above, left) and Vesta Tilley (above, right), the renowned British male-impersonator, were two of the biggest stars to shine at Pastor's 14th Street theatre. They were rivalled by the comedian W.C. Fields (right), who began his laughter-raising career as a farcical 'tramp juggler' in vaudeville.*

THE POUNDING PIANOS OF TIN PAN ALLEY

THE FAT MAN IN THE BACK ROW OF THE ROOF Garden Theatre was snoring so loudly that it was beginning to irritate the rest of the audience. They looked round in disgust at the suspected drunk. The woman with him was embarrassed. 'Wake up, Charlie,' she exclaimed shaking him. 'You're attracting attention!' People watched goggle-eyed as a policeman entered to remove him. Suddenly, the man stretched, yawned and broke into song: 'Please go 'way an' let me sleep,' he warbled. 'Ah would rather sleep than eat.' The girl on the stage and the orchestra joined in the song. So another Tin Pan Alley hit was born.

Songs from the skies
According to legend, the name 'Tin Pan Alley' originated in the 1880s from the constant pounding of the New York publishing house pianos, which was compared to the rattling of tin pans. As late as the 1920s, selling methods were crude. Boys would grab performers on the way to the theatre and give impromptu renditions of a new number. Young women stalked the department stores in search of buyers of sentimental songs. Male touts haunted racetracks and gambling houses, sheet music in hand. Stunts even included chartering aircraft and booming the latest would-be hits down on the city by megaphone.

MUSIC-MAKER *Songwriter Irving Berlin (below) began his Tin Pan Alley career in 1908, as a singing waiter who performed the latest sheet-music hits. Soon his own songs, such as 'Alexander's Ragtime Band', were all the rage.*

used to describe variety shows in the 1870s. These were colourful affairs, containing racy numbers and bawdy sketches, and played to hard-drinking, all-male audiences. But Pastor set out to clean up variety so that it was acceptable to both sexes.

'The depths of depravity'
A pleasing tenor with a repertoire of some 1500 songs – many of which he claimed to have written himself – Pastor had grown up in the tradition where shows went on the road and performed in low-grade bars and rowdy beer-halls. In these men-only places most of the receipts came from the sale of drinks, with the female entertainers doubling as waitresses. Ribald, and certainly not suitable for ladies, these shows played across the USA, closing only at daybreak or when the customers stopped buying drinks. Even in the more substantial theatres in major cities like New York, entertainers sometimes had a hard time competing with the bar.

BOTTOM OF THE BILL *Speciality acts such as acrobats and tumblers (above), and grotesque, weirdly dressed comics (left), were the very lifeblood of vaudeville.*

By the early 1860s, when Pastor made his debut, variety had sunk to what he called 'the depths of depravity'. He launched his clean-up campaign by managing a series of theatres on Broadway – including his flagship, Tony Pastor's Opera House – at which vulgar acts were outlawed and smoking and drinking were banned.

With the success some 20 years later of his theatre on 14th Street, Pastor set new standards for live family entertainment. Admired and much imitated, he ran the theatre until his death in 1908.

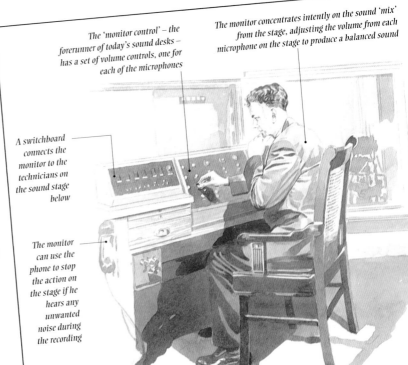

TRIAL RUN FOR SOUND *Producing a film with sound required different techniques and equipment to those used for silent movies. When Al Jolson recorded* The Singing Fool *(below), rehearsals took place with a full orchestra in the studio, so the star could hear the music and the conductor could hear and see him. The rehearsals were followed by a scene being performed for sound only, with no cameras but with the disc recorder running. Microphones suspended on wires above the orchestra and actors, some labelled 'M' and 'F' for their ability to provide the clearest recording of male and female voices, picked up the sound, while the lights were turned away from the actor to stop him getting too hot. When the scene was finished, the wax disc was played back so that adjustments could be made to improve the sound balance. Once the sound quality was assured, the cameras and recording equipment rolled for real. Most of the lighting came from above, although several high-powered incandescent lights were used.*

The 'monitor control' – the forerunner of today's sound desks – has a set of volume controls, one for each of the microphones

The monitor concentrates intently on the sound 'mix' from the stage, adjusting the volume from each microphone on the stage to produce a balanced sound

A switchboard connects the monitor to the technicians on the sound stage below

The monitor can use the phone to stop the action on the stage if he hears any unwanted noise during the recording

A second camera booth holds a seated camera operator. It is mounted on rubber wheels so that it can be moved forwards on wooden planks on the studio floor – a strip of wood at the end stops the booth going too far

The director stands close to the camera. In this position he sees the action from the same viewpoint as the audience, and the actor can see him while maintaining eye contact with the camera

Spare scenery is used as a 'baffle' to contain the sound of the orchestra and prevent it spilling into Jolson's microphone, which hangs above his head

The larger of the two soundproofed camera booths houses two cameras which film at the same time. One gives a long shot, the other a close-up

'You ain't heard nothin' yet!'

HOW THE FIRST TALKING FILMS WERE MADE

MIXING ROOM *The film's sound track was recorded in a soundproofed mixing room, its double walls treated with absorbent coverings. The engineer in charge of sound was known as the monitor. In an attempt to reproduce the acoustics of a cinema, the mixing room was very large with triple-glazed windows providing a view of the musicians below. The sound was relayed from microphones suspended over the artists' heads to loudspeakers in the mixing room. If the monitor thought it necessary, he could stop the action by calling 'cut' over the telephone.*

THE PRERECORDED BACKGROUND MUSIC OF THE FILM *The Jazz Singer* had not been on long when the star, Al Jolson, looked straight at the audience and cried: 'Wait a minute! You ain't heard nothin' yet!…You wanna hear "Toot-toot-tootsie"? All right, hold on!' People in Warners' Theater, New York, listened spellbound as Jolson launched into the song. It was October 6, 1927, and movies with sound – or 'talkies', as they were soon dubbed – were here to stay.

The story of the search for a practical sound system for the movies dated back to the earliest days of film. Various short 'novelty' films with sound had been released, as had Movietone newsreels with a soundtrack. These had made no great impression on audiences, however, and it was not until *The Jazz Singer*, with its revolutionary 'talking sequences', that full-length films began to speak and people everywhere wanted to hear them.

The soundtrack for *The Jazz Singer* was recorded in the world's first sound studio – a huge room 75 ft (23 m) wide and 100 ft (30 m) long. The walls and floors were specially soundproofed, and the doors lined with felt. The sound was recorded as it occurred on gramophone-like wax discs, with tone and volume being adjusted by an operator in a glass-fronted, soundproofed

monitor room set on a side wall some 15 ft (5 m) above the floor. The process was known as Vitaphone, or 'sound-on-disc'. Copies of the discs were then played on turntables in cinema projection rooms as the film unfolded. Matching the sound to the action depended on the skill of the projectionist, who also operated an amplifier wired to two loudspeakers, one on either side of the screen.

The film was originally meant to have synchronised music and singing, but Jolson was an unstoppable improviser. His impromptu dialogue when recording 'Toot-toot-tootsie' and a sentimental singing-and-talking scene between Jolson – who played an entertainer much like himself – and his mother, to whom he sings Irving Berlin's 'Blue Skies', was retained by Sam Warner. head of Warner studios.

On its release *The Jazz Singer* was publicised as the first 'all-talking, all-singing' feature film. It was an exaggeration, but it still achieved worldwide success.

SPONTANEOUS TALENT *The Jazz Singer was not planned as a film with recorded speech, but singer Al Jolson's talent for improvising dialogue convinced studio bosses that there was a future for the 'talkies'.*

Powerful Klieglights' have replaced the noisier arc lamps which were used in silent films

Scene hands and technicians stand silent in the wings as recording starts

Jolson stands within earshot of the orchestra and where he can see both the cameras and the director

BOX OFFICE HIT *The Jazz Singer, starring Al Jolson (above), played to more than a million Americans a week and took a record $3.5 million at the box office. The film's makers, the four Warner brothers, said it was their 'supreme achievement'.*

WARNER BROS. SUPREME TRIUMPH
AL JOLSON in "The JAZZ SINGER"

The fairest art of all

HOW THE WALT DISNEY STUDIO PRODUCED 'SNOW WHITE' – THE FIRST FEATURE-LENGTH CARTOON

THE CARTHAY CIRCLE MOVIE THEATER IN LOS ANGELES was packed. Marlene Dietrich and Charles Laughton were among the Hollywood stars gathered there on December 21, 1937, for the gala premiere of *Snow White and the Seven Dwarfs*, the world's first feature-length cartoon film. As the house lights went down, the film's creator Walt Disney was nervous; his film had cost $1 488 423 to produce – six times more than originally budgeted. If his studios were to survive, *Snow White* had to be a hit.

Unknown territory

Audiences already loved Disney's 5 to 10 minute cartoons, such as the adventures of Mickey Mouse, and the *Silly Symphonies* stories inspired by musical themes. But would fans be bored by an animated film lasting around 80 minutes?

In production, *Snow White* had presented Disney's animators with new and difficult problems. The first was how to make cartoon people move convincingly. The plot, based on the fairy tale as retold by the Brothers Grimm, called for a prince, a witch and seven dwarfs, as well as for a girl – Snow White. The studio artists were used to drawing animals, but had little or no experience in animating people, so Disney had footage of live actors – including three dwarfs – specially filmed for them to study, frame by frame.

Creating the illusion of depth

Another difficulty was caused by the complexity of some scenes in the film. At the time, animation drawings were made on paper, and then traced onto celluloid sheets – or 'cels' – of identical size. These were inked and coloured and then placed directly on top of the appropriate painted background to be filmed. The artists soon realised they would have to draw on an impossibly small scale to fit some of the bigger scenes onto the existing boards. Larger boards were installed throughout the Disney studio – which meant adapting all the camera equipment at huge expense.

With the standard equipment it was also impossible to create an impression of depth when the camera moved in from a wide shot to a detail. Because the animated drawings were laid directly on top of the background, both would increase in size at the same rate when the camera zoomed in. William Garity, the head of Disney's camera department, solved this problem with the ingenious 'multiplane camera', a massive device with several different planes of glass on which background and foreground details could be placed. This was a major innovation – for the first time, cartoon characters could move through a three-dimensional world.

Snow White took three and a half years to plan and produce. Plot lines and jokes were devised and refined in a series of story conferences, and the artists produced thousands of preliminary watercolours and sketches as the appearance of the characters developed. Each

ARTISTIC INFLUENCE *Disney's artists were influenced by popular illustrators. The finger-like branches in this early sketch of the forest scene are reminiscent of Arthur Rackham.*

NIGHTMARE SCENE
In the film, Snow White's flight through the forest is shown through her fevered imagination, with tree branches seeming to claw at her clothing as she passes.

NONSPEAKING ROLE *Walt Disney struggled to find a suitable voice for Dopey. So he decided that the dwarf should say nothing at all.*

CHARACTER SKETCH *Unlike the dwarfs in the Brothers Grimm story, those in Disney's film have distinctive personalities, with names to match.*

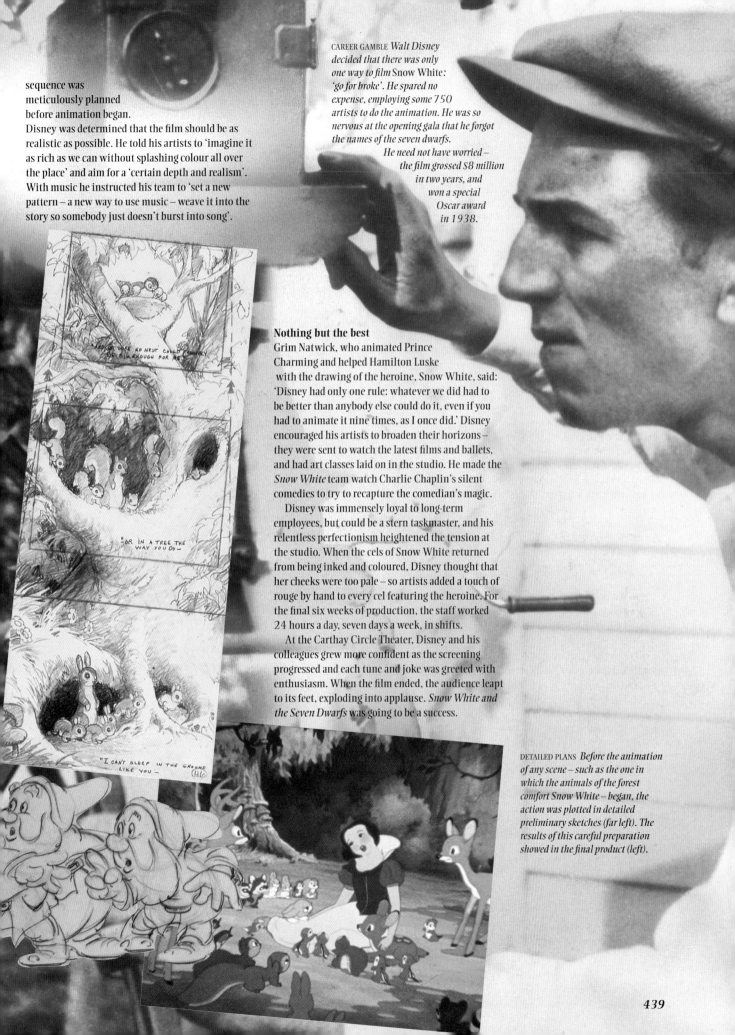

sequence was meticulously planned before animation began.

Disney was determined that the film should be as realistic as possible. He told his artists to 'imagine it as rich as we can without splashing colour all over the place' and aim for a 'certain depth and realism'. With music he instructed his team to 'set a new pattern – a new way to use music – weave it into the story so somebody just doesn't burst into song'.

Nothing but the best

Grim Natwick, who animated Prince Charming and helped Hamilton Luske with the drawing of the heroine, Snow White, said: 'Disney had only one rule: whatever we did had to be better than anybody else could do it, even if you had to animate it nine times, as I once did.' Disney encouraged his artists to broaden their horizons – they were sent to watch the latest films and ballets, and had art classes laid on in the studio. He made the *Snow White* team watch Charlie Chaplin's silent comedies to try to recapture the comedian's magic.

Disney was immensely loyal to long-term employees, but could be a stern taskmaster, and his relentless perfectionism heightened the tension at the studio. When the cels of Snow White returned from being inked and coloured, Disney thought that her cheeks were too pale – so artists added a touch of rouge by hand to every cel featuring the heroine. For the final six weeks of production, the staff worked 24 hours a day, seven days a week, in shifts.

At the Carthay Circle Theater, Disney and his colleagues grew more confident as the screening progressed and each tune and joke was greeted with enthusiasm. When the film ended, the audience leapt to its feet, exploding into applause. *Snow White and the Seven Dwarfs* was going to be a success.

"AND IM SURE NO NEST COULD POSSIBLY BE BIG ENOUGH FOR ME"

"OR IN A TREE THE WAY YOU DO –

"I CANT SLEEP IN THE GROUND LIKE YOU –

DETAILED PLANS *Before the animation of any scene – such as the one in which the animals of the forest comfort Snow White – began, the action was plotted in detailed preliminary sketches (far left). The results of this careful preparation showed in the final product (left).*

Index

Acknowledgments

The publishers wish to thank the following individuals and organisations who gave their help in compiling and checking the information contained in this book.

Prof Geoffrey Alderman; Marina de Allarcon; Alpine Club, London; Dr Bridie Andrews; Carol Andrews; Paul Bahn; Michael Barrett; Bath Postal Museum; Bill Bell; Maggie Black; Maria Blyzinski; Dr Brian Bowers; Sydney Brown; Prof Warwick Bray; Roger Bridgman; Bristol City Museum and Art Gallery; British Library; British Museum Education Service; British Telecom Museum; Malcolm Brown; Neil Brown; Sydney Brown; Cambridge and County Folk Museum; Dr Gloria Clifton; Dr Roger Cooter; Mary Coundley; Dr Andrew Cunningham; Dr Michael Daniel; Dr Andrew David; Dr Rosalie David; Mali Edmonds; English Heritage; Esperanto Centre, London; Fan Museum, London; Dr Irving Finkel; Adam Ford; Dr Lin Foxhall; Georgian Group, London; Prof John Gillingham; Richard Goman; Chris Gravett; James Griggs; Dr Willem Hackman; Dr Rick Halpern; George Hart; Hornimans Museum, London; Imperial War Museum; Ralph Jackson; Dr Stephen Jacyna; Simon James; Dr Mark Jenner; George Joffe; Peter Jones; Dr Helen King; Kingston Museum, Surrey; Lloyd's of London; Andrew Lord; Lord's Cricket Ground Museum, London; Sandy Malcolm; Bill Matthews; Dr Peter Matthias; Prof Sean McGrail; Carol Mellor; Meteorological Office; Dorothy Middleton; Museum of Costume, Bath; Museum of London; National Maritime Museum, Greenwich; National Motor Museum; Alice Naylor; News International PLC; Prof Vivian Nutton; Old Royal Observatory, Greenwich; Dr Richard Parkinson; Pitt Rivers Museum, Oxford; Pony Express National Memorial, St Joseph, Missouri, USA; Post Office Archives, London; Dr Martin Postle; Richard Postman; Dr Avril Powell; Reuters; Mark Rowland-Jones; Royal Geographical Society, London; Royal Horticultural Society, London; Dr Ellen La Rue; Donald Rumbelow; St Bride Printing Library, London; Scott Polar Research Institute, Cambridge; Dr Sonu Shamdasani; Dr Gary Sheffield; Shuttleworth Collection; Simon Stevens; Stock Exchange Press Office, London; Dr Steven Sturdy; Dr Akihito Suzuki; James Taylor; Tea Council, London; Ken Teague; Dr Annabel Thomas; Dr Simon Trew; Dr Barry Trinder; Turkish Embassy; Twining & Co Ltd; Brian Tyne; Patsy Vanags; Victoria and Albert Museum, London; John Villiers; Clive Wainwright; Dr Geoffrey Wainwright; Stephen Walsh; Karin Walton; Thomas Wiedemann; Barbara Wood; Marion Wood; David Woodcock; Dr Greg Woolf; Kevin Yhui; Yorkshire Mining Museum.

Picture Credits

The pictures in HOW WAS IT DONE? were supplied by the people and agencies listed below. Names given in *italics* are those of artists specially commissioned by Reader's Digest.

For the cover and for pages 1-9, pictures are listed from left to right and down the page. Where pictures on these pages also appear later in the book, only a partial credit is given here. For subsequent pages, the position of pictures is indicated by a description in brackets or by combinations of the following letters: [T] = top; [C] = centre; [B] = bottom; [L] = left; [R] = right.

The following short forms have been used: AAAC = Ancient Art and Architecture Collection; AKG = AKG London; AR = Ann Ronan at Image Select; BAL = The Bridgeman Art Library, London; BLI = By permission of the Board of the British Library; BLO = The Bodleian Library, Oxford; BM = By permission of the Trustees of the British Museum; BN = Bibliothèque Nationale, Paris; BPK = Bildarchiv Preussischer Kulturbesitz; CON = Library of Congress; CUP = Culver Pictures, New York; ET = ET Archive, London; HD = Hulton Deutsch Collection; JLC = Jean-Loup Charmet; KM = Kunsthistorisches Museum, Vienna; LNSW = State Library of New South Wales; MC = The Mansell Collection; MEPL = Mary Evans Picture Library; MH = Michael Holford; NMM = National Maritime Museum, London; NPG = National Portrait Gallery, London; PN = Peter Newark's Pictures; ML = Museum of London; RH = Robert Harding Picture Library; RPB = Range Pictures/Bettman Archive; RV = Roger-Viollet; SSPL = Science and Society Picture Library; V&A = By permission of the Board of Trustees of The Victoria and Albert Museum; WF = Werner Forman Archive, London; WIL = The Wellcome Institute Library, London.

COVER
Réunion des Musées Nationaux/Musée du Louvre, Paris; *Ivan Lapper*, The Image Bank/J Rajs; Sonia Halliday; ET; Trinity College Library, Cambridge; BAL; *Jonothan Potter*, RH/Adam Wolfitt; *Malcolm MacGregor*, RPB.

PAGE ONE
BLI (Ms Harl 4425 f12); BLI; CON; MH/British Museum, London; ET/Biblioteca Marciana, Venice; Ikona, Rome; Scala; Fitzwilliam Museum; AAAC; SSPL; BAL/Private Collection; Skyscan; Giraudon; MEPL; BN; Sonia Halliday.

PAGE TWO
The Image Bank/Guido Alberto Rossi; Brown Brothers; Giraudon; History Museum, Stockholm; (Cutty Sark) *Jonothan Potter*, Science Photo Library/Michael Gilbert; PN; MH/Marquis of Northampton, Castle Ashby; MAS, Barcelona; Private Collection; Sonia Halliday; V&A; The Image Bank/Nevada Weir; BLI/India Office Collection; BPK; SSPL.

PAGE FOUR
Scala; *Malcolm McGregor*, JLC; ET; RPB.

PAGE FIVE
BPK; WF/British Museum, London; JLC; *Ivan Lapper*.

PAGE SIX
BLI; BM; *Jonothan Potter*, National Geographic Society; State Library of Victoria; RPB; BAL; Archie Miles.

PAGE SEVEN
Giraudon; *Ivan Lapper*, SSPL; Permission of A.W. Freud *et al*, by arrangement with Mark Paterson & Associates; Giraudon/Chateau de Versailles.

PAGE EIGHT
Réunion des Musées Nationaux; *Ivan Lapper*, The Kendall Whaling Museum, Sharon, Massachusetts; Private Collection.

PAGE NINE
Ivan Lapper, JLC; RPB; HD; PN; RCS Libri & Grandi Opere, Milan; RPB.

CHAPTER ONE – ON THE MOVE
10-11 MH/National Maritime Museum, Greenwich. *12* [TR] © University Museum of National Antiquities, Oslo, Norway/Ove Holst/Detail from Hylestad stave church door; [BL] *Malcolm McGregor*. *12-13* By permission of the Pinxit Art Collection AB, Sweden/'The Vikings arrive in Greenland' by Oscar Wergland. *14* [TR] BN (Ms. Fr. 2810 f. 29v); [CR] BN (Ms. Fr. 2810 f. 59v) [BL] National Palace Museum, Taipei, Taiwan, Republic of China; [BC] Giraudon; [BR] BN (Ms. Fr. 2810 f. 76v). *15* [T] BLO (Ms. Bod. 264 f. 218r). *16* Scala. *16-17* Yan, Toulouse/Jean Dieuzade. *17* BAL/British Library, London. *18* William Fehr Collection, Castle, Cape Town. *18-19* Courtesy of Ikkan Art International, Inc., New York/Emanuel Leutze 'The Departure of Columbus from Palos 1492' oil on canvas 122 x 183cm, painted in 1855 (Private Collection). *20 Ivan Lapper*, [CR] Private Collection/Henry E Huntington Library, San Marino, CA; [BR] Metropolitan Museum of Art/Gift and Bequest of Alice K Bache, 1966 and 1977 (66.196.40-41). *21* [T] PN; [BL] Private collection. *22* [all] Giraudon. *22-23* Art work based on research by Louise Levathes, author of 'When China Ruled the Seas'/*Jonothan Potter*. *24* [TR] MH/Science Museum, London; [CR] BM; [BL] Philadelphia Museum of Art/Given by John T. Dorrance/Shen Tu 'Tribute Giraffe with Attendant'; [BC] Giraudon/Topkapi Museum; [BR] Scala/Museo della Scienza, Florence. *25* [TL] Bibliothèque Nationale, Paris (Ms Fr 150 f 16); [CL] [CR] MH/NMM; [remainder] NMM. *26* BLO (Ms Ash 1511 f86v). *26-27* (background) Scala/Biblioteca Estense, Modena. *27* Civico Museo Correr, Venice. *28* [TR] Scala/Galleria degli Uffizi (Giovana collection), Florence; [BL] ET/New York Public Library. *28-29 Ivan Lapper*. *29* [BR] (right) Lily Library, Indiana University (Boxer Codex f25v); [BR] (left) Lily Library, Indiana University (Boxer Codex f14). *30* [TR] BLI (Add Ms 23920 f55); [CR] LNSW/Image Library; [B] LNSW/Dixson Library.*30-31* BLI (Add Ms 7085 f16). *31* [TL] NMM; [C] [BL] [BC] LNSW/Image Library; [BR] BLI (Add 9345 16v). *32* [L] Scala/Keats and Shelley Memorial House, Rome; [B] BLI (Add Ms 36488). *33* [T] BAL/Christie's London; [BR] BAL/Roy Miles Gallery, 29 Bruton St, London. *34* [all] State Library of Victoria/La Trobe Collection. *35* Africana Museum. *36* (Lindsay) (Charles) (Jesse) Courtesy of the Douglas County Museum; (book) CON. *36-37* [T] BLI; (background) Art Resource, NY/National Museum of American Art, Washington D.C. *37* National Geographic Society/James L. Amos. *38* [CL] NPG; [R] Royal Geographical Society; [BL] Private collection; [BR] Giraudon. *39* The Granger Collection. *40-41 Jonothan Potter*. *42* [T] [BL] Sachische Landesbibliotek, Dresden; [R] Elizabeth A. Clark. *42-43* Elizabeth A. Clark. *43* (background) [TL] [TC] (Kasparek) Audrey Salkeld. *44* [TL] NMM; (poster) Private collection; [C] NPG; [B] BAL/NMM. *44-45* [T] [B] National Geographic Society/Robert E. Peary Collection; [C] RPB. *45* (background) National Geographic Society/Robert E. Peary Collection; [BR] Universitetsbiblioteket Oslo. *46 Malcolm McGregor*, [C] RV. *47 Malcolm McGregor*, [TL] [CL] Robert Opie. *48* AKG. *48-49* [T] Wright State University, Dayton, Ohio/Dunbar Library; [remainder] CON. *49* [CL] [BL] Wright State University, Dayton, Ohio/Dunbar Library.

CHAPTER TWO – EVERYDAY LIFE
50-51 ET/Marquis of Bath. *52* [L](background) Scala/Museo Egizio, Turin; [C] AAAC/Mary Jelliffe; [B] Giraudon; (scarab) BAL/Egyptian National Museum, Cairo/Giraudon; (cat) MH. *52-53* [both] BAL/Freud Museum, London. *53* [TR] [CR] ET/Egyptian National Museum, Cairo; [B] BAL/Lauros-Giraudon; [BL] MH/British Museum, London; [BC] ET/Museo della Civilta Romana, Rome. *54-55* [T] Giraudon; [B] Scala/Museo Nazionale, Naples. *55* [TL] WF; [TR] Sonia Halliday; [CL] [BR] BAL/British Museum, London; [C] Scala/Museo della Terme; [CR] MH/British Museum, London. *56* (background) BM; [TC] Giraudon/Musée des Antiquities Nationales, St Germain en Laye; [C] Giraudon/Musée des Antiquities Nationales, St Germain en Laye; [BL] BM; [BC] [BR] WF/Musée des Antiquities Nationales, St Germain en Laye. *57* [TR] Giraudon/Museo della Civiltá Romana, Rome; [C] ET/Museo Civico, Padua; [CR] ET/Archaeological Museum, Sofia; [BC] BAL/Pilkington Glass Museum, St. Helens; [BR] BM. *58-59 Ivan Lapper*. *60* Scala/Museo Pio-Clementino, Vatican. *60-61 Ivan Lapper*. *62* [T] Giraudon; [B] (both) Giraudon/Archive Nationale, Paris. *62-63* Scala/Museo Civico, Bologna. *63* Giraudon/Archive Nationale, Paris. *64-65* [T] MH/British Museum, London; [B] MEPL. *65* [T] Scott Polar Research Institute; (rest) *Paul Weston*. *66* [C] [B] AAAC. *66-67* G. Dagli Orti. 67 [TL] [CL] G. Dagli Orti; [BR] The Granger Collection. *68* [C] MAS, Barcelona/Palacio Real de Aranjuez, Madrid; [B] Giraudon/Musée du Louvre. *68-69* G. Dagli Orti/Musée du Louvre. 69 [T] [C] BAL/Museo Correr, Venice; [B] AAAC. *70* [TL] SSPL. [TR] MEPL; [CL] (lower) ET; [CL] (upper) [BL] SSPL; [CR] AR; [BR] Scala/Galleria Palatina, Florence. *71* [TL] [C] [BL] SSPL; [TR] ET; [BR] MH/Science Museum, London. *72* [CL] [B] MEPL. *72-73* JLC/Bibliothèque Mazarine, Paris. *73* [T] RV; (barometer) [B] SSPL. *74* [T] AKG; [C] [B] ET. *75* [all] MEPL. *76* [TL] HD; [CL] RV; [BL] Bulloz, Paris; [TR] [BR] MEPL. *76-77* JLC. *77* [TL] [BL] MEPL; BR (shoes) The Horniman Museum; [BR] (binding feet) JLC. *78* [L] Giraudon/Musée du Petit Palais; [TR] AKG; [B] Giraudon/Musée Carnavalet, Paris. *79* [TL] (all) JLC/Bibliothèque des Arts Decoratif, Paris; [TR] [CR] MEPL; [BL] Fine Art Photographs; [BR] RPB. *80* AKG/Albright-Knox Art Gallery, Buffalo NY. *80-81* MEPL. *81* [TL] Billie Love Collection; [TR] BAL/John Noott Galleries, Broadway, Worcs; [CL] [BL] ET/V&A; [C] *Barbara Brown*; [CR](upper) BAL/City Museum and Art Gallery, Hereford; [CR](lower) BAL/Private Collection; [BR] Fine Art Photographs/Anthony Mitchell Fine Paintings. *82* [TR] Trinity College Library, Cambridge/Munby Collection; [CR] [BL] Archie Miles; [BR] The Royal Photographic Society. 83 [TL] HD; [TR] Harry Ransom Humanities Research Center, University of Texas at Austin/Gernsheim Collection; [CL] (upper) Trinity College Library, Cambridge/Munby Collection; CL (lower) Popperfoto;[CR] [BR] Archie Miles; [BR] Harry Ransom Humanities Research Center, University of Texas at Austin/Gernsheim Collection.

CHAPTER THREE – TRIUMPHS OF BUILDING AND ENGINEERING
84-85 The Image Bank/Michael Garf. *86* [L] [BL] MEPL; [BR] National Geographic/Kenneth Garrett. *86-87* RH. *87* Giraudon. *88 Ivan Lapper*, [CR] BLI (Ms. Eg. 3028 f. 30). *89* The Image Bank/Eric Meola. *90-93 Ivan Lapper*. *93* [BR] *Terence Dally*. *94 Ivan Lapper*. *94-95* [T] Comstock Photofile Limited; [B] Scala. *95* Ikona, Rome/Biblioteca Reale, Turin/Per concessione del Ministero dei Beni Culturali ed Ambientali. *96* [TL] [BL] [BR] John Hillelson/Brian Brake; [BC] Aspect/Brian Seed. *96-97* John Hillelson/Brian Brake; *97* AKG. *98* [CL] Colorific/Lori Grinker/Contact; [BL] JLC. *98-99* Aspect/Larry Burrows. *99* [CL] Aspect/Fiona Nichols; [remainder] Aspect/Larry Burrows. *100* [TR] Aspect/Tom Nebbia; [CL] RH/Ian Tomlinson; [CR] Impact/David Reed; [BL] Gerald Cubitt; [BR] RH/Sheila Terry. *100-1* Gerald Cubitt *101* [R] WF/Private Collection; [B] Aspect/Tom Nebbia. *102* [CL] Sonia Halliday/Martine Klotz; [C] BLI. *102-3* RH/Kim Hart. *104* [CL] Scala; [C] [BL] Photograph by Jean Bernard in 'L'Univers de Chartres' © Bordas, Paris 1988. *104-5* RH/Adam Woolfitt. *105* Scala. *106* [L] *Malcolm McGregor*, [BR] Loren McIntyre. *106-7* Loren McIntyre. *107* [CR] [BR] Loren McIntyre. *108-11 Ivan Lapper*. *112-13 Malcolm McGregor*. *114* [L] Angelo Hornak/V&A; [R] BPK/Staatliche Museen, Berlin, Museum für Indische Kunst/Photo Liepe; [BC] BLI/India Office Collection. *115* [CL] Scala; [R] RH/Adam Woolfitt; [BL] Scala. *116* [TR] Giraudon/Chateau de Versailles; [CL] The Royal Collection © Her Majesty The Queen; [CR] Giraudon; [BL] JLC; [BR] AKG. *116-17* Colorific/Sylvain Grandaman. *117* [TR] Colorific/Jean Paul Nacivet; [CR] The Hutchison Picture Library/John Hatt; [BC] BN; [BR] BPK. *118* Ikona, Rome/Photo Böhm, Venice. *118-19* BPK/Museo Correr, Venice/Dagli Orti. *119* [CR] MC; [B] Gemeente Enkhuizen/Zuiderzeemuseum. *120* [CL] MEPL; [BL] Scala/Russian State Museum, St Petersburg; [BC] Novosti/Russian State Museum, St Petersburg. *120-1* [T] BN, (Cote Ge DD 894 93); (background) Novosti/Tretykov Gallery, Moscow. *121* [T] [C] [B] BLI. *122* [TR] ET/Science Museum, London; [C] MC; [BL] JLC/Bibliothèque de l'Ecole Nationale de Ponts et Chaussées. *123* [L](all) ET; [TR] SSPL; [CR] AR; [BC] [BR] University of Bristol. *124* Woodfin Camp & Associates/Bill Weems. *124-5* [T] CON; [C] [B] RPB. *126* [T] JLC/Musée Carnavalet, Paris; (Hippodrome) Giraudon; (cafe) Giraudon/© DACS 1966/Louis Anquentin, 'L'Avenue de Clichy'; (theatre) Giraudon/Musée des Beaux Arts, Lyon; (building site) RV/Musée Carnavalet, Paris; (Bois) Giraudon/Musée Carnavalet, Paris [BR] AR. *127* [TC] [TR] [CL] MEPL; [BL] AR; [BC] Popperfoto; [BR] The Image Bank/Gary Cralle. *128* [CL] Giraudon/BN; [C] BPK; [BL] BPK/Staatliche Museen. Berlin, Museum für Vor und/Fruhgeschichte; [BC] AAAC; [BR] AAAC. *128-9*. [T] Giraudon; [C] JLC; [B] Ashmolean Museum, Oxford. *129* [TR] (both) [BL] AKG/Staatliche Museen, Berlin, Museum für Vor unt Frühdgeschichte; [CR] Ullstein; [BR] BPK/Staatliche Museen, Berlin, Museum für Vor und Frühdgdschichte/Klaus Goken. *130* MEPL. *130-1 Ivan Lapper*. *132* [TL] Ullstein; [CL] BPK; [BL] Süddeutscher Verlag; [BR] RPB. *132-3* [T] Süddeutscher Verlag; [C] [B] RPB. *133* [TR] [CR] (both) [BR] RPB; [BL] RV. *134* [T](Farrington) (buggy) HD; (foot bridge) Süddeutscher Verlag; (saddle) AR; [BL] [BR] RPB; [BC] BPK. *135* RH/Adam Woolfitt. *136* [T] LNSW/Image Library; [C] (both) LNSW/Image Library; [BL] ET/British Steel Corporation; [BR] (upper) AKG; [BR] (lower) Süddeutscher Verlag. *136-7* Altitude/Baron Wolman. *137* [CR] The Hutchison Picture Library/Michael Macintyre; [BR] RPB. *138* [L] (background) AKG/Lewis Hine; [BR] RPB. *139* (background) [BR] RPB; [TL] The Image Bank/Jake Rajs; [CL] Aspect/© Vince Streano 1984; [C] Network/Bilderberg/G Fischer.

CHAPTER FOUR – FOOD AND DRINK
140-1 Sonia Halliday. *142* (stag) AAAC; (bison) © Editions d'Art Albert Skira SA, Geneva; [BR] Trustees of the British Museum (Natural History). *143* [CL] Trustees of the British Museum (Natural History); [C] G. Dagli Orti/Musée de St Germain en Laye; [BL] Musée Nationale de Prehistoire des Eyzies; [BR] Réunion des Musées Nationaux. *144* [TR] ET/Egyptian Museum, Cairo; [CL] University of Pennsylvania Museum, Philadelphia; [C] [CR] [BR] AKG/Erich Lessing. *144-5* (background) AKG/Erich Lessing; [B] Ikona, Rome/Photo Aldo Durazzi, Rome/© 1973 Time-Life Books Inc. from 'Emergence of Man: The First Farmers'. *145* [TL] ET; [TR] AKG/Drovetti Collection, Egyptian Museum, Turin; [CL] (donkey) WF/Egyptian Museum, Turin; [CL] (flock) AAAC; [BL] ET/Egyptian Museum, Cairo. *146* [C] Giraudon; [B] Ashmolean Museum, Oxford. *146-7* [T] JLC/British Museum, London; [B] Museum of Fine Arts, Boston/H.L. Pierce Fund. *147* [TR] Scala/Museo Chiaramonti, Vatican; [CR] JLC/Musée de St Germain en Laye; [BL] AKG/Museo Nazionale Archeologico, Naples/Erich Lessing; [BR] Scala/Museo Nazionale, Naples. *148* [TR] AKG/Antikensammlung, KM/Erich Lessing; (jug) AKG/Cabinet des Medailles, BN/Erich Lessing; [CR] AKG/Museo Nazionale Archeologico, Naples/Erich Lessing; (dish) AKG/Cabinet des Medailles, BN/Erich Lessing; [BL] AAAC; [BR] AKG/Antikensammlung, KM/Erich Lessing. *148-9* ET. *149* (servant) BAL/Louvre, Paris; (boy) C M Dixon; (scraps) Scala/Museo Gregoriano Profano, Vatican. *150* [TL] BLO (Ms. Maps, Notts.a.2.); (shearing wine nuts) BAL/Musée Condé, Chantilly; (ploughing) BAL/British Library, London. *150-1* BAL/Christie's, London. *151* [TL] Giraudon; [CR] ET. *152* [TL], [BC], [CR] BAL/British Library, London; [CL] BAL/BN. *153* Giraudon/Musée Condé, Chantilly; [BR] BAL/Palais des Papes, Avignon/Giraudon. *154* [T] The Trustees and Guardians of Shakespeare's Birthplace; BLI. *154-5* [T] [C] [B] BLI. *155* [BL] Sotheby's, London; [BR] BAL/Giraudon/Musée du Louvre, Paris. *156* [T] (both) [CL] [C] [BR] BLI; [BL] CUP. *157* BAL/Giraudon/Staatliche Kunstsammlungen, Dresden; [TR] BM; [C] (background) The Fotomas Index; [BC] V&A. *158* [CL] Giraudon/Musée Carnavalet, Paris [C] RV/BN; [TR] [BL] [BR] from Robin Weir and Caroline Liddell, 'Ices: the Definitive Guide' Grub Street, London 1995. *159* (background) Giraudon; [CL] Giraudon/Bibliothèque de l'Arsenal, Paris; [BL] JLC. *160* [TR] Kunstmuseum Bern/on loan from Gottfried Keller-Stiftung, Winterthur; [BL] MEPL; [BR] Robert Opie. *161* [TC] [TR] The Illustrated London News Picture Library; [BL] [BC] [BR] Robert Opie. *162* [T] The Thomas Gilcrease Institute of American History and Art, Tulsa, Oklahoma; [C] BAL. *162-3* Brown Brothers; [TL] National Cowboy Hall of Fame, Oklahoma City; [CL] Frederic Remington Art Museum, Ogdensburg, NY; [BR] RPB. *164* [CR] (Deere) [B] Brown Brothers; [C] [CR] (plough) RPB. *164-5* Brown Brothers. *165* (McCormick) CUP; [TC] Brown Brothers; [CR] [BR] RPB. *166* [TR] (both) MEPL; [BL] BLI; [BR] AR. *167* [T] Reform Club; [BL] MEPL; [CR] [BC] MC. *168* JLC. *168-9* [T] Fondation Auguste Escoffier; [B] Savoy Group. *169* [L] RV; [C] The Raymond Mander & Joe Mitchenson Theatre Collection; [BR] Fondation Auguste Escoffier.

CHAPTER FIVE – MILESTONES IN MEDICINE
170-1 BAL/Mauritshuis, The Hague. *172* (background) SSPL; [BL] [BC] Loren McIntyre; [BR] ET/Archaeological Museum, Lima (C0307078). *172-3* Art Resource, NY/George Catlin 'Medicine Man' 1832/National Museum of American Art, Washington D.C. *173* [R] RPB; [BL] ET/Archaeological Museum, Lima (54600). *174* [L] WIL; (background) MEPL. *174-5* WF/Egyptian Museum, Cairo. *175* [R] Ny Carlsberg Glyptotek; [BL] JLC; [BC] William MacQuitty. *176* [CL] JLC/Bibliothèque de l'Ancienne Faculté de Médecine, Paris; [BR] JLC/Bibliothèque Nationale, Paris. *176-7* Ikona, Rome/Biblioteca Medicea Laurenziana, Florence (Ms. Plut. 47.7 c.190r.) Per concessione del Ministero dei Beni Culturali ed Ambientali. *177* [L] RPB; [TR] Giraudon; [CR] BM; BR Kostos Kontos, Athens. *178* AKG. *178-9* WIL. *179* [TR] Zentralbibliothek Zürich (Ms. C54, f.34v.); [TL] Österreichische National Bibliothek (Cod. Med. pr 1, fol. 3); [C] Zentralbibliothek Zürich (upper – Ms. C54, f.34r) (lower – Ms. C54, f.36r); [R] Zentralbibliothek Zürich (Ms. C54, f.34v);

[BC] Giraudon/Musée Condé, Chantilly. 180 [CR] Ikona, Rome/Pavia University Library (Ms. Aldini 211, c.20)/Photo Trentani/Per concessione del Ministero dei Culturali ed Ambientali; BL BM. 180-1 Rijksmuseum, Amsterdam. 181 [TR] Scala/Museo Nazionale, Naples; [CR] Österreichisches Museum für Volkskunde.
182-3 Rijksmuseum, Amsterdam. 183 [L] BLI (Ms. Add. 42130 f.61); [R] BAL/Galleria degli Uffizi, Florence.
184 [L] Giraudon/Musée Condé, Chantilly. [TR] Giraudon/Biblioteca Civica, Padua; [BR] The Royal Collection © Her Majesty The Queen (Atlas of Anatomy Studies f.122r). 185 [L] AKG; [TC] BAL/Fratelli Fabri, Milan; [TR] ET; [BR] BAL/University Library, Glasgow.
186 [CL] NPG; [BL] JLC. 186-7 Scala/Galleria Borghese, Rome. 187 [L] [TR] By permission of the President & Council of the Royal Society; [BR] BAL/National Gallery, London. 188 [T] SSPL; [C] SSPL; [B] SSPL. 188-9 RV/Museo San Martino, Naples. 189 Historisches Museum Basel/ A. Eaton. 190 WF/British Museum, London.
190-1 [T] Scala/Ospedale degli Innocenti, Florence; [B] WIL. 191 Yale University, Cushing Medical Library, Historical Library,Clements C. Fry Collection. 192-3 NPG (detail). 193 [TR] WIL; [CL] SSPL; [C] SSPL; [CR] Philadelphia Museum of Art/The William H. Helfand Collection/Joseph Kepler; [BL] JLC/BN. 194 JLC/BN. 194-5 MAS, Barcelona/Academia de San Fernando, Madrid.
195 [T] Private Collection, copied from 'Marquis de Sade' by Maurice Lever, published by Fayard; [CR] Giraudon/Academie de Médecine, Paris. 196 [TL] [TR] WIL; [CL] Reproduced by kind permission of the President and Council of the Royal College of Surgeons of England; [BL] F.A. Countway Library of Medicine, Boston; [BR] BAL/Jefferson College, Philadelphia. 196-7 Österreichische Galerie Belvedere, Vienna/Photo, Otto. 197 (background) Reproduced by kind permission of the President and Council of the Royal College of Surgeons of England; [TL] PN; [TR] WIL; [CR] JLC. 198-9 [all] Giraudon/Chateau de Versailles. 199 [T] JLC/Musée du Val de Grace; (Mary Seacole) Courtesy of the National Library of Jamaica; (amputation) WIL; [B] BAL/Forbes Magazine Collection, New York. 200 [T] Giraudon/Albert Edelfelt 'Pasteur in his Laboratory' 1885, Musée d'Orsay, Paris. 200-1 (Patients & Pasteur) JLC. 201 [all] RPB. 202 [TR] [BR] Robert Opie; [CR] Philadelphia Museum of Art/The William H. Helfland Collection/Anon 'The celebrated oxygenated bitters'; [BL] Österreichische National Bibliothek. 203 [TL] [TC] [TR] [CL] [BL] Robert Opie; [C] Philadelphia Museum of Art/ William H. Helfland Collection/Anon, 'Ayer's Sarsaparilla' [BC] Philadelphia Museum of Art/William H. Helfland Collection/Anon, 'Le Solitaire'; [BR] PN. 204 [BL] [BC] Permission of A.W. Freud *et al*, by arrangement with Mark Paterson & Associates. 205 *Andrew Aloof* (after photography by Edmund Engelman).

CHAPTER SIX – LAW AND ORDER
206-7 BAL/Musée du Louvre, Paris/Giraudon.
208 [C] (both) BM; (Socrates) Metropolitan Museum of Art/Wolfe Fund, 1931. Catharine Lorillard Wolfe Collection (31.45); [B] (both) American School of Classical Studies at Athens: Agora Excavations. 208-9 Scala/Palazzo Madama, Rome. 209 C M Dixon/Glyptothek, Munich.
210 [C] Lesley & Roy Adkins; [BL] WF/Museo Archeologico Nazionale, Naples. 210-11 Ikona, Rome/Vasari, Rome.
211 [T] National Museum, Belgrade; [C] Lesley & Roy Adkins; [BL] Peter Clayton; [BR] Mauro Pucciarelli.
212-13 (all) National Palace Museum,Taipei, Taiwan, Republic of China. 214 *Ivan Lapper*. 214-15 Alecto Historical Editions, publishers of the Great Domesday Book facsimile. 215 *Ivan Lapper*. 216 University of Copenhagen/Det Asamagnaeanske Institut.
216-17 Bibliothèque Municipal Chateauroux/Grandes Chroniques de France, f.225v. 217; [C] BAL/Victoria and Albert Museum, London; [B] Giraudon/Biblioteca Nazionale, Marciana. 218 [T] [C] JLC; [B] Staatsbibliothek Bamberg (Ms. R.B. Misc. 120 f. 32v). 219 [T] ET/Honourable Society of the Inner Temple; [C] JLC/Musée Archaeologique, Orleans; [B] Public Record Office.
220-1 Giraudon/Academia de Bellas Artes de San Fernando, Madrid. 221 [TL] Oronoz, Madrid/Prado, Madrid; [TR] Giraudon/Musée du Louvre; [BR] Giraudon/Real Academia de Bellas Artes de San Fernando, Madrid. 222 Réunion des Musées Nationaux/Chateau de Versailles. 222-3 BAL/Chateau de Versailles/Giraudon.
223 [TL] BN/Des Heures de Louis le Grand; [CL] Réunion des Musées Nationaux; [CR] (Council) [BC] Réunion des Musées Nationaux/Chateau de Versailles. 224 [BL] Fortean Picture Library; [BC] Courtesy of the Massachusetts Historical Society. 225 (diary) [CR] Courtesy of the Massachusetts Historical Society; [CL] [B] Peabody Essex Museum Salem, MA/Photo Mark Sexton. 226 BN.
226-7 BN. 227 [CR] BAL/Musée Carnavalet, Paris; [BR] BAL/Musée Carnavalet, Paris/Giraudon. 228 [T] [B] Fortean Picture Library. 228-9 ET/Musée Carnavalet, Paris.
229 [TL] John Crabtree 'Concise History of Halifax' 1836; [CL] Giraudon/Musée Carnavalet, Paris; [CR] RV; [BR] Rijksmuseum, Amsterdam (detail). 230 Museo Condes de Castro Guimaraes, Cascais. 230-1 NMM. 231 CON.
232 BLI/India Office Collection. 233 Sotheby's, London.
234-5 National Library of Australia/Rex Nan Kivell Collection; [T] Timothy Millet, A.H. Baldwin and Sons, London. 235 [TL] (all) Timothy Millet, A. H. Baldwin and Sons, London; [TC] [TR] [C] (both) [CR] (both) [CR] Archives Office of Tasmania; [CL] [L] The Illustrated London News Picture Library; [B] Tasmanian Museum, Hobart.
236 [T] John Frost Historical Newspaper Service; [C] (upper) MEPL; [C] (lower) John Frost Historical Newspaper Service; [B] RPB. 236-7 (Rasputin) State Archives of the Russian Federation; (Yusupov) RV. 237 [TR] Harvard Law School, Cambridge, MA; [CL] State Archives of the Russian Federation; [remainder] John Frost Historical Newspaper Service. 238 CON/Alfred Stieglitz. 238-9 [T] [C] [B] CON.
239 [TR] The New York Public Library; [CR] (all) The New York Public Library/Lewis W. Hine; [BR] CON.
240 [C] MEPL; [T] [B] HD. 240-1 [T] (Niagara) (looters) MEPL; (Louisville raid) (Disguised agents) RPB.
241 (Closing speakeasy) Brown Brothers; (Speakeasy) RPB; (cards) CUP; (Texas Guinan) (bottle in book) RPB; [B] HD.
242 [BL] [BR] Royal Air Force Museum/Ley Kenyon.
242-3 Royal Air Force Museum/Ley Kenyon;
243 [CL] [CR] [BL] Royal Air Force Museum/Ley Kenyon; [BR] Royal Air Force Museum.

CHAPTER SEVEN – FEATS OF SCIENCE
244-5 BAL/Art Gallery and Museum, Derby.
246 Dr Ifigenia Dekoulakou, B'Ephoreia of Prehistoric & Classical Antiquities, Athens. 246-7 [T] *Paul Weston*; [B] Ashmolean Museum, Oxford. 247 [TR] MH; [CR] [BR] (lower) [BR] (upper) AR. 248 [C] MH/British Museum, London; [B] AR. 248-9 [T] AR; [B] Ikona, Rome/Biblioteca Ambrosiana, Milan. 249 [CR] BPK; [BR] AKG.
250 *Ivan Lapper*, (portrait) AKG/Musée du Louvre; (Ptolemy observing) Science Photo Library. 251 [TL] AR; [C] Science Photo Library/JLC; [BR] AR. 252 [TL] [TR] Courtesy of The Time Museum, Rockford, Illinois; [CL] SSPL; [BL] SSPL; [CR] Museum of the History of Science, Oxford University. 252-3 Angelo Hornak. 253 [TL] BLO (Roll 168H Frame 16); [TR] BLI (Ms. Add.19776 f. 72v.); [C] MH/Science Museum, London; [BL] [BR] [B] Courtesy of The Time Museum, Rockford, Illinois. 254 British Library.
254-5 *Ivan Lapper*. 256 [BL] MH/British Museum, London.
256-7 [T] WF/Museum für Völkerkunde, Berlin; [B] South America Pictures/Tony Morrison. 257 [C] BPK/Staatliche Museen zu Berlin – Preussischer Kulterbesitz, Museum für Völkerkunde; [B] WF/Museum für Völkerkunde, Berlin.
258 [L] AAAC; [TC] [C] [R] AR; [BC] MEPL. 259 [T] Science Photo Library/JLC; [B] ET/Palazzo Vecchio, Florence.
260 (crossbow) BAL/Biblioteca Ambrosiana, Milan; (machine guns) Ikona, Rome/Biblioteca Ambrosiana, Milan (Ms. CA 398 r.); (chain) BM. 261 (tank) Ikona, Rome/Biblioteca Ambrosiana, Milan (Ms CA 357 r.); (da Vinci) The Royal Collection © Her Majesty The Queen; (helicopter) Bulloz, Paris/Bibliothèque de L'Institut de France (Ms. Vinci f. 83 v.); (flying machine) BAL/Bibliothèque de l'Institut de France/Giraudon; (parachute) Ikona, Rome/Biblioteca Ambrosiana, Milan (Ms CA 381 v.); (wing) BN; (snorkel) Ikona, Rome/Biblioteca Ambrosiana, Milan (Ms CA 1069 r.); (diving suit) Ikona, Rome/Biblioteca Ambrosiana, Milan (Ms CA 333 v.); (lifebelt) Bulloz, Paris/Bibliothèque de L'Institut de France (Ms. Vinci B f. 81 v.); (drill) Ikona, Rome/Biblioteca Ambrosiana, Milan (Ms CA 56 v.). 262 [L] BAL/British Library, London; [R] AKG/Erich Lessing/Collegium Maius Library, Cracow; (signature) RV; [B] BAL/British Library, London. 263 [CR] MEPL; [B] JLC. 264 [L] Scala/Tribuna di Galileo, Florence; [R] Science Photo Library.
264-5 BAL/Private collection. 265 [TC] Scala/Biblioteca Nazionale, Florence; (telescope) Scala/Museo della Scienza, Florence. 266 [T] [C] [B] By permission of the President & Council of the Royal Society/Richard Valencia; [BR] Science Photo Library. 266-7 JLC. 267 [C] [CR] JLC; (flask) Institut Pasteur, Paris; (Redi) HD. 268 [L] Jeremy Whitaker/Reproduced by permission of the Portsmouth Estate; [TR] RPB; [CR] AR; [BR] BLO (Ms. New Col. 361 f. 45v) By permission of the Warden and Fellows, New College Oxford. 269 [CL] [CR] [BL] AR; [BC] ET/Royal Society; [BR] Science Photo Library. 270 [TL] JLC/Archives de l'Academie des Sciences, Paris; [CL] AKG/Metropolitan Museum of Art, New York; [C] AR; [B] JLC. 271 [CL] Courtesy of The Time Museum, Rockford, Illinois; [BC] BLI; [BL] [BR] JLC/Musée Carnavalet, Paris. 272 [L] By permission of the President & Council of the Royal Society/Richard Valencia; [R] AR; (Background) RV.
273 [TR] JLC/Archives de l'Academie des Sciences, Paris; [CR] RV; [B] Scala/Tribuna di Galileo, Florence. 274 AR; [TR] Science Photo Library; [CL] JLC; [TL] [CR] [BL] [BR] AR.
274-5 [T] Science Photo Library/Dr Jeremy Burgess; [B] AR.
275 [TL] JLC; [BR] MEPL. 276 [L] MEPL; [CL] [BL] AR.
276-7 [T] [B] (background) SSPL. 277 [T] SSPL; [R] NPG; [B] AR. 278-281 *Ivan Lapper*, 281 [CR] Private Collection.
282 [TL] MEPL; [C] C Wynn-Williams, Courtesy of the Emilio Segrè Visual Archives/Supplied by The Neils Bohr Memorial Library; [B] University of Cambridge, Cavendish Laboratory; [BR] RPB. 282-3 [T] Science Photo Library/Mehau Kulyk; [B] Science Photo Library.
283 [BC] SSPL; [BR] Manchester City Council.

CHAPTER EIGHT – THE WORLD AT WORK
284-5 BAL/Forbes Magazine Collection, New York.
285 [BL] BN (Ms. n.a. Fr. 20386 f.34); [BL] (Ptolemy) BN.
286-7 *Malcolm McGregor*. 287 [TR] Giraudon; [CR] Trustees of the British Museum (Natural History); [BR] ET.
288-9 [T] *Malcolm McGregor*, [B] BPK. 290 [BL] BM; (remainder) Private Collection. 290-1 Private Collection.
291 [R] ET/Devizes Museum; [CL] [BL] [BC] BM/Museum of Mankind; (rest) Private Collection. 292 [TL] BM; [B] Dott. Felice Torno. 292-3 Studio Koppermann/Staatliche Antiken Sammlungen, Munich. 293 [TR] BPK/Staatliche Museen zu Berlin-Preussischer Kulterbesitz, Antikenmuseum; [CR] [BR] AKG/Musée du Louvre, Paris.
294 [TL] ML; [CL] (both) BM; [BL] AAAC. 294-5 [T] Scala/Museo Archaeologico, Venice; [B] Giraudon/Museo della Civilita Romana, Rome. 295 (ship) BPK; (barge) Musée Calvet, Avignon; (cart and tavern) ET/ Diozesanmuseum, Trier; [B] ET. 296 [CL] Scala/ Biblioteca Laurenziana, Florence; [CR] BPK/ Staatsarchiv, Hamburg; [B] Scala/Catello Issogne. 297 [TC](both) Det Hanseatiske Museum; [CR] BPK/Staatsarchiv Hamburg; [BL] BPK; [BC] BPK/Gemaldegalerie SMPK Berlin. 298 [TC] Giraudon; [C] BAL/Guildhall Art Gallery, Corporation of London; [CR] The Worshipful Company of Skinners; (seal) BM; [B] Guildhall Library, Corporation of London. 298-9 BLI/India Office Collection. 299 [TR] Aspect/National Maritime Museum, Greenwich. 300 [T] Still Life Studies, Amsterdam; [B] Frans Halsmuseum – De Hallen. 300-1 ML. 301 [CL] Lloyd's of London; [BL] Gainsborough's House, Sudbury, Suffolk; [CR] [BR] Lancashire Record Office. 302 [TL] Museum and Art Gallery, Hull; [BL] Cornell University Library. 302-3 [T] MEPL; [C] (lower) BAL/Private collection; [C] (upper) BAL/Hirshorn Museum, Washington DC; [B] NMM; (chain) Museum and Art Gallery, Hull. 303 [CL] PN; [CR] RPB; [BL] Museum and Art Gallery, Hull. 304 [L] Simonin 'Mines and Miners'; [TC] MEPL; [C] Report of The Child Employment Commission 1842; [C] (both) [BC] MEPL. 305 [TR] Simonin 'Mines and Miners'; [B] MEPL. 306 (all) Manchester Central Library Local Studies Unit. 306-7 Helmshore Local History Society.
307 (all) Manchester Central Library Local Studies Unit.
308 [L] Old Dartmouth Historical Society; [BC] The Kendall Whaling Museum, Sharon, Massachusetts, USA.; [BR] Old Dartmouth Historical Society. 308-9 Old Dartmouth Historical Society; [B] The Kendall Whaling Museum, Sharon, Massachusetts, USA
310 [TL] PN; [CL] The Museums at Stony Brook, Long Island NY, Gift of Mr and Mrs Ward Melville, 1955; (nugget) [BL] University of California, Bancroft Library.
310-11 [T] PN; [C] University of California, Bancroft Library; [B] California State Library. 311 [CR] ET/Mitchell Library, Sydney; [BR] ET/Latrobe Library, Melbourne.
312 (race) NMM; [BC] William F Baker. 313 *Jonothan Potter*, [TR] Aspect; [BR] The Maritime Trust/Shell (UK).
314 [TL] V&A; [CL] MC. 314-15 (background) The Fotomas Index. 315 [TL] By permission of the Trustees of the Chatsworth Settlement; [CL] MEPL. 316 [TC] BAL/Fitzwilliam Museum, University of Cambridge; [CL] JLC/Musée Carnavalet, Paris; [CR] [BR] Albany Museum, Grahamstown, SA/Duncan Greaves; [C] Giraudon/BN; [BL] RPB; [BC] BAL/Yale University Art Gallery, New Haven, CT.
317 [T] ET; [B] AKG. 318 [CL] RPB; [C] PN; [BL] BAL/Private Collection; [BC] HD; [BR] Brown Brothers; (Store sign) F. W. Woolworth Co. 319 (all) Sears, Roebuck and Co. 320 (all) The Coca-Cola Company (Coca-Cola & Coke are registered trade marks identifying the same product). 320-1 (background) PN; [T] [B] RPB. 321 PN.

CHAPTER NINE – WAR AND WEAPONS
322-3 Leeds City Art Galleries. 324 [L] [C] [CR] [BL] BM; [BC] BPK/Kupferstichkabinett, Berlin; [BR] Scala/Museo Nazionale, Naples. 325 [T] AKG/Erich Lessing; [BR] Scala/Vatican. 326 [BL] Réunion des Musées Nationaux/Musée de Louvre, Paris. 326-7 (helmet) Antikensammlung, Munich/Studio Koppermann; (phalanx) Scala/Museo di Villa Guilla, Rome. 327 The National Gallery, London. 328 [TL] (both) BM. 328-31 *Ivan Lapper*.
332 PN. 332-3 Giraudon. 333 Musée du Louvre/Photo Josse/C M Dixon. 334 [TL] BAL/Bibliothèque Municipale de Lyon; [CL] BLI (Ms. Roy. 16 G VI f.174); [R] Giraudon/British Museum, London; [BL] Giraudon/BN; [BC] Ikona, Rome/Biblioteca Ambrosiana, Milan (Ms. S.P. 67c 41r). 335 [T] Sonia Halliday/Biblioteca Nacional, Madrid; [B] BN (Ms. Fr. 352 f.62). 336 [TR] Northamptonshire Record Office; [CR] BLI (Ms Roy 20 DXI 134v); [BL] V&A; [BC] Bibliothèque Interuniversitaire Montpellier (Ms H196 f63v); [BR] By permission of The Dean & Chapter of Durham Cathedral. 337 [TR] BM; [B] BAL/British Library, London (Ms 4397 f 23v). 338 [TL] Metropolitan Museum of Art/Bashford Dean Memorial Fund, Gift of Helen Fahnestock Hubbard, in memory of her father, Harris C Fahnestock 1929 (29.154.3); [TC] [TR] [BL] [BC] KM; [BR] Metropolitan Museum of Art/Munsey Fund,1932 (32.130.6). 338-9 Oronoz, Madrid/Ameria Real, Madrid.
339 [TL] Stadt Nürnberg/(Hausbuch der Mendelschen Zwölfbrüderstiftung); [TR] [CR] [BR] The Board of Trustees of the Royal Armouries. 340 [L] AKG/Erich Lessing; [R] BAL/BN. 340-1 Graphische Sammlung Albertina, Vienna.
341 [CR] ET/University of Gottingen; [BR] RV. 342 PN.
342-3 BN (Ms. Sup. Pers.1113 f.49). 343 [TL] RH; [BL] WF/Imperial Library, Tehran; [BC] Edinburgh University Library (EUL Or. Ms. 20 f.124v.). 344 [C] (both portions) [BR] Japan Archive; [BL] V&A. 344-5 Society for the preservation of Japanese Art Swords; The Sword Museum, Tokyo. 345 [TR] WF/Kuroda Collection, Japan; [B] JLC.
346 [L] Ikona, Rome/Biblioteca Nazionale, Florence (Ms. B.R. 232, c. 30r.) [C] Ikona, Rome/Biblioteca Medicea Laurenziana, Florence (Fl. Cod. B2 f122) Per concessione del Ministero dei Beni Culturali ed Ambientali; [R] [B] BM/Museum of Mankind. 347 [L] MAS, Barcelona/ Armeria Real, Madrid; [CL] [C] [BC] Royal Library, Copenhagen; [BR] MAS, Barcelona/Archivo de Indias, Seville. 348 BPK/Kunstbibliothek Preussischer Kulterbesitz, Berlin. 348-9 [T] BPK/Staatsbibliothek, Berlin; [B] AKG. 351 NMM. 352-3 BPK/Staatliche Museen Kunstbibliothek, Berlin. 353 [BC] RV; [BR] Bulloz, Paris/Chateau de Versailles. 354 [TL] Bulloz, Paris/Musée de l'Armée, Paris. 354-5 V&A. 355 [TR] V&A; [CR] BPK/Samuel H Kress Collection, National Gallery of Art, Washington. 356 RPB. 356-7 RPB. 357 [TL] 'The West Point Museum, United States Military Academy, West Point, New York'; [B] HD. 358 [CL] The Royal Collection © Her Majesty The Queen on loan to the Local History Museum, Durban; [B] BAL/National Army Museum, London. 359 [TR] ET/National Maritime Museum, Greenwich; [CR](lower) AR; [CR](upper) HD; [BL] MEPL; [BR] The Robert Hunt Library. 360-1 *Ivan Lapper*.
364 Imperial War Museum. 364-5 Bayerisches Armeemuseum, Ingolstadt/C. Bergen. 365 (Richthofen) Military Archive & Research Services by H. Campbell, RAF Museum, Hendon; (crash) [B] ET/Australian War Memorial, Canberra. 366-7 *Jonothan Potter*. 367 [BL] [BC] BPK.
368 Trinity College Library, Cambridge (Ms. R.17. 1.f.283v.).

CHAPTER TEN – GETTING THE MESSAGE
368-9 Trinity College Library, Cambridge (Ms. R. 17. 1 f.7r.). 370 [TR] G.Dagli Orti; (pictograms) BM; [C] (background) BLI (OR 8211/707); (paper making) Private collection; [BL] BM/Museum of Mankind; (wampum) BM/Museum of Mankind. 370-1 Réunion des Musées Nationaux/Musée du Louvre, Paris. 371 [TL] Giraudon; [TR] Private Collection; [CR] WF/Private collection; [BR] Giraudon. 372 [BC] Bulloz, Paris/Musée du Petit Palais.
372-3 *Ivan Lapper*. 374 [TR] Royal Library, Copenhagen (Ms. GKS 4II f. 183 v.); [CR] Royal Library, Copenhagen (Ms. GKS 4II f.195 r.); [BL] Ikona, Rome/Biblioteca Universitaria. Bologna (Ms. 1456 C. 4 r.) Per concessione del Ministero dei Beni Culturali ed Ambientali; [BR] Royal Library, Copenhagen (Ms. GKS 4I f.137r); [B] (both) ML.
375 [TL] [CL] Sotheby's, London; [R] KM; [BL] BLO (Ms. New Col. 121 f. 376 v.). 376 [CL] BLI; [C] Giraudon/Musée Condé, Chantilly; [CR] BLO (Ms. Douce 267 f. 36 r.); [BL] Archiv Prazskeho hradu, Prague; [BC] Sotheby's, London.
377 The Board of Trinity College, Dublin (Kells f. 34 r.).
378 TL (all) SSPL; [C] Réunion des Musées Nationaux; [BL] BLI (Ms. Or. 8210 p2); [BC] T F Carter 'The Invention of Printing' Columbia UP, 1923; [BR] BPK. 379 [TL] MEPL; [CL] St Bride Printing Library; [BR] BPK/Staatsbibliothek zu Berlin, Preussischer Kulturbesitz. 380 [TL] BPK; [C] HD; [BL] BLI. 380-1 [T] (Centre of disc) [T] [B] BLI. 381 [TR] (Cortes) BAL/Galleria degli Uffizi, Florence/Giraudon; [TR] (code) MAS, Barcelona/Archivo de Indias, Seville; [CR] (code) Public Record Office; [CR] (Mary) V&A; [BL] NPG; [BC] Master and Fellows Magdalene College Cambridge; [BR] (Louis) Giraudon/Musée Antoine Lecuyer, St Quentin; [BR] (code) BLI. 382 [CL] [BL] NPG; [remainder] BLI.
383 [TL] HD; [TC] St Bride Printing Library; [BL] [CL] (lower) RPB; [CL] (upper) BPK; [TR] [C] [BL] E.A. Smith; [BC] [BR] Times Newspapers Ltd. 384 [L] Giraudon/Musée du Louvre; [R] Giraudon/British Museum, London.
384-5 MH/British Museum, London. 385 [R] Kostos Kontos/National Museum of Athens; (Ventris) *Andrew Aloof*, [BR] Kostos Kontos/National Museum of Athens.
386 [TR] Giraudon/Musée Carnavalet; [CR] SSPL; (poster) PN. 386-7 MEPL. 387 [CL] MEPL; (Morse) PN; (Chappe) Giraudon; (Chappe Mss.) JLC/Musée de la Poste; (shutters) HD; (Murray) Ronald W. Weir/Blair Castle. 388 [TL] MEPL/Bruce Castle Museum; [TR] Pony Express National Memorial, St. Joseph, Missouri; [C] (both) [BC] MEPL; [BR] RPB. 389 [CL] Pony Express National Memorial, St. Joseph, Missouri; [B] RPB. 390 [CL] HD; [BL] Punch. 390-1 *Andrew Aloof*. [T] [C] [B] The Royal Photographic Society.
392 [L] [TR] SSPL; [L] [C] [BC] MEPL. 392-3 MEPL.
393 [BR] MEPL. 394 [TC] SSPL; [C] Science Photo Library/CON; [CR] The Illustrated London News Picture Library; [BL] PN; [BR] MEPL. 395 [T] MEPL; [TR] [B] PN.
396 RPB. 396-7 [T] PN; [C] Institution of Electrical Engineers; [B] MEPL. 397 [T] [B] MEPL; (McKay) (press clipping) Reuter Archive, Reuters Ltd.; (crowd) Guildhall Library, Corporation of London. 398 [TC] SSPL; [C] BPK; [BL] MEPL; [BC] GEC-Marconi Ltd. 398-9 Punch.
399 [TC] [BC] GEC-Marconi Ltd; [BR] MEPL. 400 *Andrew Aloof* [BL] SSPL; [BC] HD. 400-1 *Andrew Aloof*.
402 (Zamenhof) MEPL/Stuart Campbell; (letter writer) MEPL; (Ataturk) HD; (boy) Popperfoto; [remainder] Universala Esperanto-Asocio, Rotterdam. 403 (background) SSPL; [TL] Science Photo Library; [CL] Süddeutscher Verlag; [B] Museum of the History of Science, Oxford University/ National Archive for the History of Computing.

CHAPTER ELEVEN – ARTS AND LEISURE
404-5 © 1995 Board of Trustees National Gallery of Art, Washington (detail) – gift of Edgar Williams and Bernice Chrysler Garbisch. 406-7 Sygma/Jean-Marie Chauvet.
408 [TR] BM; [BL] Scala/Agora Museum, Athens; [C] [BR] Giraudon/Musée de Picardie, Amiens; 408-9 Giraudon.
409 [TR] Kostos Kontos/Epigraphical Museum, Athens.
410 [TL] MH/British Museum, London; [TR] BPK/Staatliche Museen zu Berlin-Preussischer Kulterbesitz, Antikensamlung; [CR] BM; [BL] Martin von Wagner Museum, University of Wuerzburg. 410-11 BM; (background) Kostos Kontos, Athens. 411 [TL] MH/British Museum, London; [C] (both) [BR] BM; [BC] Scala.
412 [TL] [R] RCS Libri & Grandi Opere, Milan.
412-13 RCS Libri & Grandi Opere, Milan. 413 [TL] RCS Libri & Grandi Opere, Milan; [CR] Skyscan; [BR] Mick Sharp/Jean Williamson. 414 [T] Private Collection; [B] Scala/Museo Archeologico, Palestrina. 415 [T] Scala/Museo della Civilta, Rome; [C] Scala/Museo Archaeologico, Foligno; [BL] Scala/Museo della Terme, Rome. 416 [TR] BN (Ms. 12473 f. 15 v.); [BL] Ikona, Rome/Biblioteca Ambrosiana, Milan (Ms. D75 inf. c. 6r.). 416-17 BN (Ms. 22543 f. 37 v.). 417 [TR] BN (Ms. Fr. 854 f. 121); [BL] BN (Ms. Nouv. Acq. Tr. 2973); [BC] BLI (Ms. Harley 4425 f. 12); [BR] AKG/Universitätbibliothek, Heidelberg.
418 [TR] [BL] [BC] Scala; [C] Scala/Gabinetto dei Designi e della Stampe, Florence. 419 Scala/Accademia, Florence.
420 [L] Scala/Museo Civico, Cremona; [C] [R] Ashmolean Museum, Oxford; [CR] (both) (monogram) D.R.Hill and Sons. 421 (portrait) Collection Dr N.C. Bodmen, Beethoven-Archiv; [BR] Historisches Museum der Stadt Wien, Vienna; (remainder) Staatsbibliothek zu Berlin – Preussischer Kulturbesitz/Musikabteilung mit Mendelssohn-Archiv [T] (Ms. autogr. Beethoven Art. 204 3b f.89); [CL] (left page) (Ms. autogr. Beethoven 51g f. 17 v.); [CL] (upper) (Ms. autogr. Beethoven 51g); [CL] (Ms. autogr. Beethoven 51g f. 18 r.); [BL] (Ms. autogr. Beethoven 10 f. 6 v. 422 [TR] [CL] BAL/Marylebone Cricket Club, London; [CR] [BC] Marylebone Cricket Club, London; 423 all MEPL except [BL] MC. 424 [L] © 1995 by Kunsthaus Zürich. All rights reserved, Kunsthaus Zürich, Switzerland; [TC] Harvard University Art Museum, Cambridge MA, Bequest of Grenville L. Winthrop; [BL] Giraudon/Telaru; [BR] Hubert Josse/Private collection.
424-5 [T] Musée des Beaux-Arts, Rouen; [B] Giraudon/Musée du Louvre, Paris. 426 [TR] SSZ Ltd, London; [BL] MEPL; [BC] The Dickens House Museum, London. 426-7 Guildhall Library, Corporation of London.
427 [TL] [C] [CR] BLO/John Johnson Collection; [BL] By permission of the Birmingham Museum & Art Gallery; [TR] [CL] [BR] MEPL. 428 [TR] RV; [CL] RPB; [BL] CON; [BR] RV; 429 [TL] JLC; [TR] [R] RPB. 430 (Muybridge) [TR] RV; [C] (lower) University of California, Bancroft Library; [C] (upper) [CL] [BL] Kingston Corporation. 430-1 (background) BAL/Stapleton Collection. 431 [BC] BAL/Science Museum, London; [TL] [BR] (lower) [BR](upper) Kingston Corporation.
432 [TL] Royal College of Surgeons of Edinburgh; [TR] NPG; [BL] BAL/The Sherlock Holmes Pub, London; [B] MEPL. 432-3 [T] MC; [B] MEPL. 433 [TR] MEPL; [CR] NPG; [BR] ET. 434 [TR] RPB; [TL] [CL] [C] (program) [BC] [BR] CUP. [BL] Brown Brothers. 434-5 CUP. 435 [TL] [TC] [TR] [BR] RPB; [CL] CUP; [BR] (upper) PN. 436-7 *Andrew Aloof*.
437 [CR] The Kobal Collection/Warner Brothers; [BC] RPB; [BR] The Ronald Grant Archive. 438-9 (all) © The Walt Disney Company.